NUCLEAR INSIGHTS

VOLUME 1

NUCLEAR INSIGHTS:
THE COLD WAR LEGACY

VOLUME 1: NUCLEAR WEAPONRY
(An Insider History)

Alexander DeVolpi

2009

***Dedication*s:**

— foremost to our families, especially our offspring, who have to face the Cold War legacy, good and bad.

— to public and private legions — mostly unheralded — who struggled during the Cold War to avert a nuclear catastrophe. This book is a tribute to your endeavors and success, at least so far.

— to Americans, Russians, and others now renewing friendship between peoples and countries once divided by the Iron Curtain.

— to readers: May you, your contemporaries, and your successors avoid at least some of the identified nuclear pitfalls.

All rights reserved
Copyright 2009: Alexander DeVolpi/DeVolpi, Inc.

Version V1.A3a (August 2009; Revised November 2009)

EAN-13 978-0-9777734-2-8
ISBN 0-9777734-2-6

Typesetting: WordPerfect®, PDF, and Adobe PhotoShop

Printer: CreateSpace/Amazon.Com

PRIMARY CONTENTS

VOLUME 1

Caveat Emptor
Credentials
Preface
Synopsis
Organization of the Book
Acknowledgments
Disclaimers

INTRODUCTION
I.   THE COLD WAR
II. COLD WAR NATIONAL SECURITY
III. NUCLEAR LESSONS
IV. PUBLIC INVOLVEMENT

Cold War Redux
What's in Volumes 2 and 3?
Glossary
Acronyms and Abbreviations
More About the Author
Table of Contents for Volume 1
Index of Volume 1

TABLE OF CONTENTS FOR PRELIMINARY PAGES

Dedication iv
PRIMARY CONTENTS vi
TABLE OF CONTENTS FOR PRELIMINARY PAGES vii
CAVEAT EMPTOR ix
CREDENTIALS xiii
PREFACE xvii
SYNOPSIS xxi
   Major Topics xxi
      Volume 1: Nuclear Weaponry (An Insider History) xxi
      Volume 2: Nuclear Threats and Prospects (A Knowledgeable Assessment) ... xxii
      Volume 3: Nuclear Reductions (A Technically Informed Perspective) xxii
   Major Themes xxiii
      Some Doubts xxiii
      Controversial Topics xxiii
      More Questions xxiv
      Russia's Legacy xxiv
      Shadowboxing xxv
      Recurring Themes xxv
ORGANIZATION OF THE BOOK xxvii
ACKNOWLEDGMENTS xxix
DISCLAIMERS xxxi

CAVEAT EMPTOR

When three professional colleagues and I embarked on composing what turned out to be a two-volume book, *Nuclear Shadowboxing: Contemporary Threats From Cold War Weaponry*, we found it necessary to achieve several things that we hadn't set out to do. Our original goal was to convey our unique experience during the Cold War and its immediate aftermath. However, in the 14 years that transpired before publication of Volume 1, *Cold War Redux,* and the additional year before we published Volume 2, *Legacies and Challenges,* many things changed, not the least of which was the availability of additional information and the reality of post-Cold War conditions.

We tried to keep our two *Nuclear Shadowboxing* volumes up-to-date, and we offer selections from the most up-dated version on the publically available Internet via Google Book Search (books.google.com). However, we failed in gaining widespread awareness of the lessons learned from our intense and extremely rare experience. Thus, this monograph, individually undertaken with collegial agreement, extracts from *Nuclear Shadowboxing* a simplified — yet still insightful — treatment of the Cold War experience and how it relates to modern-day problems.

These three extracted volumes of *Nuclear Insights* are both to be far more accessible, timely, and current because of new publishing technologies. The first transitioned volume is subtitled, *Nuclear Weaponry (An Insider History)*; the second, *Nuclear Threats and Expectations (A Knowledgeable Assessment)*; the third, *Nuclear Reductions (A Technically Informed Prspective)*. Each mirrors *Nuclear Shadowboxing*, but with extensive abridgement and simplification.

Omitted or Downsized in Transition from *Nuclear Shadowboxing* to *Nuclear Insights*

All appendices
All source citations (endnotes)
Many illustrations
All mathematical expressions
Many supporting details
Some figures
Many lists
Detailed footnotes
Lengthy quotations
Some recitations
Many attributions

All of these volumes and versions have benefitted from our singular credentials for the task: trained in nuclear physics, experienced in research and analysis, and first-hand witness to historic Cold War occurrences. We were often involved in events as they occurred, and some of us were also activists in relevant non-government organizations.

Many books on the nuclear arms race have been written by historians, mostly Western, almost always lacking technical backgrounds. Our long-standing professional immersion in technology, encompassing both nuclear development and international collaboration, gives us an advantage in explaining implications and limitations. Our collaboration has been synergistic — a partnership that has generated and strengthened this book because each of us contributed an understanding of nuclear complexities while providing a unique orientation from both sides of the former political, cultural, and military divide.

In gathering topical material, even more reasons were found to capture and share this experience. Although many published and Internet resources have in-depth treatments for selected aspects of the Cold War, we found it necessary to elucidate the technical and political linkage between the disruptive events and near-nuclear disaster. Nor did we find other authors as well qualified or experienced to interpret the legacy now impacting the new millennium — an impact influenced in fundamental ways by complex technical realities. From our perspective, innumerable misunderstandings, misinterpretations, and misjudgements have propagated in publications and government policies.

Indelible in the past century has surely been the Cold War, its repercussions still resonating. From that nearly cataclysmic virtual conflict, we have collated valuable lessons and policy recommendations, with emphasis on nuclear weapons and their delivery systems.

That experience carries messages for any nation that contemplates an atomic arsenal: Its neighbors and potential adversaries will react, and nuclear arms races are deceptively expensive. The diversion of national resources from urgent domestic needs challenges the best interests of any nation, often preserving and expanding the power of dictators, ideological extremists, and other vested interests.

Many actual or still-potential outcomes of the arms race were causally interlinked — one development lead to or enhanced another form of lethal escalation. With the advent of sophisticated and interlaced terrorism, *Nuclear Insights* conveys new but calibrated concerns, including the potential for nuclear blackmail. What good is overwhelming nuclear superiority? How knowledgeable are political leaders in dealing with the new threats? How well informed is the public about realistic dangers?

Having combined Soviet nuclear-weapon experience with American involvement in arms control, we present a unique insight to this episode of history. One result is new (and sometimes controversial) suggested policy alternatives. Having attained a strong technical and analytical background in nuclear reactors, arms control, nonproliferation, and weapons development — as nuclear physicists, engineers, and concerned citizens — our writings reflect personal and knowledge-based participation in countless controversies during that traumatic interval.

We emphasize this backdrop because the analysis underlying this book requires that type of expertise. Our personal experiences have coalesced to produce a book that should be valuable to serious and curious readers who lived through the Cold War or want to learn from it.

We have placed particular and selective stress on relevant history and legacies in order to derive commonsense lessons. Without systematic assessment, it would be difficult to piece together the disparate threads and arcane arguments still abiding after the decades of virtual conflict: Coming to mind are contemporary dilemmas traceable to the Cold War, such as continued dangers from weapons of mass destruction, violent struggles against transnational terrorism, risks from invading sovereign nations, and recurrent eulogies to resurrect Cold War truculence.

****

We made *Nuclear Shadowboxing* comprehensive and thoroughly indexed, with much reference data, to appeal to a broad readership: professionals in and out of government, academics, military historians, eager students, and the energetic curious. In order to reach a wider audience, *Nuclear Insights* is stripped of as much extraneous material as possible. Yet, scholarly readers of *Nuclear Insights* will benefit from its predecessor volumes.

Students of contemporary history will find just about everything they need about nuclear weapons, proliferation, and the Cold War in these volumes. The book should become a useful resource on nuclear armaments and their limitations as instruments of national policy.

We didn't start out to write such a comprehensive and lengthy treatment as *Nuclear Shadowboxing* turned out to be; in a two- decade-long effort it evolved to fill in logical gaps and to provide detailed backup. Some of our personal inclinations might have become all too evident, but compensation has been provided by presenting more than one side of issues.

Another caveat that carries over to *Nuclear Insights*: Don't assume total author's congruence in interpreting events or drawing conclusions. Presented in this book is a spectrum, rather than a singularity, of socio-political background and opinions.

Throughout our writings, individual or collective personal experiences are shared.

For the reader who wants to delve into more specifics, technical and supplemental information is available in Appendices of each *Nuclear Shadowboxing* volume. After all, *Nuclear Insights* is about half the size of its predecessor.

As researchers, participants, and consultants, we had direct or indirect experience with many of the events. Thus, our series of books provides a useful decoding of the intricate relationships, acronymic organizations, and abstruse terminologies that embellish the historical landscape.

That's partly why *Nuclear Shadowboxing* included detailed Appendices, a glossary, a list of acronyms, and ample endnotes and references. Just the task of compiling and decoding all these obscurities is a service that benefits everyone who, now or in the future, wants to find a comprehensive source of information about the development of nuclear weaponry.

Because these events affected others around the world, the international community should benefit and perhaps help fill in the gaps.

Russian-language source materials have been translated and edited for use in *Nuclear Shadowboxing*.

****

Time traipses on, but — in order to get *Nuclear Shadowboxing* published — its additions, revisions, and updates were terminated in 2004 and 2005. Likewise for *Nuclear Insights*, an early January 2009 first-edition self-imposed concurrent-events deadline evolved, but revisions will be seamless because of a new book-on-demand technology.

The reader should now keep in mind that the implicit cutoff for *Nuclear Insights* is just a week before publication. Updates are to be posted at www.NuclearShadowboxing.com, as well as a limited-edition web version available via Google Book Search (books.Google.com).

The latter web site provides a rapid, free, and sweeping means of gaining access to all the material (including citations) excised from *Nuclear Shadowboxing* in order to compile this more abridged *Nuclear Insights*.

With the publication-paradigm shift resulting from rapid book-on-demand technology (utilized for *Nuclear Insights* through the CreateSpace feature of Amazon.com), modified/corrected editions will be more convenient and timely.

Downloadable E-store and Kindle versions will be made available. If a printed copy is ordered through Amazon.com, U.S. delivery should be free. If ordered through www.NuclearShadowboxing.com, the shipped version will be personally autographed and inscribed as requested.

The copyright statement page of these printed Preliminary Pages stipulates the version number that you are reading; a comparable notation on Amazon.com indicates the most recent available version.

CREDENTIALS

This book originated from a personal endeavor of four American and Russian arms-control experts and nuclear-weapon scientists, who together bring unique insight into the history and lessons of the Cold War.

If you wanted to read an authoritative and comprehensive book about nuclear weapons and the Cold War, what qualifications would you want from the author(s)? From historians, academics and journalists you would expect a good and often substantive read. But what if you wanted to hear from the foot soldiers and field officers of the war?

And what if *you* wanted to know about present-day implications of nuclear technology, weapons, and materials?

You're in the right place here, because this is the first of three volumes derived from the experience and contributions of four Cold War scientists: all physicists and engineers, all participants in various aspects of the weaponry — especially nuclear. Together, we've had hands-on exposure (and sometimes undisclosable inside knowledge) regarding the nuclear technologies and practices on both sides of the Iron Curtain. We were close and inescapable witnesses to passing events for most of the Cold War and its aftermath.

Moreover we represent various past or present affiliations or inclinations, from Soviet Communist Party to American liberal, with a Russian-American conservative convert caught in between.

****

Nuclear Insights by itself is a three-volume monograph; it represents a fairly straightforward distillation of a coauthored book, *Nuclear Shadowboxing: Contemporary Threats from Cold War Weaponry*. These volumes differentiate themselves from other Cold War treatments, not only because of events witnessed first- and second- hand, but also because of our collective presence, capability and experience to understand and interpret crucial events that had a large nuclear technology component.

Nuclear Shadowboxing coauthors Drs. George S. Stanford and Alexander DeVolpi have been involved professionally in arms-control and nonproliferation issues for more than 40 years, not only as scientists, but also as citizens vocally objecting to excessive military buildups. Coauthor Dr. Vladimir E. Minkov, an émigré from the Soviet Union, has worked on nonproliferation programs and returned to Russia many times as part of ongoing technical collaborations. The fourth *Nuclear Shadowboxing* coauthor, Dr. Vadim A. Simonenko, is a former Soviet thermonuclear-weapons developer — now free to describe some of his Cold

War experiences, as well as continuing dilemmas in the still partially closed nuclear cities of Russia.

The four of us actively participated in events and issues during much of the Cold War. We are (now retired, for the most part) nuclear scientists with a mix of experimental and theoretical nuclear reactor, arms control, and weapons-development experience.

Plus, we are loyal citizens and family people vexed that important lessons are being ignored or distorted.

Except for Simonenko, deputy scientific director at one of the Soviet Union's two main nuclear weapons laboratories, we were largely bit players in a broad, but traumatic phase of history. We publish not as political authorities or academic historians, but as field practitioners who experienced and understood many of the events that took place, especially the more complex technical happenings.

****

DeVolpi and Stanford, before taking retirement, were researchers at Argonne National Laboratory — the first U.S. national nuclear laboratory, founded during World War II. As citizen-scientists, they have also been involved in trying to apply their unique technical experience to improve human survivability and environmental health. DeVolpi previously had non-combat duty in the U.S. Navy during the Korean War, retiring after 17 years in the Naval Reserve as a Lieutenant Commander. Stanford was born in Nova Scotia, Canada.

Minkov provides a realistic bridge to Russia: He was born in the former Soviet Union, where he gained a technical education and served in the Soviet Navy. Minkov emigrated to the United States more than 25 years ago, when he joined the technical staff at Argonne. His bilingual skills and involvement in many joint U.S.–RF governmental projects make him invaluable in giving the book a balanced perspective. Differences in language, practices, and customs must be accommodated in order to include contributions from distant Russia and to maintain a bi-cultural perspective. Dr. Minkov is able to help in that regard; he has had the linguistic and social skills for comprehensive communication with our Russian coauthor and with other sources in the former Soviet Union. He is also our in-house political conservative.

In Russia, the Institute of Technical Physics, formerly known as Chelyabinsk-70, at Snezhinsk ("Snowflake") in the Urals, was one of the two major design centers for Soviet nuclear weapons. Dr. Simonenko, a leader in design and testing of Soviet thermonuclear weapons, worked alongside many scientists who were enigmas to the West. In the midst of the Cold War, his government assigned him a direct role as a technical expert in negotiations of treaties on controlling the testing of nuclear weapons.

Simonenko is one of the Soviet military scientists who for half a century were secluded inside closed cities — their minds, bodies, and families restricted by physical, geographic, economic, cultural, and legal barriers.

Despite the ending of the Cold War, many subsisting scientists in Russia remain tacit prisoners, either by circumstances or by tradition; yet they have had much to tell the world that pertains to current policy and to an understanding of historic events. Unfortunately, those scientists and their supporting infrastructure have been largely forsaken; they have received little or no income, and became of increasing concern to all who want to see the containment of international proliferation. This lapse has been one of the problems carried over to the new millennium, and a first-hand description has been contributed by our Russian coauthor.

****

Our mode of collaboration for *Nuclear Shadowboxing* was somewhat unusual. When it came to historical context, DeVolpi and Stanford represented United States/Western viewpoints. Whenever our individual perspectives differed, Simonenko imparted the orthodox Soviet/Russian position. Minkov helped clarify how culture and tradition on the two sides affected approaches to the issues that were controversial.

During the superpower confrontation, we shared some common roles, in particular as nuclear scientists working for our respective governments. At the same time, we had different, but sometimes overlapping career experiences — in military service, in weapons design, in arms control, and with public-interest organizations. Because of this diverse background, we have been able to address a wide scope of issues, technologies, traditions, and lessons resulting from the breakup of the USSR. We often had contact in official and unofficial capacities with government authorities, policy makers, and public-interest leaders — even though we were not bureaucrats or academics, but working laboratory scientists who sometimes took on leadership roles.

Because much depends on the technical credentials, more details about the primary author are included at the end of this Volume. However, we do not rest on credentials, but rather on the information and analysis incorporated in *Nuclear Shadowboxing*, now carried over to *Nuclear Insights*.

Relevant nuclear expertise, in-depth arms control experience, diverse cultural backgrounds, and across-the-spectrum political views have been combined in writing this book. Besides offering insight into the history of the turbulent era, we disclose personal memoirs along with explanations of the underlying political, technological, and cultural turmoil that motivated the nuclear arms buildup and engendered widespread public opposition.

****

A qualitative measure of this book's topical complexity is the amount of "techspeak" that is harbored within. Being able to field this capability differentiates technically qualified authors from historians and political scientists. Not only should the presenter be comfortable with the parlance surrounding nuclear devices, one must actually understand and be able to place the terminology in proper perspective for the reader.

A more quantitative measure of topical conversance came to mind when running the spell-checker through this volume. Many terms flagged were technical/scientific terms not normally found in a standard dictionary, while others fell in the category of terminology and shorthand used among professionals. It was necessary to selectively retain unique nomenclature, accepted slang, verbal corruptions, proper nouns, military or political jargon, and "foreign" words (including transliterations from Russian).

While we supply explicit and implicit definitions of terminology, *Nuclear Insights* would be even larger if context-definition couldn't be included. That's partly why we included a detailed Glossary.

Another indication of this book's value is in the number of proper nouns flagged by the spell checker, or listed in the Table of Contents, or included in the Index. Indexing goes through a stage of creating a key-word concordance file, itself unique and comprehensive. The large number of key-word entries suggests that we were successful in retaining names of individuals, organizations, and concepts that became part of Cold War history.

Despite the risk that a credentials' allocution reads like hubris, specific qualifications have been made overt because so many presentations about the Cold War and its consequences are technically (and occasionally historically) inaccurate, misleading, or inadequate.

****

Having retired from our laboratory work and our related obligations, my original coauthors and I came into a better position to communicate our experience and our personal views. It is, frankly, difficult to disengage from the intellectual challenges that motivated each of us for about 40 years professionally and since then out of continuing curiosity. So, aside from relaying to the reader the fruits of our experience, we satisfied through *Nuclear Shadowboxing* at least our own thirst to understand the events and outcomes of the nuclear-enveloped Cold War.

Just in case the prospective purchaser, or a book reviewer, or a conscientious reader wants more specifics about the compiler/author of *Nuclear Insights*, supplementary details are provided in the Posterior Pages.

PREFACE

For almost the entire half-century of Cold War, Americans and Soviets — and most other people of our world — lived in danger of being fried to bacon crisps, for reasons still not abundantly clear.

If an all-out nuclear war had occurred, the few survivors would have been cast back into another brutal Stone Age. The civilized world now knows that it came much closer to holocaust than realized at the time. Moreover, the danger of nuclear catastrophe, although much less, still exists. Suicide **is** *an option for society, but we didn't seem to mind too much, then or now. Are all people mad? It's possible.*

In Nuclear Shadowboxing *my colleagues and I attempted to tell the story of the nuclear age and of the crazy things people did and thought (and still do); we also wrote of the many possibilities and bright achievements of the nuclear age. They, too, are real and abiding, another incentive for supplying an abridgement in this form as* Nuclear Insights.

Having had a professional career in nuclear physics and engineering, unavoidably links a person to our nuclear-weapons heritage; even specialists can be just as aware as the norm. Hopefully the average person, as distinct from political fanatics and social manipulators, will read what is presented and thoughtfully consider these matters.

This interpretive analysis is intended to de-mythicize techno-political events, incidents, and outcomes of the superpower convulsions.

Having no political axe to grind, although there will be those who think otherwise, we have had no hidden agenda. Wanted is what all but a tiny portion of mankind wants, peaceful and prosperous coexistence with nuclear energy. Hopefully this book will discourage war and the misuse of nuclear fission.

Both sides fed falsehoods to their people and to themselves. Both sides had a "nuclear priesthood," advocates of sanctified war or confrontation with potentially disastrous effects.

Other scientists — or Cold War participants — have written about this troubled era in essays or memoirs. Their perspectives are included.

Offered, in particular is invaluable first-hand experience in practically every aspect of nuclear technology, as well as insight gained from having been inside the gates and fences of just about every major nuclear program: the Soviet Union, United States, Great Britain, France, and China, as well as nuclear programs in Switzerland, Sweden, Japan, and other nations.

We have had substantial hands-on experience, a rare credential, with almost every aspect of nuclear engineering and nuclear weaponization. You will discover, if you look closely, that most individuals who have written on these complex topics approach the subjects from an academic or political background, having had little or no intimate opportunity to really get their hands greasy or radioactively contaiminated while carrying out dicey or messy experiments. Self-promotion aside, with the passing of the Cold War era ,you'll find it increasing difficult to find any other comparable experience in print.

Far from being uncritical of governments both East and West, it is noted that the two supreme antagonists piled fear upon fear until the stakes were ridiculously high. This pattern is delineated, for now there remain even fewer nuclear secrets in the world. This entire work is composed of information in the public domain.

Having personally experienced innumerable situations where government secrecy was used to curtail public discussion on controversial policies or sensitive decisions, herein are recited various examples of excessive or futile secrecy. Of special note is the well-documented success of dedicated spies. The use of secrecy still to thwart debate on questionable government policies continues to be of public concern.

The contents of Volume 1 (*Cold War Redux*) in *Nuclear Shadowboxing* (and thus *Nuclear Insights*) are essentially time-invariant — as physicists are fond of saying — meaning not much is expected to change with time. While every phase of history is subject to alternative interpretation, and this one might be affected by future declassification of interesting details, none of that is likely to significantly affect our overview, especially that of Cold War technology developments.

Volume 2 though, being largely about the aftermath if the Cold War, reflects a changing situation, mostly to the better, but with much unfinished business about nuclear reductions. This prospect is left for Volume 3, which closes with an unfinished denouement. While the postmortem is incomplete after two decades of thaw, nonetheless our treatment should need no more than some fine-tuning; it is not likely to be substantially overruled.

****

Popular advice often is to leave things up to the government, because it has intelligence data not available to ordinary citizens and state enemies. Although insiders usually are privy to details not available to the public, my experience — having had access to internal government information — is that secrecy is needed for protection of technical data but not to disguise or perpetuate most policy decisions. Governments are transient; their success in making national policy depends on career administrators and institutional memory, rarely on access to secret data.

When assembling this book, it was shocking to find the number of close calls humanity experienced, almost always without knowing it. At one point, the sun's infrared light reflected from a transient cloud formation, detected from a Russian satellite high above Montana, was construed in Moscow as the launch of five missiles. A Russian colonel had only minutes (about the time it takes to read this far in the Preface) to determine whether to launch a nuclear counterstrike. He decided, fortunately, that neither side was likely to attack with only five rockets.

The Cuban missile crisis is widely known to have brought the superpowers to the verge of nuclear war. Even after the demise of the Soviet Union, missile alerts have reached alarming levels.

Both sides have been appallingly careless, luckless, and lucky in handling their nuclear devices. Starting in 1950, dozens of nuclear weapons — some of which have never been recovered — are acknowledged to have fallen from (or crashed with) American aircraft. Some nuclear warheads from imploded submarines remain submerged on the seabed. President Kennedy was told there had been sixty accidents involving nuclear missiles, including the unintentional launching of two nuclear tipped anti-aircraft weapons. Governments have been and remain closemouthed about everything nuclear-sensitive that has happened.

Nor is there any reason to believe that these dangers to humanity have all been eliminated — eased, yes, but not undone. Massively destructive nuclear weapons remain at unneeded levels of triggerable response or forward deployment. Such weaponry, along with unpredictable leaders and preemptive and escalatory national policies, can help ratchet disputes to greater confrontational risk — raising concern about the durability of social institutions and the survival of humanity. In the nuclear age there are no immunities or sanctuaries for peoples or nations that brandish highly destructive weapons.

For the two great powers who were reluctant to dismount the Cold War tiger, the monetary cost was substantial: about $19 trillion in the United States, more than three times the post-millennium national debt. Russia and the other successor states are still trying to cope with their legacy from the Union of Soviet Socialist Republics.

The research conducted by myself and my colleagues indicates that nobody won the Cold War, any more than anybody could have won a hot war had one started. The contest was a morbid spasm of the intellect, a condition not uncommon throughout human history, but one much more perilous than heretofore.

One representational goal has been to derive useful lessons from the extended period of superpower tension, especially regarding development of nuclear weapons and their delivery systems. Hopefully nations contemplating nuclear arsenals will be discouraged from renewing such expensive arms races, thus avoiding the domestic social neglect and national spending binges that characterized a prolonged time span marked with worldwide anxiety.

Another incentive for this endeavor has been our eventful personal and professional association with both sides of what was once a civilization-threatening conflict — primarily between the United States and the Soviet Union — and most emphatically a nuclear-dominated conflict that is now unwinding to less-unreasonable levels (with numerous dangerous weapons yet to be eliminated).

****

As acknowledged on the back cover of this book, historians, political scientists, and professional journalists have ably rendered top-level compilations and personal accounts of the eventful period, its nerve-wracking transition, and now, the aftermath.

Adding to this accumulation are memoirs by former officeholders, think-tank consultants, and retired military officers. Some other scientists and former government officials are gradually filling in pieces of the Cold War puzzle.

Nuclear Insights supplements this lore by providing, first, an essential counterpoint from a technically qualified vantage; second, a chronicle of scientist and citizen involvement at the foundations, the lower tiers of participation, as witnessed by expert insiders; and, third, an integration over the subsets and interrelationships of knowledge that can be pieced together. To make this compendium, it has been necessary to link some complex technical, political, and historical elements. Cold War nuclear weaponry, its many flavors and forms of delivery, driven by political and military incentives, is largely about technological challenges in development, management, and consequence.

My colleagues and I directly and personally participated in many of the events and with many of the organizations — along with other scientists, patriots, and partisans, and volunteers. Passage of time and change of circumstances have enabled this book to be a thorough, scientifically neutral, partisan-diminished examination of consequential events around the time in which we were involved. While involvement can influence perspective (somewhat tempered with the passage of time), first-hand familiarity has the benefit of contributing something of greater value than otherwise available to historians.

Of great importance in my mind is not to lose track of minor contributors and the many other participants — movers and shakers — in the Cold War pantheon.

The treatment of events offered in this book is thus unique and supplementary. It is not comprehensive, nor intended to contradict or replace existing discourse about the Cold War, nor even about Cold War nuclear weapons. Together with other contributions, we might aim to get it as correct as something like this can be — at least in terms of understanding how nuclear confrontations occur and escalate.

****

Commentaries from readers, critics, and supporters are welcome. Originally, a website was planned to accommodate such commentaries. Aside from the reality that very few have been received to date, Google has instituted its Knol system of publication, which is a bonanza for retired professionals. Thus, updates and commentaries in connection with *Nuclear Insights* and related topics will made accessible under my identity at www.knol.google.

– Alexander DeVolpi

SYNOPSIS

Nuclear Insights *addresses the primary events and concerns that dominated the Cold War. The book does not shy away from controversial issues debated by political candidates, analyzed in the news media, discussed at home, and deliberated at the highest governmental levels. As the subtitle advises, Nuclear threats inherited from the former superpower confrontation are thoroughly examined.*

Offered in the main body of the book is a narrative, interpretive, and motivative account of nuclear weaponry. Because of the topic's enormous complexity, a good use of technical expertise and experience is to provide a panoramic view, supplemented with background information and specialized topics.

The book's post-mortem of Cold War myths, realities, legacies, and lessons is apportioned within seven substantive Chapters, bundled into three volumes. Highlights have been derived to compose this Synopsis.

For a world that often relishes sound bites, this abstract of topics and themes is offered with terseness, trepidation. and risk — and with shortcomings inherent in distilling a three-volume book into this six-page synopsis.

Major Topics

Volume 1: Nuclear Weaponry (An Insider History). Comprised of four Chapters, the first volume provides a history of nuclear weapons during the Cold War, starting with the development of fission and its explosive capability, moving on to the atomic bombings of Japan and the subsequent controversy. The next two Chapters chronicle the development of what became a fixture: the military, industrial, and laboratory linkage that supported nuclear-deterrence policies primarily through a strategy called mutual assured destruction (MAD). The last Chapter in Volume 1 recounts the awakening and influence of popular resistance to numerous unchecked nuclear-weapon systems. Without that public resistance, the frightening events might have gone on longer or been punctuated with bloody nuclear combat.

Many baby-boomers will find Volume 1 to be an apt reminder of a dicey phase in history that was under way during their youth, before they became taxpayers, family-makers, worker-bees, and voters. New generations of leaders might not have a visceral or comprehensive appreciation of the dangers and legacies from nuclear arsenals.

Chapter I throws light on origins of the nuclear action-reaction cycle, reviewing major happenings in five Sections: (A) precursor events leading to the atomic bombing of Hiroshima and Nagasaki, (B) early postwar years, beginning with the policy of "containment," (C) the coming of "détente," followed by (D) the crumbling of the Soviet Union. A final Section (E) describes the role of political leadership in the context of military buildups and efforts to achieve arms control.

The second Chapter in Volume 1 is a hindsight view of Cold War security considerations: (A) national defense claims and counterclaims, (B) weapons and military strength, (C) defense support needs, and (D) arms developments and limitations.

Chapter III derives lessons from the Cold War, first (A) evaluating myths, fallacies, and realities, (B) considering "unthinkable" doomsday and nuclear accidents, (C) identifying governmental or government-funded nuclear establishments, and (D) allocating attributable costs. Here a central theme, nuclear shadowboxing, is elaborated. Also, one pertinent observation stands out: Nuclear weapons have never really "won" anything! Although their use might have hastened the surrender of Japan in World War II, afterwards they were never militarily exploited, offensively or defensively.

The fourth and final Chapter in Volume 1 takes a detailed look at the Cold War involvement and influence of non-governmental entities and organized political factions, thereby (A) crediting public-interest organizations, (B) compiling issues of the time, (C) factoring political pressures, and (D) recognizing government intimidation of those who opposed nuclear buildups. Chapter IV thus begins the book's transition from past through present.

Volume 2: Nuclear Threats and Prospects (A Knowledgeable Assessment). Picking up after the demise of the Soviet Union, Volume 2 instructs about the radiation, chemical, biological, political, economic, and sociological "fallout" — the legacy of weapons and belligerency that new-millennium generations confront.

Chapter V discusses societal legacies — the aftermath: (A) environmental consequences, (B) institutional impacts, (C) deterioration of nuclear security, (D) existing nuclear-armed nations, and (E) beneficial applications of nuclear technology.

Chapter VI contains an update of the post-Cold War military situation: (A) inventorying major nuclear arsenals, (B) cataloging nuclear holdings of other nations, (C) listing conventional forces, (D) describing factors that sway the current balance of forces, (E) examining the impact of military development and spending, (F) taking into account nonproliferation and counterproliferation policies and measures, and (G) considering the role of technology in national security.

Volume 3: Nuclear Reductions (A Technically Informed Perspective). In light of modern-era security threats, the book's final Chapter (VII), in its own slimmer volume, addresses questions regarding nuclear reductions and disarmament: (A) the current policy debate, (B) existing treaties and negotiations, (C) prospects for reversion to non-nuclear status, (D) implications of deep cuts in arsenals, (E) disposal of nuclear materials, (F) verification of reductions, and (G) steps required for negotiated reductions. It is somewhat of a guidebook for a nuclear-weapons-free future.

Synopsis **xxiii**

Major Themes

Some Doubts. Despite huge nuclear arsenals that could wipe any nation out, both Russia and the United States are plagued with home-grown and transnational terrorism. What good, one must ask, are sophisticated nuclear warheads against elusive individuals and organizations that use simple conventional weapons with boldness, resourcefulness, and no fear of sacrifice?

It's time to pick up our marbles and move on.

Beholden to past endeavors, national and global burdens in contemporary times require safeguarding nuclear armaments, stymying proliferation of weapons, cleaning up the Cold War detritus, and reducing the dangers of accidental, unauthorized, or deliberate use of nuclear weapons.

Controversial Topics. Most controversial are policy recommendations to reduce the risks, costs, and stigma of new-millennium nuclear weapons. Many suggestions are emerging about overdue changes and improvements: making deep cuts in nuclear weapons, reducing the risk of accidental or unauthorized nuclear detonation, stabilizing Russia's nuclear deterrent, defending sensibly against ballistic-missiles, eliminating weapons-quality fissile materials, banning the testing of nuclear explosives, promoting nonproliferation, decontaminating radiation legacies, shepherding the nuclear stockpile, strengthening the protection of sensitive nuclear data, and reducing unnecessary secrecy.

Attracting too much politicized attention has been ballistic-missile defense (BMD). Finding a feasible means of guarding against missiles that can deliver nuclear weapons at intercontinental ranges has been a challenge since *Sputnik* was successfully launched in 1957 by the Soviets.

Ever-elusive defense against nuclear weapons — whether delivered by bomber, ship, or missile — has been sought since the beginning of the superpower confrontation.

We've evaluated the military, political, and technological factors that have shaped the quest for defense against long-range missiles, offering answers to questions like: What would be the most promising BMD policy? What course of action offers the best hope for thwarting an attack?

Much conventional wisdom on both sides of this question should be rejected: e.g., the notion that even a "rogue state" such as Iran or North Korea will be so irrational as to commit national suicide by launching missiles against the United States or Russia. Nor will the major powers unilaterally abandon their ability to retaliate with nuclear weapons if subjected to nuclear attack. One alternative — negotiating a multipartite ballistic-missile defense — would be much more feasible than a unilateral approach.

We advise spurning the more recent political ballyhoo about the ABM treaty: In force for 30 years, it was eminently successful in promoting nuclear stability. Also, one should discount as unrealistic the notion that defense against certain specific ballistic-missile threats is technologically impossible. In any event, it would certainly be premature to *deploy* the kind of mid-phase intercontinental-missile-intercept system contemplated by the G.W. Bush administration. Even the boost-phase intercept system advocated by the Russian Federation has pivotal reaction-time and reliability problems to overcome.

In any event, one must realistically recognize that the best affordable missile-defense system would not shield a country any better than the Maginot Line protected France in World War II. The quest for defense against ballistic missiles will undoubtedly continue, and political reality might force the eventual deployment of some sort of BMD, effective or not. When it comes to deployment, the Hippocratic admonition to doctors, "First, do no harm," could be applied to any such decision: *First, do nothing that threatens nuclear stability.* (No defense should threaten the current or future comfort zone for mutual deterrence.)

More Questions. What is the optimum size and status for post-Cold War nuclear arsenals? Because of the risk of accidental or unauthorized nuclear detonation — as well as the danger of miscalculated brinkmanship — certain questions beg for answers: How many nuclear weapons are needed for national security? (Markedly fewer than traditionally believed.)

What *is* a safe level of deterrence? Some analysts even suggest the abolition of nuclear weapons. Near-term abolition is not practicable: Treaty commitments notwithstanding, current verification inadequacies alone are sufficient to nix the idea. (For the time being, an agreed level of nuclear deterrence needs to be preserved.)

Is it feasible to put weapon delivery systems on a lower level of alert, a level that would be less susceptible to inadvertent launches? Yes, both feasible and desirable, is a good answer, because a hair-trigger alert is unnecessarily dangerous. (However, to make that possible, the United States, China, and Russia must not threaten each other's confidence in riding out and responding to nuclear attack.)

Do we still need to test nuclear weapons? There is strong evidence that nuclear warheads are stable and can be maintained without conducting proof-test detonations. (Not sufficiently informed enough to make an independent judgement on stockpile reliability, the pros and cons that have been put forth are, listed and along with an explanation why the need for more nuclear testing is not self-evident.)

Moreover, do we need new nuclear weapons with the capability to penetrate deeply into the earth? (Aside from the counterproductive aspects of betraying the nonproliferation norm, practical underground nuclear detonations will result in collateral damage — radiation well beyond the intended target area.)

What about the large inventories of fissile materials that were used, or could be used, to make weapons: Can these inventories be destroyed or safely managed? (Most of the inventories should be preserved for peaceful use and that they can be managed safely — contrary to the view of a few dissenters.)

Russia's Legacy. A serious problem facing Russia, in trying to integrate economically and politically with the rest of the world, is its nuclear heritage. Many analysts observe that Russia's protective infrastructure is deteriorating, and are fearful that non-nuclear-weapon states or terrorists might be able to get their hands

Synopsis xxv

on some nuclear materials. (Evidence is provided that this is not now the case and — with minimal Western help — Russia can handle this problem effectively.)

Regarding terrorism and weapons of potential mass casualty, a more realistic and progressive perspective is needed so that new communal dangers do not arise from theft, accidents, or misuse of residual nuclear weapons and materials still in the hands of weapons states.

Shadowboxing. As background for some of these contemporary controversies, the origins and evolution of the Cold War have been traced in this Volume, through the breakup of the Union of Soviet Socialist Republics. The rationale behind the development and use of nuclear weapons to bring World War II to a rapid close is reexamined. The subsequent evolution of a military-industrial-scientific complex, sustained by the continuation and growth of government secrecy, gave birth to thermonuclear weapons and a virtual conflict that brought the world to the brink of nuclear suicide. It is this cycle of relentless and uninhibited growth of nuclear weapons personified herein as "nuclear shadowboxing" — the sometimes furious and excessive political sparring that characterized the last half century — left vexatious and expensive problems to be addressed in the new millennium.

A major hypothesis expounded in this book is that — by inflammatory behavior, reflexive responses, and overly conservative counteractions — the United States and the Soviet Union turbocharged the nuclear arms buildup. Both nations used worst-case analysis, thrusting and jabbing with increasingly sophisticated and impersonalized technological threats, when in fact each side was jabbing primarily against its own shadow. Each image reflected one's own political, ideological, cultural, and national identity. Unfortunately, national interests were placed above the well-being of humanity.

With each technological advance inducing the other side to react, the result was a development cycle for nuclear systems having various baffling names and acronyms — fission weapons, thermonuclear weapons, ICBM, MRV, MIRV, ABM, SLCM, ASAT, etc. This action-reaction cycle, as though there were a Newton's Law of international rivalry, propelled the participants to evermore dangerous confrontations and outrageous budgetary excesses.

And what did we, could we, do to try to get out of these everlasting cycles? Was the nuclear arms race inevitable? Have the post-Cold War reductions in nuclear arms been only a move to just another level of mutual assured destruction, less expensive but still fraught with danger? Or, are arms reductions sufficient to avoid the mortal dangers engendered by the nuclear arms race?

This book has become a comprehensive assessment of the nuclear-arms race with lessons extracted from the past in order to revise future policy. These lessons, as science and technology advance, might help reduce the dangers we had meekly accepted.

Recurring Themes. The reader will find some of the following themes recurring through the Chapters:

The atomic bombing of Japan, whose military justification is debatable, was probably unavoidable because the newly vested military-industrial-scientific complex had a major stake in proving the new weapon and justifying its investment.

Political shadowboxing became a convenient means, even if unplanned, to bootstrap and nourish the concurrent arms race.

Augmenting a nuclear addiction was a new-found *persona* of government officials who, because of access to secret information, acted as though only they knew what was best for all of us. They insisted, and some still claim, that — *if you knew what they knew* — you would approve their actions. Emboldened by insider information, and strong in their convictions, they acted, notwithstanding human limitations, in what they themselves considered to be the best or patriotic interests of nations.

Countering bellicose tendencies were non-government individuals and organizations who helped keep the Cold War from becoming a hot war. Through prolonged pressure on governments to engage in negotiation and to consent in treaties, outsiders helped stave off what might very well have been a suicidal nuclear exchange.

Rigorous proof of these propositions remains to be certified, party because documentary historians lack full corroboration from archives that some governments have classified secret for another generation of two — not only to protect the living but also to perpetuate strong beliefs still deeply held.

Do you want to know more about phantom threats, exaggerated dilemmas, environmental damage, radiation cleanup, or clean-energy options? Most of that is in Volume 2, but here's a sample: So-called "dirty bombs" (improvised radiation-dispersal devices) are primarily a threat to their assemblers and handlers. They present hardly any physical threat to the public under realistic circumstances, contrary to what you might have seen on TV or read in the newspapers and magazines. But they make good press and create some transient fame.

How about nuclear abolition? If the former Cold Warriors who recently become vocal converts to nuclear abolition were to express contrition and promote unraveling egregious secrecy, this might do more for nuclear reductions than all the petitions in the world. (They would also find, especially by reading Volume 3, that the nuclear-transition road should indeed be paved with political good will, but the foundation must consist of technically viable and verifiable means of fissile demilitarization.)

ORGANIZATION OF THE BOOK

One reason the book is divided into three volumes is that one alone would have been too unwieldy. In any event, a chronological and topical separation made sense.

Each book has its own detailed Table of Contents and Index, while Volume 3 contains a consolidated Table of Contents for all three volumes.

At the beginning of each Chapter and Section — within italicized paragraphs — is a brief introduction, and at the end — also italicized — a summary and highlights.

The book is structured into many separate entities within the three volumes: Chapters, Sections, Headings, Subheadings, Boxes, and supplementary material.

Whereas *Nuclear Shadowboxing* contains detailed appendices, explanatory boxes and background notes, these are not included in *Nuclear Insights*. Also, references (endnotes), tables, and graphs had to be omitted in order to shorten *Nuclear Insights* and make it easier to read. All of this, though, can be accessed online in the full-scale mirroring volumes searchable on Google Book Search www.books.google.com. [Enter *"Nuclear Shadowboxing: Cold War Redux"* in order to access up to 20% of Volume 1, and/or *"Nuclear Shadowboxing: Legacies and Challenges"* for Volume 2.]

ACKNOWLEDGMENTS

Appreciation is acknowledged for those who have taken time to preview some or all of this book and to offer benefits of their experience in the form of valuable comments. No one else, though, is responsible for the final versions.

Nuclear Shadowboxing contains recognition for assistance in its preparation and publication.

I am especially grateful to my Washington and Lee University classmate, Gerry Lenfest, who has a keen public interest and panoptic vision about nuclear proliferation. He has supported this conversion to *Nuclear Insights*; we both hope to provide better public understanding about realistic nuclear risks.

DISCLAIMERS

Nothing herein should be construed to expressly or implicitly suggest the consent or concordance of any governmental organization.

Personal experience and advice do not replace, but are complementary to, that of government officials and arms-control negotiators.

An empirical physicist becomes especially aware of scientific and human limitations in any endeavor. Borrowing terminology from field and laboratory experience, two categories of communication error — systematic and random — are to be found in this book: systematic uncertainty is associated with errors of interpretation, while random or incidental uncertainty result from composition discrepancies, such spelling or grammar. Although mistakes have been minimized by repetitive editing and new software tools, a mechanism for periodic corrections and updates must remain in place.

All information has been taken from public sources, specifically avoiding help to any prospective terrorist or wrongdoer who might try to devise nefarious weaponry.

As explained in the Preface, *Nuclear Shadowboxing* provides attributions for quotes, paraphrases, and other obligations or courtesies.

The need for secrecy of technical details about nuclear weaponry and military strategy is respected. Above all, law and patriotism are binding in safeguarding information and data classified as sensitive or secret, as well as fear of frightful consequences for humanity if the wherewithal to make weapons of mass casualty were distributed without controls.

Fellow environmentalists, colleagues, and comrades will notice that not all points of view are represented. Some issues are divisive in themselves, especially nuclear, and especially gut-wrenching topics. Hopefully they will view the overall product as a step forward, subject to remediation as time goes on.

In the spirit of moving ahead, personal views have been curbed. Elaboration of differences might ultimately be found on the web site.

INTRODUCTION

The *Trinity* Nexus 3
　Trinity minus 50: the Nuclear Genie Emerges 4
　¡*Trinity!* Exclamation Mark for the 20 Century 5
　Deciding to Use the Bomb 6
　　[Harry S. Truman's Announcement of the Dropping of an Atomic Bomb on Hiroshima] 7
　Demonstration? 9
　Motivations 10
　Potsdam Declaration 10
　Monday-Morning Quarterbacking 11
　Ending the War 11
　　[Recollection by Wolfgang Panofsky] 12
　Nagasaki 12
　Postmortem 12
　Consequences 13
　Trinity plus 50: Getting Where We Are Now 14
　Figure 1: Full-Scale Replica of World's Largest Thermonuclear Bomb .. 15
Problems for the New Millennium 16
　Reassessing Government Secrecy 17
　Nonproliferation and Counterterrorism Issues 18

During the Cold War, and ever since,
the world has come close — more than once —
to unimaginable nuclear devastation.

In just the seven-year period from 1977 through 1984, more than 20,000 nuclear alerts were triggered — all false indications that ballistic missiles were attacking United States. And we can only guess how many times Soviet missiles were armed for massive retaliation to illusory or presumed missile assaults from the West.

Both superpowers had activated nuclear-armed ballistic missiles outside their own national boundaries when the touch-and-go Cuban missile crisis of 1962 started to escalate out of control.

From 1945 on, the United States and the Soviet Union saw themselves as enemies until the latter's collapse in 1991. That conflict is referred to as a "Cold War" because the two countries never directly fought each other in a "hot war." Instead, they engaged in political and psychological posturing, funded surrogate wars, and stockpiled conventional and nuclear weapons to outgun each other in case war did break out.

In 1995, long after the hostile decades were over, President Boris Yeltsin reportedly activated his "nuclear keys" for the first time: A scientific rocket launched from Norway was mistakenly thought by the Russian military to be a Trident missile launched from a U.S. submarine.

Although the United States and Russia signed an agreement in May of 2002 to remove strategic (nuclear) weapons from deployment, little headway yet to be made in reducing the risk of hair-trigger nuclear response to false alerts — a situation aggravated by an increase in the number of countries developing or acquiring long-range missiles.

With the East-West Cold War belligerency coming to a peaceful end, hope came that the nuclear arms race would cease and even be reversed. At first, progress was encouraging. Combined, the Soviet and American arsenals had peaked in 1985 at about 60,000 nuclear warheads, and within a few years the number deployed had dropped to about 20,000. But then the mutual reduction process stalled.

Imagine, more than 60,000 warheads! Were there that many military targets? Are there still 26,000 credible targets for the residual post-millennium arsenals, now that the self-centered conflict is over?

Half the weapons removed from deployment are still not dismantled, and the nuclear source material extracted from dismantled warheads remains available for rearming. It is not now clear that many nuclear warheads will soon be destroyed, and — worse — the nuclear competition might even resume. The number of nations with nuclear weapons is slowly growing, and global terrorism is expanding its reach.

The commonality of Russian and American interests in countering terrorism and avoiding mass destruction provide hope for some sort of joint action in mutual arms control. But eastward expansion of NATO's membership and the U.S. intention to fund a national missile defense does not inspire optimism that nuclear arms will be reduced.

The new millennium is beginning with an inauspicious burden: Enough explosive power to destroy civilization was produced and still exists in weaponized form.

How did we get into this predicament?

The *Trinity* Nexus

The first atomic explosion took place on 16 July 1945 in the New Mexico desert — a locale ominously known as *Jornada del Muerto*, Journey of Death. Code-named *Trinity*, the test stands at the approximate midpoint — the nexus — of the 20 century. The first half of that century saw theoretical and experimental physics developments that led up to *Trinity*, triggering nuclear armament and proliferation through the remainder of that millennium.

This book is written with the conviction that the lessons of the past must be understood if nuclear weapons are to be brought under control in the new millennium. Part of the current reality is that demilitarization should be global — that *unilateral* nuclear disarmament is simply unacceptable in the view of the major powers. Moreover, nuclear demilitarization has to proceed in the context of the threats perceived by major and minor nations. That's a serious contemporary challenge. A third aspect of the new reality is an internationalized threat coming from highly mobile terrorists and from state-sponsored terrorism.

From the vantage of experienced nuclear physicists and analysts, the reader will be guided through the complexities of nuclear weapons technology and the Cold War policies it spawned. First, though, let's go back fifty years and briefly trace the evolution of nuclear science.

***Trinity* minus 50: the Nuclear Genie Emerges.** At the beginning of the 20 century, given the results of earlier studies to understand the structure of the atom, scientists started to uncover mysteries of its nucleus. In 1900, physicist Max Planck introduced the theory of discrete atomic-energy states, or quanta, to represent the observed behaviour of atoms. Five years later, Albert Einstein postulated his theory of special relativity. Around 1910, Ernest Rutherford started to probe the nucleus of the atom, and Neils Bohr devised a quantum model to account for light-emission spectra. Soon afterward, concepts of quantum mechanics and the wave nature of matter were formulated by French and German scientists — Louis de Broglie, Max Born, Erwin Schroedinger, and Werner Heisenberg.

The subatomic neutron was identified in 1932 by James Chadwick in England. After Marie Joliot-Curie detected radioactivity the following year, it was realized that the atom could be broken down into smaller constituents — that each atom functionally contained a nucleus of protons and neutrons, surrounded by a cloud of electrons. Enrico Fermi bombarded substances with neutral particles (neutrons) in 1934, learning more about the nuclear substructure of elements.

Four years later, Irène and Frédéric Joliot-Curie found that low-energy neutrons induce a nuclear-breakup phenomenon later identified as fission by German scientists Otto Hahn, Lise Meitner, and Fritz Strassman. Experimental and theoretical studies quickly showed that extra neutrons were released along with a great deal of energy when fission occurred. While these startling discoveries were taking place, Adolf Hitler had started to force Jewish scientists out of the Third Reich.

Einstein, Joliot-Curie, Joseph Rotblat, Leo Szilard, Enrico Fermi, and Edward Teller soon recognized the theoretical potential of a self-sustaining, multiplicative neutron chain reaction, setting the stage for applying the newly discovered fission process to make nuclear explosives. Szilard originated and patented a theory of the nuclear chain reaction in 1933.* These developments were nearly simultaneous in several nations. Soviet physicists had an active program too on the potential of nuclear reactions for releasing usable energy.

Even earlier, at the beginning of the 20th century, imaginative physicists and writers, like Britishers Frederick Soddy and H.G. Wells, had speculated on the bright and dark implications of vast amounts of energy released from the atom.

The commandeering of nuclear physics for military applications started as the war in Europe broke out. In 1939, Nazi Germany initiated a small and secret nuclear weapons project. Next year in the United States, the first artificial element — neptunium — was produced, and from its decay, plutonium was detected a year later.

---

*The fission chain reaction was found to start with the breakup of the nucleus into energetic fission fragments and neutrons. It was quickly realized that the rare element uranium was especially favorable for the neutrons to sequentially cause additional fission in neighboring nuclei.

In response to a letter from Albert Einstein (drafted by Szilard), President Franklin Roosevelt established the Uranium Committee of the National Research Council in the fall of 1939. Less than a year later, information that Germany had begun its own research into (and was accumulating) tell-tale uranium, spurred Roosevelt to urgently expand U.S. activities; he gave the authorization just one day before the bombing of Pearl Harbor. In the meantime, British scientists had made considerable headway in specifying ways to enrich uranium and fabricate nuclear explosives.

Although Soviet scientists recognized that powerful nuclear bombs were theoretically possible, World War II interrupted their studies.

The Manhattan Project, eventually combining American and British programs, was organized in 1942 to systematically create an explosive based on the fission chain reaction. It became a resourceful and remarkable program, well funded and staffed. One of the earliest achievements was a controlled self-sustaining nuclear reactor, the first ever, assembled at the University of Chicago under the leadership of Enrico Fermi. The reactor reached its operational critical point on 2 December 1942, nearly a year after the Japanese bombing of Pearl Harbor.

¡*Trinity!* **Exclamation Mark for the 20 Century.** The 1945 *Trinity* explosion in New Mexico tested one of two nuclear-explosive designs created in the wartime crash program. It was an epochal event of the 20 century. Robert Oppenheimer chose for the test a code name based on a quotation attributed to the poet John Dunne: "Batter my heart, three-person'd God." The explosion was an inventive triumph for the large-scale Manhattan Project, which produced the many required kilograms of enriched uranium and explodable plutonium and which succeeded in devising nuclear weapons.

A month after the *Trinity* test, two atomic bombs were dropped on Japan, forcing an abrupt end to the war. In bursts lasting microseconds — millionths of a second — fission chain reactions generated explosions that killed well over hundred-thousand people at Hiroshima and Nagasaki.

The decision to use the bombs — still controversial — and the resulting human casualties were only the beginning of the broadening impact of Trinity: Yet to come would be long-term repercussions leading to the Cold War. Among the lasting consequences from nuclear devastation of two cities and their populations were widespread psychological effects. We are among those who retain a lasting imprint of those traumatic horrors.

"Little Boy," the atomic bomb exploded over Hiroshima on 6 August 1946, utilized enriched uranium: Confidence in the device was so high that no prior explosive test of that design was carried out. The "Fat Man" bomb, which destroyed Nagasaki three days later, 9 August, was based on the same plutonium configuration (nicknamed "the Gadget") that was tested at *Trinity*.

Although Germany had an early start on nuclear development in World War II, it seems that Hitler thought the war would be successfully concluded long before an atomic weapon could become available. In addition, having experienced a serious methodological error in measuring the nuclear properties of graphite,

German scientists were sidetracked from developing the uranium reactor that became an essential tool for exploiting the fission process. German progress was self-limited as a result of technical and bureaucratic factors, and the effort was further impeded by Allied bombing and internal sabotage. Until the war ended in Europe, wartime secrecy and confusion concealed the important fact that Nazi Germany had not succeeded in making nuclear explosives or radiological weapons.

Excluded from the wartime American, British, and Canadian nuclear coalition, the Soviet ally used its own espionage network to keep itself informed about nuclear developments in the West. Well-placed sympathizers and spies conveyed many essential details of fission explosive development. Through his espionage network, General Secretary Josef Stalin learned of the Manhattan Project and the results of the *Trinity* test. As the German invaders started to retreat from Soviet borders, Stalin established a secret nuclear development project jump started by prewar scientific nuclear initiatives. He also turned the vulnerabilities of American secrecy to his political advantage, notably by entering the war against Japan at the last minute in order to ensure a voice in the final settlement.

****

Humanity's nuclear destiny is slanted by past lessons and present actions. The Manhattan Project, the Nazi nuclear-weapon attempt, and the Soviet theft of plutonium bomb designs — all are events of the first fifty nuclear years that have left their legacies. Secrecy was the paramount control measure relied upon in trying to maintain an atomic-bomb monopoly. Information controls failed in that task, but excessive restraints successfully stifled what could have been a productive debate about the implications of being the first to use nuclear weapons.

Deciding to Use the Bomb. "The top news event of the 20th century," chosen by prominent journalists and scholars in 1999, is described in the adjacent clipping from files of the Associated Press. (See boxes that follow for the full radio address by President Truman.)

Use of the atomic bomb was (and to some extent continues to be) controversial. The U.S. decision (made without consulting its wartime allies) was composed of several determinations, each of which had alternatives. Would a demonstration in Tokyo Bay have had the same effect in terminating the war as the detonation over Hiroshima, which was primarily a civilian target (population of more than 300,000)? Should strictly military installations have been targeted? Was the second bomb on Nagasaki necessary, especially so soon after the first? Were long-term nuclear proliferation consequences considered at the time?

For the fortieth anniversary of Hiroshima the Smithsonian planned to have an exhibit with the Enola Gay as centerpiece in the Washington Air & Space Museum. There was intense controversy, and the proposed exhibit was canceled under political pressure, egged on by angry World-War II veterans who opposed the statement that "To this day, controversy has raged about whether dropping this weapon on Japan was necessary to end the war quickly."

> **[Harry S. Truman's Announcement of the Dropping of an Atomic Bomb on Hiroshima]**
> **(Address to the Nation, August 6, 1945)**
>
> Sixteen hours ago an American airplane dropped one bomb on Hiroshima, an important Japanese Army base. That bomb had more power than 20,000 tons of TNT. It had more than 2,000 times the blast power of the British "Grand Slam," which is the largest bomb ever yet used in the history of warfare.
>
> The Japanese began the war from the air at Pearl Harbor. They have been repaid manyfold. And the end is not yet. With this bomb we have now added a new and revolutionary increase in destruction to supplement the growing power of our armed forces. In their present form these bombs are now in production, and even more powerful forms are in development.
>
> It is an atomic bomb. It is a harnessing of the basic power of the universe. The force from which the sun draws its power has been loosed against those who brought war to the Far East.
>
> Before 1939, it was the accepted belief of scientists that it was theoretically possible to release atomic energy. But no one knew any practical method of doing it. By 1942, however, we knew that the Germans were working feverishly to find a way to add atomic energy to the other engines of war with which they hoped to enslave the world. But they failed. We may be grateful to Providence that the Germans got the V-1's and V-2's late and in limited quantities and even more grateful that they did not get the atomic bomb at all.
>
> The battle of the laboratories held fateful risks for us as well as the battles of the air, land, and sea, and we have now won the battle of the laboratories as we have won the other battles.
>
> Beginning in 1940, before Pearl Harbor, scientific knowledge useful in war was pooled between the United States and Great Britain, and many priceless helps to our victories have come from that arrangement. Under that general policy the research on the atomic bomb was begun. With American and British scientists working together we entered the race of discovery against the Germans.
>
> The United States had available the large number of scientists of distinction in the many needed areas of knowledge. It had the tremendous industrial and financial resources necessary for the project, and they could be devoted to it without undue impairment of other vital war work. In the United States the laboratory work and the production plants, on which a substantial start had already been made, would be out of reach of enemy bombing, while at that time Britain was exposed to constant air attack and was still threatened with the possibility of invasion. For these reasons Prime Minister Churchill and President Roosevelt agreed that it was wise to carry on the project here.
>
> We now have two great plants and many lesser works devoted to the production of atomic power. Employment during peak construction numbered 125,000 and over 65,000 individuals are even now engaged in operating the plants. Many have worked there for two and a half years. Few know what they have been producing. They see great quantities of material going in and they see nothing coming out of these plants, for the physical size of the explosive charge is exceedingly small. We have spent $2 billion on the greatest scientific gamble in history--and won.
>
> **(Continued on Next Page)**

The gut response of most Americans has been that these bombings ended the war almost immediately and saved many Allied lives — but persuasive arguments can be mustered that alternatives to the atomic destruction of Hiroshima and Nagasaki could have had the same outcome without the nuclear consequences. In fact, it is on record that, at the time, the U.S. Navy had opposed the atomic bombing, advising instead that Japan could have been blockaded and brought to surrender even without an invasion.

> **Harry S. Truman's Announcement (Continued):**
>
> But the greatest marvel is not the size of the enterprise, its secrecy, nor its cost, but the achievement of scientific brains in putting together infinitely complex pieces of knowledge held by many men in different fields of science into a workable plan. And hardly less marvelous has been the capacity of industry to design, and of labor to operate, the machines and methods to do things never done before so that the brainchild of many minds came forth in physical shape and performed as it was supposed to do. Both science and industry worked under the direction of the United States Army, which achieved a unique success in managing so diverse a problem in the advancement of knowledge in an amazingly short time. It is doubtful if such another combination could be got together in the world. What has been done is the greatest achievement of organized science in history. It was done under high pressure and without failure.
>
> We are now prepared to obliterate more rapidly and completely every productive enterprise the Japanese have above ground in any city. We shall destroy their docks, their factories, and their communications. Let there be no mistake; we shall completely destroy Japan's power to make war.
>
> It was to spare the Japanese people from utter destruction that the ultimatum of July 26 was issued at Potsdam. Their leaders promptly rejected that ultimatum. If they do not now accept our terms they may expect a rain of ruin from the air, the like of which has never been seen on this earth. Behind this air attack will follow sea and land forces in such numbers and power as they have not yet seen and with the fighting skill of which they are already well aware.
>
> The secretary of war, who has kept in personal touch with all phases of the project, will immediately make public a statement giving further details.
>
> His statement will give facts concerning the sites at Oak Ridge near Knoxville, Tennessee, and at Richland near Pasco, Washington, and an installation near Santa Fe, New Mexico. Although the workers at the sites have been making materials to be used in producing the greatest destructive force in history, they have not themselves been in danger beyond that of many other occupations, for the utmost care has been taken of their safety.
>
> The fact that we can release atomic energy ushers in a new era in man's understanding of nature's forces. Atomic energy may in the future supplement the power that now comes from coal, oil, and falling water, but at present it cannot be produced on a basis to compete with them commercially. Before that comes there must be a long period of intensive research.
>
> It has never been the habit of the scientists of this country or the policy of this government to withhold from the world scientific knowledge. Normally, therefore, everything about the work with atomic energy would be made public.
>
> But under present circumstances it is not intended to divulge the technical processes of production or all the military applications, pending further examination of possible methods of protecting us and the rest of the world from the danger of sudden destruction.
>
> I shall recommend that the Congress of the United States consider promptly the establishment of an appropriate commission to control the production and use of atomic power within the United States. I shall give further consideration and make further recommendations to the Congress as to how atomic power can become a powerful and forceful influence towards the maintenance of world peace.

Regarding possible alternative targets, Hiroshima was a port city with military industries, although it was not truly a "Japanese army base," nor important as such, despite Truman's assertion. (Military forces, bases, and facilities were deliberately spread throughout the islands, so that few — if any — concentrated military targets actually existed at the time.)

Demonstration? Many Manhattan Project scientists, especially those who witnessed the Trinity test, pushed for a less harmful demonstration of the bomb's awesome power. Meeting at the University of Chicago Met Lab in the Spring of 1945, scientists such as Szilard, Rabinowitz, and Franck drew up a report bearing the latter's name, advising demonstration or warning before use on a city. The petition, dated 17 July 1945, called for the bomb's power to be "adequately described and demonstrated."

The Frank committee had been appointed to review the "social and political implications" of the bomb, and the group's primary argument for restraint was based on concern that the psychological and diplomatic shock of the bomb's use would destroy chances for reaching a post-war international agreement on prevention of nuclear warfare (and none was ever reached). Favoring "direct military use" was the Manhattan Project Scientific Panel Interim Committee, consisting of Lawrence, Compton, Fermi, and Oppenheimer, which — after conferring with scientific colleagues — stated the conclusions of its own analysis: "we can propose no technical demonstration likely to bring an end to the war; we see no acceptable alternative to direct military use."

Although Fermi — hoping to keep the bomb secret and out of circulation — was originally against either its demonstration or its military use, Oppenheimer persuaded him to change his mind and accept the statement.

The overruled minority thought that a demonstration outside of Tokyo would have enabled Japanese scientists to rapidly interpret the test and its fatal implications for their government. While the Japanese Army still might have prolonged their resistance, pressure on civilian authorities to surrender could have escalated. The Emperor would have been able to "save face" by pointing to the devastating nature of a new type of demonstrated weapon that could demolish Japan.

Scientific confidence was high enough for the uranium-loaded bomb (*Little Boy*) to be committed against Hiroshima without testing. The earlier *Trinity* test had proven the plutonium-implosion design (*Fat Man*) that demolished Nagasaki. Nevertheless, the Truman administration gave greater importance to concerns that a demonstration bomb might not detonate, and that was a time when very little weapons-grade material was in the supply pipeline (In fact, no additional atomic bombs had yet been fully assembled, the schedule for replacements being several weeks to a month).

In any event, a demonstration was politically doomed because it was never as important as other factors, such as insisting on unconditional surrender or elbowing Russia out of a post-war share in Asia.

The decision to use the bomb was clouded in secrecy, with various factions and interests privately applying pressure. Some analysts consider it *un fait accompli* — once the *Trinity* test showed that the bomb worked — because it would have appeared irresponsible not to use the weapon to "end" the war

As it turned out, Japan surrendered about a week after Hiroshima was destroyed.

Motivations. While some objective considerations can be documented, subjective factors could have been quite significant in the decision to use atomic bombs, particularly the view that it was "payback time" for the vicious 7 December 1941 "sneak" attack on Pearl Harbor. Senior statesman McGeorge Bundy admitted that he would have liked to have seen more informed input to the Hiroshima bombing decision, but recognized that the decision would have been the same.

Although anxious to punish the Japanese, the U.S. Army was actually not ready to attack in August: Invasion plans could not have been put into effect for several months, allowing time to explore less catastrophic, still timely alternatives. Training and mobilization for the invasion was under way, but it is widely accepted that November of 1945 was the earliest the assault on the main Japanese islands could have been mounted. Yet, if the war had not ended when it did, additional military and civilian lives of the allies would have been lost and prisoners of war would have continued to suffer.

Because conflict on the European front had already come to an end, the Soviet army was transferring its forces East in order to enter the war against Japan; it was planning to invade Manchuria on 15 August, but rushed in on 8 August, two days after the atomic bombing of Hiroshima. Soviet involvement would have decisively hastened the Japanese surrender, but American concern about the Soviets having too much of a post-war role in Asia might explain why there was so much pressure to use the atomic bomb.

Army Chief of Staff, General George C. Marshall, was one of the generals who agreed that the atomic bomb was needed to put a quick end to hostilities. After the war, he explained:

> We [the Army Staff] regarded the matter of dropping the [atomic] bomb as exceedingly important. We had just gone through a bitter experience at Okinawa [the last major island campaign, when the Americans lost more than 12,500 men killed and missing and the Japanese more than 100,000 killed in eighty-two days of fighting]. This had been preceded by a number of similar experiences in other Pacific islands, north of Australia. The Japanese had demonstrated in each case they would not surrender and they would fight to the death.... It was expected that resistance in Japan, with their home ties, would be even more severe.... So it seemed quite necessary, if we could, to shock them into action.... We had to end the war; we had to save American lives.

Immediately after Hiroshima, other wartime generals and admirals — Dwight D. Eisenhower, William D. Leahy, William Halsey, Curtis LeMay, and Henry "Hap" Arnold — "criticized the decision to annihilate Hiroshima." General Eisenhower is quoted is saying "It was not necessary to hit them with that awful thing." Perhaps he had in mind that the firebombing of Tokyo had killed 80,000 Japanese, comparable to the number in the atomic bombing of Hiroshima. In any event, few purely military targets remained in Japan near the end of the war, and cloud cover was a major factor in target selection on the days of the atomic bombings.

Potsdam Declaration. A week before the atomic bombing of Hiroshima "an opportunity to end this war" was formally offered to Japan as a result of the three-power Potsdam summit meeting (just after President Roosevelt died), which set terms for unconditional surrender and occupation of Japan. The terms did not provide for retention of the Emperor. The Potsdam Declaration closed with the statement that "The alternative for Japan is prompt and utter destruction." (Although the means of destruction was unspecified and vague, it was a deliberate allusion to the atomic bomb.) Prior to the atomic bombing, warning leaflets were dropped over Japan, but they were not specific enough to have an influence on surrender.

Japanese government reaction to the Declaration was to publicly disregard it: *mokusatsu* — "treat it with silent contempt" — so the United States continued with plans to use its ultimate weapon. Although Japan was near defeat and was ready to surrender, it wasn't ready to surrender unconditionally; and its policy was apparently to show how costly it would be for the allies to achieve unconditional capitulation. Admiral Leahy, as did other military leaders, objected to insistence of unconditional surrender on the grounds that it would lengthen the war.

When the Potsdam conference was in recess, the Japanese sent to Russia a message through diplomatic channels — intercepted and decoded by American intelligence — indicating that they would surrender provided the emperor was retained. Because of this intercept, it was realized that the war could be ended quickly by allowing Japan to keep the Emperor, but the United States government decided against the concession (which was allowed two weeks later, following the atomic bombings).

Monday-Morning Quarterbacking. After the war, Secretary of War Stimson, who had expected one million casualties in an invasion, wrote:

> My chief purpose was to end the war in victory with the least possible cost in the lives of the men in the armies which I had helped to raise. In the light of the alternatives which, on a fair estimate, were open to us, I believe that no man, in our position and subject to our responsibilities, holding in his hands a weapon of such possibilities for accomplishing this purpose and saving those lives, could have failed to use it and afterwards looked his countrymen in the face.

President Truman, in his private diary agreed with Stimson:

> He & I are in accord. The target will be a purely military one and we will issue a warning statement [the Potsdam Declaration] asking the Japs to surrender and save lives. I'm sure they will not do that, but we will have given them the chance. It is certainly a good thing for the world that Hitler's crowd or Stalin's did not discover this atomic bomb. It seems to be the most terrible thing ever discovered, but it can be the most useful.

Regarding the three-power meeting at Potsdam, in 1945, it was not at all clear that the bomb was needed any longer to defeat Japan. The bomb would be used for another purpose, to "make Russia more manageable in Europe."

Ending the War. Almost all the U.S. generals and admirals had told President Truman that the atomic bomb was not necessary to end the war in Japan. The Allies had established complete air and sea superiority, effectively blockading the islands. In fact, General Dwight Eisenhower had angered Secretary of War Stimson with his significantly different assessment.

> The bomb was therefore dropped on Japan for the effect it had on Russia — just as Jimmy Barnes had said. The psychological effect on Stalin was twofold: the Americans had not only used a doomsday machine; they had used it when, as Stalin knew, it was not militarily necessary. It was this last chilling fact that doubtless made the greatest impression on the Russians.

Historians have since discovered that paragons of the conservative Republican establishment at the time, such as former President Herbert Hoover and *Time* magazine publisher Henry Luce, thought that dropping the bomb was unnecessary. Luce wrote:

> If, instead of our doctrine of "unconditional surrender," we had all along made our conditions clear, I have little doubt that the war with Japan would have ended no later than it did — without the bomb explosion which so jarred the Christian conscience.

It appears that most diplomatic historians, if not all military historians, now agree that there were alternatives to ending the war with the bomb and Harry Truman and his closest aides knew it.

A slightly different perspective has been offered, putting less stress on the role of the bomb to "scare" the Russians, — instead, faulting Truman's decision as being based on erroneous judgements about the situation in Japan: The military (Army) assessment was that a conventional invasion would have resulted in a million or more American casualties. Truman believed that and so acted, but some outsiders believe "the assessment was false:" In order to agree upon a unified casualty estimate, the four autonomous military services came up with an unrealistic number that did not take into account a unified attack, nor did it take fully into account the potential of diplomatic action.

> **[Recollection by Wolfgang Panofsky].** I was involved at Lost Alamos developing a device to measure the yield from a nuclear explosion with a condenser microphone quantitatively measuring the shockwave. This information was propagated by telemetry to an airplane accompanying the primary craft dropping the weapons. This telemetry device, dropped by parachute, was powered by a large battery box. Luis Alvarez, working for the Manhattan Project, taped a letter to this box addressed to Sagane, a prominent Japanese physicist who had worked at Berkeley before the war. This letter explained the nuclear nature of the weapon.
>
> Sagane indeed received this letter immediately after the Hiroshima event and delivered it to the Japanese high command. Presumably it influenced to an unknown extent the Japanese decision to surrender.

Nagasaki. Because the damage accomplished by a single atomic attack was not widely understood at the time, it was thought by some that many bombings would be required to bring the stubborn Japanese government to surrender unconditionally. This attitude fed the U.S. government's decision to use two bombs in sequence, in part to create the illusion that many more were available to fulfill the threats if peace terms were not accepted. This also fitted into the plans of General Groves, who wanted to field-test the plutonium bomb (which was one dropped on Nagasaki three days after Hiroshima was destroyed).

Postmortem. Would the war have ended promptly without the second nuclear bombing that followed only three days later? An historian, examining President Truman's diaries and other records, inferred that "President Truman did not know about the [atomic] attack on Nagasaki until after it happened."

Truman's diary, on 5 July 1945, recorded that he had given Secretary of Defense Stimson his "final order of the bomb's use" and that he wanted it used on military targets, not on women and children. Oppenheimer is reported as having been told not to ship the second bomb until he received an explicit order from the president — an order never received. Even before the bombing of Hiroshima, General Groves was pressing for speeded-up delivery of the second bomb. The historical evidence indicates that Groves gave orders that "were so carefully worded that they left considerable discretion to the field commander for the date, time of attack, and choice of target."

In March 1944 General Groves was heard to say, "You realize, of course, that the main purpose of the Project is to subdue the Russians."

A few months after the bombing, he suggested the United States should not "permit any foreign power which we are not firmly allied ... to make or possess atomic weapons." In 1954 Groves acknowledged that the Manhattan Project was conducted on the basis that Russia was our "enemy."

A present-day viewpoint sometimes expressed in Japan is that the second test supplied convincing evidence of American desire to test both the uranium and plutonium bombs on a populated area, particularly to study the effects, so as to gain superiority over post-war Russia.

Apart from military or political considerations, technical factors also were in minds of some of the Project's managers. Besides scientific curiosity to be satisfied, military interest in weapons effects might have been a factor in exploding the plutonium design over Nagasaki, to permit comparing it with the Hiroshima uranium weapon.

Consequences. The longer-range consequences of those decisions were enduring and irreversible. As pointed out by historian David Holloway,

> In August 1945, after the American attacks on Hiroshima and Nagasaki, the Soviet Union launched a full-scale effort to develop the atomic bomb ... a challenge ... to Soviet security and to Soviet interests. [This] decision came after the U.S. exhibited its willingness to use the weapon for military purposes, not earlier after the Trinity test, of which the Soviets were fully aware.

In fact, those who thought that using the atomic bomb would hold the Soviets in check made the biggest mistake of all. Holloway, in reviewing a history of the Soviet atomic enterprise, comments that "it was only after the atomic bomb was dropped on Hiroshima that Stalin made the decision to convert the project into a high-priority industrial effort." Not only did the use of the bomb fail to stop the Soviet, it seems to have accelerated the endeavor.

Whatever role these two atomic bombs had in expediting the end of Japanese resistance, Stalin's government perceived the nuclear explosions as a threat to communist expansion (or communism itself) in the postwar world.

In his Epilogue to *The Making of the Atomic Bomb*, Richard Rhodes quotes Szilard:

> The development of atomic power will provide the nations with new means of destruction. The atomic bombs at our disposal represent only the first step in this direction, and there is almost no limit to the destructive power which will become

available in the course of their future development. Thus a nation which sets the precedent of using these newly liberated forces of nature for purposes of destruction may have to bear the responsibility of opening the door to an era of devastation on an unimaginable scale.

Another way of viewing the decision is to look at how other nations have proceeded after succeeding in developing an atomic weapon. All acknowledged nuclear-weapon states (recently including India and Pakistan) have openly tested their weapons in remote locations, thus achieving their goals without bloodshed.

Two nations that did not publicly admit their success in weapon-making (and had not openly tested) are South Africa and Israel. South Africa's policy was quite telling: first, to keep the world guessing about the existence of its bombs; second, if threatened, to formally announce possession of nuclear weapons; third, to test a bomb openly as a warning. Israel seems to have secretly passed through the first stage of a similar plan, leaving the latter two stages as future options.

****

Debates about causation aside, the development and use of the atomic bomb led to four chancy decades of unstable Cold War stalemate during which the nuclear weapons states could have had their homelands obliterated.

It would be difficult to make a case that the Soviet Union, soon followed by other nations, would not have proceeded to make their own nuclear weapons had they not been used against Japan; once the nuclear genie escaped, it seemed destined to propagate. Perhaps without the bloodshed in Japan, nuclear-armed nations would have become less cautious — more cavalier in carrying out Cold War national policies.

Although the modern context differs significantly from the situation near the end of World War II, policymakers of the United States might have been able to meet their war-ending objectives without using atomic bombs on cities. By negotiating more flexible surrender conditions, the war might have ended early in August 1945 without the atomic destruction.

Originating with the deadly explosion of the atomic bomb, radiation phobia — radiophobia — is a deplorable and counterproductive outcome that has hampered acceptance of peaceful applications of nuclear power for electricity and medical treatment.

On the other hand, many who have evaluated the history of the Cold War believe — some fervently — that the existence, and possibly the use, of atomic weapons prevented a third World War.

Regrettably, though, the new-millennium conflict generated by the G.W. Bush administration's invasion of Iraq is frighteningly reminiscent of the situation in 1945: Government officials rarely, if ever, candidly acknowledge the true factors underlying their decisions.

Trinity plus 50: Getting Where We Are Now. After World War II, nuclear weapons became part of the Western political arsenal against communism. Fusion-boosted fission devices were tested as a sequel to the fission explosions that devastated Japanese cities. Thermonuclear-fusion weapons eventually became part of a nuclear arms buildup that approached doomsday scale. Just one of six 100-megaton-plus devices produced by the Soviet Union could have leveled and burned an entire metropolitan area, and, if exploded near the ground, would have spread lethal fallout over a region much larger. The Soviets made those huge aerial bombs (see photo, Figure 1) in order to counter a growing U.S. advantage in number and accuracy of smaller intercontinental nuclear missiles.

(Two Argonne colleagues, Vladimir Minkov [left] and DeVolpi [right], meeting in 1991 with the chief weapons designer Boris Litvinov in the Russian nuclear weapons museum at Chelyabinsk-70)

Figure 1: Full-Scale Replica of World's Largest Thermonuclear Bomb

To optimize performance, nuclear weapons and their delivery systems were tested in the atmosphere and underground. Most of the test information, even their occurrence, was kept secret until the mid-1990s, when some details were released. During that interlude, radiation-exposure experiments were conducted on humans, in secret and without their informed consent.

Radioactive materials left over from nuclear-materials production were simply dumped outdoors, especially in the Soviet Union, with little attention to public or environmental risk. This was done in the name of national security, the agencies doing the dumping emboldened by the knowledge that secrecy protected them from public challenge. Ensuing accidents and leaks led to devastating human and ecological problems that still persist, many of which are detailed in this book (particularly in Volume 2).

One might ask, what price national security? The secret Manhattan Project bomb program cost nearly $2 billion, a huge sum in 1945.

Problems for the New Millennium

The arms race of the past half-century was characterized by intense competition to gain military advantage or negate an adversary's favorable position. This was done through the introduction of nuclear warheads that had higher explosive yields, better yield-to-weight ratios, less throw-weight, smaller size, more reliable function, improved safety, augmented radiation, or special destructive effects. The large number of weapons-delivery platforms caused specific military standards to be established for nuclear-armed intercontinental missiles, aerial bombs, artillery shells, torpedoes, depth charges, demolitions, cruise missiles, etc. Clearly these were designed to wage war; they were not primarily defensive weapons.

The risk of nuclear devastation by accident or without authorization strongly depends, among other factors, on how many armed missiles and deployed nuclear warheads there are. Upon warning of attack, a slim margin for decision or error exists in operational policies for launching missiles; only tens of minutes are available for heads of state to make fateful command decisions. Of the 20,000 false alerts processed in the United States from 1977 to 1984, more than 1100 of them required senior Department of Defense officials to evaluate the threats; six of the warnings "escalated to senior level emergency conferences, one step short of a 'missile attack conference' involving the President." Even if nuclear arms reductions are substantial in the future, some risk will remain unless operational safeguards are put into place for all nuclear arsenals.

The military utility of nuclear weapons is being questioned more and more, and under the START I strategic-arms reduction treaty some weapons have been, and are being dismantled. To make the process irreversible, however, fissile material must be made unsuitable for re-use in weapons.

It is widely accepted that enriched uranium can be readily demilitarized by blending it with natural or depleted uranium, or by burning it in reactors. Demilitarization of weapons-grade plutonium, on the other hand, is more controversial. arms control advocates would like to render high-quality plutonium useless as a replacement material for existing nuclear-weapon designs. But the scientists and engineers who designed, developed, fabricated, and certified the nuclear weapons are understandably reluctant to face destruction of their creations. Although acquiescing to dismantlement of surplus warheads, they (and some academics) balk at demilitarization of plutonium. They support storage methods that would preserve plutonium indefinitely in a form that can be easily reused.

American, Russian, British, French, and Chinese nuclear-tipped long-range missiles, ready to fire at short notice, are re-programmable for either military or civilian targets. New participants, notably India and Pakistan, are entering the long-range ballistic-missile arena, and other nations are standing by. Nuclear and other indiscriminate weapons — chemical, biological — seem to be becoming more widely available with adequate delivery systems, despite the demise of the Cold War. Because terrorism has been getting to be somewhat more sophisticated and difficult to suppress, there is a temptation to seek a role for nuclear weapons in countering terrorism.

Reassessing Government Secrecy. Whether driven by a literal interpretation of the law or by a feeling that "it's best to err on the side of caution," excessive information control has made it difficult to have open discussion of sensitive nuclear issues. Government agencies have made futile attempts with the secrecy stamp to restrict the dissemination of information (even foreign publications) after it has become public — thereby drawing attention to, and implicitly confirming, its validity, contrary to elementary principles of information protection. Some established principles of physics are officially categorized as secret even though they have found their way into textbooks. Excessive bureaucratic zeal has led to security classification of innocuous references to fictional works.

Control of information is also essential for management of decisions, for manipulation of public opinion. For instance, rejection of alternatives to dropping the atomic bombs on Japanese cities was based, in part, on a low expected production rate for nuclear assemblies. That decision might have been different had there been a broader (albeit still security compartmentalized) discussion of its implications among government scientists and policymakers.

Events of historical importance need to be reassessed in light of an understanding of the accessible physics of reactors and fission explosives and the current risk of nuclear proliferation. The all-important success of espionage by the Soviet Union says something about over-reliance on secrecy as a barrier to proliferation. Stalin's spies uncovered much about the Manhattan Project, and his military laboratories made a weapon that appears (according to published photographs) to be nearly identical to the "Fat Man" exploded over Nagasaki. More recently there were allegations that China has succeeded in stealing critical information about the design of certain advanced American nuclear weapons.

Despite the limitations and abuses of government secrecy, technical details about weapons indeed need to be safeguarded as long as possible. Wholesale declassification of sensitive technical information that could truly damage national or international security is not advised. Moreover, "information secrecy" is distinguishable from "security procedures," the latter being the means by which restricted information is protected. While procedures for safeguarding secret information need to be maintained at a rigorous level, a balance must, and can be reached that avoids trampling on constitutional rights.

The official, operational definition of "national security" depends on who holds sway. National security is often defined to reflect the contemporary interests of special groups. It can be very promotional, as hybrid military-industrial-laboratory mutual self-interest has proven.

Secrecy was a crucial tool in sustaining the nuclear arms race. Caught up in it were many non-sensitive aspects of the peaceful application of nuclear energy that were not declassified until a decade or more after *Trinity*. With the passage of time, it's been found that deliberate, outrageous exposures of humans and the environment were carried out in the name of national security — all of this covered up until recently by secrecy restrictions. This leaves open the question of how to strike the right balance between data that needs continued protection and information that should be disclosed to benefit discussions of policy.

As explained in several Chapters, unnecessary government secrecy continues today to interfere with public understanding of, and participation in the formulation of nuclear policy.

Nonproliferation and Counterterrorism Issues. For nuclear weapons, very-high-quality plutonium and uranium are essential ingredients. Any insufficiently safeguarded inventories of these materials pose a global danger. To reduce this danger, demilitarization of nuclear materials used for weapons must remain high on national and international agendas.

Will nuclear proliferation be rolled back in the new millennium? This will surely depend on more progress in implementing the Non-Proliferation Treaty, in cutting off production of weapons materials, and in curtailing tests of nuclear explosives.

Under Article VI of the Non-Proliferation Treaty, the United States and other signatories obligated themselves to pursue negotiations in good faith on effective measures to stop the build up of nuclear arsenals at an early date. They also agreed to achieve nuclear disarmament; thus, while nuclear elimination might seem to be highly implausible, it is already a commitment of the treaty parties. Although the arms race for a while was winding down, the quantity of nuclear weaponry still in functional form is huge. The policies that have led to this situation need critical and impartial evaluation.

One can visualize a world in which reciprocal verification of reductions — even deep cuts in armaments — by the United States and Russia would be militarily effective and budget-friendly. Safeguarding nuclear materials in storage, based on realistic assessment of the sensitivity of the information, would create mutual confidence and security. The cost of safeguarding fissile materials could be held in check by applying measures commensurate with realistic chances and consequences for illicit diversion of nuclear materials. Sufficient data could be released to fashion a domestic and international nuclear-safeguards policy that would be cost-effective while preserving the national security of all parties. Meanwhile, other forms of destructive weaponry are being sought by malefactors, creating problems that cannot be simply dealt with by traditional negotiated arms control.

Just some of the more pressing issues and goals to be discussed here have been highlighted. There are also hidden costs associated with policies that favor storage rather than elimination of surplus nuclear weapons materials. Delay in dismembering institutions created to promote nuclear weapons will continue to bloat national budgets. Weapon scientists assert that they know how to keep their nuclear weapons in a state of readiness. So, given the more-than-sufficient stockpile of stable certified warheads, doesn't an immense nuclear arsenal seem unnecessary?

A major challenge for security in the new millennium is to lessen the still-present danger of deliberate, accidental, or unauthorized use of nuclear explosives. While the extended state of East-West belligerency is over, the quantitative and qualitative nuclear arms race has not died out, nor are weapons being eliminated at a pace consistent with the newfound security that comes with the end of superpower confrontation. Scary situations have occurred, one of which was the 1995 incident mentioned earlier: the scientific-research rocket that was initially registered by a Russian early-warning system as a U.S. missile which could blind Russian radars to a surprise attack. Of the ten scarce minutes available for Russian officers to decide whether it was a hostile missile, it is said that eight of those minutes were used to alert the Russian command, all the way up to President Yeltsin.

For this new century to benefit from nuclear demilitarization, several things must be considered:

Finishing the task of controlling and curtailing nuclear arsenals will have to be given high priority, but we cannot expect progress to be rapid. Elimination of weapons would take place stepwise, an early stage being at least partial demilitarization of the nuclear materials created in the 20th century.

If vested interests, taking advantage of counterproductive security rules and practices, are the force behind delays in demilitarization, the public should be made more aware of it.

Absent verifiable, reciprocal, and binding arms control accords, the potential for another type of arms race (for example, between the United States and China) is imaginable. Reinforcing worldwide nonproliferation norms and systematically backing away from excessive nuclear inventories and dependencies will reduce all aspects of the nuclear risk. The nuclear superpowers should lead by setting better examples of restraint, if they want to minimize the expansionist plans and capabilities of such nations as India, Pakistan, North Korea, and Iran.

In addition to the ticking nuclear time-bomb, other weapons — like chemical and biological that were developed during the Cold War — are coming back to haunt humanity and disrupt peaceful coexistence.

Chapter I:

THE COLD WAR

(Two Scorpions in a Bottle)

Chapter I: The Cold War

A. NUCLEAR EVENTS THROUGH WORLD WAR II 23
   [Enriching Uranium] 24
   [The Frisch-Peierls Memorandum] 24
   [Nuclear Chain Reaction] 25
The First Nuclear Weapons 25
   [Nuclear-Weapon Configurations] 26
   [German Atomic-Bomb Research] 26
   [Soviet Pre-War Nuclear Research] 27
   Figure 2: Mushroom Cloud After Atomic Bombing of Hiroshima 28
Hiroshima and Nagasaki 29

B. POST-WAR YEARS: *MANO A MANO* [1945–1960s] 30
   [The Iron Curtain] 30
   [Nuclear Secrecy] 31
   [Dropshot] 34
Berlin 34
End of Nuclear Monopoly 36
   [RDS-1] 36
   [The Nuclear Aftermath] 36
NATO 37
Korean War 38
   [Strategic vs. Tactical] 38
Atoms for Peace 39
   [The Smyth Report] 39
Testing Nuclear Weapons 40
   [Nuclear-Weapon Testing and Secrecy] 41
Developing Thermonuclear Weapons 41
   [Boosted Weapons] 41
   [Fusion Weapons] 42
Long-Range Bombers 44
Theater Weapons and the Neutron Bomb 45
Intercontinental Ballistic Missiles 46
Bomber and Missile Defenses 47
Surviving Nuclear War 48
Cuban Missile Crisis 49
   [Events Leading Up to the Cuban Missile Crisis] 50
Vietnam War 52
   [Vietnam War Narrative] 53
   [Gulf of Tonkin Incident] 54
Changing Nuclear Balance 54

C. PHASES OF *DÉTENTE* [1970s-1989] 57
The 1970s 57
   [ABM] 58
   [MIRV] 58
   [SALT I] 58
   [CSCE/MBFR] 59
   [Backfire] 60
   [B-1] 60
   [Cruise Missiles] 60
   [SALT II] 61
Afghanistan 62
The 1980s 62
   PD-59 62
   [Neutron Weapons] 63
   [INF] 65

D. COLLAPSE OF THE SOVIET UNION [1989-1991] 66
  [START I] 67
  [Dzerzhinsky Toppled] 67
  [Nunn-Lugar Act] 68
Cold War Reprise 68
  Winners and Losers 69
Russia as Successor 70

E. COLD WAR LEADERSHIP 71
  Superpower Leadership 73
  Superpower Leaders 73
  Non-Aligned Nations 74
  Treaties and Agreements 74
  [START II] 76

When the atomic bomb was dropped on Hiroshima in 1945, the most revolutionary weapon ever to be introduced into warfare destroyed not only a city and much of its population, but also all classical concepts of warfare. It was no longer possible to defend industrial centers, territories, or populations. The concept of defense became an anachronism, to be replaced by deterrence — the threat of (nuclear) retaliation. In this Chapter, we trace the course of nuclear weaponry from its beginnings in World War II to the unceremonious end of the Cold War. During that span, nearly a hundred-thousand nuclear weapons were fabricated, and the world came close to self destruction.

A. NUCLEAR EVENTS THROUGH WORLD WAR II

In this initial Section, the development of nuclear weapons is followed from the pre-war discovery of fission to their horrific use against Japan. A novel combination of experimental, theoretical, and developmental activities were successfully accomplished by Allied scientists and engineers in a sustained and cooperative project.

During World War II the Western Allies undertook to develop an atomic bomb before Nazi Germany could get one.

The fundamentals of nuclear fission had been discovered well before the war. Scientists in many countries, notably including Germany, Russia and Japan, were aware of that discovery and its implications.

In May 1939, Joliot-Curie's team in France deposited three patents on exploiting nuclear energy and fission explosion. Moreover, they had calculated the necessary size of a nuclear reactor using heavy water and were thinking about preparing a test site for an experimental explosion in the Sahara.

British contributions to the concept and realization of the explosive chain reaction are often slighted. James Chadwick and Joseph Rotblat were among the first to appreciate the potential for weaponization. Fundamental insights coherently assembled by Frisch and Peierls were that a bomb would depend exclusively on the rapid multiplication of fast neutrons, that the fissile materials must be pure uranium-235, and that mere kilograms of fissile uranium were needed. This theoretical foundation for developing a nuclear explosive was passed on to the Americans. A secret study by the British MAUD Committee in July 1941 concluded that uranium-235 could be separated from natural uranium "within three years" and that both weapons and energy production from uranium-235 were scientifically feasible and should be pursued under high priority; this report helped Roosevelt decide to allocate huge resources to a U.S. effort to make an atomic bomb.

[The Frisch-Peierls Memorandum]. If any published document can be said to be the first convincing technical argument about making an atomic bomb, it was a UK memorandum of March 1940 by O.R. Frisch and R.E. Peierls of Birmingham University. Circulated among UK and U.S. scientists, the memo indicates the reasoning that scientists of that day could deduce about development of a fission explosive. It was titled "On the Construction of a 'Super-bomb' based on a Nuclear Chain Reaction in Uranium."

Their ideas included key aspects that later turned out to be correct: the importance of isotope uranium-235 to a rapidly multiplying chain reaction; the number of neutrons produced per fission; the negative effects of hydrogen for a fast reaction; the exponential increases in neutrons; the temperatures reached and energies released; and the necessity to have a critical size of dense material.

Frisch and Peierls estimated that "The energy liberated by a 5-kilogram bomb would be equivalent to that of several thousand tons of dynamite...." The memo contained suggestions for rapid assembly of the uranium parts in order to produce a fission-induced explosion.

In order to produce large amounts of concentrated uranium-235, the two pioneering authors advised a multi-tube thermal-diffusion uranium-enrichment system of 100,000 tubes.

Even the destructive effects of the explosion were qualitatively understood, including prompt radiation and fallout.

[Enriching Uranium]. It was soon discovered that isotopes of natural elements could be separated from each other, e.g., uranium-235 could be extracted out of natural uranium, in a process called *enrichment,* by any of several very complex methods.

Isotope enrichment is very important for weaponization because uranium-235 is *fissile* — that is, when bombarded with neutrons it breaks up (fissions) into smaller components accompanied by the all-important release of nuclear energy and neutrons.

Germany initiated a competent atomic program led by renowned nuclear physicist Werner Heisenberg, but internal and external factors served to impede a favorable outcome.

As the war got under way, Japanese scientists, who followed the pre-war physics publications of the West, initiated their own small nuclear project. They made critical-mass calculations and pinpointed uranium mines in northern Korea, then under Japanese occupation. Although the initial goal was to provide nuclear propulsion for their Navy, an aeronautic institute decided that an atomic bomb was feasible. Uranium separation was started on a small scale, but the facilities in mainland Japan were bombed out.

The western Allies (United States, Canada, and Great Britain) did not share nuclear information with the Soviets, who, hampered by the relentless German invasion, tried to keep up with developments by setting up an espionage network. Foremost among Russian spies was the British-naturalized German physicist, Klaus Fuchs, who had been assigned to the top-secret U.S. atom-bomb development project. Guided partly by purloined information, the Russian scientists developed their own program — which, however, was limited by higher wartime priorities.

[**Nuclear Chain Reaction**]. Chain reactions to create explosions are fueled primarily by fissioning fissile isotopes. Although a nuclear reactor can be made to operate in a controlled fashion with natural or slightly enriched fuels (such as uranium), a nuclear explosive requires highly enriched uranium or weapons-grade plutonium.

Because plutonium doesn't exist in nature, it had to be made in a nuclear reactor. Under the leadership of Enrico Fermi at the University of Chicago, the first self-sustaining nuclear-fission chain reaction was accomplished in the CP-1 (Chicago Pile 1) reactor at the end of 1942. Creating this reaction was an essential step in developing nuclear weapons, and it also turned out to be the first phase in the ultimate development of nuclear reactors for power, research, propulsion, and radioisotope production.

CP-1 was fueled with natural uranium — uranium separated chemically from mined ore ("yellow cake"). It had already been recognized that either a form of the natural element uranium or of the artificial element plutonium would be an essential component of nuclear explosives.

The CP-1 nuclear "pile" consisted of more than 6 tons of uranium metal embedded in a stack of 350 tons of purified graphite assembled at a squash court beneath Stagg Field on the University of Chicago campus.

The First Nuclear Weapons

While the feasibility of a fission chain reaction was being established in Chicago, a crash program called the "Manhattan Project" or the "Manhattan Engineering Project" was initiated to create a nuclear explosive. Formed in September 1942 under the leadership of Brigadier General Leslie Groves, the atomic-bomb project remained in existence until 1947.

One of Groves' first actions was to choose Oak Ridge, Tennessee, as the location for production of enriched uranium. Also, while physicists at the University of

Chicago (Met Lab) under Enrico Fermi worked on assembling a nuclear reactor, a remote location (Site Y) at Los Alamos, New Mexico, was selected in late 1942 to engineer weaponization of the fission reaction.

In the meantime, Berkeley physicist J. Robert Oppenheimer gathered a small group of theoretical physicists — including Hans Bethe and Edward Teller — to discuss designs for an atomic bomb.

> **[Nuclear-Weapon Configurations].** Two workable configurations, *gun barrel* and *implosion*, were conceived for making nuclear explosions.
>
> The gun-barrel design, so-called because fissile material was rapidly consolidated by compressing it in a tube, worked only with enriched uranium.
>
> The implosion technique was more complex, requiring inward-directed explosive-driven uniform compression of fissile material. This was the only method that worked with plutonium.

The gun-barrel method for achieving an atomic explosion was devised at Los Alamos, and isotopic separation to produce the enriched uranium required for its fissile core was undertaken by several methods. Enriched uranium was needed because naturally occurring uranium had been found to be inadequate for the explosive chain reaction (in a *critical mass*, the smallest amount needed).

Eventually, for plutonium, the alternative *implosion* method was developed to trigger the nuclear detonation. Implosion was necessary to overcome erratic initiation of the chain reaction due to the quality of the plutonium then being produced in the nuclear reactors. An implosion design is still considered necessary for plutonium-based nuclear weapons.

> **[German Atomic-Bomb Research].** Not all salient conclusions about fast-neutron multiplication were reached correctly by Werner Heisenberg and his German team. They were sidetracked — not by lack of zeal — but by some consequential theoretical and experimental errors, coupled with resource limitations. Walter Bothe at Heidelberg in January 1941 made a measurement of the neutron-absorption probability in graphite that mistakenly led them to conclude that it was not an effective medium. In addition, high purity graphite on an industrial scale was expensive in the wartime environment.
>
> By October of 1941, experimental results were obtained in Leipzig from a reactor composed of uranium oxide and heavy water. As a result, Heisenberg realized by 1942 that uranium-235 could be used to make a bomb and that a reactor could generate plutonium; however, he was hampered by overestimates — a factor of 1000 — of the quantity needed to make a nuclear explosive. (This was not the case for the Manhattan Project.)
>
> The German (war-constrained) decision to use heavy water left the program vulnerable to sabotage and aerial bombardment of the production plant in occupied Norway. Nuclear reactors would have been easier to hide.
>
> The entire project was thwarted also by German arrogance, complacency, rivalries, and the consequences of losing talented Jewish scientists.
>
> Meanwhile, the German military campaign to destroy communism was beginning to take its toll, the British were standing up to the aerial bombardment, and the United States was about to enter the war. Hitler needed Von Braun's missiles sooner than he needed a theoretical weapon.

Section A: Nuclear Events

After achieving practical proof of the nuclear chain reaction, Fermi personally guided essential steps to design, develop, and construct weapons-grade plutonium production reactors. (Following the World War II Manhattan Project, this initiative lead to the development by Argonne National Laboratory of nuclear reactors for power, propulsion, and materials testing.)

> **[Soviet Pre-War Nuclear Research].** Scientists in the Soviet Union closely monitored European developments before World War II, and they maintained their own nuclear-research programs, producing original results in nuclear isometry, spontaneous fission, and nuclear interactions. Other initiatives, dating back to World War I, had led to basic and applied research in nuclear physics, radiochemistry, and accelerator physics.
>
> As the World War II approached, Soviet scientists studied conditions for nuclear chain reactions, separation of isotopes, and concepts for realizing controlled or explosive energy releases. In 1940, as combat for national survival raged, resources were limited to essential research rather than development of a nuclear-explosive device.

While the intensive espionage conducted by the Soviets during World War II seems to have been of benefit to them, they would have developed their own nuclear weapons without it: According to atomic historian Richard Rhodes,

> the {U.S. and Soviet] bomb programs ran in parallel because the raw materials, the processing and the technology depended upon universal physical fundamentals that both sides could determine independently. At that basic level, there was never any "secret" of how to make an atom bomb. Knowledge derived from espionage could only speed up the process, not determine it, and in fact every nation that has attempted to build an atomic weapon in the half-century since the discovery of nuclear fission has succeeded on the first try.

Gregg Herken, also an historian, estimates that Klaus Fuchs' stolen information might have advanced the Soviet A-bomb by 18 months. David Greenglass, through the Rosenbergs, supplied sketches and drawings of the detonating mechanism. Another spy who worked at Los Alamos, Theodore Hall (cover name "Mlad") was able to give to the Soviets conceptual information — but probably not detailed designs — on the implosion method. Although Stalin formed a "Uranium Committee" in 1939, he did not convert the Soviet atomic project into a crash program until after the Hiroshima detonation.

Sergei Kapitza, son of physicist Peter Kapitza, has estimated that the intelligence information might have gained the Soviets about one year in advancing their atomic bomb. Actually they developed two bombs, their own being about half the size and twice the yield of the *Trinity* device, but — under pressure from Beria — they tested the imitated design first.

The European phase of World War II ended on 7 May 1945. Although the Allies knew by the end of 1944 that the Germans were no longer working on the atomic bomb, the Manhattan Project continued unabated. Only Joseph Rotblat left the project at that point, stating that his original rationale and obligation were fulfilled.

Figure 2: Mushroom Cloud After Atomic Bombing of Hiroshima
(from *Newsweek*)

Hiroshima and Nagasaki

On 6 August 1945, the Enola Gay — a modified B-29 bomber — delivered the first atomic bomb to destroy Hiroshima and directly kill about 70,000 people. Three days later a second atomic weapon demolished Nagasaki, putting to death 35,000 in the city. Each of the two bombs dropped on Japan weighed about 10,000 pounds (4550 kilograms) and were delivered to targets located more than 1000 miles (1600 kilometers) away from American bases in the Pacific.

These bombings hastened capitulation of the Japanese government. World War II came to an end without an invasion of the main islands of Japan. Many lives would have been lost in an invasion, had one been necessary.

The atomic strikes had a lasting impact on nuclear policy during and since the Cold War.

The decision to use atomic bombs on Japanese cities, vis-a-vis other options, remains controversial. Although the Imperial Government was responsible for initiating the war, with the resulting devastation throughout the Pacific, long-term implications of using atomic weapons were recognized by only a few insiders. Secrecy concerning the development and existence of the atomic bombs was paramount. Some scientists who helped develop the new weapon made a plea for a harmless demonstration, but President Truman, with the advice of the cabinet, decided to bomb cities directly.

The war then ended, and major fatalities for the allies quickly came to an end.

> A demonstration of the awesome new power could have been carried out, for example, in Tokyo Bay. Although wartime casualties and atrocities were still occurring, the risk of massive losses of Allied soldiers was not imminent, since an invasion of Japan could not have been mounted for several months.
>
> Given the powerful effect of the first bombing, the wisdom and ethics of destroying Nagasaki three days after Hiroshima have also been questioned. Unconditional surrender might have been attained by allowing more time for the effect of Hiroshima's devastation to sink in, and by threatening in advance to destroy another unspecified city.

Other alternatives for early termination of hostilities also existed: Having broken the Japanese diplomatic code, the United States was aware of, but would not then agree to, their conditions for capitulation — principally that the Emperor be allowed to remain. Ironically, the United States did in the end accept those same conditions for surrender.

Often considered one of the strongest reasons for the nuclear attack was the emotional residue from the devastating surprise attack by the Japanese on Pearl Harbor (a day that "will live in infamy"), which brought the United States into the war. Moreover, racial biases were often verbalized and acted upon during the war.

Military commanders of World War II have regarded the bombings as more political than military in purpose.

"[A] top-secret 1946 study found in the War Department's files suggested that Truman knew before the bombing of Hiroshima and Nagasaki that such action was not required to force Japan's surrender." Much current thinking characterizes the decision as primarily a warning shot to discourage Soviet expansion — to make the Soviet Union "more manageable," particularly in Eastern Europe.

Many analysts consider the atomic destruction in Japan to have triggered the nuclear arms race.

****

Nuclear fission and the atomic bomb became good-news/bad-news highlights of the last century. It was not planned that way: Events in World War II led to the exploitation of scientific discoveries in the development of nuclear explosives as a tool for staving off aggression. Given the institutional pressures, it was probably inevitable that atomic bombs would be used to put a prompt end to the bloody war with Japan.

B. POST-WAR YEARS: *MANO A MANO* [1945–1960s]

With the end of the World War II (the Great Patriotic War, as it is known in Russia) came the first steps by the Western Allies toward containing Soviet expansionism in Europe — an early stage in the prolonged belligerency. These steps included the improvement and deployment of nuclear weapons, with the associated political decisions to incorporate them into military strategies.

The next quarter century saw the retrenchment of the Soviet Union behind its "iron curtain" — Churchill's phrase — followed by the Berlin blockade, the formation of NATO and the counterbalancing Warsaw Pact, the Korean War, and the Cuban missile crisis.

> **[The Iron Curtain].** By 1946 the whiffs of Cold War were becoming evident, a result of the continued Soviet occupation of eastern European nations. At Fulton, Missouri, on 5 March 1946, in one of his defining moments, Winston Churchill stated that "an iron curtain has descended across the Continent."

Significant nuclear developments in this period included the testing of an atomic bomb by the Soviets, the start of extensive nuclear testing in the atmosphere, the development of thermonuclear weapons and intercontinental ballistic missiles, and attempts to find defenses against those missiles. The initial nuclear superiority enjoyed by the West eroded to a level of parity as the two sides reached a condition of mutual assured destruction (aptly, MAD). Efforts to control the proliferation of nuclear weapons began.

The first five years after World War II are characterized by historian Gregg Herken as a period of "diplomacy and deterrence." An early American showcase for deterrence was a nuclear-weapon test on Bikini Atoll in the summer of 1946, to which a Soviet observer was invited (whether the invitation was accepted is not clear). The test, having the mission of evaluating the effects of atomic bombs on ships, was actually a product of American military inter-service rivalry.

George Kennan, the senior American diplomat in the USSR, had been monitoring official Russian views, particularly those of Stalin. His "Long Telegram" — 8000 words in February 1946 from Moscow to the State Department — analyzed Soviet foreign policies, observing that their sense of being encircled by capitalist nations led to a "neurotic view or world affairs." Given Leninist beliefs about the inevitability of war between capitalist and socialist nations, this "neurotic" viewpoint made peaceful coexistence difficult. Kennan's analysis led to the American policy of *containment*.

In response to the deterioration of Soviet-American relations, the Truman Doctrine was announced by the President on 12 March 1947. Its essence was to provide economic and military assistance to those nations threatened by communism, particularly Greece and Turkey. A few weeks later, what became known as the Marshall Plan was presented by the Secretary of State to meet the needs of war-devastated Europe.

One of the greatest miscalculations at the beginning of the Cold War was the American belief that the United States would have an enduring atomic monopoly (and that it was safeguarded by government secrecy). This belief was repeatedly expressed by the military head of the atomic-bomb project, General Leslie Groves, who dominated security thinking in the early days after World War II. In trying to keep the Soviets from making an atomic bomb, Groves attempted to maintain a strict mantle of secrecy, gain a monopoly in the key raw material — uranium ore — and keep German nuclear scientists out of the hands of the Soviets. This effort to oversee technology by controlling information, materials, and scientists turned out to cause only a small delay).

> **[Nuclear Secrecy]**. One month after the atomic bombings of Japan a secret U.S. planning document was issued embodying a nuclear first-use policy in case of war with the Soviet Union. Thus began a sequence of post-Cold War events.
>
> Secrecy became a tool for sustaining the nuclear arms race. Fear-generating tactics, hidden behind a veil of secrecy, permeated governments and their judicial systems. The overly trusting publics were kept unaware of the implications (and even the existence) of nuclear decisions. Many aspects of the peaceful application of nuclear energy were not released until a decade after Trinity.
>
> Besides hiding strategic decisions of major public impact — affecting the very survival of humanity — official secrecy by nuclear-weapon aspirants provided the cover to conceal radiation exposures to unaware humans and contamination of the environment.

The Soviets managed to circumvent all of the embargoes. Unbeknownst to Groves, they had put their first reactor, which would produce plutonium, in operation before the end of 1946. In any event, failure of the United States during World War II to officially inform their ally of the Manhattan Project helped to provoke comparable allocations in Soviet nuclear activities. One Soviet response to the atomic bombing of Japan was to immediately accelerate their nuclear program, to make an atomic bomb as soon as possible.

Even while much American post-war effort went into maintaining a nuclear-weapon monopoly, the government embarked on peaceful applications of nuclear energy, such as civilian power and the medical radioisotopes that come from nuclear reactors. Proposals were also made for sharing and controlling the international development of nuclear fission.

These proposals were influenced by the 11 June 1945 Franck Report, written in part by Eugene Rabinowitch, later the founding editor of the journal that became known as *The Bulletin of the Atomic Scientists*. Not only did the report convey to President Roosevelt a plea against dropping the atomic bomb on Japan, it also warned of a post-war nuclear arms race and possible destructive nuclear war. The report's underlying message about the future of nuclear weapons and atomic energy was summarized in a 1947 letter (partially quoted in the box) signed by Albert Einstein.

> This basic power of the universe cannot be fitted into the outdated concept of narrow nationalisms. For there is no secret and there is no defense; there is no possibility of control except through the aroused understanding and insistence of the peoples of the world.

However, the post-war years were not without Soviet diplomatic and military machinations that frightened the West. "In February 1948 the Kremlin engineered a coup which brought the Czech Communists to full power...." This coup,

> stunned the West. It reinforced the belief that the Soviets would stop at nothing to expand their power and produced a general hardening of attitudes toward the [Soviet Union] throughout Western Europe and America.

After evaluating post-war conditions, the U.S. National Security Council, in a secret 1948 document (NSC-20), identified the Soviet threat to American security as both "dangerous and immediate." It recommended a comprehensive national policy on atomic warfare that became "America's grand strategy for dealing with the Soviet Union and communism." While the American military services were divided on use of the atomic bomb in warfare, the Strategic Air Command was designating major Soviet urban-industrial concentrations as targets. Had such an attack happened, it could have foreclosed political settlements because Moscow and other cities would have been obliterated. The prospect of initiating such an attack when the Soviets could not respond in kind was not consistent with a conciliatory environment envisioned by NSC civilian members.

On 24 June 1948, the Soviet military blockaded Berlin by halting incoming rail traffic, the major source of food and fuel into the city. Two days later an elaborate airlift began carrying food and supplies into Berlin. In the wake of the Berlin crisis,

Section B: Post-War Years

a new policy on atomic warfare was issued as NSC-30; it confirmed that American plans for the defense and economic recovery of Europe depended foremost on using the atomic bomb in the event of war.

The first Soviet A-bomb test, in August 1949, was detected by American scientists. On 23 September, President Truman announced the findings, and two days later the Soviets acknowledged that they had the atomic bomb. This, says Herken, "signaled not only the end of America's nuclear hegemony but the start of the Soviet-American [nuclear] arms race." (Of course, *races* don't begin spontaneously; the preceding *warmup* can make them inevitable.) At the time, the United States had a stockpile of about 100 nuclear weapons and long-range delivery capability with the B-36 bomber.

Derived from NSC-20 was the war plan *Dropshot*, which is described in the box on the next page. The United States saw itself as an "aggressor for peace." Unbeknownst to American policymakers, the Soviets were aware (through their spies) of these internal decisions.

A sequence of related international events inviting reaction then unfolded: The Soviet Union's swift consolidation of control over Eastern Europe and the first Soviet atomic test, victory of China's Communist party in October 1949, the Soviet-Chinese alliance, the Korean War, McCarthyism episodes, and revelations of atomic espionage.

United States response to these perceived threats was to develop the more powerful hydrogen bomb. In January 1950 Harry Truman, spurred on by the Atomic Energy Commission (AEC) and the small (eighteen-member), but singularly powerful, congressional Joint Committee on Atomic Energy, approved the development of a thermonuclear hydrogen weapon (H-bomb). This had the effect of increasing U.S. reliance on nuclear weapons. On 30 November 1952, the United States tested its inaugural hydrogen-bomb device.

At about the time of the first Soviet atom-bomb test, the policy of *first-use* of nuclear weapons worked itself through the American bureaucracy: Sanctioning U.S. first-use of nuclear weapons, NSC-57 was approved in 1949. Under *first-use* or *no-first-use* doctrines, the conditions were specified that would justify (or prohibit) employment of nuclear weapons under various anticipated military confrontation. The first-use policy was intended to deter a massive conventional attack; a no-first-use policy would still allow nuclear retaliation to an atomic attack.

Reliance on nuclear weapons became a legacy of the Truman administration. In April 1950, a new policy statement, NSC 68, provided (besides conventional rearming) the rationale for an "all-out effort" to build an H-bomb. Paul Nitze, well-known as a hawkish[*] presidential advisor, was instrumental in its formulation. He predicted that by 1954 the Soviet Union would be capable of a nuclear attack on the United States, so he advised an increase in both conventional and nuclear arsenals. Not long afterwards the Korean War began, at which time President Truman began to implement NSC-68's recommendations.

---

[*] Advocate of aggressive foreign policy

> **[Dropshot].** A U.S. government war plan *Dropshot*, drafted early in 1949, scripted preventive and pre-emptive war with the Soviet Union, a conflict that was postulated to occur by 1957. Russia was assumed to have by then at most 250 atomic bombs, but it was also taken for granted that U.S. nuclear supremacy would be enduring.
>
> Regardless of Soviet ability to retaliate, the U.S. air force was committed to atomic attack upon the Soviet urban areas that were inseparable from industry; the plan called for destruction of noncombatants on a great scale.
>
> Private doubts about this premeditated carnage were held by civilian and military officials, and both secret and public forums challenged the assumptions that led to the plans for an atomic air blitz.
>
> As an adjunct to *Dropshot*, a plan was drafted to deal with domestic threats in the event of war. "Intelligent people" who had a weakness for "causes," and "various youth and women's organizations," were among those considered vulnerable to enlistment for Soviet espionage, subversion, and sabotage. The plan had visions of bacteriological agents introduced into U.S. reservoirs and atomic bombs smuggled in cargo holds of ships. *Dropshot* contained provisions for mass arrest of Americans suspected of disloyalty.
>
> The Soviet nuclear test in August of 1949 quashed the myth of an enduring atomic advantage for the United States.
>
> Nevertheless, because of *Dropshot*, increased production of atomic bombs was requested and carried out. By 1950 the United States had more than 300 nuclear warheads (and the Soviets had 5). By 1960, the American arsenal had built up to over 20,000, a 6-to-1 advantage.

In 1954, the Eisenhower administration policy to become known as "massive retaliation" was explained by Secretary of State John Foster Dulles:

> to depend primarily upon a great capacity to retaliate, instantly, by means and at places of our choosing [which meant that] it is now possible to get, and share, basic security at less cost.

On 5 May 1950, an American U-2 high-altitude surveillance aircraft was shot down over Russia, with major diplomatic repercussions. Two months later the State Department, intensifying the tension, accused the Soviets of violating international treaties whenever it was in their interest.

Berlin

A Soviet blockade of Berlin (for almost a year, starting mid-1948) was one of the first major international crises of the Cold War, the first major East-West confrontation.

After the victory in Europe, Allied Powers reached the Potsdam Agreement, outlining the division of defeated Germany into four occupation zones, one each for the Soviets, French, Americans and British. Additionally, its capitol Berlin was to be divided into four zones. But Berlin was deep inside Soviet-occupied eastern Germany; so the French, American, and British sectors were isolated by the outer Soviet occupation zone.

Section B: Post-War Years

Having been twice invaded by Germany, the Soviets were alarmed at the prospect of a strong Germany. To prevent its military resurgence, the Soviet-zone administration imposed tight controls on commerce, currency, and freedom of movement within its zone of occupation, including Berlin. Western efforts to install a uniform currency was seen by the USSR as part of a systematic effort to resuscitate their worst fears. An increasing trickle of defectors, some through Berlin, was another problem for the Soviets.

Consequently, they installed their own new currency in East Berlin, and as of 24 June 1948, proceeded to blockade railroad, street, and barge access to the western sectors of Berlin,. This left the western-occupied sectors, having a civilian population of about 2.5 million, without resupply of provisions and goods.

Being a divided city occupied by four powers under the 1945 peace agreement, Berlin's unique location within Soviet-dominated Eastern Europe made it vital for intelligence and ripe for intrigue. The Allies were determined to remain in Berlin also for political and military reasons. However, passage rights for highway, railroad, and rivers had never been negotiated.

The siege, which lasted for 320 days, was relieved by a massive Allied airlift. Unarmed cargo aircraft, not a blatant military threat, could not be challenged by the Soviet administration without using military force. An imaginative and sustained Allied airlift, mostly from the adjacent British and American sectors, kept Berlin supplied and humiliated the Soviets. About 70 British and American pilots lost their lives.

International diplomatic and popular pressure had been rising against the Soviets. The Western relief of Berlin became one of the first major Cold War political defeats for the Soviets.

Also as a result of the blockade,

> The emotional, political, and moral effect in the United States was so great that for many years thereafter it was almost impossible for any American political leader to see the USSR as anything but a cynical aggressor.... The Soviets have not always understood how aggressive and how threatening Westerners felt this act of coercion (and to a lesser degree the Czech coup) to be.... It was in the wake of the Berlin blockade that the United States and most of the Western European countries created the North Atlantic Treaty Organization — NATO.

As a gesture of Western resolve, the airlift was a major plus in political investment. Meanwhile, a hemorrhage of refugees from the oppressive and economically poor East continued, peaking at about 360,000 people per year in 1960. Moreover, presence of a Western showcase within the heart of a communist zone remained an irritant to the Soviets and a threat to the stability and survival of East Germany. In mid-August of 1961, barbed-wire barriers were erected around Berlin. The Western Allies did little to protest, hoping that tensions might subside if East Germans had to remain locked inside. By 1980 the Berlin Wall was more than 850 miles long.

Still, some harrowing escapes were successful. A Soviet blockade of alternate land routes from East Germany in 1961 was probably an attempt to stem the embarrassing escape of refugees from Stalinist persecution.

The building of the Berlin Wall in mid-1961 further symbolized the oppressive consequences of the standoff between East and West, and its toppling 28 years later signaled the end of superpower confrontation.

End of Nuclear Monopoly

As a backdrop to East-West clashes after World War II, the Soviet Union was urgently engaged in developing its own nuclear deterrence. It was simply a matter of time, and in 1949 the first Soviet nuclear-explosive test was conducted. Radioactive debris carried in the atmosphere was detected and quickly recognized in the West for what it was. It shattered the American vision of an enduring nuclear monopoly. Additional tests and improved weaponization continued through 1953, when the Soviet Union started to stockpile deliverable weapons.

> **[RDS-1].** Spurred on by the notorious KGB head, L.P. Beria, and guided by a harvest of espionage data, the Soviet nuclear weapons program in mid-1945 shifted into high gear. A year or so later, Laboratory No. 2 (now known for its founder, I.V. Kurchatov) became the leading institute for fissile material production. Both gun-type and implosion-type atomic weapons were initially pursued. Plutonium was produced at the Chelyabinsk-40 complex in Siberia.
>
> The first (and successful) nuclear explosion RDS-1 took place in 29 August 1949, after four years of intense activity, at a test site chosen in the Semipalatinsk area of Kazakhstan. A Fat-Man lookalike, the initials RDS stood for "Stalin's rocket engine" (It was designated Joe-1 by the United States). Physicists at the Soviet nuclear center Arzamas-16 had already devised a smaller, lighter design, RDS-2, which had twice the yield.

During its nearly decade-long atomic dominance, the United States had produced hundreds of weapons and had more than enough strategic bombers left over from World War II to deliver nuclear bombs to the Soviet land mass. In addition, technology for making nuclear weapons had been shared with the British, who tested their first fission explosive in 1952.

> **[The Nuclear Aftermath].** Following World War II, nuclear weapons became part of the American and eventually Western arsenal against perceived Communist threats. The massive military forces of the Soviets continued to occupy the easternmost parts of Europe.
>
> Improvements in the military utility of nuclear weapons soon became a major objective of the United States. The original bombs were large and heavy. To reduce their size, fusion-boosted devices were tested; these required less of the heavy fissile material for a given explosive yield because some light-weight thermonuclear fusion made up for the difference. Later, with better yield-to-weight ratios (a bigger bang per pound and per buck) thermonuclear fusion weapons became part of a nuclear arms race that approached doomsday scale.
>
> A single 100-megaton-plus device eventually developed by the Soviet Union (see Figure 1 in Introduction) could have leveled, burned, and/or irradiated an entire metropolitan area — New York, London, or Paris. The Soviets made six of these huge aerial bombs to counter the U.S. advantage in intercontinental nuclear missiles.

Perhaps, in the long term, this balancing transition might be seen as the most significant occurrence of the time. Soviet neutralization of Western nuclear dominance permanently changed the calculus of international affairs.

Eventually the in-your-face stage was to be emphatically influenced by the advent of intercontinental missiles that could deliver nuclear weapons to the continental United States.

NATO

A pact was signed in April 1949 establishing the North Atlantic Treaty Organization (NATO). North American members were the United States and Canada; European members were Belgium, Britain, France, Germany, Netherlands, Luxembourg, Denmark, Iceland, Italy, Norway, and Portugal. Greece and Turkey joined in 1952. NATO's purpose was to provide physical security for Western European allies of the United States.

By 1948, preparatory talks for the organization had begun in London, where "the atomic bomb had become the key to U.S. guarantees of European security, as a counterweight to Russia's massive military manpower." Even before the Soviet Union had tested a nuclear weapon, Article 5 of the North Atlantic Treaty committed the Allies to come to the defense of all members in the event of an attack. This was understood to be a "nuclear guarantee."

NATO agreed to integrate tactical nuclear weapons into its own defensive strategy in 1954, and there were 2,500 U.S. tactical nuclear weapons deployed in Western Europe by the end of 1960.

It was not until 1967 that the doctrine of *flexible response* (anything short of a nuclear attack) was adopted, abandoning the idea of massive nuclear retaliation. But the first-use option was retained by NATO, even though many of NATO's nuclear weapons would have been exploded on its own territories.

To counterbalance NATO, the Warsaw Treaty Organization (WTO) was established in May 1955, to come to the assistance of any member that was a victim of aggression.

The NATO tactical nuclear stockpile in Western Europe grew to about 7400 weapons in the early 1970s. Nuclear explosives were incorporated in artillery shells, missiles, gravity bombs, landmines, and depth bombs.

Information about nuclear weapons was shared with NATO allies. For instance, in May 1959 the United States signed an agreement permitting secret information about atomic weapons to be transmitted to Germany and allowing their personnel to be trained in the use of those weapons. A few weeks later Congress was informed that Netherlands and Turkish troops could also be trained to handle nuclear weapons. (In 1965 the White House publically confirmed that nuclear warheads were available to NATO nations, but said that the warheads could not be used without the President's authorization.)

Korean War

The first "hot" war began June 1950 on the Korean peninsula with land battles between North and South Korean forces, the latter assisted by U.S. military forces under treaty obligations. Even when the war was going badly for the South Koreans and the Americans, targets in North Korea were not considered useful enough for atomic bombs. As a result, the A-bomb already became viewed as a "wasting asset" rather than a "winning weapon."

China entered the war in November 1950. Later that month, President Truman caused a controversy when he stated at a news conference that the United States was considering using the atomic bomb in Korea in order to compensate for its losing conventional war. Although commentators consider it doubtful that the threat to use the atomic bomb elicited concessions from China and North Korea, President Eisenhower and his Secretary of State, John Foster Dulles, "claimed to believe it." The Korean war ended three years later with an enduring armistice on 27 July 1953.

Stalin, who had largely kept the USSR out of the Korean War, died before the armistice, after nearly 30 years of being in power. He was succeeded by Georgy Malenkov.

> [Strategic vs. Tactical]. "Strategic policy" is a long-term plan for achieving a defined purpose, such as destroying military, industrial, or political targets. It is often contrasted with tactical operations, which have the aim of gaining a specific military end in immediate support of a given operation. Usually the weapons employed to achieve these goals are different.

In 1954, the doctrine of *massive (nuclear) retaliation* to a conventional attack was announced as official policy by Secretary of State Dulles. He said that the existing policy of meeting communist challenges head-on at time and place of their choosing caused the United States "grave budgetary, economic and social consequences." His alternative was a policy that allowed the United States "to retaliate instantly and at places of our own choosing." Not long afterwards the Soviets spoke out for a broader international renunciation of weapons of mass destruction, including atomic and hydrogen weapons, a proposal rejected by the Western powers.

By the end of 1955, a tacit understanding had emerged between the United States and the Soviet Union that "nuclear war was ultimately unacceptable." Gradually U.S. strategic policy changed to (1) an assured second-strike capability (through the triad of air, land, and sea forces), (2) a single integrated strategic operations plan (SIOP) that did not directly target cities, and (3) a flexible response in place of massive retaliation.

But "flexible response" had a serious drawback: It narrowed the gap that separated nuclear from conventional weapons. In fact, Dulles earlier had alluded to tactical nuclear weapons that could destroy enemy targets while keeping civilian casualties to a minimum. Eisenhower had said a week later that one could use tactical nuclear weapons "just exactly as you would use a bullet or anything else."

Sputnik, the first human-made satellite, was launched by the Soviet Union on 4 October 1957. Like the day John Kennedy was assassinated, the beep-beep-beep of Sputnik is the type of event that becomes engraved in memory and history. Earlier, in August of that year, the Soviets had announced that they had successfully tested an intercontinental ballistic missile (ICBM) with a real nuclear warhead (Actually they had an earlier test in February 1956). The United States launched its first successful ICBM from Cape Canaveral in December 1957 and a satellite — *Explorer 1* — at the end of January 1958.

Atoms for Peace

A detailed history of the Manhattan Project, the *Smyth Report,* was published immediately after the atomic bombings.

> **[The Smyth Report].** The official government publication, named after its compiler, presented to the public the first chronicle of atomic-bomb development: "A General Account of the Development of Methods of Using Atomic Energy for Military Purposes under the Auspices of the Government 1940-1945."
>
> The Report was published by the Superintendent of Documents in Washington, D.C., on 12 August 1945 (six days after the atomic attack on Hiroshima and three days before the Japanese surrender). It is a remarkably full and detailed disclosure of development work carried out between 1940 and 1945 by the American-directed team of allied physicists under the Manhattan District code name. The project culminated with production of the first atomic bombs. It was priced at one dollar by the U.S. Superintendent of Documents.
>
> From the foreword by General Groves:
>
> The story of the development of the atomic bomb by the combined efforts of many groups in the United States is a fascinating but highly technical account of an enormous enterprise. Obviously military security prevents this story from being told in full at this time. However, there is no reason why the administrative history of the Atomic Bomb project and the basic scientific knowledge on which the several developments were based should not be available now to the general public. To this end this account by Professor H. D. Smyth is presented.

Throughout the post-war years, efforts were made to reach agreements on how to manage the new nuclear technologies. Trying to head off foreseeable temptations by other nations to acquire the technology for making atomic weapons, an historic U.S. State Department document known as the "Acheson-Lilienthal Report" was released in March 1946. It called for the development of atomic energy to be assigned to an international organization. The subsequent "Baruch Plan," presented by the United States to the United Nations from June through October 1946, was a compendium supporting a proposal for international control of atomic development and beneficial applications of atomic energy. The plan required the Soviets to be the first to turn over their nuclear technology, and they rejected it.

At the December 1953 UN General Assembly, President Eisenhower proposed a plan for nonmilitary applications of atomic energy. He advocated establishment of an International Atomic Energy Agency. Four years later, the IAEA was formed as an autonomous agency of the UN to encourage and monitor peaceful uses and to help prevent the spread of nuclear weapons.

In June 1968 a Non-Proliferation Treaty (NPT) was approved by the UN General Assembly. The treaty was promptly signed by the United Kingdom, United States, and Soviet Union. The signatory nations agreed not to transfer nuclear weapons to other nations nor to assist or encourage the others in developing their own. In 1972 the IAEA was formally assigned the task of establishing safeguards for facilities of nations that signed the treaty. Nations that did not sign could still enter into safeguards agreements with the IAEA. In any event, only *nonmilitary* facilities were subject to IAEA safeguards and inspections, a limitation that persists.

Testing Nuclear Weapons

The testing of nuclear weapons and explosives has always been a contentious issue. Nations with them (the *haves*) defended their actions on national-security grounds; others (the *have-nots*) expressed growing antipathy to the political and environmental consequences.

The nuclear weapons states also took issue with each other. There were frequent Soviet accusations that the United States and United Kingdom inherently opposed treaties limiting testing because they wanted to continue developing more lethal weapons. The holdouts countered, as they did in 1959 (and mostly still do), that a test-ban treaty would not have controls adequate to prevent violations.

In the United States the pace of nuclear testing was increased by funding authorizations from the Joint [Congressional] Committee on Atomic Energy. The AEC thus accelerated the testing rate by opening proving grounds in Nevada and creating an extensive set of production facilities. These developments were strongly encouraged by scientists working within weapons programs.

Altogether, the Soviet Union eventually conducted a total of 715 nuclear tests, detonating 969 explosive devices. For the United States, the number of nuclear detonations revealed was 1030. Several other nuclear weapons states also carried out atmospheric and underground proof tests. Very few tests were ever announced publically at the time.

During the 1950s, various pauses in testing took place while negotiations proceeded on treaties that would limit or ban testing. A three-year joint moratorium began in 1958. In considering a permanent ban, Eisenhower insisted that there be an adequate number of inspections to detect possible violations, while the Soviets feared that inspections would be used to gather intelligence. Since the Soviets would not agree to inspections of test sites, Eisenhower suggested a treaty simply banning atmospheric nuclear explosions, which could be detected by national technical means.

In August 1961, after the three-year moratorium, the Soviets resumed testing, giving as a reason aggressive policies of NATO. However, the United States had previously announced that it would resume testing, which it did with an underground explosion a few days later. Soon afterwards, the Soviets detonated a nuclear device of more than 50 *megatons*, and Premier Khrushchev warned that if Western nations unleashed a war, the Soviet Union would retaliate with 50- and 100-megaton bombs. The following February, President Kennedy said that the United States was prepared to resume atmospheric nuclear testing.

It was almost two years (August 1963) before a partial nuclear test ban treaty was signed, prohibiting testing in the atmosphere, outer space, and under water. Ostensibly because the United States and Soviet Union could not agree on verification, no ban on underground nuclear explosions was reached.

> [Nuclear-Weapon Testing and Secrecy]. To optimize performance, nuclear weapons and their delivery systems were tested in the atmosphere and underground. The experiments were also used to study the effects of radiation, blast, and fire on various targets, including military vehicles, housing structures, ships, and personnel. Much non-sensitive information, even the occurrence of some tests, was kept secret for at least 10 years after the Cold War ended. Keeping the information secret for so long had the appearance of blunting public opposition to future testing.

Developing Thermonuclear Weapons

In May 1951 the United States performed a nuclear test, code-named "George," which successfully demonstrated fusion in an atomic explosion. The first test to get appreciable energy from the fusion reaction was the November 1952 U.S. *Mike* shot. The unwieldy Mike assembly, however, did not have the practical military potential of a thermonuclear device exploded by the Soviets in August 1953. That breakthrough in the development of deliverable thermonuclear weapons was followed by a huge (15-megaton) U.S. explosion in March 1954. The first H-bomb to be dropped from an airplane was tested by the Soviet Union in November 1955.

During the transition from atomic to thermonuclear bombs, boosted-fission devices were developed. Even after megaton hydrogen bombs were developed, fusion-boosted fission weapons continued to be important in nuclear arsenals.

> [Boosted Weapons]. Fission weapons t are said to be "boosted" if they have an enhanced yield due to the presence of gaseous deuterium and tritium, which support thermonuclear fusion reactions. Fusion reactions contribute to the explosive energy by magnifying the fission rate. For a given yield, fusion-boosted devices can be made somewhat smaller and lighter than fission weapons. They therefore become weapons of choice for limited warfare, so-called "tactical" nuclear explosives. Progress with fusion-boosted fission devices encouraged development of more powerful multistage thermonuclear weapons.

Thermonuclear, or fusion, weapons increased the devastation that could be accomplished by nuclear weapons. A fusion weapon requires an initiating stage — a fission trigger (a small atomic bomb) — in order to start the thermonuclear reaction; this method of triggering was suggested by Enrico Fermi in 1942 to Edward Teller, who then became the chief advocate of the "Super," as that early design of a thermonuclear bomb became known.

The evolution of thermonuclear weapons became a "crucial turning point" in the arms race because mutual deterrence began to emerge. The U.S. *Mike* shot in 1952 gave powerful proof that thermonuclear weapons worked, and it inspired the Soviets to proceed vigorously with their development program.

A year earlier there had been the U.S. *George* test. It was a demonstration that fusion of deuterium and tritium would indeed occur if the substances were heated to a very high temperature in an atomic explosion — an idea registered with a still-secret patent granted in 1946 to John von Neumann and Klaus Fuchs (who was later unmasked as a Soviet spy). The *George* experiment, however, did not contribute much to the development of a practical hydrogen bomb.

> **[Fusion Weapons].** Fusion weapons (thermonuclear, or hydrogen, bombs), are nuclear devices in which an uncontrolled, self-sustaining fusion reaction takes place — resembling the large-scale nuclear reactions that power the sun. In a fusion reaction, two energy-rich nuclei collide and combine ("fuse"), rearranging their neutrons and protons into reaction products and releasing energy.
>
> Thermonuclear weapons are multistage devices quite different from fission and fusion-boosted bombs, and far more destructive. They contain three essential components or stages: (1) a fission-based primary "trigger," (2) separated fusion fuel (such as lithium deuteride), and (3) a massive casing. Each stage is important in obtaining a high explosive yield. Typical multistage fusion weapons have yields in megatons. In 1961 the Soviet Union detonated the 50-megaton bomb which, with a straightforward design change, could have had a yield at least twice as great.
>
> Relatively small and lightweight thermonuclear warheads were eventually made that fit into small re-entry vehicles for ballistic missiles.

Fuchs, while at Los Alamos (he returned to England in 1946), had himself originated an idea for initiating a thermonuclear explosion, although not the all-important concept of radiation implosion that made the H-bomb possible. In any event, the Soviets, gratified by their own successful 1949 atomic detonation, were already aware of theoretical possibilities for a thermonuclear weapon.

Although Teller worked almost exclusively on the Super, the AEC General Advisory Committee in the late 1940s, with Oppenheimer as chairman and James Conant, Enrico Fermi, and I.I. Rabi as members, opposed deliberate development of the superbomb. Instead, they

> approved buildup and diversification of fission weapons (including tactical weapons), endorsed preparation for radiological warfare, and supported the development of boosted fission weapons.

Despite the reluctance of that committee, a group consisting of AEC commissioners, senators, and military chiefs pressured President Truman to go ahead with the Super, and early in 1950 the *New York Times* carried the headline,

"Truman Orders Hydrogen Bomb Built." That decision impelled the Soviets to accelerate their own thermonuclear program.

It was another year before Los Alamos scientists Stanislaw Ulam and Teller were able to work out a configuration that resulted in two-stage thermonuclear weapons, namely "the confinement and utilization of radiant energy from a primary atomic bomb to compress and ignite a secondary, physically isolated core containing thermonuclear fuel." This was the configuration, in a bulky, undeliverable assembly, that was tested at the 1952 *Mike* shot.

The second hydrogen-fusion bomb designed in the United States was exploded at Bikini on 1 March 1954; at a yield of 15 megatons, the *Castle Bravo* thermonuclear explosion was almost 1000 times as powerful as the early atomic bombs, and resulted in "unexpectedly widespread radioactive fallout." That's largely because its yield was about 2½ times larger than calculated.

Japanese fishermen about 80 miles away on their ship, the *Lucky Dragon*, were seriously burned by radiation. Residents of an island 176 miles from Bikini had to be evacuated because of radiation fallout, and for years they were medically treated.

Robert Oppenheimer, who led the technical development of the atomic bomb, anticipated dire consequences of mutually destructive deterrence because of hydrogen bombs and long-range bombers:

> We may anticipate a state of affairs in which two Great Powers will each be in a position to put an end to the civilization and life of the other, though not without risking its own. We may be likened to two scorpions in a bottle, each capable of killing the other, but only at the risk of his own life.

The Soviet thermonuclear program began in earnest when they received intelligence data about U.S. work on the Super. Igor Tamm set up a group of younger physicists, including Andrei Sakharov, to work on the project, and a "layer cake" (*sloika*) design was initially proposed. Although their first successful atomic bomb test in 1949 stimulated thermonuclear weapons development, it was Truman's announcement in 1950 that caused the Soviet government to make a formal decision to proceed. The 1953 test at Semipalatinsk was measured at 400 kilotons; unlike the *Mike* assembly, the Soviet device — using lithium deuteride and tritium as thermonuclear fuel — could be made into a deliverable bomb.

Although sometimes disputed, it is evident that the Soviet fusion concept grew out of the then-existing foundation of physical knowledge and theory, because Fuchs' information did not provide the Soviet Union with the basis for a workable design. Sakharov and Yuri A. Trutnev proposed a new ("third") idea, which was based on separation of the fission primary and the fusion secondary; this resulted in a successful test of an air-dropped bomb (RDS-37) in November 1955, with a yield of 1.7 megatons.

Unwittingly, Edward Teller's obsession with his original "Super" design might have wasted years "working on the wrong thing," thus putting the United States behind the Soviets in development of a militarily useful thermonuclear weapon.

The largest thermonuclear weapon ever was tested in the atmosphere by the Soviets at the end of October 1961. This bomb had an estimated yield of between

50 and 60 megatons, but its capability with additional fissionable materials was much larger — perhaps up to triple the nuclear-energy release. Conveyed by a Tu-95 strategic bomber and exploded at an altitude of 4 kilometers over Novaya Zemlya in the Arctic, it has never been surpassed in yield. Deriving its explosive power from fission and fusion and fission reactions, its success proved the feasibility of devices of virtually unlimited power. The bomb, sometimes labeled a "political" weapon, was tested half a year after the U.S. Bay of Pigs invasion in Cuba.

Both the U.S. and Soviet thermonuclear programs were highly secret, with little practical information available from espionage, and no input from the public. American scientists viewed the immense thermonuclear fusion weapon as qualitatively different because its explosive power was so very much greater than a fission bomb. Apparently, however, the Soviet political leaders and scientists initially deemed it to be just another weapon of war.

After the successful *Castle Bravo* shot, political leaders "gained a new appreciation of the destructive power of nuclear weapons." Nuclear war slowly became less acceptable, and nuclear deterrence was understood to mean that retaliation could inflict an unacceptable level of damage: "the shared understanding of the destructiveness of nuclear war created a common interest in avoiding such a war, and this was a factor for stability in the political rivalry between the two countries."

But getting the big weapons halfway around the world needed delivery systems.

Long-Range Bombers

Because thermonuclear weapons were large and heavy, both the United States and Soviet Union started in the mid-50s to deploy long-range bombers (ICBMs were not yet developed). First was the Soviet M-4 Bison, later joined by the Tupolov Bear. The United States continued to maintain its superior B-52. Great Britain developed its V series of bombers (Valiant, Vulcan, and Victor), but after around 1960 depended primarily on ballistic missiles to deliver nuclear warheads.

In World War II, the Soviet Union had used long-range aircraft primarily to destroy specific military targets, whereas Great Britain and the United States employed them as weapons against industrialized cities in Germany and Japan. The bombing operations of the Allies thus "motivated the Soviet Union to begin working on the creation of its own heavy bomber for strategic bombing," starting with an engineering copy of the American B-29 bomber. The Tu-4, based on Soviet modifications of the B-29, went into production in 1947, with 847 of them being built by 1952. After nuclear bombs were developed, the Soviets modified the Tu-4 into the nuclear capable Tu-4A, an intermediate-range aircraft that could not reach beyond nearby theaters of war, except for a few that could be refueled in midair.

From 1947 to 1957, the commander of the Strategic Air Command (SAC) was General Curtis E. LeMay (who earlier had organized the Berlin airlift). His audacity in the early 1950s in deploying the SAC long-range bomber fleet — which was mid-air-refuelable and could be stationed at forward bases — often led to controversy and provocation whenever his reconnaissance crews flew B-47s deep into the Soviet Union. Although SAC was subject to presidential authority, General LeMay felt that he could override the authority under special circumstances. While he was SAC commander, LeMay had considerable access to, and control over, nuclear bombs — generating unease in some quarters.

By 1962, the Soviets had close to 160 long-range bombers that could deliver about 270 nuclear weapons to U.S. territory. The Tu-95 had entered service in 1954 and the 3M bomber a few years later. Concerned about being caught unawares, the Soviets kept them on permanent alert at bases deep within their borders.

SAC, also fearing a surprise attack, sustained many bombers airborne at any one time; in 1961 the United States was keeping about a dozen B-52s in flight; each bomber was capable of carrying several thermonuclear weapons.

Theater Weapons and the Neutron Bomb

Although some very efficient nuclear weapons were designed and distributed, many of the ones initially fielded by NATO and the U.S. naval fleet were boosted-fission weapons. Being smaller and lighter, these were more deployable,. Theater nuclear weapons are intended for tactical — in contrast to strategic — warfare. They were for limited use in what was expected to be a contained battlefield area, such as unfortunate boundary nations situated between NATO and WTO. Tactical nuclear weapons are also the type and size suitable for naval engagements.

The Soviet Union emerged in 1955 from development and testing of a compact tactical weapon (RDS-9) that had a yield of about 5 kilotons. It was mated with a torpedo. A nuclear-artillery-shell concept followed in the next year. A tactical-missile warhead entered the Soviet stockpile in 1960, to be followed by nuclear air-defense and cruise missiles.

The *neutron bomb* is a tactical nuclear weapon, developed first in the United States. All nuclear weapons cause casualties by three mechanisms: radiation, blast, and heat. In comparison with a conventional nuclear explosive, more of the neutron bomb's energy output is in the form of neutrons and other radiation, and less as blast and heat; thus it is more formally known as an *enhanced radiation weapon* (ERW). The idea was that in combat near civilian targets, soldiers would be irradiated, but equipment, buildings, and landscape would receive minimum damage. Depending on their distance from the detonation epicenter, personnel would receive lethal or debilitating doses of radiation. Sometimes the propaganda got out of hand, as when the weapon was referred to as a "clean" bomb, even though it had lethal and destructive features common to all nuclear weapons.

The United States also developed intermediate-range ballistic missiles (IRBMs) to deliver lightweight nuclear weapons from countries close to the Soviet Union. By late 1957 the Thor and Jupiter IRBM programs were well along, both types of missile having been authorized for deployment, even before consultation with America's allies. Thor missiles were deployed in the United Kingdom in 1958, and

Jupiters went to Italy in 1960 and Turkey in 1961. These rocket batteries were not deactivated until 1963.

Soviet SS-20 IRBMs were fielded in increasing numbers, causing considerable public outrage in Western Europe.

Intercontinental Ballistic Missiles

Having carried out a full-scale test of a ballistic-missile, the R-5M (SS-3 Shyster) in 1956,the Soviet Union was actually a couple of years ahead of the United States. Its nuclear warhead detonated near Lake Balkhash, about 750 miles (1200 kilometers) from the launch site.

However, the USSR was hamstrung by its early decision to boost large (10,000-lb) thermonuclear payloads. Eventually this paid off for them, but in the meantime they were behind in the race for small missiles with comparably smaller payloads. For the time being, though, in the mid-1950s, the United States lagged the Soviet Union in ballistic-missile development.

The technology for intercontinental-range missiles evolved rapidly, the Soviets developing the R-7 (SS-6) ICBM starting with flight tests in May of 1957. The R-7 had parallel staging, consisting of four peripheral boosters symmetrically mounted around the central module. Not only was the R-7 successful as an ICBM, it was the first space-launch vehicle. An R-7 rocket placed into orbit the original artificial satellites PS-1 (Sputnik) and PS-2 in October and November of 1957, respectively.

After the launch of Sputnik, the payload of ballistic missiles increased until it became possible to carry thermonuclear warheads about 3000 miles — across continents. The first U.S. ICBM was the Atlas D, which became operational in 1959; the Soviets fielded a fully operational ICBM, the two-stage modified R-7A (SS-6 Sapwood), a year later. These were fueled by volatile liquid propellants that were highly unstable and unstorable. In time, improvements were made in propellants, the ultimate being solid fuel.

Although the Soviets had started official development of rocket-driven projectiles in the 1920s, both the Soviets and the Americans made use of captured World War II German technology, scientists, and technicians in producing their missiles,. The Germans had begun long-range rocket research in 1944, based on prior development of the intermediate-range V-1 and V-2 rockets. While the earliest deployed Soviet and U.S. ICBMs were unreliable, inaccurate and expensive, they were capable of delivering nuclear warheads to cities at great distances without being intercepted by air defenses.

Originally the land-based ICBMs were located at launch pads that were not protected against attack. This disposition allowed easier fueling, maintenance and reloading, but the sites were vulnerable, especially to nuclear missiles. Eventually, the technology was developed to put liquid and solid-fueled rockets underground in hardened silos. Minuteman I, introduced in 1962, was the first ICBM with solid fuel; it was also more accurate, faster responding, improved in reliability, and less expensive.

Six major parameters drove ICBM design: warhead yield and weight; reentry-vehicle (RV) velocity on approach to target; guidance system and targeting accuracy (CEP); rocket-engine technology; basing method; and vulnerability to attack. Optimizing these parameters was a complex and expensive task.

Because ICBMs came to have such significant military value and each side was defenseless against them, they symbolized the nuclear arms race more than any other weapon system.

Bomber and Missile Defenses

Efforts were made to defend against strategic bombers and missiles. In the early 1950s, air defenses were set up in the United States, consisting of a network of detection and tracking radars and surface-to-air interceptors. In 1958 the more advanced Nike-Hercules bomber-intercept missiles became operational, specifically in sites around the cities of New York, Washington, and Chicago; these missiles were armed with nuclear warheads. Eventually the system was expanded to cover more of the continental United States, and missile batteries were also shipped to Europe and Asia.

In terms of percentage of attacking force destroyed, air- and ground-based defenses can be very effective against bombers, which is part of the reason why aircraft are not good first-strike, counterforce weapons. Air defenses, though, have less value against a nuclear attack on cities because even one bomber getting through can cause horrendous damage. A major role of aircraft in a nuclear exchange would be for mopping up — destroying any remaining targets, hardened or otherwise, after ICBMs and SLBMs eliminated most of the defenses. Other factors favor aircraft as part of a stable deterrent.

Because air defenses could have some success against bombers, it was natural to think of trying to intercept and destroy ballistic missiles. However, the ICBMs much faster than aircraft, and much of their trajectory is outside the atmosphere; so attempts to find a ballistic-missile defense have met with little success.

The Nike-X anti-ballistic system, consisting of short-range Sprint and longer-range interceptor missiles, along with the radars to guide them, was developed in 1966 — intended as a defense against Chinese and Soviet ICBMs. An ensuing, more advanced, system, called Sentinel, was approved in 1967 by President Johnson's Secretary of Defense, Robert McNamara, who said that its primary purpose was to defend against a Chinese nuclear attack or an accidental launch. Short-range Sprint and long-range Spartan ICBM interceptors were to be part of that system.

The ABM systems were fatally flawed, however, in a number of ways:

(1) The computer programs were enormously complex and therefore impractical to debug reliably;

(2) Even by high-altitude interception, defense of all the cities and other targets in the contiguous 48 states was hopeless; and,

(3) Although the Sprint terminal-defense missile was incredibly spry, the trajectory-analyzing system could not give it enough time to intercept an incoming warhead.

Starting in 1967, the Soviets deployed an ABM system, called *Galosh*, around Moscow; an earlier version had been intended for defense against bombers. Both the Soviet and U.S. ABM systems at the time used a "shotgun" approach: the nuclear-armed interceptors needed only to get close enough to detonate their nuclear explosives, which created a huge burst of radiation to disable the incoming missile.

According to McNamara, the U.S. counter-response to the Galosh was to deploy MIRVs (missiles with multiple warheads), which could penetrate Soviet defenses.

In March 1969, the U.S. government announced that a modified Sentinel system (called Safeguard) was being designed to protect missile sites, rather than cities. This was a belated acknowledgment that the American people could not be protected against a Soviet nuclear attack.

Surviving Nuclear War

Civil defense became synonymous with the construction of fallout shelters in an effort to protect civilians during and after a nuclear assault.

The 1957 launch of *Sputnik* sparked fears of a "missile gap," whereby the United States was behind in producing long-range missiles. As an immediate result, the U.S. government embarked on an effort to formulate mass-evacuation plans and to persuade people to build nuclear-fallout shelters for their families. The government built shelters to protect military and civilian decisionmakers, and some civilians dug their own.

The stakes were huge: The U.S. nuclear arsenal stood at 2400 weapons in 1955, when the United States began stationing nuclear weapons in NATO countries. A U.S. government agency concluded that

> the currently planned [U.S.] nuclear offensive against the Soviet Union would result in 77 million casualties in the Soviet bloc; of the 134 major cities, 118 would be bombed, and 75 to 84 percent of the populations in those cities would be killed.

Scientists and decisionmakers were divided on the wisdom and potential effectiveness of fallout shelters. Shelter or no, survival near a nuclear explosion was quite unlikely, although it took some time for scientists to convince decisionmakers of this. However, for people far from ground zero, underground shelters did offer some possibility of surviving the cloud of radiation that would be carried by prevailing winds and might be lethal for weeks. The problem was to find ways of subsisting long enough to outlast the worst of the radiation while living, breathing, and eating without severe contamination.

Another problematic aspect of civil defense was emergency-management preparedness. While prudence called for such preparedness, including massive evacuations of cities if warning of nuclear attack were received, much controversy erupted over the effectiveness of the various measures considered or implemented.

A controversial aspect of civil defense was the risk that it could lull people and decisionmakers into a false sense of security. The theory behind mutual assured destruction was that a rational attack was deterred because there would be no realistic prospect of avoiding horrendous, unprecedented damage in a counterattack. But risk of miscalculation from nuclear-policy brinkmanship was increased if leadership:

(1) perceived that civil-defense would protect them, their families, and others enough to justify risking an all-out war,
or, perhaps more likely,
(2) thought that the other side sensed that it could survive a nuclear attack,
or, perhaps even more likely,
(3) thought that the other side believed they were going to be attacked, and therefore felt the need to blunt that attack by striking first.

The most egregious soft-pedaling of radiation hazards occurred in the early stages of East-West hostilities, when misguided officials treated civil-defense as though atomic detonations were simply an extension of conventional warfare. Propaganda was lavished on civilians and armed-service personnel, trying to persuade them to prepare for nuclear attack by digging shelters or foxholes, and otherwise by taking shelter ("duck and cover") or trying to move out of harm's way ("run like the wind").

Civil-defense exercises, such as Operation Alert, were conducted in major cities; people were told they had only themselves to blame if they were not prepared for atomic war. Operation Cue at the Nevada test site was carried out with manikin families arrayed in the direct path of a nuclear blast. These futile exercises soon began to encounter public resistance, notably from some outspoken mothers in New York City who realized that they did not, in the first place, want their children being sacrificed to national-security policy.

And indeed, nuclear weapons were being touted and brandished as tools of a forceful national policy. For example, SAC General Curtis LeMay let it known that he wanted two nuclear bombs available for every city in the Soviet Union — arousing hostile reactions from their citizens and leaders. Even after the Korean War truce in 1953, the United States had 30 nuclear weapons stationed at a forward base in Okinawa for delivery to the USSR by B-36 bombers.

Civil defense was really not designed for population survival, but rather to accustom constituents and civilian decision-makers to the idea of atomic warfare. However, people began to realize that the winner of a nuclear war might very well be decimated along with the loser — that an underlying motivation for civil defense was to get civilians to accept the notion that people could be sacrificed. To survivors would go the hollow "victory."

Cuban Missile Crisis

In April 1961, s U.S.-backed Bay of Pigs invasion of Cuba occurred. Despite its failure, the attack caused considerable concern not only to Fidel Castro, but also to his Soviet allies. By October 1962, the United States had discovered that the Soviets had secretly installed nuclear-capable missiles on Cuba, just 90 miles from Florida.

This Soviet action was partly in retaliation for 15 intermediate-range Jupiter nuclear-armed missiles the United States had set up in Turkey. In the words of Premier Khrushchev, the United States "had already surrounded the Soviet Union with its own bomber bases and missiles [in Turkey, Italy, and West Germany]." More important, the Soviets were convinced that the United States would not let Cuba alone. Khrushchev feared that if Cuba fell, other Latin American countries would reject Soviet Communism — a "domino effect," to borrow a phrase often used later in the Vietnam era.

> [Events Leading Up to the Cuban Missile Crisis]. At the end of 1958, General Fulgencio Batista's dictatorial regime was overthrown by revolutionary elements within Cuba. Some of the elements, including the movement lead by Fidel and Raul Castro, had been supplied by the CIA.
>
> The new Cuban government began expropriating land and private property under the auspices of the agrarian reform. By the end of 1960, the revolutionary government had nationalized more than $25 billion of private property owned by Cubans. Cuba also nationalized all United States and other foreign-owned property in the nation. The United States, in turn, responded by freezing all Cuban assets in the United States and by tightening an embargo on Cuba, still in place at least a half-century later.
>
> Cuba was a major focus of the new J.F. Kennedy administration when it assumed power in January 1961. In Havana, fear prevailed of U.S. military intervention. The threat of invasion became real when, in April 1961, a force of CIA-trained Cuban exiles opposed to Castro landed at the Bay of Pigs. The invasion was quickly terminated by Cuba's military forces.
>
> Fidel Castro was convinced the United States would eventually mount a full-scale invasion of Cuba. Shortly after routing the Bay of Pigs invasion, he declared Cuba a socialist republic, established formal ties with the Soviet Union, and began to modernize Cuba's military.
>
> The United States feared any country's adoption of communism or socialism, but for a Latin American country to openly ally with the USSR was regarded as unacceptable, given the Cold War enmity.
>
> In late 1961, Kennedy authorized Operation Mongoose against Castro's government, a series of covert operations which were to prove unsuccessful. More overtly, in February 1962, the United States launched an economic embargo against Cuba.
>
> The United States also considered direct military attack. Air Force Gen. Curtis Lemay presented to Kennedy a pre-invasion bombing plan, while spy flights and minor military harassment from the United States Guantanamo Naval Base were the subject of continual Cuban diplomatic complaints to the U.S. government.
>
> By September 1962, observers would have seen increasing signs of American preparations for a possible confrontation, including a joint Congressional resolution authorizing the use of military force in Cuba if American interests were threatened, and the announcement of an American military exercise in the Caribbean planned for the following month.
>
> The Cuban Missile Crisis in October 1962 itself was a confrontation between the United States, the Soviet Union, and Cuba during the Cold War. In Russia (and most of Europe), it is termed the "Caribbean Crisis," while in Cuba it is called the "October Crisis." The crisis ranks with the Berlin Blockade as one of the major confrontations of the Cold War, and is often regarded as the moment in which the Cold War came closest to a nuclear war.

Another factor motivating the Soviets at the beginning of the 1960s was the strategic superiority that the United States appeared to have. At the end of the Cuban missile crisis, the Soviets were at a severe strategic disadvantage. They could not send more than 300 nuclear weapons onto U.S. territory, while the United States had "more than 1300 strategic bombers, capable of conveying more than 3000 weapons to targets in the Soviet Union [as well as 183 ICBMs] and 144 missiles on nine Polaris submarines." In addition, early in that very month of October 1962, the United States began deploying the solid-propellant Minuteman ICBM that had been previously and openly tested. The Soviets thought that they could counter the apparent strategic imbalance by moving medium- and intermediate-range ballistic missiles to Cuba.

In an address to the nation in October 1962, President Kennedy announced a quarantine of offensive weapons going to Cuba, and warned that if any missiles were launched from Cuba the United States would retaliate by attacking the Soviet Union. A few days later Khrushchev said he would remove the missiles if the United States promised not to invade or support an invasion of Cuba. This decision defused the most dangerous of the confrontations.

On 2 November 1962, President Kennedy went on radio and TV to announce that the Soviet missile bases in Cuba were being dismantled, thus ending the tense two-week international crisis.

Both sides had shown restraint. Only after the crisis was over did the United States learn that (98) nuclear warheads had already been slipped into Cuba. It was said that 20 warheads were in place for the medium-range R-12 (SS-4) missiles, which could reach as far as Washington, D.C., and that nine tactical nuclear weapons were in the hands of Soviet commanders in Cuba, who — for the first (and only) time — had been delegated field authority for their use.

This crisis was apparently the sole occasion during the Cold War that the Soviet Union raised its strategic-force alert level.

> Historians still debate whether U.S. conventional superiority ... or overall U.S. nuclear superiority was responsible for the outcome of the Cuban missile crisis. The outcome was also the product of both sides' intense desire to avoid nuclear war, of skillful diplomacy, and of some compromise: the Soviets removed their missiles from Cuba, and the U.S. promised not to invade Cuba and implemented a previous decision to remove medium-range missiles from Turkey.

A somewhat conflicting account suggests that the United States had no previous intention of removing the missiles from Turkey, but that in order for President Kennedy to "save face," he made a secret agreement with Chairman Khrushchev for their withdrawal. It was later declared publicly that the United States had planned to remove the missiles before the Cuban crisis. Indeed, Kennedy had twice before ordered removal of the missiles, but they were still in Turkey when the crisis broke out. Turkey was reluctant to part with them; however, Turkish leaders realized they might become a target if the crisis degraded into an actual nuclear exchange. The missiles were removed from Turkey several months later, in March 1963.

Robert McNamara is reported to have said that Kennedy and his advisors never considered using nuclear weapons during the Cuban entanglement because of U.S. conventional superiority in air and sea around Cuba. That, however, would not have prevented an irreversible chain of events, each side with their finger close to the nuclear button. In fact, documents declassified 40 years later are reported to indicate that a particularly dangerous moment occurred when "a U.S. Navy destroyer dropped depth charges on a Soviet submarine without realizing the sub was carrying a nuclear [torpedo] weapon [as were four other Soviet subs near Cuba]." Although the Soviet submarine commander "ordered a nuclear-tipped torpedo readied for launch," he decided — after consulting with other officers — "to hold fire and surface."

The Carribean missile crisis, aside from its frightening aspects, was a watershed in military and diplomatic policy. The bitter lessons of the confrontation caused the Soviets to place greater emphasis on nuclear potency: They believed that their humiliation — and lack of success in protecting socialist Cuba — was due in part to Soviet inferiority in strategic arms. The Soviet indignity was followed by a massive expansion of their nuclear arsenal — development of the R-36 (SS-9) and UR-100 (SS-11) beginning in 1962 and 1963 respectively — reaching rough nuclear parity with the United States by the end of the decade.

Although many attempts were made by the United States to overthrow Cuba's government, the United Sates promised after the Cuban Missile Crisis to never invade the island.

Getting so close to the brink caused diplomats to focus on negotiations for a nuclear test ban and for the Moscow-Washington "hotline" to be installed. American "success" in the Cuban/Carribean/October missile predicament also led to increased U.S. governmental arrogance, which helps explain growing involvement in Vietnam.

Vietnam War

In the early 1950s the Truman administration supported France's colonial war against communist-led indigenous coalitions seeking independence in the Indochina peninsula. President Eisenhower's administration had supplied substantial military aid to the French, who withdrew in 1954 from northern Vietnam after losing a decisive battle, leaving the country divided into two parts.

Guerrilla resistance commenced in South Vietnam, as corrupt American-supported governments attempted to suppress the civil war with CIA assistance.

The potential independence of Vietnam and Laos gave rise to the Western "domino theory" — the notion that the fall of those countries would trigger the toppling of other nations into the communist sphere of influence. President Kennedy raised the number of American military personnel in South Vietnam, but resisted Pentagon proposals for major escalation. When Lyndon Johnson came into office, he committed a large force of American combat troops after the debatable Gulf of Tonkin incident.

American bombing of North Vietnam started in 1965; it had little military impact, but turned international public opinion against the United States. A vigorous antiwar public movement became a major brake on U.S. support of the corrupt, despotic, and unpopular regime in South Vietnam.

At the war's height, President Johnson had over half a million troops in Vietnam. The mission was doomed because the communists, fighting as nationalists, were able to recruit South Vietnamese peasants to join their side in a lengthy war of attrition. As the Viet Cong and North Vietnamese began to succeed militarily in the South, Johnson — who was facing growing public disenchantment — lost confidence in the Pentagon's ability to hold back the communist incursion. He decided to stop the bombing and de-escalated American involvement, trying to turn the war over to a propped-up South Vietnamese Army.

[Vietnam War Narrative]. The Vietnam War was the longest and most unpopular war in which Americans ever fought. And there is no reckoning the cost. The toll in suffering, sorrow, in rancorous national turmoil can never be tabulated. And for many of the more than two million American veterans of the war, the wounds of Vietnam will never heal.

Direct American involvement began in 1955 with the arrival of the first advisors. The first combat troops arrived in 1965 and we fought the war until the cease-fire of January 1973.

The end of World War II opened the way for the return of French rule to Indochina. Despite the ties he had forged within the American Intelligence community, and his professed respect for democratic ideals, Ho Chi Minh was unable to convince Washington to recognize the legitimacy of his independence movement against the French. French generals and their American advisors expected Ho's ragtag Vietminh guerrillas to be defeated easily. But after eight years of fighting and $2.5 billion in U.S. aid, the French lost a crucial battle at Dien Bien Phu — and with it, their Asian empire.

With a goal of stopping the spread of communism in Southeast Asia, America replaced France in South Vietnam — supporting autocratic President Ngo Dinh Diem until his own generals turned against him in a coup that brought political chaos to Saigon.

With Ho Chi Minh determined to reunite Vietnam, Lyndon Baines Johnson determined to prevent it, and South Vietnam on the verge of collapse, the stage was set for massive escalation of the undeclared Vietnam War. In two years, the Johnson Administration's troop build-up dispatched 1.5 million Americans to Vietnam to fight a war they found baffling, tedious, exciting, deadly and unforgettable.

The Vietnam War was seen from different perspectives by Vietcong guerrillas and sympathizers, by North Vietnamese leaders and rank and file, and by Americans held prisoner in Hanoi.

The massive Tet, 1968, enemy offensive at the lunar new year decimated the Vietcong and failed to topple the Saigon government but led to the beginning of America's military withdrawal from Vietnam.

Richard Nixon's program of troop pullouts, stepped-up bombing and huge arms shipments to Saigon changed the war and left GIs wondering which of them would be the last to die in Vietnam.

Despite technical neutrality, both of Vietnam's smaller neighbors [Cambodia and Laos] were drawn into the war, suffered massive bombings, and, in the case of Cambodia, endured a post-war holocaust of nightmarish proportions.

While American and Vietnamese soldiers continued to clash in battle, diplomats in Paris argued about making peace. After more than four years, they reached an accord that proved to be a preface to further bloodshed.

Through troubled years of controversy and violence, U.S. casualties mounted, victory remained elusive, and American opinion moved from general approval to general dissatisfaction with the Vietnam War. South Vietnamese leaders believed that America would never let them go down to defeat — a belief that died as North Vietnamese tanks smashed into Saigon on April 30, 1975, and the long war ended with South Vietnam's surrender.

Johnson's successor, President Richard Nixon, who had campaigned under the slogan "peace with honor," decided to attack communist supply lines and bases in Cambodia, Laos, and North Vietnam. These military actions, besides being counterproductive, led to new rounds of vigorous public anti-war protests. In 1973 Nixon reluctantly withdrew the American forces who were facing defeat, and the South Vietnamese government fought on until it was overwhelmed in 1975, excluding the gruesome toll on non-combatants.

America's biggest Cold War setback thus took place in Southeast Asia, particularly Vietnam. Over the years, more than 8.5 million Americans fought there; 58,000 Americans (and more than 1 million Vietnamese soldiers) were killed.

The financial cost for the United States has been estimated at $155 billion (1970 dollars), and the heavy expenditure helped stimulate worldwide inflation in ensuing years. The losses to the Vietnamese people were appalling.

[Gulf of Tonkin Incident]. The Incident prompted the first large-scale involvement of U.S. armed forces in Vietnam. It was a pair of attacks allegedly carried out by naval forces of the Democratic Republic of Vietnam (DRV — North Vietnam) against two American destroyers, the USS Maddox and the USS Turner Joy on August 2 and 4, 1964 in the Gulf of Tonkin.

The outcome of the incident was the passage by Congress of the Gulf of Tonkin Resolution, which granted President Lyndon B. Johnson the authority to assist any Southeast Asian country whose government was considered to be jeopardized by "communist aggression". The resolution served as Johnson's legal justification for escalating American involvement in the Vietnam War, which lasted until 1975.

In 2005, it was revealed in an official NSA declassified report that the USS Maddox first fired warning shots on the August 2 incident and that there may not have been North Vietnamese boats at the August 4 incident.

Not only did the military campaign fail to hold back the spread of communism to South Vietnam, the floundering effort destroyed myth-like projections of American military might and undermined cherished domestic views of moral superiority.

Changing Nuclear Balance

The years from 1957 to 1962 were a period when U.S. invulnerability ended and its superiority was challenged. Here is a summary of some key events that affected strategic and political policy.

The 1958 U.S. dispatch of nuclear-capable artillery to the islands of Quemoy and Matsu, in coastal waters very near the Chinese mainland, was part of an exercise in the then-prevailing concepts of nuclear diplomacy. Gradually, it no longer became feasible to so closely confront China.

Nuclear stakes began to change significantly as the Soviet Air Force deployed their Bison and Bear bombers (capable of one-way trips to the United States), and when the Soviet Rocket Forces tested a ballistic missile at full intercontinental range.

American ability to gain reconnaissance information about the closed Soviet landmass was impaired when the high-flying U2 airplane was downed on 5 May, 1960. Up to then, the United States was able to partially compensate for the closed nature of the Soviet Union.

In the late 1950s, critics of Eisenhower's administration had contended that there was a developing *missile gap*; Democrats claimed that the Republicans had "allowed the American deterrent posture to deteriorate." The claims were later shown to be based on incorrect, perhaps falsified data. The alleged gap became a major topic in the 1960 presidential election campaign, when John F. Kennedy used the issue to his political advantage. After Kennedy was elected, his defense secretary, Robert McNamara, dismissed the missile gap as illusory.

In 1961 the U.S. Arms Control and Disarmament Agency (ACDA) was created by Congress, as a result of considerable grass-roots pressure stemming from faulty missile-gap information and other misgivings about U.S. policy.

Massive retaliation was considered by many to be a nuclear addiction, but the comparatively inexpensive nuclear defense helped facilitate the post-World-War-II economic recovery of the United States. Extending the nuclear umbrella to allies enhanced stability, allowing economic progress in Western Europe and Japan.

Nuclear weapons had multiple roles: NATO wanted a formula for neutralizing the Soviet military threat while avoiding costly remilitarization; the answer was tactical nuclear weapons lent by the United States. America itself had been invulnerable to Soviet nuclear retaliation until *Sputnik*, but felt the need for a growing nuclear arsenal when it became fearful that a Soviet ICBM could be a "winning weapon."

Eventually the Soviet Union's growing nuclear capability caused NATO members to suspect that the United States might be deterred from defending Europe; at the time, the Joint Chief of Staff's only operational plan was "cataclysmic." Thus, a new strategy emerged: flexible response; conventional forces would be used before nuclear counterattack.

Nuclear-powered Polaris submarines with ballistic missiles (SLBMs) became operational in 1962, and by the end of the year were made available to the United Kingdom. Also, both the United States and the United Kingdom agreed to provide some of their nuclear weapons to the NATO multilateral force.

From 1962 to 1970, the strategic position of superiority moved to parity for the United States *vis-a-vis* the Soviet Union. Although the specific types and mix of delivery systems chosen by the United States and Soviet Union differed, there was an overall standoff: Each side had a secure deterrent because of a dependable retaliation capability, and both sides understood it.

A major factor in the transition to parity was the deployment of photo-capable satellites, first by the United States in 1960 and then by the Soviet Union in 1962. The images obtained allowed both sides to gain some confidence about strategic stability, particularly the absence of major military redeployment.

Kennan's "containment" notion is said by one historian to have lasted from 1945 to 1969, at which point détente became a better description of the superpower relationship.

In the late 1960s, American reliance on mutual assured destruction (MAD) was based on the belief that the Soviets knew that 50 percent of their population could be killed and 80 percent of its industry destroyed in massive retaliation. Needless to say, it had not escaped the attention of public-interest groups that nuclear war would also inflict insufferable devastation on the continental United States. Public concern over excessive militarism, nuclear arsenals, and radiation risks began to grow even greater. Pressure was brought to bear against the arms race by individuals and public-interest groups.

Not widely known in the 1960s was that radiation-exposure experiments were being conducted by governments on humans, in secret and without their informed consent. Radioactive materials were being incautiously dumped, especially in the former Soviet Union, with little attention to public or environmental risk. These heedless actions were carried out in the name of national security, with the responsible agencies emboldened by secrecy that precluded public challenge. Accidents and hazardous leaks have had devastating human and ecological consequences, and they are still a problem in some locales.

****

After World War II, the East-West political conflict began in earnest, symbolized by the Iron Curtain, and marked by armed hostilities throughout the world, some of which came close to involving nuclear weapons. Regardless of the devastating nature of nuclear arms, their development continued: Thermonuclear devices were tested, and the means of delivery progressed to unstoppable ICBMs.

An atoms-for-peace program was attempted, but it was inherently flawed because it preserved nuclear weapons states as an upper caste that has never fulfilled its obligations to reduce dependence on nuclear armaments.

Almost-transcendental experiences like the Berlin and Cuban crises, which brought nuclear-armed nations into apprehensive conflict — potentially to the verge of war — resulted in policy shifts that led to dangerous global strategies of pre-emption and to large increases in armaments. The earliest U.S. national-security policies and war plans depended on first-use of atomic bombs to wipe out military, industrial, and civilian centers in the Soviet Union.

Perhaps, in the long term, the most significant event in this period was neutralization of the American atomic monopoly. It started in 1949 with the first Soviet nuclear test — its fallout detected in the West (literally and figuratively) — and continued with additional tests and improved weaponization through 1953, when the Soviet Union started to stockpile deliverable weapons. Other nations began developing nuclear forces too.

In order to minimize the number of nuclear weapons states, an international regime under the NPT and the IAEA was set up. Showing little self-restraint, the weapons-states were only able to discourage growing atomic militarization by those nations that had little reason to choose that path.

As the Soviet Union displayed a capability for massive atomic retaliation, first with long-range bombers and later with long-range ballistic missiles, the United States tried to find a defense against missiles, but came up against nearly insurmountable technological obstacles associated with the high speed and destructive power of nuclear-armed ICBMs. Passive measures, such as civil defense, also showed little hope of protecting populations from the effects of nuclear war.

Needless to say, while nuclear arsenals created a strategic military standoff, they did not help nations prevail in regional conflicts like those in Berlin, Korea, Vietnam, and Afghanistan.

Overall, on the strategic side, it was mano-a-mano, *a risky direct confrontation, largely between the Soviet Union and the United Sates — symbolizing East and West — but mostly through surrogates on the ground at a tactical level..*

The U.S. hard-line approach for Vietnam notably failed to achieve success for both sides. The colonial era was over, and the "domino" theory never got anywhere. The Cuban missile crisis was a "success" only in the sense that it didn't flare up into a nuclear exchange.

Many of these preceding topics will be examined in more detail in the remainder of this Chapter and in other Chapters.

C. PHASES OF *DÉTENTE* [1970s-1989]

The 1970s can be characterized as a period of **détente** *— a normalization of relations between the Soviet Union and the United States. Ironically, it took a reversal by hard-line Republican presidents (Nixon and Reagan) to obtain the necessary congressional support for several treaties that would otherwise have been difficult to ratify. Although the pace of détente itself suffered a setback upon the 1980 election of Ronald Reagan, the groundwork for a renewed relaxation of tension was laid by Mikhail Gorbachev, with his* **perestroika** *(economic restructuring), accompanied by* **glasnost** *(openness).*

The 1970s

Two months after Richard M. Nixon took office as President in January 1969, the Senate gave its consent to the Non-Proliferation Treaty (NPT), and in November the United States and Soviet Union started the strategic arms limitation talks known as SALT. Nixon in February 1971 signed the Seabed Treaty, which prohibited placement of nuclear weapons on the ocean floor.

One of Nixon's first policy pronouncements linked arms control with the resolution of political issues, regardless of the fact that an excessive buildup of nuclear arms and the associated risk of accidental war were concerns that burdened both sides. While he emphasized U.S. military strength as a prerequisite to arms control negotiations, Nixon recognized that parity could also be assured by means of agreements with the Soviets.

Although the U.S. administration endorsed the concept of *nuclear sufficiency,* it proceeded with both ABM and MIRV systems, partly as *bargaining chips* in negotiations with the Soviets. Under the *Nixon doctrine*, all treaty commitments were to be honored while American allies were to be include within an effectual nuclear shield.

The year 1972 was a major one for détente (a French word for "relaxation"). Henry Kissinger, Nixon's Secretary of State, believed that détente might bring about an element of stability to the problematic political relationship of the superpowers.

[ABM]. In an effort to defend Moscow, the Soviets in the late 1960s deployed the Galosh anti-ballistic missile system. When the Chinese detonated a hydrogen bomb in June 1967, pressure arose for the United States to develop its own ABM system; so that year, the United States announced the Sentinel system (later renamed Safeguard), to be built around and supposedly protect 15 major American cities.

[MIRV]. The development of ever-lighter warheads and more accurate delivery made it possible to mount from 3 to 15 nuclear warheads on an ICBM, and moreover to aim each one at a separate target. The United States had a significant lead in this development, and was deploying MIRVs (multiple independently targetable re-entry vehicles) while the Soviets were still at the MRV stage (multiple re-entry vehicles that could not be aimed independently).

As the war in Vietnam continued, President Nixon visited China, opening new trade and political arrangements. The Strategic Arms Limitation Treaty (SALT) talks made progress as a result of several factors: reduced tensions, increasing cost of nuclear-weapon systems, popular pressure for arms control, existence of nuclear parity, and technological improvements in ABM and MIRV. The SALT I agreement and the ABM Treaty were signed in 1972, placing limits and controls on long-range ballistic missiles and nuclear defenses.

Section C: Détente

> [SALT I]. Because the 1972 Strategic Arms Limitation Treaty was the first major arms control agreement between the Soviet Union and the United States, it was one of the high points of détente. It enshrined the principle of mutual deterrence. Both sides agreed to limit their offensive and defensive weapon systems, including submarine-launched ballistic missiles, ballistic-missile submarines, fixed land-based ICBM launchers, and anti-ballistic-missile systems.

SALT I was a big step in nuclear arms control, but it omitted controls on new strategic systems, notably (and deliberately) MIRVs, and was silent on Soviet deployment of newer, larger missiles, such as the SS-19. Somewhat to American surprise, the Soviets tested their own MIRV in 1973 (having successfully tested an MRV in 1970).

1974 was a year when international affairs were eclipsed by U.S. domestic politics (especially the Watergate scandal following the reelection of President Nixon). At a news conference early in the year, President Nixon defended, against critics from his own party, his policy of détente in world affairs, including arms control agreements. Indeed, he signed the Threshold Test Ban Treaty (TTBT) in July 1974. A month later Nixon had to resign as a result of the Watergate scandal, to be succeeded by his vice-president, Gerald Ford.

Détente had reached its high point during the Nixon administration; afterwards, tensions began to rise. This was partly due to American hardline conservatives questioning whether the United States was benefitting from eased tensions with the Soviets; they saw economic agreements as subsidizing an inefficient Soviet economy, "thereby enabling the Soviet Union to augment the size and quality of its military establishment." The conservatives blocked ratification of a U.S.-USSR trade agreement and imposed additional restrictions on SALT II. At the same time, both covert and open combat operations in Indochina increased despite American liberal opposition.

Kissinger maintained continuity in policy and leadership after Nixon's resignation (8 August 1974), and nearly a year later the Helsinki Final Act was signed. This pact set the stage for agreements on security, human rights, economics, and other issues. The Helsinki meetings also represented the start of the Conference on Security and Cooperation in Europe.

> [CSCE/MBFR]. While the SALT I and SALT II negotiations were under way, two other important sets of wider-ranging East-West talks were being conducted. Those talks, which covered conventional — rather than nuclear — weapons, also involved all of the European states and Canada. The Conference on Security and Cooperation in Europe (CSCE) started in 1972, and the Mutual and Balanced Force Reductions (MBFR) meetings began a year later.
>
> The MBFR discussions ended in 1989 without an agreement, after almost two decades of widely politicized encounters (In fact, Ambassador Jonathan Dean characterized those negotiations as a "train without a locomotive"). The CSCE negotiations, though, led to a pact in 1975 to settle disputes peacefully, build confidence, and avoid accidental alerts when large military maneuvers were being practiced.

To supplement the threshold test ban, a Peaceful Nuclear Explosions (PNE) Treaty was signed in 1976, prohibiting underground nuclear explosions for peaceful purposes if they exceeded 150 kilotons.

Although another phase of strategic-arms talks (SALT II) began in 1977 under President Carter, the arms race continued. In a delayed reaction to NATO's deployment of Pershing and cruise-missiles in the 1960s, the Soviet Union kept was deploying intermediate-range mobile SS-20 missiles in Eastern Europe.

Did the SS-20 provide the Soviets a margin of usable superiority in Europe? In retrospect, they simply heightened the level of an already existing nuclear stalemate. From their viewpoint, the SS-20s advantages were more operational in nature: (1) damage limitation [greater accuracy, lower yield], (2) ease of operation [solid fuel and reloading capability, and limited numbers to coordinate], and (3) unambiguous limitation to theater application [because of their short range]. Longer-range missiles, even if launched toward Europe, might appear at first to be headed to the continental United States, thereby triggering a counterattack. The U.S. strategic command might have held back if a nuclear war were confined to Europe.

Meanwhile, the Soviets objected to NATO's plans for introducing anti-personnel neutron bombs (enhanced-radiation weapons — ERWs) in Europe. Other weapons developments that complicated the SALT II discussions included the Soviet MIRV, their large ICBMs, the *Backfire* bomber, the B-1 bomber, and U.S. cruise missiles.

[Backfire]. The Soviet medium-range bomber Tu-22M, (called the *Backfire* by the United States) was deployed in 1974; it became a source of disagreement during the SALT negotiations and public debates over whether it had a range sufficient to reach the continental United States without midair refueling.

Hawkish groups such as the Committee on the Present Danger (formed in 1976), often alleging military disparities, put pressure on President Carter to strengthen the U.S. defense position. That particular Committee included Democratic foreign policy "experts" such as Paul Nitze and Eugene Rostow. When the B-1 bomber program was canceled by Carter, the Committee charged him with being "soft on defense."

[B-1]. The U.S. B-1 stealth bomber was developed to replace aging B-52 for delivering strategic nuclear warheads. However the B-1 proved to be extremely costly and was abandoned by President Carter in 1977.

[Cruise Missiles]. Air-breathing jet-engined cruise missiles gained considerable accuracy in the 1970s when the U.S. Tomahawk was fitted with guidance based on matching terrain contours. Being able to fly at very low altitudes, these missiles can avoid radar detection. Cruise missiles could be launched from ground (GLCMs), sea (SLCMs) or air (ALCMs). They were relatively inexpensive, capable of delivering a highly destructive nuclear warhead, and comparatively easy to conceal.

Section C: Détente

In early June 1979, President Carter approved production of the large, 10-warhead ICBM known as the MX, although its basing mode had not yet been decided. A few weeks later the second Strategic Arms Limitation Treaty (SALT II) was signed by Carter and General Secretary Brezhnev. Later in the year, Brezhnev said he was willing to consider reducing SS-20s deployed in Europe.

Even though Carter had signed SALT II, he stopped supporting it when the Soviet Union invaded Afghanistan, and it was never ratified — having been opposed very effectively by Richard Perle, who was one of the negotiators and an aide to hawkish Senator Henry "Scoop" Jackson.

Carter left office in 1981 with American-Soviet relations at their lowest point since the Cuban missile crisis. Although Carter was partially to blame, one analyst concluded that the inflexibility of Brezhnev's leadership made the Soviets "primarily responsible for the collapse of détente."

The 1970s, through 1977, were characterized by parity among the superpowers, the introduction of serious arms control measures, and divergence in weapons-development programs. In the mid-70s, the Soviets introduced the heavy missiles SS-17, -18, and -19 (U.S. designations), called "heavy" because of their relatively large payloads. The United States went deliberately "light," concentrating on better accuracy and smaller, more efficient warheads.

It can be argued that through the 1970s, the longtime *modus vivendi* of containment was working: Soviet expansionist threats had subsided. (Concluding that containment had broken down would have required acceptance of the dubious proposition that Western Europe had been *Finlandized.*[*]) Also failing to materialize yet was the supposed USSR "breakdown" that had been declared by Reagan, Nixon, and Carter.

> **[SALT II].** Because of the ever-intensifying nuclear arms race, the second SALT Treaty moved ahead to deal with issues that had been sidelined in SALT I. Equal ceilings of long-range weapons were agreed, with sub-ceilings for MIRVs. Limits were also placed on heavy ICBMs, on deployment of new ICBMs, and on the number of warheads on MIRVed missiles. Also, cruise missile limits were set. SALT II was not been brought up for ratification by the U.S. Senate nor ratified by the Soviet Union or Russian Federation; however, both parties have been complying as though it were in effect.

From 1977 on, parity between the superpowers was often a topic of political dispute: With each side deploying new, often different types of nuclear delivery systems, it was difficult to arrive at clear-cut comparisons. The SS-20 IRBM was being routinely produced and fielded. On the U.S. side, cruise missiles became available. In 1980 the Ohio-class Poseidon submarines, armed with Trident-I C-4 long-range missiles, began to be launched.

---

[*] dominated without a fight — that is, cowed in international and economic affairs.

Repeated Soviet attempts to influence West German and NATO decisions with military threats invariably failed.

Afghanistan

The Soviet invasion of Afghanistan in 1979 negatively impacted détente. That unwinnable war embroiled Moscow in its own "Vietnam," with disastrous consequences to the Soviet Union's prestige and economy. The Soviets, attempting to bolster Afghanistan's puppet atheistic communist regime, ran into fierce resistance from guerrilla fighters called the Mujahideen in the tribal Muslim nation.

Worldwide condemnation of the invasion followed immediately, including a boycott of the 1980 Olympics in Moscow, and President Carter's cancellation of a U.S.-USSR grain agreement. A quick increase in the Western aid came to Pakistan, which became home to millions of Afghan refugees.

Although Soviet troops were able to take control of major urban centers in Afghanistan, the Mujahideen attacked supply convoys. After the United States (through the CIA) and Britain supplied the guerrillas (which included foreign volunteers, like Osama bin Ladin) with anti-aircraft weapons, operations by Soviet transport planes and helicopters were seriously disrupted.

A large death toll was suffered by Afghans and the invading Soviet army, which had little success in attaining control of the nation. Gradually the Soviets sought ways of backing out of their quagmire. It was under this dark cloud that the 1980s got under way, and it was not until 1989 that Soviet withdrawal of forces was completed.

The failure in Afghanistan contributed to changes in leadership and ultimate demise of the Soviet Union.

The 1980s

When Mikhail Gorbachev came into office in 1985, he shifted Soviet policy toward finding a negotiated solution in Afghanistan. But the damage to détente had already been done, and the beginning of the 1980s represented a major escalation of the superpower confrontation. In particular, the NATO-Warsaw Pack stalemate was boosted to a higher, more dangerous level. A resurgence of nuclear arms in the United States came about through decisions made by Presidents Carter and Reagan.

U.S. executive-branch policy, which evolved through an interagency process, was usually embodied in a secret document (a presidential directive). This directive is for all governmental agencies, particularly the Departments of Defense and Energy, in order to shape their budget requests and strategic armaments in accordance with a common goal — at least in principle. Each president usually begins a term in office with a new policy survey, which sometimes results in significant changes, but more often was an evolutionary document that reflected a compromise between competing interests.

PD-59. In mid-1980, President Carter approved Presidential Directive 59 (PD-59), which placed emphasis on a counterforce nuclear doctrine — the targeting of military installations. It also authorized the largest arms-production program in 30 years. PD-59 came on the heels of decisions to deploy Pershing-II and Tomahawk missiles in Europe, to construct an MX missile system, and to shelve the SALT II Treaty. The new policy shocked the Soviets; they regarded this directive as tantamount to rejection of nuclear parity — the very foundation of détente — and the Soviets threatened to accelerate their own nuclear weapons programs, which they did. NATO allies also were dismayed. One result of PD-59 was a long hiatus in arms control progress — lasting eight years or so, well through President Reagan's second term.

In 1980 the U.S. Department of Defense had five categories of Soviet targets: nuclear forces, conventional military, industry of war, industry of recovery, and command and control. One of many problems apparently overlooked in PD-59 was that if an opponent's command and control were targeted, it would be very difficult to put an early halt to a nuclear war with no one left to negotiate and end reflexive retaliation.

President Ronald Reagan took office in 1981. He faced a more difficult situation than his predecessors, which was partly his own doing; he touted weapons and uttered threats, along with unfounded accusations that the Soviets were violating international treaties. Not long after taking office, President Reagan said the United States would store neutron weapons in Europe. He also repeated his controversial opinion that tactical nuclear weapons, including the neutron bomb, could be used in Europe without igniting an all-out war.

Reagan was not inclined to be conciliatory. Strategic arms reductions talks (renamed S T A R T) resumed in 1982, but for a long time they made no appreciable progress. It was long after he left office, just before the collapse of the Soviet Union, that the resulting treaty, START I, was signed.

[Neutron Weapons]. The neutron-bomb concept, credited to RAND weapons-scientist Sam Cohen, was an enhanced-radiation weapon (ERW) with an on-again, off-again development history starting in the late 1950s. The design emphasized lethal neutron radiation, with a reduced radius for destruction by blast and decreased fission-product production. Because of these modifications, it could kill humans without as much property damage as ordinary nuclear weapons that had the same explosive yield.

Cohen considered it ideal for killing Viet Cong soldiers in the jungles of Vietnam. While never used there, an ERW was considered by military commanders to be a way to destroy Soviet tank formations in Europe without collateral damage (although separating tanks by more than 100 yards was an effective counter-tactic).

Even though the relative amounts of radiation, blast, and heat had been modified, the ERW was still a nuclear weapon based on an explosive fission-chain reaction, and it aroused controversy. The debate about its deployment was especially lively in Europe. The controversy about their deployment didn't subside, particularly in Europe, until 1981, when President Reagan conceded to pull back to the United States the stockpile of neutron weapons he had sent to Europe at the beginning of his term.

Several hundred neutron warheads were produced in the 1970s and 1980s. Neutron warheads were developed for aerial bombs and artillery projectiles.

In 1983 Reagan publicly called the Soviet Union an "evil empire," and continued his opposition to a freeze on nuclear expansion. A little later he announced his intention to develop the Strategic Defense Initiative (SDI), which opponents soon derisively dubbed "Star Wars" because of the movie-like visions it evoked both for the President and for the public.

Reagan's dream of a defense umbrella led him to assail mutual assured destruction (MAD) as immoral and unreliable — which had the curious effect of causing those interested in arms control to uncomfortably support reliance on MAD. The premise was that mutual assured destruction was stable; whereas, a missile-defense system perceived to have a chance of working would lead to strategic instability: In time of crisis; one side or both might feel the need to "use it or lose it" — to launch a massive strike on warning of an attack. That was the reasoning American supporters had used earlier to overcome Soviet opposition to the ABM Treaty. It now became the Soviets' turn to use that argument on the United States, observing that SDI coupled with the Carter-Reagan nuclear buildup made arms control much more difficult.

One of the specific treaty-violation accusations leveled by the Reagan administration was made in May 1983 when the Soviets had tested a new ICBM (the SS-25) that violated limitations of the SALT II agreement (which, however, had not been ratified). The charge hinged on technicalities and was never substantiated, although the dispute dragged on for several years.

In July 1983 President Reagan asked Congress to support production of the MX missile (which he wistfully dubbed the "Peacekeeper") because it would let the United States "negotiate through strength" and make the Soviet Union more willing to compromise. Later that same year he used a similar argument in contending that the Soviets might be more willing to negotiate if the Pershing-II and cruise missile deployments were to proceed on schedule. (Four years later the INF Treaty was agreed upon.)

Low-orbit satellites were (and still are) particularly vulnerable to many feasible forms of attack. In 1984 the Soviets indicated a willingness to have a moratorium on anti-satellite weapon testing until a treaty was negotiated under which the parties would use national means of verification. Despite the fact that reconnaissance satellites provided indispensable information over inaccessible lands of the Soviet Union, the treaty never came about: U.S. hardliners wanted the ability to destroy Soviet satellites, and they expected that the Soviet Union would not be able to muster the resources to similarly threaten U.S. "eyes-in-the-sky."

An event that foreshadowed the end of the Cold War took place in March 1985 when Mikhail Gorbachev became General Secretary of the Communist Party. Five months later his government offered and initiated a nuclear-test moratorium; President Reagan rejected U.S. participation, pointing out that the Soviets had just completed a series of nuclear tests.

Although Reagan's election signaled the ending of détente, the mid-1980s produced the *perestroika* and *glasnost* of Gorbachev. In the Soviet Union the press became freer, industry was made self-managing, and government bodies were opened to pluralistic voting. During Gorbachev's tenure, "the Soviet Union

Section C: Détente

announced a revision of its military doctrine to emphasize defensive operations, implemented a unilateral reduction of a half-million troops and their equipment, and renounced military intervention in neighboring countries...." The doctrine moved toward "reasonable sufficiency," with conventional forces restructured to defeat an invasion but not carry out a large-scale offensive action.

> **[INF].** U.S. intermediate-range nuclear forces in Europe were "modernized" in 1979, in response to Soviet deployment of their mobile, triple-warhead SS-20. And in 1983, NATO added ground-launched cruise missiles (GLCMs) and Pershing-II intermediate-range ballistic missiles. A decade later the INF Treaty was negotiated, banning all such missiles, resulting in the eventual withdrawal, retirement, or destruction of about 2400 tactical nuclear-weapon delivery systems.
>
> One of the risks associated with having short-range and intermediate-range nuclear weapons within Europe was that they could devastate and contaminate the very nations and peoples that were to be defended. Adding provocation, Western intermediate-range missiles could reach well into Russia.
>
> Considering the fact that the high-tech conventional forces of NATO were adequate to deter the massive military of the Warsaw Pact, tactical nuclear weapons were at best superfluous. Nevertheless the nuclear threat in Europe spiraled — contributing to the tradition of nuclear shadowboxing.
>
> The INF Treaty, signed on 8 December 1987, after seven years of discussions, was a major *disarmament* (in contrast to *arms control*) treaty. It resulted in elimination of all INF systems in Europe (missiles with a range between 500 and 5500 kilometers). On-site inspection, despite its comparative intrusiveness, was accepted and proved successful as a means of verifying INF disarmament by both sides.
>
> In 1991, as the Soviet Union was collapsing, Presidents George H. Bush and Mikhail Gorbachev announced reciprocal withdrawals of about 4000 more tactical nuclear weapons from Europe.

In October 1986 the two superpower leaders met in Reykjavik, where President Reagan rejected Gorbachev's proposals that included reductions in strategic nuclear weapons, a comprehensive test ban, elimination of INF weapons in Europe, and reinforcement of the ABM Treaty. Yet, a month later, the United States deliberately exceeded the limits imposed in the unratified SALT II agreement by deploying a B-52 bomber capable of carrying cruise missiles.

It didn't take long, just a few months, before the Soviets ended their year-and-a-half voluntary nuclear-test moratorium, in February 1987, citing "American policies." In September, Secretary of State George P. Shultz charged that the Soviet radar site at Krasnoyarsk violated the ABM Treaty (it would have, if completed, but it never was). Regardless of these policy conflicts, the INF Treaty was signed on 8 December 1987. The following year, the Soviets agreed to dismantle what there was of the Krasnoyarsk radar.

During the Reagan years, the influential Committee on the Present Danger continued its pressure for increased defense spending. Nearly half of the new expenditures under Reagan were for a massive military buildup. Yet, because of countervailing Soviet buildups, the perceived window of vulnerability (for land-based nuclear missiles) was never closed. In a sense, the expenditures did not matter to neoconservatives on the Committee because they believed that the Kremlin did not view nuclear war as the worst of all outcomes — a belief that paralleled their own — and so whatever it took to stay ahead was justified.

Nevertheless, as a result — and an indication — of a thaw, President Reagan visited Moscow in June 1988. Economic, political, and sociological events were then already under way to dismember the Union of Soviet Socialist Republics.

****

Détente was a long time in coming, and it didn't last very long. Warlike leaders vacillated, pushed by combative constituencies and militant institutions, exacerbating the arms race until it became nearly unmanageable. Provocative deployments led to comparable responses. Extramural military adventures were costly and unsuccessful, eventually unleashing public-activist uprisings. Economic imperatives were beginning to intervene, bringing domestic pressure to limit military acquisitions.

D. COLLAPSE OF THE SOVIET UNION [1989-1991]

By the late 1980s, the Soviet internal and external hegemony was already starting to decay — the system was racked by various disturbances and overburdened by domestic and international pressures. Gradually the rot spread and the structure itself crumbled.

Some think that the beginning of the end for Soviet-style communism occurred when Paul John Paul II went to Poland in 1979, which was when the Solidarity movement took hold t here. In any event, communism in Eastern Europe was finished by the end of 1989. During that decade, and arms buildup had taken place, costly in many ways to both East and West.

In the summer of 1989, Hungary lifted its border controls with Austria, an action that effectively allowed transient East Germans an exit route to the West. By October, the East German government began to fall apart, especially when the Soviet leader, Mikhail Gorbachev, refused to use his troops to put down the ongoing unrest. After the resignation of East German Communist Party head, Erich Honecker, a mass exodus of East Berliners began a confusing sequence of events that resulted in a breach of the Berlin Wall on 9 November 1989.

Section D: Soviet Collapse

A month after rupture in the Berlin Wall, Romania and Czechoslovakia ousted their Communist Party and its leaders (achieved by Poland earlier in the year).

Although not directly relevant to the end of the Cold War, a major military conflict, running from mid-1990 through early-1991, was occurring on a different battleground — the Middle East — with Iraq's invasion of Kuwait and an allied counter-invasion that brought big changes to that oil-rich area.

Another memorable milestone was the reunification of Germany, whose partition had been aggravating the superpower relationship for nearly half a century. The East and West German governments signed terms for unification on 31 August 1990, and the four allied powers (Britain, France, Soviet Union, and United States) ratified them on 12 September (an act that brought a formal end to World War II). The following day, Germany and the Soviet Union signed a treaty pledging 20 years of friendship. The reunification took formal effect on 3 October 1990. At the beginning of December, "the first freely elected all-German government since the 1920s took office."

Early in 1990, even the Soviet Communist Party had agreed to a proposal by President Gorbachev for a multiparty system. As the year went on, Lithuania declared its independence. Ending 1990 was the Conventional Forces in Europe (CFE) Treaty being signed by NATO and WTO. Some consider this event the "formal" conclusion of the Cold War.

Although the CFE treaty did not undermine the Soviet Union's ability to defend itself against attack, the perception that Gorbachev had backed down under Western pressure reduced his domestic political support.

Certainly 1991 was the final year for the Soviet Union as a political entity. Fortunately, the START I Treaty had been signed just before the breakup (Negotiations, which began in 1982, had dragged on during the Reagan and Bush administrations).

> [START I]. The United States and Soviet Union signed the first S T A R T accord in July 1991. Each side was to curtail its deployed strategic nuclear warheads and launchers (missile and bombers). Certain categories of heavy ICBMs were reduced in number deployed and in throw-weight (payload) capacity.

Political events came intensified and rapidly for the Soviet Union in the latter half of 1991, commencing with a coup temporarily ousting President Gorbachev, who was promptly restored to power with the assistance of Boris Yeltsin, then President of Russia.

By December 1991 the republics of the Soviet Union had declared their independence. The end of that month marked the official termination of the Union of Soviet Socialist Republics, with the

> [Dzerzhinsky Toppled]. The statute of Felix Dzerzhinsky erected across from the KGB headquarters, which he had founded, was knocked over by demonstrators on 22 August 1991. Perhaps as much as any other symbolism, this event marked the demise of both Communism and Stalinist repression.

formation of the Commonwealth of Independent States (CIS), consisting of Armenia, Azerbaijan, Belarus, Georgia, Kazakhstan, Kyrgyzstan, Moldova, Russia, Tajikistan, Turkmenistan, Ukraine, and Uzbekistan. Soon thereafter, the leaders of the republics that inherited nuclear forces pledged to honor nuclear agreements between the former Soviet Union and the United States. With the resignation of President Gorbachev on 25 December, the Soviet Union no longer existed, and Russia became the dominant political remnant. The bipolar pattern of confrontation came to an end.

Because the Communist hegemony fell apart, the vast Soviet and American nuclear holdings were not needed and their dismantlement began, resulting in abatement of the world's abiding fear of nuclear holocaust. Some of the first, most forward-looking moves came as a series of unilateral measures.

The Nunn-Lugar Act of 1991 was a major piece of legislation, signed by President Bush, marking the start of financial aid for the dismantling of Soviet weaponry.

Some feel that the superpower confrontation was not actually over until the last cold-warrior American President (G.H. Bush) left office in 1993, to be succeeded by Bill Clinton. In any event, Clinton's election signaled a shift away from foreign entanglements and toward the nation's neglected domestic problems.

> [Nunn-Lugar Act]. In November 1991, the U.S. Congress (lobbied by the Federation of American Scientists and other public-interest organizations) passed a bipartisan bill (sponsored by Senators Sam Nunn and Richard Lugar) authorizing the government to spend up to $400 million in fiscal-year 1992 to help the Soviet Union destroy nuclear, chemical, and other weapons. Verifiable safeguards were to be set up against proliferation (transfer) of such weapons. Financial and technical assistance was granted to Russia, Ukraine, Belarus, and Kazakhstan, thereby making possible implementation of both S T A R T treaties.

Cold War Reprise

When the Soviet Union collapsed at the end of December 1991, President G.H. Bush could claim that the West had "won" the drawn-out superpower showdown. "Not only was Soviet communism gone," observed one analyst, "but under Boris Yeltsin a new Russia had emerged that was committed to democracy and the free market system." However, it was not just that the price was high for sustaining the half-century of virtual conflict: To this day, it is not clear that Russia will remain a democratic nation, nor is it far-fetched to suspect that nuclear armaments might grow and spread worldwide.

Among the costs of Cold War confrontation were military and civilian deaths and casualties in the limited hot conflicts of that period, especially in Korea and Vietnam. Another cost was to American political institutions, including abuse of presidential powers (Watergate being a sordid revelation). Perceptions of national-security needs endangered the most cherished principles of American constitutional democracy.

In dollars alone, the conflict was very expensive. For the United States, national debt reached $4 trillion, and economic infrastructure — such as transportation, education, nonmilitary R&D — suffered from neglect (a conclusion supported by subsequent recovery and infrastructure reinvestment). Social problems became rampant and are still visible in the inner cities, with their slums, serious unemployment, and dependence on welfare.

Some damaging consequences were psychological — paranoiac fears and suspicions generated during the superpower engagement. In the Soviet Union, such feelings helped maintain Joseph Stalin's rule, and sustained brutal dictatorships there and elsewhere, long after Stalin was gone. In the United States, citizens were intimidated and careers were ruined by mere suspicion and accusation of Communist sympathy. And the military ventures in Czechoslovakia, Vietnam and Afghanistan divided governments from their people.

Most dangerous was the risk of nuclear war, exemplified by the Cuban/Carribean missile crisis. An adverse outcome would have been shared by the entire world.

During the contentious half-century, the United States alone produced more than 70,000 nuclear weapons, a legacy that still haunts humanity.

Winners and Losers. Who "won"? Mikhail Gorbachev has said that "both the Soviet Union and the United States *lost* the Cold War." Unfortunately, there are still some who *miss* it.

When the Soviet Union disintegrated, President Bush claimed that it was a "victory for democracy and freedom." But Anatoly Dobrynin, Soviet Ambassador to the United States during 24 years of the intransigency, decided that "internal contradictions" had doomed the Soviet Union; likewise, an analyst attributes the "victory" to changes within the Soviet Union, especially economic deficiency. Even so, G.H. Bush is credited with facilitating the dramatic concessions that Gorbachev felt compelled to make. Prior arms control initiatives, which aided the cooperative transfer to Yeltsin's control of nuclear weapons in the former Soviet Union, were also important. That peaceful transition averted a potentially catastrophic failure — a nuclear incident or confrontation.

Although the West can easily be considered to have won the ideological clash, there remains debate as to who or what else can take credit. Conservatives argue that the day was saved by their steadfast, hard-line anti-communism, particularly the massive arms buildup under President Reagan. But early on it was President Kennedy, a Democrat, who accelerated the nuclear arms race by ordering an expansion of the strategic arsenal in response to a nonexistent missile gap, prompting the Soviets to try to catch up. Democratic President Carter began a major arms build-up that was continued and expanded by Republican President Reagan. The U.S. arms increases had bipartisan support.

Since 1982, Soviet economists and management analysts had been simulating the effects of possible economic reform. *Perestroika* (the Russian word for *restructuring*) was "the most comprehensive, in-depth, management-science project ever implemented in the world," says economist Thomas H. Naylor, one of the few Westerners who had frequently visited the Soviet Union. He observed technological deficiencies in Soviet computer hardware and software, process-control systems, and energy-related equipment. Naylor considered that the "biggest impediments ... to Soviet economic reforms [were] lack of experience [with] market-oriented management practices [and] no experience with such basic managerial functions as accounting, finance, marketing, or organizational development." That was prophetic: These have turned out to be major impediments to post-Cold War economic advancement of the former Soviet Union.

"The Cold War was also the first total war between economic and social systems.... The West prevailed because its economy proved able to supply guns as well as butter..." says a Western historian, who explained

> Every time the West fought [in a hot war] after 1945, it lost or at least failed to win. [Yet] while suffering reverses on the battlefields of Korea, Dien Bien Phu, Suez and Vietnam, the West managed to build up regional economic bases in Europe and in Asia which were to prove decisive in the long run....

Serious economic problems in the Soviet Union included a stagnant economy, inefficient agriculture, an inadequate supply of consumer goods and services, a serious technology gap with the West, a rigid political structure, a police-state mentality, a high death rate, and an increasingly alienated population. Even so, as Naylor has observed, Reagan's administration had "...no interest whatsoever in [Soviet economic reform — giving them] information that could potentially undermine the rationale on which the greatest [U.S.] peacetime military buildup in history was based." Moreover, the Jackson-Vannik amendment to the Trade Reform Act of 1974 kept the Export-Import bank from participating in U.S./USSR trade as long as the Soviet Union denied its citizens (especially Jewish) the right to emigrate.

According to Naylor, despite the inflexibility of the Soviet economy, "between August and December 1989, every Eastern European government made significant shifts [away from economic rigidity]." "China's foreign trade responded dramatically [to an] open door [policy for export controls]."

Some economists argue that the high levels of U. S. defense spending were dictated by domestic economic considerations, not the Soviet threat. American business and government leaders realized that it had taken a world war to put an end to the depression of the 1930s. Thus, putting the U.S. economy on a permanent war footing was good for business and therefore good for the nation. This hypothesis has been reinforced by the continuation of high U.S. defense budgets ever since.

Russia as Successor

The international community quickly accepted Russia as the primary successor state to the Soviet Union. The newly constituted nation was accorded a seat on the UN Security Council and inherited non-proliferation responsibilities of the NPT (and Soviet international debts). In January 1992, Russia explicitly assumed Soviet arms control obligations, with President Yeltsin agreeing that:

> Russia regards itself as the legal successor to the USSR in the field of responsibility for fulfilling international obligations. We confirm all obligations under bilateral and

multilateral agreements in the field of arms limitations and disarmament which were signed by the Soviet Union and are in effect at present.

****

The opening of the Berlin Wall in November 1989 turned out to be the first and most dramatic indicator of imminent political change in the East-West conflict. Reverberations of the breakdown quickly reached all of Eastern Europe. Within a few weeks, Czechoslovak and Romanian communist regimes fell.

The Cold War ended rather abruptly, unexpectedly, and peacefully in 1991 with the disassembly of the Union of Soviet Socialist Republics. The presidency of George H. Bush witnessed all of this — the rupture of Communism in Europe, disintegration of the Soviet Union, disengagement of nuclear forces, expiration of the ideological clash, and the beginning of postmortem policy assessments.

An important reason for the apparent inevitability of the original bitter East-West conflict was incompatibility of ideologies. Now maybe — with some luck, perseverance, and support — democracy will take hold in Russia and other former Soviet republics. Perhaps that will be one of the most beneficial outcomes (along with the freeing of Eastern Europe). However, at this time there is in Russia not much public support for, or understanding of, market capitalism and democratic processes, although well-connected individuals are profiting. Moreover, no effective system or culture for collecting taxes had existed in the Soviet Union; this has turned out to be a major complication and obstacle in the development of Russia's economic independence. Russian nationalism has revived, and communism is hanging on, while worries continue in the West that the demise of the current Russian state "and replacement by a more aggressively nationalist, and less democratic and [less] capitalistic Russia could again revive the Cold War."

E. COLD WAR LEADERSHIP

Throughout the extended superpower conflict, various parallel efforts to defuse or harness the nuclear arms race were undertaken. Some pacification attempts were the result of personal commitment by government leaders. But other efforts were purely for political gain, and some were driven by ideology. In all cases, leadership came under pressure from embedded as well as public interests.

In this final Section of the Chapter, we tabulate some key aspects, leadership, and treaties of the Cold War, leaving it to subsequent Chapters to provide more background and to explain in more detail.

Immediately after World War II, President Truman tried a somewhat conciliatory approach toward the Soviet Union, although the latter's expansionist and intimidatory tendencies made it difficult. At the same time, Cold War analyst Ronald Powaski finds, "the Truman administration exaggerated its estimate of Soviet military strength to win public support for its commitment to NATO." Actually, the Soviets had demobilized the vast majority of their forces, and approximate parity with NATO existed at the end of World War II. While there was not then a realistic possibility for a Soviet invasion of Western Europe, the explosion of the first Soviet atom bomb in 1949 caused Western leaders to consider exaggeration a more prudent reaction than underestimation. Attempts by communist parties to become dominant in governments were real, as the 1948 coup in Czechoslovakia demonstrated.

Further fueling an American sense of vulnerability was the rise of *McCarthyism*, with Senator Joseph McCarthy charging that Truman was "soft" on both internal and external communist threats. Powaski determined that "[Any] realistic assessment of the Soviet threat was politically impossible, indeed almost treasonable, during the Truman era, and long after."

During Eisenhower's administration, the nuclear arms race "intensified and prolonged the Cold War," concludes Powaski. The United States started to rely heavily on nuclear weapons "primarily because they were relatively inexpensive to construct and deploy compared with conventional forces." As the saying went, they gave "more bang for the buck."

With the Soviets developing their own nuclear arsenal and means to attack the American homeland, nuclear weapons "were considered necessary not only to deter a Soviet nuclear attack," says Powaski, "but also to counter a wide variety of other communist challenges." The United States "made nuclear threats during the offshore-islands crises with China in the 1950s, the Berlin crises of 1948 and 1961, the Cuban missile crisis in 1962, and the Arab-Israeli war of 1973." The overwhelming nuclear superiority of the United States lasted until the 1970s.

> With the exception of ICBMs and ABMs, the United States led the Soviet Union in developing and deploying every major strategic nuclear weapons system. To Americans, nuclear superiority was not only a requirement of national security but a matter of national supremacy. To the Soviets, U.S. nuclear superiority was a threat to their continued existence.

The military-industrial-laboratory complex added self-serving pressure to build more nuclear weapon systems. This, Powaski goes on to say, "included the Pentagon, defense contractors, scientists in the nation's nuclear weapons labs, and politicians with industries in their districts or an ax to grind with the administration in power." He concludes from his study of the superpower conflict:

> To raise congressional support for increased defense spending, the military-industrial[-scientific] complex capitalized repeatedly on the public's fear of the Soviet Union with exaggerated estimates of Soviet capabilities. The result was one weapons gap scare

Section E: Cold War Leadership

after another: the bomber gap and then the missile gap in the late 1950s, the ABM gap in the late 1960s, the throw-weight gap in the 1970s, and the space-weapons gap in the 1980s.

The Soviets felt compelled to keep up with their more technologically advanced adversary, and, in time, they succeeded in matching in number if not quality virtually every major U.S. nuclear weapon. Paradoxically, the ultimate and inevitable result of this action-reaction cycle was an increase in both American and Soviet insecurity. The more nuclear weapons the Americans targeted on the Soviet Union, the more nuclear weapons the Soviets aimed at the United States.

Superpower Leaders. Who were the individuals in charge during this extended unstable status of hostility?

Superpower Leadership

Beginning Date	U.S. Leadership	Beginning Date	USSR Leadership
4 Mar. 1933	Franklin D. Roosevelt	~1924 (various positions)	Josef I. Stalin
12 Apr. 1945	Harry S. Truman		
20 Jan. 1953	Dwight D. Eisenhower	5 Mar. 1953	Georgi M. Malenkov
		Feb. 1955	Nikolai A. Bulganin
20 Jan. 1961	John F. Kennedy	Mar. 1958	Nikita S. Khrushchev
22 Nov. 1963	Lyndon B. Johnson		
20 Jan. 1969	Richard M. Nixon	14 Oct. 1964	Leonid I. Brezhnev
9 Aug. 1974	Gerald R. Ford		
20 Jan. 1977	Jimmy Carter		
20 Jan. 1981	Ronald W. Reagan	11 Nov. 1982	Yuri V. Andropov
		13 Feb. 1984	Konstantin Chernenko
		11 Mar. 1985	Mikhail S. Gorbachev
20 Jan. 1989	George H. Bush	Nov.-Dec. 1991	Boris N. Yeltsin
20 Jan. 1993	William J. Clinton		
20 Jan. 2001	George W. Bush	31 Dec. 1999	Vladimir V. Putin

Throughout President Nixon's term of office, Henry Kissinger was his most influential foreign-policy advisor, often striving to develop a working relationship with the Soviet Union. However, his insistence on "linkage" of the SALT process with Soviet actions in other parts of the world did not bear fruit until May 1972, when Nixon and the Soviet leader Leonid Brezhnev signed the SALT I and ABM Treaties.

Non-Aligned Nations. From the earliest stages of the Cold War, many nations chose (or had no choice) to stay out of the political and military engagements of the superpowers. Many poorer states became part of an international non-alignment movement, later joined by independent oil-producing nations. They all took a strategic and political position of neutrality with regard to the Cold War struggle.

Mostly composed of nations that had gained independence from the European empires in the period after World War II, the movement tried to act as a stabilizing force — between the two superpower blocs *vis-a-vis* the non-aligned members — as well as a more powerful voice through unity.

Dating back to 1955, the movement was founded by the leaders of India (Jawaharlal Nehru), Ghana (Kwame Nkrumah), Egypt (Gamal Abdel Nasser), Indonesia (Achmed Sukarno), and Yugoslavia (Josep Broz Tito).

Both the United States and the Soviet Union tried to attract non-aligned countries into their respective camps through monetary aid. Though never a strongly unified group like NATO or the Warsaw Pact, its members were able to increase their status through unity and cooperative voting in the UN General Assembly. None of them had nuclear weapons, but some had nuclear ambitions.

Treaties and Agreements. Many arms control treaties and agreements between or including the superpowers were put in place.

During public demonstrations in the 1980s after Reagan was elected President, he was frequently criticized for resisting arms control agreements.

Long before he was President in 1981, Reagan's record was one of rigid opposition. In 1963 he was against the Limited Test Ban Treaty. In 1968 he objected to the Non-Proliferation Treaty, saying it was not "any of our business." (The NPT has since become the universal mainstay against worldwide nuclear proliferation.) In 1972, Reagan opposed SALT I and the ABM Treaty negotiated by President Nixon. Two years later he spoke against the SALT II understandings reached by Presidents Ford and Brezhnev. In another two years he was in opposition to the PNE Treaty. In 1979 Reagan contested Senate ratification of SALT II, calling it "fatally flawed." He openly advocated a nuclear arms race in his 1980 campaign.

Even after becoming President, in addressing the British House of Commons in June 1982, Reagan deplored what he derisively called "accommodation," asking the rhetorical question, "Must freedom wither in a quiet, deadening accommodation with totalitarian evil?" That address, which became known as "The Evil Empire Speech," met with applause in some circles and consternation in others. A year later, he explicitly referred to the Soviet Union as an "evil empire" and described Soviet Communism as the "focus of evil."

President Reagan's efforts to "negotiate through strength," while pleasing his conservative supporters, engendered considerable public opposition; moreover, it was counterproductive — leading, in due time, to a comparable arms buildup by the Soviet Union rather than a negotiated outcome. After NATO's decision in 1983 to deploy U.S. nuclear missiles within range of the Soviet Union, riots developed in Europe and governments eventually toppled. As soon as West Germany's parliament voted to accept deployment, Pershing-II intermediate-range missiles quickly began to arrive. That prompted the Soviet Union to walk out of arms talks. Moreover, it was feared that Western Europe might become "increasingly vulnerable to a conventional blitzkrieg" because NATO had fallen into "such a lopsided reliance on nuclear defense."

The Soviet strategic response followed in three arenas: expediting deployment of its SS-20 missiles, (2) accelerating installation of new tactical nuclear weapons in East Germany and Czechoslovakia, and (3) deploying nuclear-armed cruise missiles in submarines off the coasts of the United States. It is difficult to argue that Western security was being improved when both the qualitative and quantitative threat was being amplified.

All of this came soon after a landmark TV program, "The Day After," drew an audience of 100 million Americans to what one commentator called a "searing version of nuclear annihilation."

Reagan's support of the MX missile, widely recognized as a first-strike weapon because of its accuracy and vulnerability to pre-emption, further inflamed public opposition.

In a 1983 address to the nation, President Reagan alleged (1) that Soviet defense spending in 1982 was nearly double that of the United States, and (2) that the Soviet Union had about 50 percent more strategic ballistic missiles, nearly twice as many land-based intermediate-range missile warheads, and 30 percent more strategic missiles and bombers. This was a selective use of one-sided statistical and intelligence data; Reagan failed to give a realistic assessment of relative military strength, neglecting other important considerations, particularly weapons quality and the fact that mutual vulnerability made weapon quantities largely irrelevant.

Many prescriptions to halt the arms race were offered by those in and out of government. For example, a group of non-partisan corporate leaders in 1984 became "apprehensive about a costly and dangerous competition neither superpower can win." To stop the arms race, they proposed five steps: suspend nuclear testing, negotiate a test-ban treaty, halt deployment of new nuclear weapons systems, ban production of nuclear warheads, and carry out "stable, orderly, balanced" reductions. The businesspeople recognized the "deadly — and costly — competition" that was under way among the superpowers. They also had no illusions about the need for arms agreements to be mutual and verifiable, having had their proposed steps reviewed by scientists, specialists, retired military officers, and in particular by former CIA director William Colby, who became a vocal proponent of reducing nuclear arms.

At the end of his second term, President Reagan and Premier Gorbachev had a summit meeting in Iceland, where they came close to ending the arms race. But the 1986 Reykjavik meeting embarrassed Reagan because he and Gorbachev agreed to eliminate all ballistic missiles and nuclear bombs within 10 years, a proposal quickly recanted by Reagan's advisors. Even so, agreements were reached on INF and strategic reductions, but not on SDI restrictions (which remained a thorn in relations).

> **[START II].** Just before leaving office in January 1993, President Bush signed the START II agreement with President Yeltsin. This agreement called for both sides to cut their nuclear arsenals by about one-third in a decade. Although it has been ratified by legislative bodies of both nations, stipulations that were added delayed and even thwarted implementation.

As historian Powaski reviews the Reagan presidency,

The "full-court press" launched during Reagan's first term — which included a military buildup capped by the Strategic Defense Initiative, the denial of economic assistance to the Soviet Union, and a willingness to challenge the Soviet Union in the Third World — did force the Soviets to consider the prospect of having to expend resources they did not have. But the Cold War came to an end primarily because of the inherent weaknesses in the Soviet system, not the pressure the Reagan administration brought to bear upon it. Simply put, the Soviet economic system had failed to support the Soviet empire.

****

The occupation of Eastern Europe and the export of communism was anathema to the West, especially the United States. Nuclear weaponry was the primary deterrent through which the Cold War was kept under control by leaders of both sides.

The course and nature of bilateral relationships were kept in check at times largely by the profound influence of sensible military and political leaders. Some of them carefully shepherded their constituencies, evading slippery slopes; others eagerly sought confrontational paths, egged on by their hawkish military-industrial-scientific complexes.

All of the supreme leaders sooner or later realized the importance of diplomatic approaches and agreements in holding nuclear warfare at bay. As a result, a framework of arms control treaties emerged during what turned out to be an extended period of East-West tension. Normalized by this formalistic framework were mutual strategies of deterrence, the Soviet de-facto hegemonic control over Eastern Europe, and a Western policy resigned to containment.

Eastern Europe simply remained under Soviet communist shackles.

****

The devastation of Hiroshima and Nagasaki in 1945 stands out as the most historic and significant event of the atomic century. Although an East-West hostility was already in the making, the nuclear arms race thereupon became inevitable. The bombing also signaled a willingness to use nuclear weapons, and thus added credibility to threats to use them in other conflicts.

After World War II, the Soviet imposition of the virtual "iron curtain" in Europe marked an intractable phase, stirring up further development of nuclear weapons, first by the Americans and later by the Soviets, who rapidly caught up. Many of these developments were shrouded in secrecy, helping to sustain some questionable policies that were not subject to public scrutiny. The strategy of containment was invented in Washington and pursued for decades by driving forces within the DC beltway.

NATO began in 1949 as a conventional response to insecurity sensed by Western allies of the United States, as well as reaction to the Berlin blockade, which had started a year earlier. NATO eventually acquired nuclear weapons and a doctrine for their use in Europe. The late 1950s brought Sputnik and intercontinental missiles that were capable of reliably delivering nuclear weapons around the world.

New nuclear designs, including thermonuclear fusion weapons, were tested extensively in the atmosphere during the 1950s and 1960s, eventually being tested underground in response to public opposition to the release of radioactivity into the atmosphere.

Attempts in the 1960s to find defenses against nuclear attack led to controversial ballistic-missile defense systems and promulgation of civil-defense measures; these too attracted public opposition.

The Cuban/Carribean missile crisis of 1962 was patently the closest brush with mutual nuclear destruction. That crisis resulted in the withdrawal of missiles from Cuba (and Turkey), but also in a massive expansion of the Soviet nuclear arsenal. U.S. nuclear superiority evaporated, and a period of détente developed. Facilitated by fears engendered by the existence of mutual assured destruction and by the excessive economic toll of the nuclear arms race, several treaties were negotiated to reduce the tempo of the competition. (However, before the buildups were curtailed, the number of nuclear weapons in the world reached a peak of more than 60,000 — enough to destroy every aspect of civilization.)

In 1989 the Berlin Wall was torn apart, the first of a sequence of events that culminated in the disintegration of the Soviet Union itself two years later, ending the bipolar hostilities. Only then did an uncertain process begin: of standing down from the capacity for mutual destruction, accompanied by hesitant dismantling of nuclear weapons and their delivery systems. In the early 1990s a more-focused national and international process commenced for coping with the nuclear-proliferation and environmental-pollution legacies of the arms race.

In essence, the Cold War was a period of incessant antagonism — a political, economic, sociological, psychological, and militaristic struggle — largely between the two superpowers. Having begun just as World War II was ending, it persisted through escalating degrees of conflict until the Soviet Union collapsed. Although nuclear weapons were not detonated in the Cold War, they were brandished via various proxies and subtleties. Despite exaggerated rhetoric, oversimplified villainy, counterproductive secrecy, and so-called "worst-case analysis," a combination of restraint and luck prevented the outbreak of nuclear war.

To outsiders, the struggles could be likened to compulsive sparring between two scorpions trapped in a bottle. Although exhausted, both scorpions survived and one gained bragging rights.

Chapter II:
COLD WAR NATIONAL SECURITY

(Hypothermia and Hyperthermia)

Chapter II: National Security

A. NATIONAL DEFENSE	82
Deterrence	83
[X=Y≠Z]	86
Figure 3: Mutual Assured Destruction	87
Multiple Warhead Missiles	90
Survivability	91
Nuclear War Strategies	92
U.S. Strategies	92
Soviet Strategies	94
[Hawks and Doves]	95
British Global Strategy	97
Germany's Nuclear-Capable Delivery Systems	98
France and China	98
Nuclear-Explosion Phenomena	99
Fireball	100
Blast	100
Radiation	101
Fallout	101
[Radiation Units]	102
EMP	103
Examples of Nuclear Devastation	104
[Nuclear-Explosion Effects]	105
B. THE WEAPONS	106
Categories and Types of Weapon Systems	106
PALs	107
Strategic Systems	107
Triad	108
Bombers	109
U.S. Intercontinental Ballistic Missiles (ICBMs)	110
Soviet Strategic Modernization	111
Submarine-Launched Missiles	111
Shorter-Range Missiles	112
Soviet Nuclear Weapon Systems	112
Theater Weapons	113
[Tactical Nuclear Weapons]	113
U.S. Overseas Deployments	114
Soviet Deployments	116
INF Missiles	117
Enhanced-Radiation Warhead	117
British and French Deployments	118
Nuclear Warheads	118
Warhead Safety	119
Comparative Military Strength	120
British, French, and Chinese Nuclear Systems	121
C. DETERRENCE-SUPPORT TECHNOLOGY	122
Production of Nuclear Weapons	122
[Oak Ridge]	123
Nuclear weapons Testing	124
[A Sea Drill]	125
Nuclear Propulsion	126
U.S. Naval Vessels	126
Soviet Nuclear-Powered Vessels	127

Section A: National Defense

Early Warning 127
Canceled Programs 128
Intelligence Collection and Satellites 129
   U-2 Flights 129
     [U-2 Memoir] 130
   Satellite Reconnaissance 131
Command, Control, and Communication 133
   C³I 133
   Command 134
     [The Football] 135
   Control 136
   Communications 137

D. ARMS DEVELOPMENTS AND LIMITATIONS 139
Advances in Delivery Systems 140
   Stealth 140
   Technology Creep 140
Civil Defense 141
   The 1960s 143
   The 1970s 143
   Emergency Management 143
   Fallout Shelters 144
Strategic Defenses 145
   Tallinn 147
   Galosh 147
   Sentinel 148
   Safeguard 148
     [Nixon's ABM Decision] 149
   SDI 151
   Directed-Energy Weapons 152
Antisatellite Technology 153
Overkill 154
   [Nuclear Warheads Galore] 154
Improved Weapons Safety 156
Arms Control 157
   SALT I 157
   Nuclear Testing Limitations 158
   SALT II 159
   INF 160
   START I 161

In this Chapter we examine, with more detail, the armaments that sustained the Cold War and underlay the two sides' postures and notions of national security or insecurity. Since the end of that traumatic period, much information has been declassified, giving us the opportunity to examine the validity of various comparisons of military strength. Also, the passage of time permits a less impassioned view of the role rendered by negotiated arms control in tempering the arm race.

In a sense, the bilateral competition caused a chilling effect (institutional hypothermia) on international and bilateral relations. This, in turn, led to political and ideological expressions that were rhetorically and audaciously overheated (hyperthermia).

First basic strategic concepts are outlined, and changes in them, during the ups and downs of the virtual conflict. These policies encouraged the growth of nuclear weaponry — their quantity, explosive yield, means of delivery, and support systems. Later on, we discuss sub-strategic (theater or tactical) weapons.

In the first Chapter atomic weaponry was treated superficially within the context of historical events, while here more particulars are provided about the underlying policies, strategies, technologies, and influences that steered the arms buildup.

Some details about the nuts and bolts of weapon systems are provided. This lays the foundation for understanding not only some major aspects of the superpower confrontation but also its aftermath. While not essential reading for everyone, this helps remove some mystery from subsequent events and causes. Even if the reader skips specifics, the background information and summaries will be useful, especially in terms of lessons spotlighted.

Most important are the factors that led to the United States having, at one point, more than 32,000 nuclear weapons in its arsenal. Well before that peak was reached, the nation had attained an explosive power of 1.4 million Hiroshima-sized bombs. Needless to say, the other nuclear powers, especially the Soviet Union, also countered with their own highly destructive nuclear arsenals.

A. NATIONAL DEFENSE

In reviewing strategies and armaments of military establishments, we must keep in mind the background underlying Soviet feelings of near-paranoia about outlanders. Invasions from east and west pervade Russian history; fresh and persistent was the memory of German's World War II invasion — an attack repulsed by military commanders who became the post-war Soviet leaders. After the war the Soviet politburo strongly felt the need for unassailable military strength, to offset perceived vulnerabilities. After all, some 30 million of their citizens had been killed, their country had been devastated, and their economy had been shredded.

At the same time, the United States and the balance of Europe felt threatened by the reality and the rhetoric of international communism being exported from the Soviet Union. As noted in Chapter I, one of the underlying reasons for the United States to use the atomic bomb against Japan was to intimidate Stalin and forestall further Soviet aggrandizement. Success in that regard, however, was limited: The subsequent Soviet takeover of Manchuria and the vassalage of Eastern Europe went notably undeterred, notwithstanding the short-term American nuclear monopoly.

On each side of the ideological divide there were several possible post-war strategies for preserving national security, ranging from insular retrenchment to overt pre-emptive aggression. The United States chose to rely on deterrence through overwhelming nuclear strength, expanding the capability that had devastated two Japanese cities. However, because of accompanying rhetoric, that buildup looked to the Soviets not so much like deterrence than a threat to their survival.

While the United States could have assumed a less threatening stance, it also could have done worse: Another option — more available to it, because of its superior resources, than to the Soviet Union — would have been a vigorous pursuit of nuclear war-fighting capability. Aspects of such a hard-line posture were often implicit in U.S. policy — (or internally explicit, but kept secret from the public because of their alarming implications).

Three contrasting policies for strategic-force utilization were conceivable, especially as delivery systems changed from bombers to missiles: (1) pre-emptive (or first-strike) attack, (2) launch-on-warning of an attack, and (3) retaliatory (deterrent) response. Ostensibly, the preferred policy became deterrence, around which forces were configured.

Western strategists considered themselves realists, and virtually all believed the Soviet Union to be "a formidable adversary that will exploit every foreign-policy opportunity presented to it." That did not mean the Westerners expected nuclear war, but most were inclined to think that a strong military posture would induce Soviet leaders to act less aggressively.

In pages that follow, these postures are dissected and analyzed, with some detail about why nuclear weapons have so profoundly changed some of the concepts of warfare.

Deterrence

Deterrence is a national strategy for displaying an image of formidable military strength, intended to convince a potential adversary that an armed attack would be self-destructive. While the threat of massive and certain retaliation is a powerful dissuasion, such a policy provides no guarantee against being attacked.

A cohesive and up-to-date conventional military stance is itself a strong deterrent. Because of human factors, what is not clear in abstract analyses of strategic posture is exactly what is necessary or sufficient to deter. If an adversary poses a relatively minor military threat, the arsenal for deterrence need not be large. Against a major adversary, the retaliatory damage that could be inflicted even by a modest nuclear arsenal is devastating.

Moreover, other institutional capabilities, such as economic and political muscle, contribute to deterrence. They should be factored in, along with positive or negative societal factors, such as public zeal.

Negotiated arms control provided the means to gradually achieve mutual limitations that tempered some arms-race excesses. Nevertheless, during the Cold War, the magnitude of armaments required to exercise effective deterrence was often publicly and privately disputed.

Deterrence depended on some factors that were not particularly objective, as acknowledged in 1959 by strategist Bernard Brodie (using the genderized language of his profession at the time):

> Deterrence after all depends upon a subjective feeling which we are trying to create in the opponent's mind, a feeling compounded of respect and fear, and we have to ask ourselves whether it is not possible to overshoot the mark. It is possible to make him fear us too much, especially if what we make him fear is our over-readiness to react, whether or not he translates it into clear evidence of our aggressive intent.

Not only is deterrence a motivator of strategy, it is also an outcome of the actual acquisition, deployment, and operation of military forces. Like other possible strategic policies, it is a construct that can have a positive or negative influence in a feedback loop between policy and implementation.

Freedom from fear of traumatic surprise is probably the essence of stable deterrence. Because advances in military technology often look as though they will destabilize the "balance of terror," they tend to stimulate an escalatory response in anticipation of the worst, even though many weapon systems fail to live up to their billing when implemented. It nominally had taken 15 years for a new type of weapon to be fully integrated into service: An adversary typically had more time than usually assumed to respond when it became clear that the new system should be countered. Some warfare systems (such as aircraft carriers) can remain in service for as long as 40 years, so implementation represents a long-term commitment not to be undertaken lightly.

Acquisition of atomic weapons has engendered military policies that could destabilize a coexisting balance of power. When two nuclear powers confront one another, the situation is stable only when each has an offsetting *second-strike capability* (the ability to ride out an attack and still retaliate with a devastating response). A bilateral second-strike posture fosters the perception that deterrence is mutual and stable. Obviously, in such a situation neither side has a *first-strike capability* (the ability to destroy the other side's retaliatory capacity in a first blow), the existence of which would create a condition of *crisis instability*.

A first-strike capability, on one or both sides, is inherently unstable — disrupting mutual-assured destruction (MAD) and leading to hair-trigger *launch-on-warning* policies. Ominous as it might seem, many hawkish individuals and organizations favored such a capability.

Even with mutual assured destruction, one or both sides might feel that a significant portion of their forces was vulnerable to a first strike. In that case, in a time of crisis, hawkish elements considered it reason enough to risk a pre-emptive attack while still possible ("use it or lose it").

Crisis stability prevails when neither side has an incentive to strike first during a crisis.

Nuclear weapons thus became conceptually *useful* only to the extent that they deterred war. An important corollary inherent in crisis stability is confidence to ride out a single missile launch that might be accidental or unauthorized, so that war does not inevitably follow.

Even the possibility that the opponent *might* be moving toward a first-strike capability is considered unacceptable. For example, each side during the Cold War felt threatened by the prospect that the other might be developing an effective missile defense that could let them mount a first strike with impunity. President Reagan's Secretary of Defense, Caspar Weinberger, testifying before Congress, said that one of the most frightening situations he could imagine would be for the Soviets to acquire an effective missile defense (thereby supposedly denying the United States a second-strike capability).

In this context, Israel's 1981 destruction of the Iraqi Osirak nuclear reactor before it could go into operation is telling. There was no possibility that Iraq would in the foreseeable future have been able to get a first-strike nuclear capability, even against Israel. But the possibility that Iraq could get a *second*-strike, retaliatory (deterring) capability might have also been considered unacceptable to Israel's policymakers.

Apparent challenges to stability, deterrence, or power projection have often been invoked as excuses for pre-emptive actions.

Early strategies were rooted strongly in the perceived threat to Western Europe from the large standing Soviet Army. The initial American response and the subsequent Soviet counter-response was an expanded strategic air force, armed with atomic bombs. A key factor in establishing a convincing aura of deterrence has been the ability to send warheads over great distances; as a result, short- and long-range bombers became the initial mainstay of strategic forces on both sides.

Late in the 1950s, intermediate- and long-range ballistic missiles were developed that could be guided to targets the size of cities; inevitably their guidance systems were so improved that they could be aimed at much smaller targets like specific military installations.

Eventually there evolved arsenals of hydrogen bombs that could be delivered by missiles. That in turn led to hardening of *silos* (underground facilities for storing and launching missiles) to minimize their vulnerability to a first strike. The logical next step was to try to devise an anti-ballistic missile system, even though it threatened to destabilize mutual deterrence. In time, submarine-launched ballistic missiles (SLBMs) were developed; these had the advantage of being nearly immune to a pre-emptive strike, thus increasing the stability of the deterrence.

Based on concepts articulated by George Kennan, the policy of *containment* was devised to thwart Soviet expansionism without going to war. The Truman Doctrine and the Marshall Plan were developed as part of the plan to check the spread of Communist militancy. Although it took about fifteen years for the Soviets to tone down their aggressive stance, the perception that a major war was imminent slowly diminished. But the repressive Soviet system did not change as much as Kennan and his fellow planners expected, and the Soviet Union somehow maintained a workable economy along with a giant military machine. What was expected to be short-term turned out to be a long-term, threat-and-counterthreat competition between East and West.

A U.S. governmental structure for national security affairs was created by the National Security Act of 1947, and with minor changes this structure has continued. Under that Act, the military services were unified, at least in principle, within a new Department of Defense (DOD). The institution of the Joint Chiefs of Staff (JCS) was established, the World-War-II Office of Strategic Services (OSS) became the CIA, and the National Security Council (NSC) was created. In 1948, an agreement clarified the roles of the U.S. services, authorizing the Navy to conduct tactical air operations, and making the Air Force responsible for strategic air warfare.

The Atomic Energy Act was passed at about the same time, placing the atomic-bomb program under civilian control.

In the Soviet Union, the Defense Council — consisting of members of the civilian and military leadership — was the agency that subordinated the armed forces to party leadership. The Defense Council was empowered to make decisions about nuclear weapons. Headed by the General Secretary of the Central Committee of the Communist Party, it was accountable to the Politburo. But in defense policy, the political leadership of the Soviet Union relied almost entirely on recommendations and evaluations by officials of the armed forces. The Military-Industrial Commission established basic guidelines and coordinated development of military equipment and weapons.

> [X=Y ≠ Z].. When George F. Kennan (Mr. X), the father of containment, realized that his idea was misunderstood, he renounced "fatherhood" of the strategy because "it had come to mean mostly military containment," whereas he had expected "mostly political and diplomatic pressure." Kennan wanted the strategy of containment (Y) to avoid a war, not equate to military action (Z).

Rendering deterrence to avoid a third world war became the principal function of American military power. Alternative strategies would have been to launch a preventive war, or to position a massive American conventional military presence in Europe, but neither of these alternatives was as feasible, cheap, nor acceptable to the public as deterrence.

In 1949 two events severely shocked the American sense of security: the final victory of the Chinese Communists in their civil war, and the first atomic test explosion by the Soviet Union. These events were instrumental in energizing U.S. work on a fusion bomb and a governmental study that led in 1950 to a new strategic policy decision, NSC-68.

Massive retaliation became the official U.S. military doctrine, lasting at least until 1961. It was a strategy aimed at deterring both non-conventional and conventional attack. Immediately after World War II, a new big war was arguably precluded by America's atomic monopoly. However, with that monopoly having been lost, NSC-68 acknowledged that deterrence was more complex, requiring a solution for each level of threat. When Secretary Dulles spoke about instant and massive retaliation, he introduced the idea that deterrence was a tool to be used vigorously in support of policy aims throughout the American sphere of interest. It became the means for preventing not only big wars, but small ones as well, and the United States issued threats of using nuclear weapons at least three times: to encourage Korean War peace negotiations, to warn against resumption of that war, and to keep the Chinese from intervening in the war in Vietnam.

NSC-68, according to an historian of the war in Vietnam "created an almost apocalyptic vision of the international communist menace, one that fundamentally obscured the political, economic and social complexities of a decolonizing Vietnam."

Mutual assured destruction, a situation that materialized as the nuclear arsenals grew, became an acronymic symbol of their counterproductive risk.

A sense of mutual dependence for survival is captured in the cartoon reproduced in Figure 3.

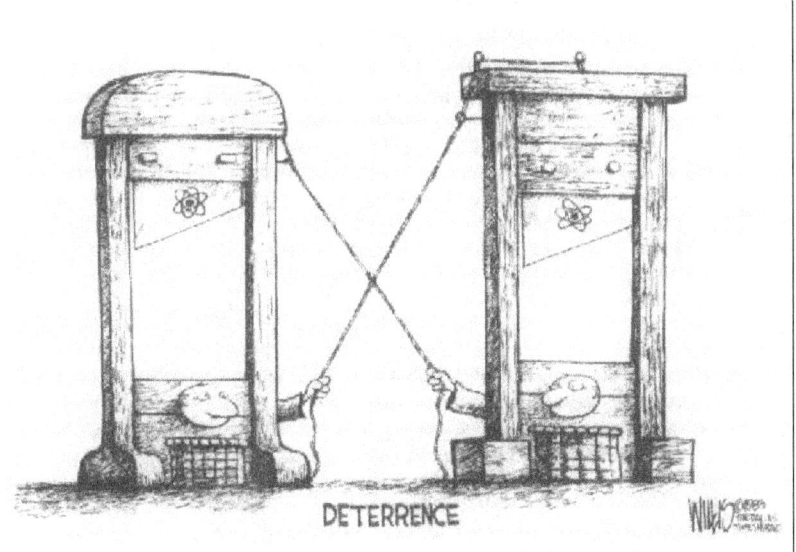

Figure 3: Mutual Assured Destruction

Two major problems developed with the policy of massive retaliation as Soviet strategic power grew to counterbalance the West: (1) the policy (challenged by the public) depended on assured destruction of the homeland, and (2) the standing American security umbrella for Europe was no longer seen as credible. Some strategists doubted that "one can rationally — that is to say, in one's own interest —

carry out the threat of a nuclear counterattack once the threat has failed to deter." Would the United States place its own cities in jeopardy if the Soviets attacked in Europe? One answer came with a new doctrine, *flexible response,* which gave rise to an arsenal of sub-strategic and strategic weapons, as well as to maintenance of sizable conventional forces.

Because nuclear-war initiation could be tantamount to national suicide, deterrence has been characterized as "prudent, reciprocal self-deterrence." An apt analogy, attributed to J. Robert Oppenheimer, is two scorpions in a bottle — if one attacks, they both die. Although rational self-deterrence favors survival, history does not assure rationality.

Assured destruction became truly mutual in the early 1960s, when the Soviet Union reached nuclear sufficiency. MAD, though, has stayed on as a fact of life. In the mid-1960s Defense Secretary McNamara advocated relying on MAD to prevent nuclear war, a posture that he eventually repudiated. Destructive capability peaked around 1968, when the United States decided to freeze the size of its arsenal and to concentrate on technological improvements to its weapons systems. The Nixon administration accepted the concept of "sufficiency" in atomic weapons, and began to negotiate treaties to limit the arms race.

Nevertheless, the arms race continued through the 1970s and 80s. One analyst and author of treatises on superpower politics — the "War Game" — expressed concern about the escalation in 1983:

> American strategic planners have been thinking the unthinkable for more than 25 years: trying to make nuclear warfare controllable and even winnable. The result has been an escalating nuclear arms race that has increasingly reduced our security and endangered our survival.... [The danger is increasing because] a war-fighting doctrine is fundamentally incompatible with a policy of nuclear deterrence: The greater the capacity to fight a nuclear war, the more likely it is that deterrence will fail because of the increased chances that a war will start through accident, miscalculation or pre-emption in a time of crisis.

The struggle between *war fighters* and *advocates of stable deterrence* has been traced to the late 1950s. War fighters ridiculed the notion of relying on mutual assured destruction, claiming it was immoral because it would mean destroying cities; moreover, they derided it because they did not think the Soviets would be deterred, because it left no alternative to extensive devastation of civilian targets if deterrence failed, and because preservation of MAD precluded deployment of a first-strike force. The war fighters wanted to be able to coerce the Soviet Union; on the other hand, advocates of deterrence were uncomfortably gratified with simply an assured capacity to destroy one another.

Britain and France had no moral qualms about relying on MAD: They too aimed their strategic missiles at Soviet cities, expecting to deter an attack. In any event, official Soviet propaganda impressed their public with the notion that their citizens, not their military, were targeted by Western nations.

In 1962 Secretary of Defense McNamara adopted a strategy of "controlled and flexible nuclear response aimed at military targets rather than cities — a counterforce strategy." This plan for limited atomic war might have made sense then, but by 1982 it did not. For one thing, the Kremlin made it clear that while it would not be the first to use nuclear weapons: Its response to an attack would not be self-limited.

Much U.S. policy was based on what turned out to be an inaccurate interpretation of Soviet views — notably the notion that the Soviet military leadership thought nuclear victory was feasible. In the 1970s the Soviet leadership publicly acknowledged the holocaust that unconventional war would bring. McNamara, for one, eventually changed his own views, rejecting the concepts of flexible response and limited warfare. He has since converted to the position that "nuclear weapons serve no military purpose whatsoever. They are totally useless — except only to deter one's opponent from using them."

During the Johnson administration, *gradualism* became a strategic concept: measured application of force so as to minimize the risk of Soviet or Chinese intervention (into the U.S. war in Vietnam). Former Secretary of the Air Force, Thomas C. Reed, faults these trends for abandoning "the principles of surprise, massed force, and timely execution, all fundamental to military success." He laments the "misguided allocation of resources" that "left the U.S. Air Force focused on nuclear war, to the exclusion of other duties." He adds that "any American use of nukes could only trigger a broader exchange that would be lethal to both sides, having particularly in mind the concept of "assured destruction."

On the other hand, when Caspar Weinberger became Ronald Reagan's Secretary of Defense in 1981, he wanted preparations to fight a protracted nuclear war and to prevail. (Only later did he acknowledge that such wars are not winnable.) The Reagan administration, in which Reed held prominent positions, vigorously pursued a *decapitation* doctrine — a capability to destroy Soviet civilian and military leadership and communications — undeterred by the fact that decapitation left nobody with whom to negotiate an end to the war. Decapitation risked a furious spasmodic all-out reflexive nuclear-doomsday death blow from a doomed antagonist.

A published 1983 analysis of superpower policies argued that deterrence must be made "as stable as possible," favoring a reformed basis for deterrence, consisting of "genuine arms control negotiations," and moving away from multiple-warhead weapons toward bombers, submarines, and single-warhead missiles.

Another strategist articulated the perception that "the system of deterrence as it is presently [1985] constituted is extremely stable, that the so-called balance of terror is not delicate, and that it is not likely to be significantly affected by changes in the level of armament in either direction for the foreseeable future." Others held different views: " it is morally untenable to base deterrence on the threat to destroy large portions of each side's population.... Thus ... counterforce weapons and strategic defensive systems [should be developed]."

An alternative to threatening unmitigated retaliation in response to a conventional attack is to place greater reliance on non-nuclear forces (a *finite deterrent*), de-emphasizing dependence on non-conventional weapons. In the absence of counterforce weapons, MAD can be preserved by having only enough weapons to destroy a significant fraction of the opponent's population. This approach would minimize concern that a limited conflict might escalate to an all-out war, which under nuclear-only deterrence could be the outcome of a policy game of chicken. Finite deterrence, based on a strong conventional force, also defuses the concern that an atomic holocaust might occur.

One can find an apt analogy for deterrence in nature: some substances are metastable. Certain natural glasses are thermodynamically unstable but nevertheless have persisted in equilibrium for thousands of years. That's because a crucial triggering event (such as sufficient rise in temperature) which would cause them to change hasn't happened yet. During the half century of Cold War, the international temperature remained low enough that deterrence — fundamentally metastable — remained in uneasy equilibrium.

Multiple Warhead Missiles. The original multiple-warhead missile system was designated MRV, for multiple-reentry vehicle. With MRV, more than one reentry vehicle could be released in flight by the upper stage of a single missile. The RVs, containing nuclear warheads, would be spread apart somewhat, in order to inflict more damage on a target area than a single large warhead. They could not be accurately aimed.

Mutual deterrence could be destabilized by developments that permit a pre-emptive attack which would inflict considerable damage to retaliatory forces. Just such a situation loomed when the United States developed missiles, such as the Minuteman and the MX, armed with multiple, accurately targetable warheads (an outgrowth of a NASA program that launched four satellites from a single missile).

MRV was followed by MIRV, which has the ability to aim each warhead individually, allowing the warheads to be dispersed somewhat more widely, and directed at individual targets with high accuracy.

MIRV immediately upset the balance of potential destruction: The Soviets, faced with the possibility of losing several unlaunched missiles to each attacking missile, found themselves deprived of confidence in their second-strike capability. They thus felt compelled to place their largely land-based attack force on "launch on warning" — the hair-trigger alert status that is the antithesis of nuclear stability. While MIRV seemed to gain the United States a momentary edge in offense, it added to the risk of accidental or unauthorized launches.

According to Senator Daniel Moynihan, "The decisive technological event that led to the shift in balance of power was the deployment of MIRVs — a term first used in public in 1967." Viewed in hindsight, it should have been recognized that MIRV technology would work ultimately to U.S. disadvantage.

Herbert Scoville, a former deputy director of the CIA, explained MIRV's operation and its destabilizing effect in 1981 as follows:

> The basic principle [of MIRVs] was to make the final stage of the missile, which contained in the U.S. case from three to fourteen warheads, a sort of "bus," which successively released its warheads on trajectories toward separate targets. With large numbers of warheads on a single missile, it was possible to overwhelm an ABM [antiballistic missile] system and guarantee that a warhead would reach the specific target.... It had a corollary characteristic that is now profoundly destabilizing the arms race. Since each missile could fire several warheads at different targets, in theory a single missile with MIRVs could threaten several missiles on the other side....

Thus, now there could be a real gain by launching a first strike against the opposing side's land-based ICBM force. MIRVs now created incentives to initiate a nuclear strike....

The U.S. first tested MIRVs in August 1968.... The Soviet Union, running about five years behind the United States in MIRV technology, tested its first MIRVed ICBM in the summer of 1973 [and] by 1980 it was obvious that Soviet MIRVed missiles would soon threaten our land-based ICBMs.

The next key arms development in multiple-warhead delivery was the MX, a highly accurate U.S. missile that could be transported by railcar or truck. A troublesome aspect of the new missile was that it invited pre-emptive attack: destroying one MX would eliminate 10 warheads at once.

The original MX proposal called for 200 missiles to be deployed in 4600 hardened (blast-resistant) widely separated structures. Various controversial schemes were cooked up to reduce vulnerability, such as transporting the missiles from one launch tube to another, as in a shell game, using rail and/or road loops. About 10,000 miles of load-bearing roadway would have been paved throughout a large area of the western United States. The proposal was scuttled when it became apparent that those states would become a potential target for nuclear attack, and that eastern states would become a deposit zone for extensive fallout.

The ten-warhead MIRVed MX missile was first proposed by President Nixon in 1973 as a means of improving U.S. ability to strike hard targets in the Soviet Union. The MX elicited an equivalent response from the Soviets in the late 1970s, although the accuracy of their MIRVs did not approach the 100-meter CEP ("circular error, probable") claimed for the MX. The missile system was enormously expensive, and its launch sites immediately became prime targets because a large number of warheads could be eliminated by just a few missiles in a first strike.

No survivable, politically acceptable basing mode was ever found for the MX, despite a study of 35 different schemes — including mobile road- and rail-based carriers. The Reagan-proposed "dense pack," with 100 of the MX missiles clustered near Warren Air Force Base, Wyoming, was not a survivable configuration, being vulnerable to pre-emptive attack.

Although Congress finally did cap the MX program at 50 missiles (deployed at Warren), every other strategic program proceeded largely as proposed by the Defense Department.

Survivability. Strategic stability depends in part on having a retaliatory force that can outlast a surprise attack. This calls for hardened, dispersed, and diversified methods of retaliation. A strategic triad of ICBMs, long-range aircraft, and submarines can be designed to ride out a pre-emptive attack, assuring survivability by dispersing forces in such a way that they cannot all be eliminated or disabled by a single, coordinated counterforce strike.

Hardening is a way to shield against damage. Land-based ICBMs are placed in underground silos that have blast-resistant, removable covers. A near-direct hit is needed to destroy or disable the protected missile. For effective resistance to attack, all of the command, control, and communications for the missile systems also are hardened.

Bombers also are elements of a stable force: They fly much slower than missiles, they can be kept airborne for long periods during times of crisis, and there is enough time and communication capability to recall them if an attack were ordered on misconstrued evidence.

Wide-ranging ballistic-missile submarines (called *boomers*) are more stabilizing than land-based missiles because they can ride out an attack, allowing time for the situation to be assessed before retaliation is ordered. They are virtually undetectable, and are designed for patrolling quietly in the oceans for long intervals without surfacing. Their diminished underwater communication would make it difficult to undertake a coordinated preemptive, spur-of-the-moment strike. Nuclear-armed and -powered submarines do not have (or need to be placed on) a hair-trigger response, because their comparative invulnerability makes quick response unnecessary. For many years their rockets were not as accurate as the land-based missiles; they were inefficient for a counterforce attack because several warheads would have had to be assigned to each enemy missile silo. But that changed with the deployment, starting in 1990, of the very accurate Trident-II (D5) missile, reputed to have a CEP (accuracy) of less than 125 yards — the size of a football field.

Mobile land-basing modes, such as those used by the Soviets (rail and road) and originally envisioned by the United States for the MX system, can also give missiles a degree of immunity from total destruction.

Nuclear War Strategies

While deterrence may sound like a purely passive posture (what is overtly threatening about an assured capability to retaliate?), in reality the point where a deterrent force takes on offensive characteristics is not clear-cut, and there was an ongoing dispute between those who favored finite (restrained) deterrence and those who pushed to maintain a policy of being prepared to use unlimited nuclear weapons. Advocates of the latter were essentially endorsing *war-fighting* strategies (sometimes called *damage limitation*, although, as was clear to many, damage would not have been significantly limited).

Although strategic doctrine focused on the broader features of national and international security, the battlefield (sub-strategic) use of atomic weapons was both a bounded domain of its own (limited nuclear war) and a potential instigator of strategic war.

Fortunately, the Cold War never erupted into an actual nuclear exchange, and some argue that the standoff in atomic weaponry helped maintain the peace. As one commentator expressed it, "the Cold War became a form of global insurance against catastrophe, a system of political control [that] prevented local wars from getting out of hand."

U.S. Strategies. Regarding U.S. strategic policy, a pair of analysts concluded,

> There never was a consensus in Washington about what was necessary to deter the Soviet Union.... Military superiority offered little comfort to the administration [in the

Cuban missile crisis] ... because [quoting McGeorge Bundy] "if one Soviet weapon landed on an American target, we would all be losers."

The U.S. Defense Department allowed ambiguity to permeate its strategic planning. Nuclear targeting and war planning became the basis of both warfighting and deterrence strategies.

It didn't start out that way. President Truman, although he chose to use atomic bombs to expedite unconditional surrender by Japan, recognized that they were not "military" weapons: Women, children, and unarmed people were most of the victims.

The Atomic Energy Act of 1946, which Truman supported, gave responsibility — over strenuous objections of the military — to the civilian-run Atomic Energy Commission for manufacturing and maintaining a nuclear arsenal. The Act also made it clear that atomic weapons could only be used with the explicit authority of the President.

A subsequent 1948 atomic-policy directive, NSC-30, instructed the military services to formulate options for the first time, including massive nuclear air strikes against the Soviet Union; the directive reserved for the president the decision of when and how to use the bombs. Under NSC-30, deterrence, based on military readiness, was adopted as U.S. policy.

NSC-20 (issued in 1948 also) had postulated an immediate and dangerous threat from the Soviet Union. This was taken seriously enough to justify planning for all-out atomic war scenarios. To deal with the strategic issues of Soviet aggrandizement and the political issues of communism, SAC was prepared to obliterate Russia's cities, industrial centers, and command sites. NSC-20 gave rise to the war plan *Dropshot*, which envisioned premeditated attack on the Soviet Union and was forthwith implemented by substantially scaling up production of nuclear warheads.

In 1953, during the Eisenhower administration, national-security council strategy directive NSC-162/1 was drawn up, and it dominated early U.S. thinking. It established the policy that nuclear weapons could be used to actively promote the interests of the United States, rather than simply discourage their use by others. The directive, approved by Eisenhower, reads: "In the event of hostilities, the United States will consider nuclear weapons to be as available for use as other munitions." During the same period, the Atomic Energy Commission was ordered to gradually transfer custody of certified nuclear munitions to the military. These policy initiatives had the effect of increasing the danger of nuclear war because they changed the American military establishment into an organization that equipped and trained itself on the assumption that such devices would be used in combat.

When the Carter administration came into office, another ominous change took place: A controversial war-fighting policy directive, PD-59, decreed that U.S. forces must be able to respond in a flexible manner to Soviet strikes and to fight, if necessary, a protracted nuclear war. It was signed by President Carter just before he left office, over the objections of his Secretary of State, Cyrus Vance.

Later, according to a Brookings Institute *Atomic Audit*,

> The election of Ronald Reagan in 1980 brought into the White House and Capitol Hill a large number of conservative policymakers with a deep dislike of arms control, a strong distrust of the Soviet Union, and a desire to rebuild American military strength following what they considered a period of laxity and weakness in the 1970s. To that end, the administration proposed the largest peacetime military program in U.S. history.... "Peace through strength" became a shorthand rationale for the program.... The administration leaked word of a new strategy to fight and prevail in a protracted nuclear war with the Soviet Union.... [This] precipitated widespread antinuclear demonstrations and the founding of the nuclear freeze movement.

All of these epochal decisions were made in strict secrecy, without public debate or public knowledge. The U.S. government had, as early as the Eisenhower administration, secretly decided to use nuclear ordnance in any future conflict, to disperse it throughout the world, and to choose enough targets to obliterate the Soviet Union. Similar secret and momentous decisions were made by the Soviet Politburo.

The *Atomic Audit* ascertained that, to formulate an expanding catalog of targets to match the Reagan emphasis,

> Strategic war planning ... was carried out by the Joint Strategic Target Planning Staff, [which] drew up a long list of potential Soviet and Chinese targets, along with plans to strike them, known as the Single Integrated Operational Plan (SIOP). Tens of thousands of targets made the list during the height of the Cold War. SAC staff assigned the most important targets to the U.S. strategic bomber, submarine, and land-based missile forces and scheduled the launch and delivery of U.S. weapons to ensure maximum target destruction in wartime.
>
> Because the SIOP was shrouded in secrecy (all information pertaining to it has its own classification code, SIOP-ESI, "extremely sensitive information"), oversight of it was poor to nonexistent.... Even today, the overriding goal of the SIOP — ensuring the coverage and hence destruction of virtually all possible targets — drives the entire U.S. nuclear infrastructure, including targeting assignments, force deployments, warhead production and retirement, alert postures, and launch response times.... In 1966 the stockpile peaked at an estimated 32,200 warheads and bombs — the idea was to find targets for excess weapons.

By the late 1960s, Soviet strategic forces had reached a point where they required little or no advance preparation to mount a nuclear strike on a massive scale; this forced U.S. strategic planners to assume that a Soviet attack could come as a complete surprise. The standard worst-case scenario thus drove the United States to deploy missiles in hardened underground silos, to disperse stealthy submarines, to put bombers on airborne alert (until 1968, then on runway alert until 1991), and to enable a good portion of these forces to survive a massive attack launched without warning.

By mid-1980, the National Strategic Target Database had grown to 50,000 targets. The number of SIOP targets evolved along with the number of Soviet strategic warheads, many of which were stored at the same target site. The United States usually had more than enough of its own strategic warheads to wipe out the designated targets; only well after the Cold War did the number of targeted Russian sites decrease significantly.

Soviet Strategies. While Soviet military strategies were rarely made public, some indications reached the West suggesting what was *not* part of their strategy. For example, M.A. Milstein, formerly a General of the Soviet Army, who traveled extensively out of the Soviet Union, belittled the logic of limited nuclear war. Referring to an analysis by Henry Kissinger, General Milstein agreed that "it would be nearly impossible to get any reasonable definition of what should be understood as [limited nuclear war]." If each branch of the armed forces had permission to destroy whatever they deemed necessary for performing their tasks, "limited nuclear war waged in such a manner could quite well become indistinguishable from a total war."

General Milstein publicly asserted that "the possibility of using nuclear weapons as a 'flexible' and 'acceptable' means of foreign policy should be rejected once and for all." As a case in point, he cited a NATO war game, code named *Carte Blanche*, held in Western Europe in 1955. Theoretically, 335 sub-strategic weapons were exploded by NATO forces in East and West Germany. Assuming WTO retaliation, 1.7 million civilians would have been killed and 3.5 million wounded. Thus, the number of people hypothetically killed was about five times the number that died in air raids over Germany in World War II.

Using similar examples, a respected British strategy analyst emphasized the implausibility of the majority of civilians surviving a limited atomic war. Two notorious mock operations that he cited were (1) *Operation Sage Brush*, a war game in Louisiana, after which the umpires ruled that all life in the state had ceased to exist after some 70 tactical nuclear bombs were dropped on military targets, and (2) *Carte Blanche*, the exercise in West Germany mentioned above. Clearly the use of nuclear weapons could never be left to a tactical decision by a local commander during battle; the consequences dictate that it requires a strategic decision at the highest levels.

In the Soviet Union, the primacy of political leadership in determining military goals was well established. Military doctrine was not originated by the military but by the Communist Party. The doctrine had socio-political and military-technical aspects. Subordinate to, and consistent with, military doctrine was military science (the science of war), for which professional military expertise played an important role in formulating. Military strategy, which implemented military doctrine, was the basis for war planning, force structuring, weapons development, and deployment.

Chairman L.I. Brezhnev contradicted bellicose assumptions of Western warhawks, saying (as quoted by General Milstein): "... there is no Soviet threat either to the West or the East. It is all a monstrous lie from beginning to end. The Soviet Union has not the slightest

[Hawks and Doves]. Although labeling people and their attitudes is simplistic, individual viewpoints were often contrasted with terms like *hawks* or *doves* with regard to militant or peaceable viewpoints about the state of hostilities, about the superpower adversaries, and about arms control negotiations. Sometimes *hardliners* (warhawks — strong on defense) linguistically confronted *peace-niks* (soft on defense). *Cold-warriors* were the stereotyped opposites of *arms controllers*.

Conservative and liberal were, and are, terms applied to political positions on opposite sides of the American center.

intention of attacking anyone." These assertions were reflexively discounted by hardliners in the West.

In 1960, First Secretary Khrushchev outlined a new military doctrine whose essence was that the nuclear rocket would be the centerpiece of Soviet military strategy. The newly formed Strategic Rocket Forces became the premier branch of the armed services. Civil defense was underscored, and the Moscow ABM system was begun.

However, by 1963 (after the Cuban missile crisis) Soviet overemphasis on nuclear weapons was criticized and attitudes shifted: Conventional response would be used against conventional attacks. The corresponding change in NATO to "flexible response" (no longer insisting on nuclear reaction to a conventional-weapons attack) led in 1967 to a Soviet strategy that allowed for initial conventional neutralization of a Western assault against the WTO.

The hawks in the United States did not trust the ability of Americans to see through Communist propaganda; hard-liners demonized Soviet intentions, exaggerated American vulnerabilities, and articulated a generally paranoiac mind-set about Soviet strategies. That rhetoric reinforced prevailing Soviet political attitudes, which precluded any dependence on an adversary for even a hint of Soviet security. Of course, there was a continuum of opinion between the extremes, and there were many agnostics — undecided, indifferent, or ambivalent.

Although Soviet literature spoke of "contemporary wars that may be unleashed by the imperialists," their strategic thinking about war was more realistic and defensive than in the West. The Soviet strategic outlook also recognized that nuclear weapons did not make war completely unthinkable — that it was necessary to consider the possibility that deterrence might fail. Thus, a flexible concept of *combined arms* was indispensable, and Soviet forces would not be expected to absorb an enemy strike without lashing back. In Soviet thinking, defense and deterrence went hand-in-hand.

Soviet military writing also assumed that rapid and total escalation was the most likely consequence of any initially limited war between the major powers. While some Westerners used the Soviet losses in World War II as evidence that they were somehow inured to a high level of damage, informed Soviets realized that the scale of damage and loss in an atomic war would be unprecedented and intolerable. The role of Soviet military power was seen as the major instrument for impressing on the "imperialist camp that military means cannot solve the historical struggle between the two opposing social systems...."

While the United States had embraced a concept of *deterrence by punishment*, the Soviet position was one of *deterrence by denial*. The Soviets were not inclined to accept the Western metaphysics of deterrence, which appeared then to be part of an imperialist policy of containment. Soviet deterrence policies were designed to prevent the outbreak of hostilities by minimizing incentives for being attacked — in essence, *denial* to the opponent of the opportunity to mount a successful attack, rather than avoidance of war by threatening *punishment* if attacked. In the Soviet Union, war-denial had a stronger footing than war-avoidance.

To the Soviets, American policy had the appearance of "blackmail." In the 1980s, Western strategic developments appeared unequivocally headed toward counterforce capability: the MX missile, SLBMs, forward-based systems, and PD-59. Modernization of NATO's sub-strategic forces seemed to be an effort to regain military superiority. Pershing-II missiles, first stationed in Western Europe in 1984, had a range of about 1100 miles; they threatened ICBM silos and control centers in the western part of the Soviet Union.* With only a four-minute flight time, the Pershing-II nullified the Soviet option to withhold launch until warning of an attack. Nuclear-armed cruise (air-breathing) missiles based in Europe deepened the Soviet sense of vulnerability. The American buildup belied the notion of "sufficiency," looking instead more like counterforce for the purpose of coercion.

The idea of a limited nuclear war leading to military superiority over the Soviet Union was out of tune with the Soviet view that *political* objectives — rather than the West's emphasis on weapons performance — were the essence of war. Consequently, any war operation — such as the tactical (battlefield) use of small atomic weapons — could not conceivably be limited. The Soviet military used the term "correlation of forces" when analyzing the outcome of a postulated struggle, recognizing an unfavorable outcome for the Soviet Union whether or not NATO had nuclear forces.

While the Soviets proclaimed the "unthinkability" of war, in fact they understood the possibility of atomic war and chose a strategy of war-prevention through strength. In the West this was considered an implacable attitude and a quest for military superiority. The dispute over their intentions notwithstanding, it is clear that the Soviets did not plan to initiate an armed conflict — a state of hostilities that, in their thinking, would almost certainly become nuclear.

British Global Strategy. The basis for British strategy seems to have been prepared in a 1952 paper by the Chiefs of Staff. Because of changing economic and strategic circumstances, Winston Churchill had determined that

> deterioration in Britain's economic position over the past year required savings to be made in the defence budget.... [He] believed that insufficient account was being taken in current British defence planning of the value of nuclear weapons.... [There was fear that if] Western European economies were to collapse, then the Soviet Union could achieve a "bloodless victory."

Other analysis in the 1952 paper indicated that because "there was no effective defence against atomic attack in prospect ... British and NATO strategy had to be modified to place greater emphasis on deterrence." In particular,

---

*U.S. nuclear missiles went to western Europe to satisfy Western Germany's desire to be under a nuclear "umbrella." The idea was that the missiles would constitute a "doomsday machine" — an incursion from the East would immediately go nuclear. That sequence would then escalate to a world-wide nuclear conflagration, the prospect of which would deter the Soviets from invading. In other words, the United States was risking national suicide to protect West Germany.

the intention of the allies to use atomic weapons immediately in the event of aggression had to be made "unmistakenly clear to the Russians." [And] the allies had to have the capability to carry out their own atomic air strike.

The British strategy paper thus concluded that "it would be necessary for the West to keep one step ahead of the Russians in a wide variety of fields of scientific [military] development."

Moreover, the Chiefs of Staff

> stressed that it would be "most unwise" for Britain to become "completely dependent" on the United States for the provision of atomic weapons....

The significance of the paper lies in its impact on the long-term strategy of Britain and NATO. The recommendations of the Chiefs were expected to produce "substantial savings in the United Kingdom rearmament program...." Meanwhile, "[the American Chiefs of Staff] had a clear understanding of the British desire to change Alliance strategy in a more nuclear direction from the summer of 1952 onwards."

The views of the British staff, in the context of the beginnings of the Cold War, are typical of the time:

> We consider it likely that Russia, appreciating that she could not defeat the United States, would initially concentrate on trying to overrun Western Europe.... [Allied military aims are to] establish firmly the anti-Communist front within countries outside the Iron Curtain [and] intensify measures designed to weaken the Russian grip on the satellite states.

Germany's Nuclear-Capable Delivery Systems. To keep West Germany from acquiring atomic weapons, that nation was the object of *extended deterrence* — under the *nuclear umbrella* to discourage either a conventional or unconventional attack on allies of the United States. Yet, often not realized and deliberately unadvertised, Germany was requested by NATO after World War II to rearm with nuclear-capable delivery systems. While not ending up with sole-use authority, Germany had access to weapons that were nominally in the control of U.S. custodial units. The stated rationale was that "[German] forces had to be equipped with nuclear weapons in order to compensate for [NATO's] reduced conventional capability."

Six types of sub-strategic short- and intermediate-range ballistic missiles (Little John, Lacrosse, Honest John, Corporal, Sergeant, and Redstone) and many nuclear-capable aircraft were deployed. To expedite availability in time of crisis, Germany "asked for the pre-delegation of nuclear authority to military commanders." German NATO officers in the field then technically would have been able to activate them in responding to a conventional attack. Nevertheless, their use (nuclear release) required "a positive decision by the U.S. President."

In addition, West Germany gained "access to information on nuclear capable delivery means, nuclear warheads, and nuclear targeting procedures." Also, the Federal Republic "successfully resisted the perceived attempt by the United States to withdraw from (or at least reduce) its commitment to extended deterrence...."

France and China. Both France and China developed independent retaliatory forces based on their own regional concerns. Each of these nations accepted some assistance from a nuclear weapons state, unofficial and deniable enough to avoid blatant violation of the NPT.

France, which was fully capable of developing its own nuclear weapons — having the scientific and engineering talent, as well as resources — is widely believed to have received from the United States some type of *sub rosa* validation of its thermonuclear designs, in order to expedite its program. Enduring a long history of invasions, France chose at an early stage to develop its *force de frappe*, a deterrent capability based on aircraft, missiles, and submarine platforms.

China also had been subjected to many foreign incursions. Their military scientists acknowledge that "In the 1950s, the Soviet Government provided important assistance in the development of defence science and technology," but deny that there was "substantial" help in the development of atomic weapons. The Soviet assistance that had been received, namely, "teaching model and drawings and information," was interrupted in the early 1960s by "the sharpening of differences between the two countries and the deterioration of their relationship."

China evidently decided to develop its own nuclear weapons and missiles in 1955. Their first atomic-bomb test occurred in October 1964, five years after the interruption of Soviet assistance. Altogether, they have conducted 45 fission and fusion tests, about half of them in the atmosphere.

China created an independent retaliatory force, small in the context of the nation's size and population, to deter an attack; its weapons are deployed so as to "have a good chance of survival from first strike and be capable of penetrating missile defences, as well as adequate safety and reliability."* China has adopted and advocates a policy of "unconditional no-first-use of nuclear weapons ... not to be used as conventional war-fighting weapons nor as offensive weapons."

Nuclear -Explosion Phenomena

Deterrence in the Cold War largely depended on threats to inflict destruction by nuclear detonations. To help readers understand the policy and strategy options, some effects of nuclear explosions are described.

The devastation that nuclear weapons can cause has radically changed the nature of contemplated or inadvertent war. The (17-kiloton) Hiroshima bomb was about 10,000 times more powerful than a conventional high-explosive of the same weight. Large thermonuclear weapons magnify explosive yields by another factor of 1000.

A rapidly expanding fireball created by an atomic explosion spreads destruction via heat, blast, and radiation. Heat sets flammable objects on fire; blast consists of destructive shock waves that damage structures, equipment, and people; and radiation duces ionization in animal and human cells. The effects combined are devastating.

---

* The survivability claim is highly questionable, but there is no reason to doubt their ability to penetrate missile defenses.

The dimensional profile of the explosive range — within in the atmosphere, above ground, on the surface, under the surface, under water, or in space — significantly influences the human, military, structural, and environmental impact of the fireball.

A nuclear detonation lasts only a small fraction of a second (a microsecond or less), although a tiny fraction of the energy release is delayed (up to hours or more). Initially, the energy is transported by short-wavelength radiation (gamma rays and x-rays) and atomic particles (mainly fission fragments and neutrons). Much of it is rapidly converted to heat as it is absorbed in surrounding air and materials.

Fireball. A fireball of superheated air and bomb materials is the first visible evidence of the explosion; it expands faster than the speed of sound, with waves of heat and blast that crush, destroy, or burn practically everything nearby.

The high temperature (tens of millions of degrees) immediately produces soft x-rays, which are largely absorbed within a short distance, heating the surrounding bomb materials and air to form the fireball: Its expansion causes a shock wave, and some of the initial high-temperature radiation is absorbed and re-radiated at lower temperatures. The energy released ends up about evenly divided between blast and heat.

Heat causes two kinds of harmful effects on people: "flash" burns from direct exposure of skin, and "contact" burns from flammable clothing or other fires.

Ignition of various common substances occurs at radiant-exposure levels that depend on the material. About ten calories per square centimeter will ignite newspaper or wood, while clothing might require five times as much. For perspective, a hand-held lens or reflector can concentrate ordinary sunlight to one or two calories per square centimeter per second, igniting newspaper or tinder in a few seconds.

Because typical atmospheric heat-transmission factors for low-altitude bursts are as high as 60 to 70 percent at distances of 5 miles, a 2-megaton thermonuclear detonation could produce a radiant thermal exposure of 50 calories per square centimeter on a clear day at that distance — more than enough to ignite clothing. Permanent blindness is likely in eyes that happen to be looking toward the fireball, even at large distances. A firestorm, creating rapid and widespread incendiary damage, occurred in Hiroshima (but not Nagasaki, where the terrain was hilly).

Clouds, smog, and dampness will impede the transmission of radiant heat. Prior preparations to eliminate exposed flammables and to construct shelters can mitigate somewhat the incendiary effects of an nuclear burst.

Blast. Most of the militarily significant effects of nuclear weapons are caused by blast — the primary phenomenon of significance in targeting hardened installations. Sudden overpressure from the shock wave crushes objects and collapses structures. If exploded at an altitude of about a mile, a 1-megaton burst would demolish a reinforced concrete multistory building almost 2 miles from ground zero, and destroy brick and wooden houses at 4 miles. Within a 3-mile radius, almost every structure would be destroyed and no humans would survive.

A 1-megaton bomb will produce, at a distance ten times farther, the same overpressure as a 1-kiloton bomb. The area of damage would be about 100 times larger.

If a missile silo were hardened to withstand a blast overpressure of 1000 psi (pounds/square-inch), a 1-megaton weapon would probably have to detonate within a few hundred feet to impair a retaliatory missile launch. An overpressure of just 5 psi is sufficient to destroy normal residential structures and kill about half of the people within.

A wave of wind (up to hundreds or thousands of miles per hour) and high pressure (hundreds or thousands of psi) delivers the blast to distant structures. There will be flying objects and falling buildings.

Possible defensive measures include rapid evacuation on warning of an impending attack (or early evacuation in time of crisis), construction of blast shelters, and reinforcement of buildings.

Radiation. Ionization within living cells is the principal mechanism of biological damage from radiation. The ionization process removes electrons from atoms or molecules, a process that can harm cell constituents. This can lead to impairment of reproduction of a cell or even destruction of its nucleus.

Near-term human physiological effects from acute radiation exposure include sickness, sterility, hemorrhage, and death. For survivors of heavy doses, long-term aftermaths include blood disorders, cataracts, malignant tumors, hypothyroidism, fetal malformation, and genetic aberrations. To produce such effects in detectable numbers, the doses must be relatively large and sudden: After all, we are continually and naturally exposed to a low level of background radiation — each of our cells being struck many times a day — and we survive because cellular repair occurs rapidly and continually.

Neutrons and gamma rays are the most important ionizing radiations created in nuclear explosions. Most of their harmful effects are from ionization in human and animal cells. Neutrons and gamma rays will penetrate about 8 in. (20 cm) through living tissue, just enough to cause cellular damage in human-size organs.

There is also a long-term risk of cancer and of birth defects in children exposed *in-utero* at the time of the explosion. While severe doses of radiation cause death by massive irreversible cell damage, sublethal doses can impair functioning of the immune system.

For people not very far from ground zero of a nuclear explosion, the initial (*prompt*) ionizing radiation that occurs at the time of the detonation can be fatal. For large, thermonuclear weapons, the prompt-radiation effects are outstripped by lethal blast and fire damage that extends to greater distances.

Evasive actions, such as taking shelter — within a second or two — might, in theory, help against the slightly delayed portions of the gamma and thermal radiations. However, in practice, you would have to take such action instinctively in response to a bright flash of unknown cause, without looking in the direction of the flash. (Typically there are two flashes close together, the first not as bright as the second: The initial one gets your attention, so you are likely to look in that direction, just in time to be blinded by the next one.)

Fallout. After an atomic detonation, residual radioactivity — emitted primarily from the fission fragments created in the explosion — lingers, with decreasing intensity, for many months. If it is a *surface burst* (one in which the fireball touches the ground), a lot of earth (and whatever else is present) is vaporized and rises in the mushroom cloud, where it soon condenses. The resulting dust — radioactive because of the fission products — gradually descends to the ground, the larger particles first, as the cloud drifts downwind, creating *local fallout*.

A 1-megaton weapon detonated at the surface could produce a time-integrated fallout dose of more than 9 Sv (900 rem) in a region of 1000 square miles around the target. The *footprint* of the fallout (the area affected) is typically elongated, extending perhaps 100 miles downwind from ground zero, depending on the size of the explosion.

Local fallout will occur within a day of the explosion, but *delayed fallout* continues for days or years, especially if the mushroom cloud reaches very high altitudes (which will happen for all but the smallest explosions). Much radioactivity (all of it, if from an air burst) comes from extremely fine particles that rise with the mushroom cloud to very high altitudes and drift with air currents. Most fission products that rise to the stratosphere decay before coming down; nevertheless, some radioactivity will persist over the ensuing months or years, descending worldwide as *global fallout* in slightly contaminated rain, snow, or dust. There is no clinical evidence that the global fallout generated during the era of atmospheric testing ever caused any physical health effects. However, global fallout from a massive nuclear war would not be so harmless.

[Radiation Units]. One *gray* (Gy) of radiation dose deposits an energy of 1 joule/kilogram (1 rad = 100 ergs/gram = 0.01 Gy). Each type of radiation has a different "relative biological effectiveness" (RBE). A *sievert* (Sv), which is a gray multiplied by the RBE, gives the effective biological dose (1 rem = 1 rad multiplied by the RBE).

Humans are, on the average, exposed to about 0.003–0.004 Sv (0.3–0.4 rem) of radiation annually from natural sources. However, the rate of absorption, whether acute (rapid) or slow (chronic), makes a big difference in the effects of radiation.

Serious radiation-induced illness in humans begins at an acute exposure of about 2 Sv (200 rem). Death occurs (despite prompt medical care) with about half the people suddenly exposed to 6 Sv, and in most people above 10 Sv. A 1-megaton explosion would give a prompt radiation dose of 6 Sv at a range of about 1.5 mi. from the epicenter, but someone that close would probably be killed first by blast or fire.

If the explosion were an *air burst* (one whose fireball does not touch the ground), there is virtually no local fallout (but still plenty of direct radiation, blast, and heat); the fission products remain airborne, becoming global fallout. The atomic bombs dropped by parachute at Hiroshima and Nagasaki were detonated well above ground; so their destructive effects and radiation doses were direct, not from fallout. For attacking soft targets like industrial areas or airfields, air bursts do far-reaching damage.

Modern strategic weapons are designed with hard targets in mind, which means that the fireball is intended to touch the ground if the target — perhaps a missile silo or command bunker — is hardened. This type of attack would produce considerable airborne plumes of aerosols and radioactive debris, and eventually fallout.

Fallout particles range in size from microns to millimeters (the larger ones coming down sooner as local fallout). The radioactivity in fallout is due to fission fragments, with hundreds of radioactive products in the mix. Overall, these products decay at a rate roughly proportional to time. One hour after an explosion the dose rate is less by an order of magnitude (a factor of ten) than it was at one minute, and in two weeks it's down by another factor of 1,000.

For unsheltered, unprotected organisms, the local fallout lying around can be hazardous even weeks after the explosion. Eventually the radiation decays, but some significant activity could remain for a long time. Because plants can become contaminated, the food chain will be somewhat more radioactive than usual for years.

Concentrations of radiation near ground zero and downwind along the "footprint" of the local fallout are likely to cause serious inhalation or ingestion consequences for those who are unable to evacuate or find safe shelter. For people outside the concentrated downwind footprint, the radiation is much less intense and, thus, less likely to induce physical illness; however, panic and other forms of societal disruption are probable, aggravated by radiation phobia, which increases severe anxiety and other emotional impact.

Well-provisioned shelters that sustain life for days or weeks could diminish the physical and emotional harm done by wind-borne fallout: Filtered ventilation, food, water, radiation monitors, and other emergency systems and supplies would be needed. Radiation-qualified crews (taking advantage of natural radiation decay over time) could significantly reduce contamination of individuals and structures.

EMP. Impelled by gamma rays from a nuclear explosion, charged particles (electrons knocked from air molecules) moving rapidly in the Earth's magnetic field can create a significant electromagnetic pulse (EMP) at high altitudes. Such a pulse can induce excessive voltages and currents in electrical circuitry, causing severe damage to electronic equipment.

An EMP-producing explosion should take place from tens to hundreds of miles above ground level.

For explosions near the earth's surface, the EMP is localized, but the range increases with altitude. If the explosion is above the atmosphere, the damage to electronics can extend for great distances (hundreds of miles), and to power stations for shorter distances. Generally there is no effect on living things. Modern solid-state circuits are especially sensitive to EMP, but special shielding can provide very good protection, and all U.S. military electronic equipment is said to be hardened against EMP.

The charged particles — electrons — are from ionization of air molecules by gamma rays. A high-altitude nuclear explosion over Johnson Island by the United States in the early 1960s produced significant electronic effects 800 miles away.

The primary impact of a burst at higher altitudes would be from EMP generated during the explosion. Unshielded Earth-orbiting satellites are especially vulnerable; the collateral effects might be substantial — enough to interrupt communications, impair reconnaissance, cause aircraft to crash, cars to stall, and electric power to shut down.

Loss of military surveillance satellites would be highly destabilizing, seriously jeopardizing any strategy of riding out the first wave of an attack. Fearing that a communications blackout to would be a forerunner of a first strike, an EMP blast could very well trip an immediate, all-out, retaliatory response.

Examples of Nuclear Devastation. To impress upon a less-familiar generation some consequences of nuclear retaliatory exchanges, here are some damage reckonings.

A single nighttime 1-megaton explosion over downtown Detroit was estimated by a U.S. government agency to cause 470,000 immediate deaths and 630,000 casualties. Many fatalities (up to 190,000) could also occur in the metropolitan area due to flash burns and fires. There was no reason to expect the effects to be any less on a comparable city in the Soviet Union.

The indirect effects on civilians of a counterforce exchange (limited by each side targeting only ballistic-missile launchers and strategic bomber bases) would still be devastating: Fallout from 2000 Soviet warheads directed at ICBM silos (necessarily ground bursts) would expose a large fraction of the United States, particularly in the Midwest and East, to high levels of radiation. A U.S. counterforce attack on the Soviet Union would be expected to cause comparable casualties.

An all-out exchange, in which urban-industrial targets are hit as well as military facilities, would involve thousands of high-yield warheads. In this scenario, most large cities would be destroyed and the inhabitants killed by blast and thermal effects. Fifty to one-hundred million deaths in both the United States and Soviet Union would be expected from direct effects of the explosions.

Making ghastly reading are the effects of a nuclear explosion. Many injuries and deaths are the result of synergisms in the multiplicity of direct and indirect effects from nuclear detonation. Deaths occurring some time after the explosion could well be due to multiple causes, such as nuclear radiation combined with burns and mechanical injuries (flying glass and other objects, collapse of buildings, etc.). Moreover, the widespread destruction or impairment of emergency services and the radioactive contamination of food and water would aggravate the consequences of a large-scale nuclear attack. Pandemonium is likely to prevail.

> **[Nuclear-Explosion Effects].** Out of a combined population of about 450,000 in Hiroshima and Nagasaki in August 1945, approximately one-fourth were killed or missing and another one-fourth were injured by the atomic bombings.
>
> Injuries resulted from three main sources of trauma: blast, burns, and nuclear radiation.
>
> Direct blast injuries resulted from the positive phase of the airborne shock wave acting on the body so as to cause damage to the lungs, stomach, intestines, and eardrums, and also internal hemorrhage. Compression and subsequent decompression of blast waves caused damage to the body mainly at junctions between tissue and air-containing organs, and at areas where bone and cartilaginous tissue join soft tissue. The chief consequences were: damage to the central nervous system, heart failure due to direct disturbance of the heart, and suffocation caused by lung hemorrhage or liquid extrusion into the lung tissue. The indirect or secondary effects due to collapsing buildings and the great quantity and variety of the debris flung about by air blast contributed strongly to injury and death.
>
> Most of the burns were flash burns, although individuals trapped by spreading fires were subjected to flame burns. The effects of thermal radiation on the eyes fell into two categories: retinal burns and flash blindness.
>
> About 30 percent of those who died at Hiroshima had received lethal doses of nuclear radiation, although this was not always the immediate cause of death. Many individuals suffered multiple injuries. Acute whole-body radiation effects depended on the dose: 5000 roentgens led to almost immediate incapacitation and fatality within a week; 500 roentgens caused vomiting and nausea on the first day, followed by radiation sickness with about 50 percent fatalities in 1 month; 50 roentgens had no obvious short-term effect.

****

With so much inherent destructive potential, nuclear weapons have changed relationships among the nations of the world forever. Atomic weapons brought an abrupt end to the war with Japan — a militarily and politically desirable effect — but their subsequent presence in Europe was a mixed blessing: While deterring Soviet military aggression, nuclear weapons failed to stop or reverse political subjugation in East Europe.

The United States lost its atomic monopoly early in the Cold War. Nevertheless, it felt compelled to brandish atomic weapons ostensibly to deter the Soviets and their allies from using their conventional forces to attack American and NATO interests. As such, nuclear-explosive devices were really a demonstration of military will and determination. They were not weapons that could be used with real advantage in a battlefield without devastating the region or nation they were intended to defend. Nevertheless, strategic policies of NATO and the United States increased the chances of nuclear war, especially as the incentive to use-it-or-lose-it grew. The need to make quick decisions — with serious, long-term importance — magnified the pressure on leaders to plan the use of last-resort weapons in response to any attack — conventional or nuclear. The danger of accidental or unauthorized use grew as more and more weapons were produced and deployed on both sides.

The MX missile concept was part of a fruitless quest for the holy grail of security through strength. Indeed, MX (and MIRV before it) gave a transient superiority to the United States. However, as had other new U.S. weapon systems, MX induced a reflexive response on the part of the Soviets. With their retaliatory capability undermined, they simply placed missiles on a launch-on-warning status, thus adding to the risk of accidental, unauthorized, or impulsive launch. The next step in the Soviet response was to develop their own MIRVs. The nearly inevitable reaction made the United States vulnerable to a first strike against its land-based multiple-warhead missiles. As a result, the MX was particularly inviting as a target for pre-emption, a risk that alarmed residents in the western and midwestern United States.

Assured deterrence and stability suffered as the two sides ratcheted up the stakes.

B. THE WEAPONS

The possibility was quite real that nuclear weapons would be used in combat if deterrence failed. If that's the case, what were the specific characteristics of these weapons, and what was required to dispatch them to targets on the field of battle? One rudimentary impression about nuclear warheads, delivery systems, and control systems was that of a bewildering array of devices, with many bells and whistles. In this Section the reader is introduced to nuclear weapons-delivery systems, their deployment, and primary military function. Also, the Soviet and U.S. arsenals are compared.

Nuclear munitions must be exploded at a distance from the launch point because of the range of lethality from blast, heat, and radiation — including wind-carried fallout. The original A-bombs were very large and heavy, requiring long-range strategic aircraft for delivery. Eventually nuclear ordnance was reduced in size so that even a recoilless rifle could be used to propel a low-yield nuclear explosive far enough away for battlefield use.

Categories and Types of Weapon Systems

Weapon systems can be of three types — strategic, tactical, or surveillance — with applications in several military categories. The nuclear-capable weapon-delivery systems have applications in one or more of the four battle environments: space, air, land, and sea. In selecting nomenclature for U.S. nuclear weapons, the AEC used the prefixes *B* for bombs and *W* for warheads.

Overhead surveillance systems are not used for delivering weapons although they play an essential role in planning and evaluating military operations. Many types of carriers could deliver other "weapons of mass destruction," a loosely defined category often considered to include chemical and biological agents.

PALs. Aside from custodial controls, the first line of defense against unauthorized activation of a nuclear weapon was a secure locking device. The earliest were simple mechanical combination locks. Since the early 1960s a more sophisticated U.S. system called a "permissive action link" (PAL) has been increasingly employed. It is an electronic (originally electro-mechanical) device that prevents arming of the weapon unless the correct codes, consisting of quite a few digits, are inserted into it. Two different codes must be inserted, simultaneously or close together. This follows the "two-person rule," which dictates that no single individual should be able to arm any nuclear weapon. The codes are usually changed on a regular schedule.

PALs were installed from the late 1960s to the mid-1970s on U.S. bombers and ICBMs, requiring the individual crews to receive the unlocking codes from a higher authority in order to launch. The first crucial step was to ensure that the code *authorizing* launch could be transmitted to ICBM, bomber, and submarine crews. Eventually, however, concern developed over whether the codes *enabling* launch could become an Achilles heel, making it possible for an adversary who intercepted the code to prevent the launch.

Strategic Systems

The means of delivering strategic ordnance developed in directions that depended on perceived military requirements and improved technology for both the weapons themselves and their application to the growing set of targets. Altogether, the United States produced 116 different nuclear-weapon delivery systems, including 6125 ballistic missiles of 11 types, 59 SLBMs, and many shorter-range missile systems.

The atomic bombs detonated over Japan were very heavy (10,000 pounds), and were carried about 1000 miles by specially fitted B-29 bombers. In the United States, production of practical and economical weapons was initially impeded by technical problems and by a strong bias toward bombers on the part of the Air Force. "By 1953 the situation was beginning to change."

> Breakthroughs in thermonuclear designs made light-weight, high-yield warheads practical. Fission warheads as well were becoming smaller in diameter and significantly lighter than the first-generation atomic bombs. The steady advances in missile technology being pursued by the army, navy, and air force showed promise of increased range and greater accuracy.

Smaller bombs and nuclear warheads permitted more delivery options, such as land, sea, and air platforms for ballistic missiles. Sub-strategic artillery, torpedoes, depth charges, and landmines were deployed by the U.S. Army and Navy. Nuclear weapons could be transported and utilized at intercontinental ranges and in theaters of war throughout the world.

In the late 1970s and early 1980s, new systems for strategic warfare were developed in the United States. These included the B-1 bomber, the MX missile, the Trident submarine, and cruise missiles that could be launched from air, sea, and ground. All of these were a threat to the Soviet mainland. Moreover, there was enough built-in technical sophistication to seriously undermine mutual deterrence: The new systems had the characteristics of pre-emptive first-strike weapons, designed to quickly disable the opponent's retaliatory capability.

Triad. An underlying concept for the secure allocation of strategic forces was the *triad*, consisting of three types of reliable systems ("legs of the triad") for delivering weapons: ICBMs, SLBMs, and long-range bombers. Together, the threesome provided redundancy and survivability, while each leg had its own distinct characteristics and implications for military planning and arms control.

Pre-launch survivability, controllability, and readiness are all important for strategic-weapons. After launch, especially in time-urgent counterforce operations, time-to-target, aim-point accuracy, and its damage-infliction potential are significant attributes. An ICBM can reach a target halfway around the world in about 30 minutes, while strategic bombers require many hours. Thus bombers are inherently stabilizing in their ability to be recalled before reaching a target. On the other hand, to abort a missile attack in flight, the ICBM projectiles would have to be destroyed remotely by flight controllers (a capability unlikely to be installed, because such a option would make the missile vulnerable to electronic countermeasures).

The accuracy of a missile-delivery system is generally represented in terms of the *circular error, probable* (CEP), the radius of the circle centered on the target, within which a warhead has a 50 percent chance of impact. CEPs of U.S. ballistic-missile models were reported to be a hundred meters or less (about the size of a football stadium). That's sufficient to destroy a hardened target such as a missile-launch silo or an underground concrete- and steel-protected command-and-control center.

Because survivability began to be emphasized in the early 1960s as a critical characteristic of a balanced strategic force, the United States undertook a comprehensive missile-replacement program. The more vulnerable Atlas and Titan I batteries were replaced with 1000 Minuteman and 54 Titan II missiles in blast-hardened underground silos. The United States also built 41 nuclear-powered submarines, each armed with sixteen ballistic missiles, and 600 B-52 intercontinental bombers to complete the strategic triad.

Each part of the triad has its own strengths and weaknesses:

> The land-based ICBMs had the best command and control because of the simplicity of communications between the highest authority and the officers with access to the firing circuits.... High accuracy [small CEP] is most easily achieved with fixed, land-based ICBMs. They can reach their targets within thirty minutes of launch, so they provide for rapid response if required. But once they are out of their silos, they cannot be recalled, so a nuclear war could not be avoided [if they were launched in error].

Because of their known, fixed location, [ICBMs are vulnerable in their silos and] can be the target of accurate missile warheads [in a pre-emptive attack].

The bombers have a slower reaction time (six to eight hours to reach their targets), but they do not have to be committed irrevocably to a nuclear attack when they take off from their airfields.

The submarine ballistic missile force is by far the most survivable of all elements of the triad.... There appears no way in which a fleet of about thirty submarines that would be at sea at any one time can be destroyed on short notice. The greatest disadvantage of the submarine force is the problem of maintaining two-way communications without disclosing the location of the ship. This difficulty creates problems for command and control.

By 1966 this well-balanced U.S. strategic force ... was completely operational. The Soviets in the meantime were far behind.

While both superpowers depended on their own triads of nuclear-delivery systems, the mix of components hinged on the respective national-security strategies. In 1989 Soviet intercontinental rocket forces carried 60 percent of their strategic warheads, compared with 20 percent for the United States (which relied much more on highly survivable sea-going SLBMs). The Soviets took advantage of their huge land mass to deploy road- and rail-mobile ICBMs, reducing their vulnerability to surprise attack.

Bombers. Another means of avoiding vulnerability was to keep bombers either constantly aloft or on "airstrip alert" — able to take off within minutes of warning that an attack is coming. The airplane fleet of the Strategic Air Command (SAC) was the mainstay of U.S. nuclear-bomb delivery during the early stages. (When World War II came to an end, 2900 B-29s were operational, but by 1947 only about 300 remained in service with SAC. At the time of the Berlin crisis (1948), SAC had 32 nuclear-qualified bombers.)

New aircraft entered the force in the 1950s, including the B-50 *Superfortress*, of which 370 were built; the B-36 *Peacemaker*, of which 385 were constructed; the B-45 *Tornado*, the first jet bomber; and the B-47 *Stratojet*, with 2041 built in what was the largest jet program ever, with the possible exception of the Soviet Union's Tu-16 *Badger*.

The B-52 *Stratofortress* was the first intercontinental jet-powered bomber, entering service in the mid-1950s. It was updated several times, the G and H versions being capable of launching cruise missiles. Boeing built 744 of the them, and the Air Force still had 94 B-52Hs at the end of 1996.

The B-1, originally intended to replace the B-52, promised increased ability to elude Soviet air defense. Searching for a new intercontinental bomber with a large payload to replace the B-52, the Air Force in June 1970 awarded North American Rockwell a contract to produce five B-1A test aircraft, later reduced to three. Rockwell "subcontracted the work for the plane to more than five thousand corporations in forty-eight states ... creating a formidable lobbying coalition with considerable clout in Congress."

The Soviet Union began in the late 1960s or early 1970s to develop a supersonic strategic bomber. It wasn't until the 1980s, however, that development of the Tu-160 Blackjack was completed. In the meantime, the Soviet Air Force relied on the intermediate-range Tu-22M Backfire, which was accepted for service in 1976. Its actual range was a contentious issue in treaty negotiations during the 1970s, with the United States insisting that it could reach its territory (albeit on a one-way suicide mission or with in-flight refueling) and thus should be counted as a strategic bomber.

In 1977 President Carter canceled production of the B-1A and focused on cruise missiles that could be launched from B-52 bombers. They could carry as many as 20 nuclear-armed cruise missiles. The B-1, however, wasn't dead: It was revived as the B-1B *Lancer* by President Reagan, who ordered procurement of 100, at a cost that turned out to be about $330 million per aircraft. Eventually their nuclear mission was rescinded, and they were converted to conventional bombing missions.

In 1980 the Carter administration publicly announced another reason the B-1A had been canceled: The Air Force already had another bomber under development — the B-2A *Spirit*, which incorporated *stealth* technology, meaning it had a very low profile for detection by radar. President Clinton reduced the Air Force request from 132 to 75 of these dual-capable "flying-wing" B-2As. President G.W. Bush, under budgetary pressure, further reduced it to 20. It became the most expensive bomber ever — about $2.6 billion per aircraft, five times its weight in gold, although unit cost of additional aircraft would have been about half of that.

U.S. Intercontinental Ballistic Missiles (ICBMs). By the late 1950s airplanes were beginning to be supplanted, for delivering strategic warheads, by various kinds of missiles. The very first ICBMs were based on German wartime research, the American versions becoming the *Atlas D* and *Titan I*. Armed with thermonuclear warheads, these and the Soviet equivalents were considered "city busters" (They were not accurate enough to be used specifically against hardened military or governmental sites). The Titan I was particularly significant, being the first big multistage missile.

The first generation of ICBMs had liquid propellants that required cryogenic storage. They could not be kept fueled, except for a very brief period before launch, without running the risk of being seriously damaged. Also, because of the time required for their fuel tanks to be filled prior to launch, and because the launch sites were unhardened, these missiles were considered to be vulnerable to a pre-emptive strike.

The United States has produced, since World War II, many types of ballistic-missiles, including thousands of ICBMs and SLBMs, loaded with eight classes of warheads having explosive yields ranging from 170 kilotons to 10 megatons.

The United States also fielded the world's first solid-fueled rocket.

The number of U.S. land-based missiles remained relatively constant from 1967 to 1991, but their military effectiveness was continuously improved.

The MX (*Peacekeeper*) was originally designed for either air- or ground-mobile basing. It incorporated several new design technologies, including extendable exit cones for thrust enhancement, motor casings wound with Kevlar fiber for reduced weight, and compressed-gas cold launching (as done by the Navy to launch its SLBMs under water).

In a plan called "multiple-protective shelters," 200 MXs were to be shuttled among 4600 unhardened shelters in Nevada and Utah. Criticism of the high cost ($37 billion) and extensive land use was instrumental in its cancellation. A subsequent idea, supported by President Reagan, for putting them in Minuteman silos also drew criticism because close spacing led to possible concurrent destruction from a single large thermonuclear blast.

Soviet Strategic Modernization. During drawn-out SALT I negotiations in the 1970s, both sides engaged in strategic modernization, not expecting a quick consummation of the treaty. The Soviets gained a *launch-on-warning* capability by constructing early-warning radars and orbiting satellites. They also protected their ability to *launch-under-attack* by improving radiation hardness and other facets of combat readiness.

The most important element of the Soviet modernization program was deployment of MIRVed land-based missiles, thus catching up with the United States.

Submarine-Launched Missiles. When it became possible to launch missiles directly from submarines, a new dimension in national security and deterrence stability came into effect. Nuclear-powered submarines can range and hide underwater in the more-than-half of the earth that is covered by deep seas. Because they were designed to patrol with minimal acoustic noise, the nuclear-powered (SSBN) "boomers" became very difficult to detect or track. As a result, the SLBM force became the least vulnerable leg of strategic triads.

The first U.S. submarine-based missile was the *Regulus*, with a 500-mile range, originally deployed in 1955 aboard a cruiser and an aircraft carrier. Later it was secured in watertight hangers atop submarines. The turbojet-driven cruise-missile *Regulus* — derived from the German World War II "buzz bomb" — was armed with a 120-kiloton warhead that could be launched only when the submarine surfaced.

The Soviet Navy armed their submarines with cruise missiles too, beginning in the 1950s. They developed the means to launch directly from a container. Both diesel- and nuclear-powered submarines were equipped with nuclear-armed cruise missiles, starting late in the 1950s. At the time, cruise missiles were largely effective only against stationary coastal targets, although later models were intended to attack ships at sea. (The original focus of Soviet military doctrine was seaward operations against enemy ships and naval bases.)

The U.S. Navy initiated the *Polaris* ballistic-missile submarine program at the close of 1956, and these boomers began their patrols in 1964. The *George Washington* nuclear-powered submarine, carrying 16 Polaris missiles with a range of about 2200 kilometers, entered service in 1960. In early 1971, MIRVed *Poseidon* SLBMs were introduced, bringing the number of U.S. submarined-based warheads from about 1500 to 5500.

As early as 1949, the Soviet Union began to study ways to install ballistic missiles on submarines, but it was not until 1954 that its development was officially approved. When their Navy fired missiles from a converted submarine in 1955, it was the first test launch of a ballistic-missile from a submarine. A full-scale nuclear-armed missile test took place in 1961, resulting in a yield of about 1.5 megatons at the Novaya Zemlya test range. A few years later, the Soviet Navy had submarines that could launch missiles under water.

Delta-class SLBMs, developed in 1965 (just after the U.S. Polaris submarines were deployed), became the second generation of Soviet missile submarines. Armed with the single-warhead , these subs had a cruising range of 7800 kilometers; this made them much less vulnerable to antisubmarine warfare operations because they could make lengthy patrols in safe seas close to their own coast, with targets in the United States still in range.

The third Soviet SLBM generation consisted of Typhoon-class submarines, which carried 20 solid-propellant missiles with a range of 5000 miles. Each missile held up to 10 MIRVed RVs. This generation of submarines had a significantly lower noise level, and they were adapted to patrolling under the Arctic ice.

In the mid-1980s, U.S. Poseidon SSBNs were retrofitted with the Trident-I SLBM. The more modern Trident-II SLBM is now deployed on a fleet of 14 modern Ohio-class SSBNs, each of which carries up to 120 W88 warheads.

Shorter-Range Missiles. Ballistic and air-breathing intermediate- and short-range missiles were developed to be launched from specific land, sea, and air platforms. They were primarily for delivering sub-strategic warheads throughout NATO battle areas.

Some missiles had an inherent strategic capacity also, because they could reach the Soviet Union from ships, naval aircraft, or sites in Western Europe.

Unpiloted cruise missiles with either conventional or non-conventional explosives can be used in theater operations. Also, they have been deployed on launch platforms (aircraft and submarines) that give them a strategic reach. With a range of 1500 nautical miles, subsonic speed, and a 200-kiloton warhead, nuclear-armed cruise missiles are a potent threat. In addition, with better fuels, terrain-matching guidance, and improved radar-evasion technology, they can have the speed, accuracy, and kill capability to attack hard targets (such as ICBM silos). Nuclear-armed cruise missiles have become a potentially destabilizing weapon system because of their first-strike implications, combined with difficulty in devising verifiable arms control procedures.

The U.S. Army developed its own surface-to-surface missiles, unguided artillery rockets, suborbital ballistic missiles, liquid-propelled guided missiles, and also a number of surface-to-air missiles. The Navy acquired its own cruise and air-to-air missiles. The Air Force brought out a variety of winged missiles.

Cruise missiles that are submarine-launched (SLCMs) and aircraft-launched (ALCMs) can fly long distances, are difficult to intercept, and can be guided with great accuracy. Their cost is small compared with other types of delivery systems. The United States equipped its Los Angeles-class submarines with the Tomahawk cruise missile.

Soviet Nuclear Weapon Systems. The Soviet Union placed cruise missiles that could be launched from torpedo tubes on their third-generation submarines.

The best available estimate of total Soviet nuclear weapons in mid-1988 was 33,000, comprising 13,000 strategic offensive weapons, 4,200 ABM and surface-to-air defensive warheads, 11,800 tactical weapons, and 3600 naval weapons. (At that time, the United States had about 25,000 nuclear warheads.)

In mid-1988, the Soviets had 2516 strategic bombers and missiles in service, and their ICBMs were loaded with 6500 warheads. Sixty-three submarines were loaded with 3400 warheads. One-hundred-sixty long-range bombers could transport 950 warheads.

In 1989, Soviet surface-to-surface missiles could release up to 2900 warheads, and their short-range aircraft could deliver up to 5100 warheads. Soviet naval aviation had aircraft that could load up to 1700 nuclear weapons. Four-hundred sea-launched cruise and anti-ship missiles were deployed.

Sometimes the throw-weight (payload plus auxiliary mechanisms) for Soviet weapons — but more often their total explosive megatonnage — was used as a parameter for comparison with the U.S. arsenal. Officially, DOE estimated about 5000 deliverable explosive megatons for the Soviets in 1985, while U.S. hawks estimated they had about twice as much.

While throw-weight and megatonnage had their own significance, they were not indicative of the actual military value of nuclear holdings in time of war because qualitative factors such as accuracy, reliability, or yield allocation were not taken into account.

Spreading a given megatonnage among more warheads increases the potential area of destruction: so a 1-megaton warhead could destroy an area of about 80 square miles. If, instead, the 1-megaton total were to be apportioned among eight warheads, each with 125-kiloton warheads, the area destroyed would be twice as large. Similarly, twenty 50-kiloton warheads, with the same megatonnage, could destroy 216 square miles. Bigger was not necessarily better. Improved CEPs permitted U.S. warheads to be smaller in size for the same hard-target kill capability; as a consequence, from a peak of 19,000 megatons in 1959, U.S. megatonnage decreased, reaching approximate parity with the Soviets' 5000 megatons by 1985.

Theater Weapons

In the early 1950s, the main instruments of nuclear warfare in Europe were U.S. and Soviet bomber squadrons, but smaller ordnance was already under development for field use. *Theater* weapons, also called *tactical* or *sub-strategic*, consisted of a variety of relatively small nuclear explosives and bombs. Some could be carried in a backpack (with yields as low as 22 tons or less TNT-equivalent).

Whereas weapons of strategic deterrence are those that would be used in retaliation against major military installations and metropolitan centers, theater weapons were designed for actual use in forward battle zones such as West Germany, Poland, or Czechoslovakia. In order to be forward-based, custom modifications of weapons custody, command, and control were required. Because of radioactive fallout and political consequences, the zones designated for nuclear combat were outside the possessor's national boundaries. (A possible exception would be a nation with a small nuclear arsenal intended for last-resort self-defense.)

> [Tactical Nuclear Weapons]. Some nuclear weapons were specially designed and deployed for battlefield use. For the most part, they had explosive yields less than the bombs used against Japan. Often their outputs could be changed in the field to suit the intended application. These smaller warheads were intended for short-range aircraft gravity bombs, antisubmarine depth charges, artillery and howitzer projectiles, ground-, air-, and sea-launched missiles, underwater torpedoes, and land mines.

Regarding method of delivery,

the vast majority of theater and tactical nuclear weapons built during the Cold War — in the West, more than 25,000 warheads and bombs, 36.5 percent of all U.S. nuclear weapons — were designed to be delivered by "conventional" [dual-capable] military systems, such as air force and navy tactical fighters, army ground-based and navy shipborne surface-to-air missiles (SAMs), naval antisubmarine warfare (ASW) systems, and army and Marine Corps artillery pieces.

The backpack-carried weapon, *Davy Crockett*, could be launched in the battlefield. It was the lightest and smallest fission weapon (implosion type) ever deployed by the United States.

The Army also fielded a series of improved nuclear artillery rounds and atomic demolition mines (ADMs) — including a special mine, weighing 58.6 pounds, that could be carried by a single soldier. Needless to say, considerable training and precaution were essential to avoid physical trauma from being too close to the explosion of these smaller atomic munitions, especially since detonation on the surface created significant local fallout.

Starting in 1962, the U.S. Army deployed surface-to-surface missiles: in 1972 the *Lance*, and in 1983 the *Pershing-II* missile. The Navy began deploying nuclear depth charges in 1961.

U.S. Overseas Deployments. Nuclear weapons deployed outside the continental United States were stockpiled on naval vessels, on U.S. overseas bases, or at sites authorized by a foreign government. NATO had an blanket arrangement, and some nations specific secret agreements. As a rule, strategic weapons were based in the United States, while some intermediate and short-range delivery systems were assigned overseas. NATO nations hosted most U.S. forward-based nuclear weapons.

Information on Cold War deployments of U.S. sub-strategic weapons outside the United States is slowly being declassified. This was the "nuclear umbrella" that the United States placed over Europe.

The first known overseas deployment of U.S. nuclear weapons was to Britain in 1950 — warhead assemblies without the "physics package" (thus only the non-nuclear components). Later that year, when the Korean War broke out, foreign deployment of complete assemblies was accelerated. Naval ships carried complete warheads. Non-nuclear components had been stored at the U.S. possession of Guam; in 1951, the physics package s were added.

The earliest known introduction of atomic artillery to NATO nations was in 1955, when the 280-mm gun was positioned in West Germany. Germany since then hosted the most U.S. nuclear warheads, about 3500 of 21 different types.

During President Eisenhower's term, West German Luftwaffe fighter-bomber pilots had virtual control of nuclear bombs during times of alert. After President Kennedy came into office, he ordered installation of coded permissive-action links (PALs).

Altogether, by 1960 NATO was assigned about 3000 U.S. nuclear weapons; in the mid-60s the total number of U.S. sub-strategic weapons in Europe reached about 7000, and in 1971 the number peaked at about 7300. By 1983, just before the deployment of cruise and Pershing-II, the total had dropped to 5845. About half of these were "dual-key," that is, assigned to delivery systems owned and garrisoned by the armed services of an allied nation; those weapons could not be used without approval of the host country. The remainder, almost 3000 nuclear weapons, were "single-key" devices allocated to delivery systems under U.S. control.

Fearing the possibility of being overwhelmed by Red Army armor and artillery, NATO positioned its forward-based systems close to possible lines of advance. Storage bunkers were adjacent to borders and at NATO airbases. Tactical aircraft were ready to deliver small H-bombs to the rear echelons of Soviet invaders. The ADM ("backpack" munitions) were intended for placement near bridges and other choke points.

Complete atomic bombs were shipped to Morocco early in 1954, and to Britain later that year. From 1955 through the early 1960s, various fission and fusion weapons arrived in Italy, France, Turkey, the Netherlands, Greece, and Belgium. Intermediate-range nuclear-armed Thor and Jupiter missiles were deployed in England, Italy, and Turkey.

While under American occupation, the Japanese island of Okinawa was the storage site for about 1000 warheads of 18 different types, practically the full range of capability in tactical nuclear weapons. Post-Cold War disclosures indicate that the United States deployed nuclear warheads on two other occupied Japanese islands as well: Iwo Jima and Chichi Jima. This forward-basing in the Pacific was not particularly surprising to those who had grown accustomed to American official secrecy.

Although not officially acknowledged, declassified information indicates that for about ten years, through 1966, non-nuclear components, i.e., nuclear weapons without the physics packages, were stored on U.S. bases in mainland Japan (in addition to the complete warheads on U.S. bases in the above-mentioned islands). Also, U.S. bombers and warships used the bases and port facilities in Japan for routine transfer of nuclear weapons and components.

In 1968 the United States and Japan are believed to have signed a secret agreement allowing reintroduction of nuclear weapons on Chichi Jima and Iwo Jima in the event of an emergency. Although Japan's aversion to nuclear weapons was more a diplomatic sentiment than a strict reality, the nation sought to exempt itself from being a nuclear target, and its leaders understood that compromises were necessary for Japanese security during the dark days of the superpower confrontation. There is no reason to believe that the government of Japan itself ever had possession or control of nuclear weapons or their delivery systems.

Nuclear-armed *Matador* surface-to-surface ground-launched cruise missiles (GLCMs) were deployed in Taiwan in 1958. Several types of nuclear weapons were also deployed in South Korea that year; eventually as many as ten types of sub-strategic nuclear weapons were shipped to Korea.

From 1963 to 1966 the U.S. Army stationed a *Nike-Zeus* nuclear-armed antiballistic missile system on Kwajalein Atoll of the Marshall Islands.

Outside of NATO, locations for U.S. atomic weapons included Greenland, Cuba (Guantanamo Bay), Iceland, Philippines, and Spain. Adding Hawaii, Midway, Johnston Island, Alaska, and Puerto Rico to those already mentioned, all-in-all an estimated 27 countries or bases outside the lower 48 states were sites for their deployment. From a report made public that deleted some destination names, it is possible to infer that Canada might have been a host in the interval of 1950 through 1977; if not, some other unidentified and still-secret nation harbored U.S. nuclear weapons. Unclear are the reasons for continuing to keep this information buried, inaccessible to the public of both the United States and the host nation.

At the beginning of the 1980s, the United States upgraded its intermediate-range nuclear force (INF) deployment by installing 108 Pershing-II missiles and 464 GLCMs in Europe.

By 1983, the United States had assigned substantial nuclear forces for use in NATO contingencies: 400 Poseidon SLBMs, 156 fighter-bombers, 66 long-range aircraft, and 40 carrier-based aircraft.

A particularly heavy veil of secrecy was drawn over information about stateside production and overseas deployment of U.S. nuclear weapons. Senator Stuart Symington (Missouri) was a member of the Senate Foreign Relations Committee. In 1970, he complained that the American public often knew much less than the rest of the world about these deployments and the related foreign-policy commitments. Senator Symington found that the executive branch failed to reveal, explain or justify the true dimensions of deployments — his Committee having never been given any details about the positioning of more than 7000 nuclear warheads in Europe, or the reasons for doing it.

Soviet Deployments. According to one estimate, just before the breakup of the Soviet Union there were about 15,000 Soviet sub-strategic weapons, divided among ground-based air defense, aircraft wings, and naval forces. They were assigned to 13 Soviet republics and the three Baltic republics. The bulk of them were in Russia (about 10,000), the remainder in Ukraine, Kazakhstan, and Belarus, with 700–800 allotted to the Baltic states.

Although the Soviet nuclear design bureau had been directed in 1952 to develop a nuclear artillery shell, it wasn't until 1954 that the government authorized a tactical nuclear weapons program. Their first compact device — intended for a nuclear torpedo — was tested in 1955. Designers completed the concept development of an artillery shell in 1956.

Around 1959 or later, the Soviets started to deploy SS-4 and SS-5 intermediate-range nuclear-armed missiles in their western region. Because these missiles were liquid fueled, preparation for firing required eight hours, and standby to launch could last for only five hours.

In the mid-1970s the Soviets began to upgrade their intermediate-range forces, including the Backfire bomber and the SS-20. The latter was a significant military advance, being a mobile solid-fueled ballistic missile with three 150-kiloton MRVs and a range of 2700 miles. By 1983 about 350 SS-20s had been deployed, two-thirds of them able to reach targets in Western Europe.

The notorious mid-60s misadventure in Cuba began when sub-strategic missiles were transferred there from the Soviet Union. They were noticed by the United States, and their withdrawal was negotiated under stressful conditions. At some undisclosed time, the Soviet Union moved tactical weapons into WTO-allied nations, but evidently kept them under tight central control.

Before the Soviet Union completely dissolved, all Soviet nuclear weapons were withdrawn to the new Russian Republic. Also, tactical weapons that had been in WTO states, such as East Germany and Czechoslovakia, had already been moved eastward to the republics that eventually constituted the CIS.

INF Missiles. Typically, intermediate-range nuclear force (INF) missiles could travel between 300 and 3000 nautical miles. In 1975 the United States had 324 Pershing-I missiles (range 460 miles) within Europe. The Pershing-II, deployed in 1983, was a mobile solid-fueled rocket with a range somewhat more than 1000 miles; 108 were deployed in Europe. American GLCMs had a single nuclear warhead and a range of 1500 miles.

The Soviet Union deployed within its boundaries and in affiliated WTO nations, the SS-20 (range 2700 miles), which could hit targets in Western Europe and China — but not in the United States.

The 1987 INF Treaty eliminated all nuclear-armed ground-launched ballistic and cruise missiles with ranges between 500 and 5500 kilometers (about 300 to 3400 miles), as well as their infrastructure. Included were all Soviet SS-4, SS-5, SS-12, SS-20, and SS-23 missiles, and all American Pershing and nuclear-armed cruise missiles (no matter where they were located — anywhere in the world).

Enhanced-Radiation Warhead. Popularly called the "neutron bomb," a specialized enhanced-radiation warhead (ERW) was designed to kill or disable enemy soldiers while minimizing damage to nonliving things. Its primary purpose was to help defend Western Europe against the feared onslaught of Soviet armor and artillery.

The ERW is a small nuclear weapon modified to produce more neutrons per kiloton than a fission weapon. By reducing the proportions of blast and heat from the explosion, the "collateral" damage to nearby property can be minimized.

The ERW is intended for use in theater warfare. Such warheads could be delivered to a battlefield by rockets, howitzers, or cruise missiles.

If detonated high enough in the air above troops or armored tanks, its primary effects could be direct lethal radiation, with reduced blast and heat. Local radioactive residue would be relatively low-level, mostly from short-lived elements activated by neutrons. However, if the fireball touched the ground, which would be quite possible, fission products would be spread about too. In any event, an ERW *is* a nuclear weapon, with effects not qualitatively different from those of any other nuclear explosive of the same yield.

Armored tanks offered resistance to radiation from ERWs, depending on their distance from the epicenter of explosion and upon the amount of shielding provided by its armor; the crew would not be immediately incapacitated. The fact that some crew members could continue in battle for days raised skepticism about the neutron bomb's tactical utility; although doomed they would be able to fight on for as much as a week.

Aside from increasing the armor plating on tanks, the most effective way to counter the neutron bomb is to disperse formations. With tanks more than 1000 feet apart, one ERW might irradiate only three tanks.

The idea of reducing collateral (nonmilitary) damage is meaningful mainly to military tacticians: Nearby civilians — who would receive high doses of radiation while being subjected to less blast and heat — are likely to react to the thought more irrationally than analytically.

President Reagan, pushed by his Secretary of Defense, Caspar Weinberger, decided in 1981 to deploy neutron weapons in Europe after the Soviets occupied Afghanistan and continued to produce SS-20 intermediate-range missiles aimed at European cities. The underlying idea was to equip the U.S. Army with capabilities to fight with conventional, nuclear, and chemical weapons all at once in an "integrated battlefield." (To be prepared to function under such conditions is mind-boggling.)

Meanwhile, the United States had embarked on its most ambitious nuclear weapons-building program, which included the Trident-I submarine-launched missile; a new warhead for the land-based Minuteman III; new strategic and sub-strategic bombs; a new long-range, air-launched cruise missile; and plans for the proposed MX missile. In addition, NATO was planning to deploy cruise and Pershing-II ballistic missiles that could reach targets in the Soviet Union.

British and French Deployments. Nuclear weapons and delivery systems developed by Britain and France were important elements for the balance of forces in Europe. Both nations had nuclear-equipped bombers and nuclear missiles on submarines. France also had its own land-based nuclear missiles.

By 1983 the two countries each had 300 nuclear warheads, give or take 50 or 100, depending mostly on the year chosen for accountability.

The French announced in 1980 that they too had begun development of a neutron bomb, even though at the time the United States had temporarily shelved its plans. Both France and Britain developed sub-strategic warheads to complement their strategic arsenals.

Nuclear Warheads

Military strength was often gauged during the Cold War by the number and variety of nuclear warheads, especially strategic weapons.

The United States designed and built 65 types of nuclear bombs (B-series) and warheads (W-series) for 116 different delivery systems, ranging from battlefield to strategic bombs and warheads home-based in the United States or assigned to naval vessels. The U.S. Air Force has, over the years, deployed 42 types of nuclear weapons, the Navy and Marine Corps 34 types, and the Army 21 types. Another 25 warhead designs were canceled before production. Some models have had wide applicability, used in one configuration as a bomb, and in another as the payload for one or perhaps several kinds of missiles.

The last completely new type of warhead designed and produced was the W88, assembled at the Pantex Plant, intended for a Trident-II D-5 missile. The W88 has been reported to have the highest explosive-yield-to-volume ratio of any U.S. nuclear ordnance. Production was halted in 1990, but resumed in 2003.

The historic peak for the U.S. stockpile was reached in 1966 when about 32,000 nuclear warheads were simultaneously in service.

In 1954, thermonuclear (hydrogen) bombs — hundreds of times more powerful than their fission predecessors — began to enter the U.S. stockpile in great numbers, and the total stockpiled megatonnage (yield) increased sixty-fold in five years. It peaked in 1960 at almost 20.5 billion tons of TNT — equivalent to nearly 1.4 million Hiroshima-sized bombs. The explosive power of the arsenal today is less by a factor of ten, but still equals some 120,000-130,000 Hiroshima-sized bombs.

Before their designs were retired, 14 Little Boy and Fat Man bombs were made.

Warhead Safety. Along with carefully regulated and controlled handling, well-thought-out designs are vital in assuring the safety of nuclear weapons. Three specific improvements have been: (1) enhanced nuclear detonation system, which reduces the chance of a warhead's detonators being fired electrically in an accident; (2) insensitive high-explosive (IHE), which is much less prone than conventional HE to detonation by fire or impact; and (3) a fire-resistant pit,* which can withstand prolonged exposure to a jet-fuel fire.

Additional safety features include mechanical safing, which greatly diminishes the possibility of having a significant uncontrolled fission chain reaction; separable components, where the fissile material is physically separated from the high explosive until the weapon is about to be used; and one-point-safe design, which ensures that no significant chain reaction will result if the HE is detonated at any single point of impact.

---

* The *pit* is the central fissile component — the core — of a nuclear explosive. It is primarily composed of fissile materials arranged in a geometric configuration that sustains the explosive nuclear reaction.

Insensitive HE was first put into service with U.S. nuclear ordnance that entered the stockpile in 1978. After that, timely retirement of older weapons improved not only reliability but also overall safety for devices that lacked the newer features.

In addition to possible accidental or malevolent detonation of the explosive material in warheads, inadvertent detonation or ignition of a nuclear-armed missile's propellant could occur, leading to dispersal of particulate plutonium in a populated area. (Although that would cause consternation and cleanup activity, the plutonium would not constitute a major safety or health hazard.) In the missile's blowup, the possibility of a small but highly destructive nuclear burst (a few tens of tons) cannot be completely ruled out.

A Trident missile submarine uses W76 and W88 thermonuclear warheads. The W88 is a high-yield warhead that was extensively tested before the Threshold Test Ban Treaty came into effect. But, instead of IHE, the W88 uses HE to minimize the weight and diameter of the re-entry vehicle. Also, all three stages of the D-5 missile used in the Trident have high-specific-impulse propellants to achieve maximum range; these propellants are more likely to detonate during impact accidents than propellants used in many other ICBMs.

A combination of shortcomings — a warhead using a comparatively sensitive explosive, a missile propellant that is more prone to detonation in an accident, and a unique through-deck missile-tube design that is not without problems — has led to suggestions that Trident safety could and should be improved.

Some of those concerns have been addressed through operational procedures for handling and transporting RVs and missiles. Other problems could be reduced by adapting existing IHE warheads, retrofitting the W76 and W88 with IHE, or developing new, safer warheads (testing of which, however, would be prohibited by the CTBT — which is another reason that right-wing pressure has been generated to forsake the treaty).

Comparative Military Strength

Meaningful comparisons of overall military strength are difficult. For example, the armed forces required for a conventional military conflict between NATO and WTO would have been different from that needed in a distant surrogate war (such as in or near Cuba, Angola, Vietnam, or Afghanistan). In an extended conflict, the competence of support forces and military technicians could be important. Moreover, simple tabulations of the numbers of weapons and armed personnel do not reflect the quality of armaments and training.

Just before Reagan came into office as President, the Soviet Union has about twice as many personnel in their armed services. They also had up to three times as many armored tanks. The Soviets also had more ICBMs, SLBMs, and missile submarines; they probably had about 24 percent more nuclear warhead. At the same time, the United States "led" in the number of strategic bombers and deliverable nuclear warheads.

On balance, neither side could survive an all-out nuclear attack by the other; neither could "win" without suffering comparable mortality and damage. Even the smaller nuclear-weapon states had enough capability to inflict devastating, and therefore deterring retaliation against any first strike. "Parity" existed at least in terms of equivalent devastation, if not in each numerical weapon category.

Allies of the Soviet Union and the United States are excluded from balance-of-forces comparisons, particularly because the numbers for Warsaw Treaty Organization members are suspect. Anyway, their contribution in a conflict would have been limited mainly to military personnel and conventional arms. At that time China could not have been considered an ally of the Soviet Union.

During the Cuban missile crisis, 1960–61, the Soviet Union had a much larger standing army than the United States, but only half as many strategic bombers.

While the Soviet Union might never have caught up in strategic warheads, it probably exceeded the United States in total warheads — strategic and tactical.

British, French, and Chinese Nuclear Systems. Great Britain, France, and the People's Republic of China had substantial inventories of nuclear weapons and the means to deliver them.

In 1987 Britain had aircraft and ballistic-missile submarines armed with nuclear warheads: About 190 bombs could be carried by Tornado bombers, 25 on the older Buccaneer bomber, and 195 warheads on Polaris-type submarines. In addition, carrier aircraft (Harriers) could deploy 23 bombs, and ASW helicopters could have dropped 160 depth charges.

France had a larger nuclear arsenal than the UK: about 250 warheads on three submarines and 160 bombs that could be carried by Mirage and Jaguar aircraft. Their short-range and long-range land-based missiles could launch as many as 90 warheads. France had one aircraft carrier, a platform for delivery of about 40 atomic bombs.

China is reported to have had about 400 nuclear bombs and about 200 warheads that could have been released from land- and submarine-based missiles of medium, intermediate, and intercontinental ranges. Their long-range aircraft, primarily of Soviet origin, could transport more than 100 bombs, and their land-based intermediate- and long-range missiles could send aloft up to 200 RVs with warheads. They also had a submarine that could launch two or three dozen nuclear-tipped missiles.

****

When the North Atlantic Treaty Organization decided to deploy nuclear battlefield weapons in Europe, it stimulated another reactive arms cycle. The Soviet Union thereupon began to develop its own tactical nuclear weapons, and a few years later it deployed them in Warsaw Treaty Organization nations.

During most of the Cold War arms buildup, the Americans were six to ten years ahead of the Soviets in quantity of nuclear weapons. Because the United States chose to develop smaller, more efficient, and more numerous weapons, its total megatonnage declined steadily from the mid-1960s, remaining below that of the Soviet Union even to the end.

Because of the better quality and accuracy of U.S. nuclear weapons, American military chiefs, when asked, reaffirmed they would not want to swap places with their Soviet counterparts.

C. DETERRENCE-SUPPORT TECHNOLOGY

As any military officer would advise, force projection cannot be particularly effective without production capacity and logistics support to generate and sustain the implements of warfare. Because of the unique and menacing nature of nuclear weapons, they had to be produced and handled under very special conditions. Airplanes, missiles, and ships qualified to launch special weapons required careful management by trained crews. Some naval vessels were powered by nuclear reactors. The potential use of special weapons required an investment in leading-edge technology for reliable command, control, communications, and intelligence (C^3I) support.

In the United States, governmental support of the deterrence infrastructure came from DOD agencies, DOE's national laboratories, intelligence collection organizations (such as the Defense Intelligence Agency, CIA and National Security Agency), and the related military/civilian system for controlling nuclear weapons and their use. The corresponding Soviet agencies included the Ministry of Defense, KGB, and Ministry of Medium Machine Building (later designated Ministry of Atomic Power and Industry, now MINATOM). These U.S. and USSR agencies conducted research and development, operated facilities for weapons production, nuclear propulsion and testing, and administered the CI network.

Production of Nuclear Weapons

Many, now historical, events led to the production of the nuclear weapons, including the "Super" thermonuclear explosive. Even while the breakthroughs needed for development of fission and fusion explosives were taking place, preparations were initiated to produce the necessary core ingredients — primarily exotic materials such as plutonium, enriched uranium, and tritium. These materials, not naturally available in sufficient purity or quantity, required specialized manufacturing processes and facilities.

The Manhattan Project, officially known as the Manhattan Engineering District, or simply the Manhattan District, was in charge. It undertook mining of yellowcake ore. The uranium extracted from the ore is about 0.7 percent of the fissile isotope uranium-235 and 99.3 percent of the non-fissile uranium-238. From natural uranium, the Manhattan Project produced HEU (highly enriched uranium), which had been augmented by a factor of a hundred in fissile U-235. Three different enrichment methods were originated: electromagnetic isotopic separation (with a *calutron*), gaseous diffusion (cascaded stages), and thermal diffusion (which was abandoned early in the project). Another isotope-separation method, using high-speed centrifuges, was developed later in the Cold War and has ever since continued to be important.

> [Oak Ridge]. Here we were, a DOE technical advisory team, at Oak Ridge, Tennessee, amid the rolling, forested hills.... Nuclear reality was right in front of us.
>
> We were there in 1987 to witness procedures for possible verification of a treaty on warhead dismantlement. (I was a representative of one of the "peace" labs [Argonne, Brookhaven], while other team members were from the weapons labs [Los Alamos, Livermore, Hanford]). We were shepherded by DOE officials, and hosted by Oak Ridge weaponeers.
>
> It seemed so simple: a technician took a warhead apart matter-of-factly with a hammer and chisel. Yes, that's all he needed. I reflexively cringed when he pounded on it.
>
> When it was stripped down to a few bare essentials, this soccer-ball-sized creation of civilization was all it took to incinerate a smallish city and its inhabitants.
>
> Imagine my feeling of helplessness, so much potential devastation right in front of me — the reality, not the pictures, not the theory — the real thing. Just a metal ball you could cradle in your arms.
>
> Verdant scenery greeted us outside; the earth's sun thankfully was still shining.

Uranium was enriched mainly because the weapons program needed HEU (enriched to more than 90 percent) for warhead *pits* (and later for thermonuclear *secondaries)*; the program also used HEU as fuel for naval reactors, some plutonium-production reactors, and developmental fast-breeder reactors.

Power reactors can be fueled with uranium or thorium, but most U.S. electricity-producing reactors burn with uranium enriched merely to the range of 2–5 percent. About 725 metric tons of HEU were separated by the United States.

Unlike uranium, plutonium exists naturally only in trace (insignificant) quantities. All plutonium used to build nuclear weapons has been derived from uranium-fueled reactors: It is created when the abundant uranium-238, in or surrounding a reactor core, absorbs neutrons released in the nuclear chain reaction. Plutonium was later chemically separated from the reactor fuel by Purex (Plutonium-URanium EXtraction) *reprocessing*, which has to be conducted in facilities that protected the operators from high radioactivity. Just over 100 metric tons of weapons-grade plutonium was produced in the United States during the Cold War.

The weapon-development laboratories for new U.S. designs were Los Alamos (where the first work was done) and Lawrence Livermore (starting in 1952), with engineering implementation assigned to Sandia Laboratories in Albuquerque, New Mexico. Other facilities in the weapons complex included Mound Laboratory (Miamisburg, Ohio), the Kansas City Plant (Kansas City, Missouri), Rocky Flats (Jefferson County, Colorado), and Pantex (Amarillo, Texas). Advanced methods for more efficient weapons were always being sought, but "Such concepts as levitated cores, composite pits, and 'boosted' weapons were considered during the Manhattan Project but not acted upon until [World War II] ended."

DOE's production and assembly of warheads involved many facilities. Plutonium parts (*pits*) were produced at the Rocky Flats plant near Denver, Colorado. Thermonuclear *secondaries* were made at the Y-12 plant in Oak Ridge. The fuse and electrical system came from the Kansas City Plant. Warheads were assembled at the DOE Pantex facility. Completed warheads and re-entry vehicles (RVs) were shipped to military bases and staging areas in special truck-trailer rigs.

In the Soviet Union, the nuclear-warhead production complex was concentrated at twelve principal sites — all in Russia. Ten of these sites (and their closed support cities) were not marked on maps at the time. The two secret weapon-design laboratories were later acknowledged to be Arzamas-16 (now Sarov) and Chelyabinsk-70 (now Snezhinsk); both were responsible for full life-cycle technical support of nuclear weapons. The Moscow Institute of Automatics had a role in designing components of warheads.

The Kurchatov Institute in Moscow was the original center for atomic-bomb research; in fact, their first reactor (F-1) was still operating long after the Cold War with its initial uranium fuel. Weapons production was distributed among four locations, particularly Sverdlovsk-45. Materials production took place largely at Mayak (Chelyabinsk-65), Tomsk-7, Sverdlovsk-44, and Krasnoyarsk-45. Weapons-production was carried out at Sverdlovsk-45, Penza-19, and Zlatoust-36. The Soviet production complex was reported to employ about 100,000 people.

Nuclear weapons Testing

Testing nuclear weapons with full- or partial-scale explosions has gone on ever since the Trinity experiment. Most tests have been for weapons development, that is, to evaluate the correctness and effectiveness of a design; others were for studying destructiveness — the effects on inanimate targets and living. Some proof trials of weapons in service have been conducted for safety, reliability, and operational diagnostics. In addition, there was considerable testing for so-called "peaceful" purposes.

The United States carried out atmospheric atomic detonations in the Atlantic and Pacific Oceans, in Nevada, and in other states, including Alaska. The Marshall Island atolls of the Pacific Proving Ground became the site of 66 atmospheric and underground tests, ending in 1958. In the Atlantic, closer to South Africa than any other land mass, three secret explosions were conducted in a single series ("Argus" in 1958, for EMP excitation); they were the first shipboard launches of short-range ballistic missiles with nuclear warheads. A total of 1030 U.S. nuclear devices were discharged altogether, 904 of them at the Nevada Test Site.

The Soviet Union had an extensive testing program, both in the atmosphere and underground, their military sites being Novaya Zemlya (near the Arctic Circle) and Semipalatinsk (in Kazakhstan). The Soviets are known to have carried out 715 proof tests. About three-quarters of them were in support of military programs: weapons development, troop maneuvers, weapons safety, weapons effects, and fundamental research. The remaining tests, most of which were took place outside the military test ranges, were for peaceful purposes: industrial blasts, oil and gas release, and ground excavations.

> [A Sea Drill]. In 1952, a second-year cadet in the Soviet Engineering Navy Academy was assigned to a month-long naval duty on a destroyer in the Black Sea. A directive was received to initiate sea drills imitating a nuclear attack. The exercise was to restore the destroyer to its duties after the assumed explosion of an atomic bomb 200 meters away. Of course, weapon developers and military leaders knew that nothing would be left of the ship after a nuclear detonation that close. However, government policy was not to provide field personnel with real information for fear of paralyzing their ability to fight, even if they didn't know what was happening to them.

Before being were banned, most Soviet tests were airbursts. The number and yield of surface detonations were kept relatively low because of the large amount of fallout hurled into the air. Some nuclear weapons were detonated under water or just above the water. Four weapons were exploded in space, one at high altitude. Nineteen of the underground Soviet blasts were said to have (accidentally) ejected a substantial quantity of radioactive gases into the atmosphere.

The superbomb (mentioned and pictured in the Introduction) was created by the Soviet Union for political (intimidation) purposes; it was tested in 1961 — at about half its 100-megaton explosive rating because the full bomb would have caused much more radioactive fallout.

Other nations also have conducted explosive trials above and below the Earth's surface: Great Britain (45 tests), France (210), and China (45). The British tests were in the Indian Ocean and Australia, the French experiments in Algeria and the South Pacific. The first Chinese nuclear detonation was in October 1964, at their site in the Lop Nur region of Xinjang; code-named 596, it yielded around 20 kilotons. Their first H-bomb experiment followed by only two years, and in June 1967 they achieved a yield of 3 megatons. After 1981, China's nuclear explosions were carried out underground.

At various times, treaties and voluntary moratoria narrowed the arena for nuclear testing, interrupted experiment continuity, or limited permissible yield. For instance, an understanding between the superpowers gained a moratorium from 1958 into 1961. There was a second Soviet moratorium from mid-1985 to the beginning of 1987.

After the Limited Test Ban Treaty of 1963, both nations discontinued experiments above ground, under water, and in outer space. Altogether, 825 underground tests, some of which had nonmilitary objectives, were conducted in Nevada between 1957 and 1992. The U.S. program between 1966 and 1996 cost approximately $20 billion. Under presidents Reagan and Bush, expenditures doubled compared to the previous decade.

Combined explosive generation of all known Soviet tests is estimated to be 470 megatons, equivalent to 30,000 Hiroshima bombs, and more than twice the U.S. total. A catalog of worldwide nuclear testing accounted for 2049 detonations by the five nuclear powers and several recent salvos by India and Pakistan.

To detect and identify nuclear detonations under the scope of the 1963 Limited Test Ban Treaty, a U.S. program for earth-based sensors was created for its three weapons laboratories. In 1966 the first prototype, an unattended seismic station, was put into service in Alaska; eventually a worldwide network of seismic monitors was developed. Years later, satellites became space-based platform for detecting nuclear tests.

Nuclear Propulsion

Naval nuclear propulsion became an important strategic element, providing global "projection" of military power. It also helped stabilize deterrence because nuclear-powered submarines and surface vessels were mobile, lethal, and comparatively invulnerable to surprise attack. Each superpower had developed submarine-launched long-range missiles (SLBMs), so nuclear submarines became the most robust of the three legs in the strategic triad.

Prolonged cruising ranges and remote stationing of naval forces allowed military power to be projected with minimal dependence on overseas bases, refueling provisions, or supportive alliances — all especially important in time of crisis. Several of the nuclear weapons states took advantage of this improved naval mobility and range.

Aircraft propulsion by nuclear means, on the other hand, failed to become viable. Between 1946 and 1961 more than $7 billion was poured into the U.S. Aircraft Nuclear Propulsion program. Although no aircraft has ever been propelled by nuclear power, the Air Force risked flying ran a three-megawatt, air-cooled, operating reactor 47 times aboard a converted B-36 bomber.

U.S. Naval Vessels. The American Navy's reliance on nuclear propulsion began with the 1954 commissioning of the attack submarine Nautilus.

Most reactors on naval ships and submarines are fueled by highly enriched uranium (permitting a more compact engine compartment than natural or low-enrichment uranium). A single reactor is usually sufficient for a submarine; twin reactors wee installed in cruiser-class ships, and up to eight reactors in larger vessels.

In addition to long cruising ranges, elimination of refueling outages that require frequent returns to port, is particularly advantageous under wartime conditions. Modern naval-reactor cores lasted fifteen years between refuelings.

Four new nuclear-powered aircraft carriers were funded from 1983 to 1988 as part of the Reagan-administration's military buildup. In 1988 the number of operating naval reactors peaked at 169. Nuclear-powered cruisers are gradually being retired as they reach the end of their service life. At the end of 1997, the United States had 131 operational seagoing naval reactors (compared with 108 operating civilian power reactors).

The most powerful and eminently successful leader in U.S. naval nuclear propulsion was Admiral Hyman Rickover. His strong emphasis on quality in manufacture and operation led to safe and reliable reactors, leaving the Navy an enduring legacy.

Soviet Nuclear-Powered Vessels. At the peak of its nuclear-propulsion evolution, the Soviet Navy had more than 300 nuclear-powered vessels. They consisted of submarines, surface combatants, and a small commercial fleet. The Soviets also had several nuclear-powered icebreakers.

In mid-1988 the Soviet Union, with about 1000 surface ships and 370 submarines, had the largest navy in the world. They had nuclear-powered submarines armed with cruise and ballistic missiles, and many of their attack submarines were similarly powered. Some of the submarines were powered by dual reactors. One aircraft carrier and at least one Kirov-class cruiser were reactor-powered.

The Delta- and Typhoon-class submarines had 50,000 and 100,000 shaft-horsepower nuclear-propulsion units, respectively; the reactor capacities would therefore have been in the range of 185 to 370 megawatts (Mw). Some Soviet submarine reactors were of a high-energy neutron variety, cooled by molten metals rather than water as in American (low-energy-neutron) reactors. It was after the United States deployed Polaris SLBMs that the Soviet Union responded with its new "Yankee-class" generation of ballistic and cruise-missile submarines.

As of 1994, the Russian commercial nuclear fleet consisted of six icebreakers (including one pending) with dual 62.5-Mw reactors, two icebreakers with single 90-Mw reactors, and one container ship with a 79-Mw reactor (power ratings approximate).

Early Warning

After the Soviet Union tested its first nuclear weapon in 1949, the United States realized that its homeland was vulnerable to strategic attack by long-range (piston-powered) bombers. Early warning of attack was needed to activate the relatively primitive air-defense systems of the time, which were assigned to protect priority facilities for producing nuclear weapons. Batteries of radars were deployed for point defense. By 1952 the line of radar stations posted in Canada provided an additional two hours of warning.

By 1957, because additional warning time was needed against Soviet supersonic jet bombers, the DEW (distant early-warning) radar line was completed, stretching along the Aleutians, across far-northern Alaska and Canada, extending across Greenland to end in Iceland. The U.S. Air Force also flew radar-equipped aircraft.

In the course of time, the Navy contributed to early warning with its fleet of picket ships, radar aircraft, and blimps.

At the high point of the Cold War, the United States eventually had assigned more than 100,000 military personnel and 350,000 volunteer ground observers to warning and defense; Canada contributed 80,000 volunteer observers.

The launch of *Sputnik* in 1957, as well as fading of the "bomber gap," brought about a major shift in the strategic situation and mode of defense. *Sputnik* led the way to a capability for satellite reconnaissance that has endured and thrived to this day — tolerated, in part because it is visually inconspicuous.

To track missile launches, the United State placed radars (ground-, air-, and sea-based) outside the periphery of the Soviet Union. Over-the-horizon ionospheric-backscatter radars were installed by both nations. The Soviets were able in the early 1970s to detect ballistic missiles soon after launch at distances of 2400 kilometers.

Canceled Programs

About two dozen U.S. nuclear bomber and missile programs never reached deployment. One of the ill-fated programs, called "Skybolt" was to use airborne B-52 bombers to launch ballistic missiles. The United Kingdom was also interested in the concept, but success was elusive, costs were mounting, and the alternative of submarine-launched ballistic missiles became more attractive.

Undertaking many development programs in an effort to catch up and build a massive deterrent force, the Soviets too suffered program casualties. They sometimes willingly risked failure if eventual success would mean a high payoff.

In the West there was a lot of reliance on "worst-case analysis" — unfounded conjecture — that led to expensive aircraft nuclear-propulsion programs, all terminated without consummation. Serious U.S. efforts to achieve airborne nuclear propulsion were stimulated as early as 1947 by speculation about Soviet nuclear-powered aircraft — an example of nuclear shadowboxing with an imaginary foe.

Another canceled program was Project PLUTO, a remote-controlled ramjet aircraft that could crash its lethal cargo consisting of its own reactor and a nuclear weapon into a target. Alternately, the remotely piloted aircraft could release atomic bombs as it flew around at supersonic speeds. After reaching its final target, the material from the reactor core that vaporized in the nuclear explosion would have dispersed downwind to become radioactive fallout. In any event, there was risk that the unpiloted craft would crash onto United States or allied territory with its highly radioactive reactor and one or more nuclear warheads.

And then there was Project Orion, intended to propel a spacecraft by detonating atomic bombs behind it. An unsolved safety question was how to launch such a vehicle reliably.

The Aircraft Nuclear Propulsion project was directed toward a high-performance airplane with objectives of dropping atomic bombs and performing aerial reconnaissance. The $1-billion project was described as "one of the silliest airborne ideas ever ... every engine start could be an environmental disaster."

Another ill-fated program was ASTOR, a submarine-launched nuclear torpedo guided by wire to its target, an enemy submarine. But the underwater detonation would also sacrifice the U.S. vessel — eventually judged too high a cost to pay.

No doubt the Soviet Union spent time and resources on unconsummated programs too; in fact, their major rocket and weapon programs were usually hedged by having two parallel avenues of development.

Intelligence Collection and Satellites

A perception of national security, even with mutual assured destruction, requires confidence that decisive military surprises are improbable. Helpful in this regard are the lengthy time and extensive resources an adversary would require to develop a strategic weapon system, thereby giving opportunity for intelligence processes to discover the program and monitor its progress.

Military and civilian data collection were prominent aspects of the Cold War. In the United States, both the CIA and the Defense Intelligence Agency were well funded, and the Soviets had comparable KGB and military (GRU) espionage operations.

Information gathering by humans (HUMINT) has always been a foundation of intelligence collection. Also growing in importance were technology-assisted means, including aerial surveillance, signals intelligence (SIGINT), and eventually satellite reconnaissance.

> The decision to base national security on nuclear weapons opened a Pandora's box of complex and expensive operational burdens. To deter Soviet attack by credibly threatening to inflict apocalyptic destruction, the United States needed spy systems capable of gauging the strength of enemy nuclear forces, warning systems capable of detecting an imminent enemy attack, command systems capable of protecting the president and other key nuclear decisionmakers during wartime, and communication networks capable of transmitting the "go-code" to the dispersed nuclear forces.

One of the earliest American electronic intelligence (ELINT) surveillance programs required aircraft to fly along the Soviet periphery, or even trespass across borders, in order to elicit and detect radar signals. Started in the 1950s, the program had always been risky and provocative. One ELINT plane, a C-130 Hercules aircraft that violated Soviet borders, was shot down over Armenia in 1958. President Eisenhower denied, in a public statement, that the American plane had deliberately crossed Soviet territory in order to monitor their air defenses by triggering radars and scrambling fighters.

An April 2001 incident, involving an ELINT plane flying close to territorial waters claimed by China, resulted in diplomatic difficulties.

British and American patrols over the no-fly zones of Iraq under UN mandate were part of a military exercise to trigger radar beacons.

U-2 Flights. Starting in July 1956, aerial surveillance inside the Soviet Union was carried out through overflights by the specially designed high-altitude reconnaissance aircraft. The U-2 was developed partly to supplant low-altitude aircraft that could, and sometimes did, trigger dangerous responses during overflight of national territories.

The U-2 had an operational altitude of more than 75,000 feet (22,860 meters), a range of 3700 miles, and a top speed of 400 knots (nautical miles per hour). Although its flight speed was relatively slow, its long range gave it an important role in surveillance flights, and its high altitude capability reduced susceptibility to surface-to-air missiles (SAMs) and to interceptor aircraft.

> [U-2 Memoir]. It was an air-defense detachment defending the highly secret Soviet nuclear weapons development site Chelyabinsk-70 that shot down the U-2. It happened during a May-Day parade about 10 am local time, when many people were on the streets in Chelyabinsk: The bright flash in the clear sky was first thought to be part of the day's demonstration. Although the flash was widely noted, it wasn't until the next day that people were informed that a U.S. spy plane had been brought down (and that the pilot had been captured) — a great victory for Soviet air defense.
>
> [Powers had been assigned on 1 May 1960 to carry out photo-reconnaissance of missile-launch facilities near Chelyabinsk. Unbeknownst to U.S. intelligence, a new longer-range SAM-2 battery had been deployed to protect the high-priority Soviet facilities near Sverdlovsk and Chelyabinsk.]

Its 750-pound payload was devoted exclusively to photo reconnaissance, using a special camera and high-resolution 9-inch-wide film that could resolve 2.5 feet over a ground-swath beneath the plane almost 100 miles wide. Because of its low radar visibility, the U-2 "was initially expected ... to go undetected by Soviet air defenses.... But, in fact, the U-2 was detected immediately...."

Intense Soviet protest caused President Eisenhower to order a hiatus in the flights. But to help resolve the increasingly sensitive matter of Soviet ICBM development, he re-authorized reconnaissance missions. On May Day 1960 they had to be temporarily suspended again when the U-2 piloted by Francis Gary Powers was brought down over the nuclear weapons laboratory Chelyabinsk-70 (now Snezhinsk). At the time there was a raging public (and private) debate over the extent of the "missile gap." Actually, the U-2 missions had failed to detect operational ICBM launch sites that were in place in the southern Soviet Union.

The U-2 overflights furnished a strong incentive for the Soviet Union to intensify its ballistic-missile and aircraft defense programs. Until Power's aircraft was downed, the Soviet military was frustrated by its inability to intercept an aircraft that was illegally intruding on their airspace.

The high-flying aircraft were also sold or transferred to other nations. U-2 surveillance of China began in 1960, continuing for a decade. Four U-2s belonging or assigned to the Chinese Nationalist government were reportedly shot down over Communist China during the 60s.

It was the American U-2 coverage of Cuba in mid-October 1962 that first revealed Soviet nuclear missiles being installed there.

Satellite Reconnaissance. Stimulated partly by limitations on aerial overflights, technical progress with rockets that could put payloads into earth-orbit made satellites the mainstay of technical intelligence. At first, high-altitude photo-reconnaissance led the way, but eventually more sophisticated sensors were deployed in the satellites, using both visible and invisible spectral regions, along with active radar scanning.

The earliest practical satellite reconnaissance program was named CORONA. A panoramic camera with 70-mm film eventually gave a ground resolution reaching six feet. To recover the film, the reentry vehicle was ejected high over a neutral zone; it deployed a parachute in the atmosphere and was recovered in mid air by a C-119 aircraft.

The first successful CORONA mission, launched on August 18, 1960, returned imagery covering more than 1.6 million square miles on 20 pounds of film. CORONA film-return reconnaissance satellites were perfected by the CIA; 95 of them were successfully orbited between 1960 and 1972.

Before information from CORONA was available, U.S. National Intelligence Estimates predicted that the Soviets would have 140 to 200 ICBMs deployed before 1961. After film recovery from CORONA, the projection was reduced to 10 to 25 ICBM. The highly classified CORONA program was a major success for technical intelligence collection.

During the late 1970s and early 1980s, U.S. photographic satellites followed a fairly standard pattern. Two KH-11s circled the earth at all times, transmitting pictures to ground stations during every orbit. U.S. photographic-imaging intelligence satellites operated film and electronic cameras to produce high-resolution images of objects on the ground at ranges of up to 622 miles (1000 kilometers) with resolutions better than 4 inches (10 centimeters).

Soviet development of space-based early warning systems started in 1973 with the Cosmos-series of satellites on two tiers: highly elliptical orbits (*Molniya* launch vehicles) and geostationary satellites.

Low-earth-orbit (LEO) satellites, at altitudes ranging from 150 to 3000 km, have been used for military reconnaissance and arms control verification. They make an orbit around the earth about every hour-and-a-half; a typical U.S. LEO satellite made 11 passes per day over the former Soviet Union, dwelling about an hour there altogether.

Mid-earth-orbit (MEO) satellites are at an intermediate altitude (3000 to 30,000 km), and are used for navigation and missile guidance. The U.S. Global Positioning System (GPS) consists of dozens of satellites, each of which circles the earth every 12 hours. The Soviet Union began deploying its comparable Glonass system in 1982, but it was not completed until 1996.

Highly Elliptical Orbit (HEO) satellites have orbits with large eccentricities (stretched ellipses). *Molnyia* orbits are a common example. HEO satellites typically dip below 3000 km and rise above 30,000 km during an orbit.

Geosynchronous (GEO) satellites are orbited at around 36,000 km altitude, which keeps them above a fixed point on the earth's equator. They warn of missile launches from land or sea.

Since the Cold War, more and more and more have nonmilitary functions. In particular, civilian reconnaissance spacecraft have recorded in unmatched detail the recent environmental history of our planet — such as changes in the earth's surface, shrinkage of freshwater lakes, pollution in rivers and coastlines, deforestation, expansion of deserts, and evolution of volcanoes.

Signals intelligence (SIGINT) satellites are designed to detect transmission from broadcast communications such as radios, radars, and other electronics. Besides the satellites, telemetric monitoring of radio signals is useful for determining missile flight data and electronic activity that might be associated with military preparations. Nations have used ships and other detector platforms to determine missile test parameters, such as distances traveled and impact patterns of multiple warheads.

In addition to overhead surveillance of the Earth's surface, operations were carried out underwater in the oceans.

> Since 1965 the [U.S.] Navy has operated a submarine-based deep-ocean reconnaissance and retrieval program.... Among the missions these submarines have accomplished are the retrieval of Soviet nuclear warheads and tapping into Soviet underwater communications cables.

In June 1974 the *Hughes Glomar Explorer* (a ship specially built for the mission) sailed to the site of a Soviet Golf-2 submarine sinking. Two nuclear torpedo warheads were reportedly salvaged. The loss of that attack submarine has been attributed to a collision during maneuvers in a naval war game.

The CIA tried to determine deployments and budgets of Soviet forces. The estimates proved to be a mixed bag, rated as "highly accurate" for current Soviet strategic deployments, "inconsistently prescient" for future deployments (including frequent under- and over-estimation), and "quite inconclusive" for Soviet deployment of tactical and theater nuclear arsenals. Estimates were often biased as a result of domestic political influences and the insular nature of the intelligence community.

According to one CIA insider, U.S. technical intelligence capabilities, particularly observation satellites whose photographs reveal the location of deployed systems, increased confidence in estimates of foreign strategic weapons, past and current. Conversely, strategic stability was jeopardized when anti-satellite development was undertaken — first by the Soviets starting in 1968, and later by the Americans to destroy surveillance satellites in orbit.

Although surveillance and espionage enabled the United States to verify adherence to strategic arms control agreements, they failed to produce accurate and timely intelligence on many other important nuclear issues, such as the size of the Soviet sub-strategic arsenal and the increases in alert level of Soviet strategic forces during crises.

Command, Control, and Communication

Three integrated military structures — command, control, and communication — combining explicit procedural rules with technological restraints, have evolved for managing and using nuclear weapons.

Elaborate resources were (and are) devoted to ensuring that nuclear weapons could be detonated only upon orders of the respective supreme governmental authority. That authority devolves down a designated chain of *command*, which — through multiple, redundant channels of *communication* — exercises *control* over their launch and detonation. A fourth essential element of this regime is definitive information (usually *intelligence*), often gathered by national technical means and by human intelligence. This nuclear-management assemblage was abbreviated variously as C&C, C, or CI, depending on how many elements were being considered.

Decapitation is the term applied to hypothetical destruction by first strike of a nation's infrastructure for commanding and controlling strategic forces. Such an attack would at best be of questionable benefit to the attacker, because there would then be no coherent authority left to negotiate with for ending the retaliation and conflict. Nevertheless, the Soviet government (and many U.S. citizens) often feared that decapitation was an underlying American government strategy.

C^3I (Command, Control, Communication, and Intelligence). The infrastructure commonly abbreviated C^3I — pronounced "C-cubed *I*" — is the organizational and technical network that provides coherent direction and targeting for strategic forces.

Command refers to the military chain of command, which, in the case of authority to release nuclear weapons, extends from the head of state to the weapons officer.

Control is the means, either procedural (two officers with different keys) or mechanical/electronic (such as a PAL) for ensuring that the weapons are not armed without proper command.

Communication consists of secure, encrypted methods, usually redundant, to maintain links between headquarters and the field.

Intelligence information linked to the process is needed for evaluating apparent threats that might lead to defense alerts — real, practice, or false.

Conversely, the U.S. C structure was in principle vulnerable, at least through the early 1980s, to a decapitating strike. The strategic command system probably could not have survived the kind of deliberate attack that was then within the capability of the Soviet Union: Fewer than 100 delivered nuclear warheads could have disrupted the U.S. military chain of command, rendering it incapable of a coherent response. Even so, very damaging non-coherent retaliation could have taken place — under responses practiced by trained bomber crews, submarine officers, and ICBM launch personnel.

While both superpowers were vulnerable at various times to a decapitating strike, an inescapable and devastating retaliation capability always existed on both sides. With all this embedded reflexive response, sensible deterrence of a rational attack was never in doubt. Described below are some of the major features and weaknesses of the respective Soviet and American C structures and cultures.

Command. At the apex of nuclear decision-making were the highest appointed or elected officials: for the USSR, the General Secretary of Central Committee; and for the United States, the President. A chain of succession was established to cope with the absence or incapacity of the top official. Given adequate time, the Soviet Defense Council and the U.S. National Security Council would be involved in the decision-making process. If the entire civilian chain of succession were wiped out, responsibility for nuclear decision in each nation would pass down the military pyramid, starting with the armed-services commanding officer.

In order to retaliate reliably in case of a surprise attack, emergency command posts were set up at remote and/or secure locations. To maintain continuity of government under the threat or actuality of nuclear war, large underground shelter complexes were dug to protect the chain of authority. The center of the U.S. C&C system was the National Military Command Authority in Washington, D.C., with secondary centers at hardened sites like the NORAD Cheyenne Mountain Complex, in Colorado Springs, Colorado, and aboard specially equipped aircraft. NORAD's primary mission has been, "to detect and assess a Soviet nuclear attack, notify senior military commanders, and coordinate the launching of retaliatory strikes before the first Soviet warheads [detonate]." (Note that this was explicit implementation of a "launch-under-attack" policy).

Adding to a state of continuous American military preparedness was the National Airborne Operations Center, which consisted of aircraft that either were always in the air or could be sent aloft on short notice. An Alternate National Military Command Center was installed under Raven Rock Mountain in Pennsylvania.

For civilian government workers in the executive branch, an unpublicized continuity-of-government facility was established at Mount Weather, Berryville, Virginia. Other sites included Mount Pony, in Culpeper, Virginia, which was operated by the Federal Reserve Board as the central computer node for the transfer of all American electronic funds. The Congress could evacuate to prepared underground complexes at the Greenbrier resort in West Virginia.

Because an ICBM requires only a half-hour to go from one continent to another, there is very little time for human decisions on whether to retaliate. The Cheyenne Mountain command center, after receiving information from their monitoring instruments, would have three or four minutes to make a reality assessment. An emergency teleconference would take place with the duty officer at the White House, and perhaps 12 minutes would be available to contact and obtain a decision from the President, who would have to access and use the "Football" for ordering a retaliatory strike. As we mention later, nowadays even less time is available.

The Soviet Union displayed great sensitivity to the vulnerability of command structure. Because their management of military forces was more highly centralized than that of the United States, a decapitating strike against the USSR would have had even greater consequences. Soviet strategic forces were kept under tight operational restrictions, with significant numbers of bombers and submarines retained at base.

> **[The Football].** A constant companion to the U.S. President is a military aide who carries a briefcase nicknamed the "Football." Contrary to popular belief, the Football does not contain the famous "go-code" — the codes needed by the launch crews to unlock and fire their ground, sea, and airborne weapons. The actual authorizing and enabling codes remain in military custody.
>
> The Football contains information the President would need in deciding whether to launch nuclear weapons. The president personally carries a card with special codes for positively personal identification to the key commanders who might be ordered to initiate a nuclear attack. The military, although constrained by regulations and procedures, does not really need these codes to launch a strike.
>
> A comparable Soviet briefcase system was designated the *Cheget*.

To reduce the risk of an "erroneous decision," the supreme Soviet military commander could not issue an order to launch retaliatory forces unless their early-warning system gave a "missile-attack" signal. The missile-attack finding would have to follow a succession of ominous events, starting with ballistic-missile launches being detected by the space-based early-warning system and progressing to target tracking by the radar surveillance.

The Soviets deployed a network of reserve command centers, some airborne and some rail-based, which could deliver retaliatory strikes, if necessary. Members of the top leadership could proceed to an underground command center near Moscow. For the general staff, a huge, superhardened, underground command center was reportedly been under construction for two or three decades at Yamantau Mountain near Beloretsk in the Urals.

Inasmuch as American and Soviet strategic forces shared a common vulnerability, the actual balance of power (or terror) was precarious and changing. It was crucial to avoid a complete, simultaneous alert of strategic forces during political crises (such as the Cuban missile scare). Until a drawdown of nuclear forces and until a more stable, distributed-force posture could be attained with reliable C^3I, it was necessary to avoid letting secondary issues rise to the level of political crisis.

An assessment presented to the National Security Council in February 1950 found that sixteen properly targeted nuclear weapons could "most seriously disrupt" U.S. government operations. (The same conclusion emerged from authoritative, detailed reassessments conducted in the early 1960s, late 1960s, late 1970s, and mid-1980s.) "Acutely vulnerable to sudden destruction by a few tens of Soviet weapons" were the top political leaders, the senior military commanders, their fixed and mobile command posts, and their communication links.

The historical and budgetary significance of this persistent finding of vulnerability is hard to overstate. "It pressured the United States to emphasize heavily the ability to strike first or launch on warning: U.S. commanders needed to be able to authorize and implement the activation of U.S. forces *immediately* after detecting the imminent or actual launch of Soviet missiles. That is why launch authority was delegated to so-called "prepositioned national command authorities," namely, senior military officers in the nuclear chain of command. However, this facilitation of hair-trigger reaction "increased the risk of launching on false warning or of improper exercise of launch authority."

Launch-on-warning as a strategy was only as good as the timeliness and reliability of the notification systems. Three ballistic-missile early-warning system (BMEWS) radars (in Alaska, Greenland, and England) covered all attack corridors for Soviet ICBMs (not SLBMs), but provided only about fifteen minutes' notice. Under the best of circumstances, the sequence of radar detection, presidential notification, decision making, code authorization, and weapon firing would have taken longer than fifteen minutes. With the advent of early-warning satellites, the United States gained an additional ten or fifteen minutes for verifying that a Soviet ICBM attack was on the way.

After investing much more than the United States in protecting its command and control network, the Soviet Union also eventually gravitated to a quick-launch policy. In the 1980s it established and tested a backup system known as "dead hand," which would allow the quasi-automatic launch of thousands of nuclear weapons in the event senior commanders were killed or otherwise neutralized before they could issue the launch commands. The dead-hand system is active today.

Nuclear practice has departed in fundamental ways from the doctrines that were supposed to guide it. Moscow's (Soviet and now Russian) command and control system has been even more vulnerable than Washington's. This led to the conclusion that "retaliation after rideout was an abstract idea in the theory of stability but not a viable option in the real world." (*Rideout* is the concept that one side would absorb some destruction before determining the means and extent of retaliation.)

As accuracy of ICBMs increases, potential decapitation strikes create a menace for command and control facilities. The launch-on-warning posture forced on both superpowers is considered to be a "half-way house" between "pre-emption" and "rideout." While launch-on-warning bypassed some uncertainty in command and control, it also drastically compressed (to a few minutes, or largely to a computer evaluation) the time available for deciding whether to retaliate. The better strategy would be a policy rejected by both sides: to rely on invulnerable, de-alerted forces — submarines being the strongest candidates.

Control. The following three steps are undertaken in launching a missile:

(1) Target coordinates are punched into the missile computer or console after receipt of a validated launch-preparation message (emanating from the President);

(2) Another enabling message must be received, either to remotely release the PAL or — for the crew using two keys — to unlock the controls of a warhead/missile; and

(3) The actual launch command would be the final message from designated authority.

As many as 2000 U.S. land-based missiles could be fired in a minute or two. Submarine-launched missiles would require five to ten minutes to be launched because the emergency-action message to the underwater craft takes longer to be transmitted and validated. The United States keeps two SSBNs on alert in each ocean, so about 500 warheads could be quickly sent aloft.

The Soviets are said to have maintained control of nuclear firing through the use of specific authorization and release codes, coupled with carefully vetted authentication requirements. Not much has been published about their mechanical or electronic controls over access to or release of special weapons.

The United States and the Soviet Union (and Russia, soon after its creation) could each have launched within ten minutes about 5000 warheads, a 10-fold hedge ensuring obliteration of 250 major targets in each nation.

Because nuclear weapons are so frightful, not only must their use be securely managed, but their physical possession must be carefully tracked. The chain of custody for each U.S. warhead is said to be strictly maintained from the moment a DOE production facility turns it over to DOD. Besides the paper trail, each weapon is assigned a unique serial number, engraved or etched on the weapon; the numbers are used to track changes in custody, configuration, and location. Only a portion of the stockpile is actually examined and directly inventoried each year. Some weapons on alert or in inaccessible containers do not have frequent physical inspections.

Communications. Channels for communicating military information and authenticating firing orders are varied and secure. To coordinate forces in time of attack, special networks have been established. The networks use the entire electromagnetic bandwidth for transmission and reception, and make use of land-, sea-, and space-based messaging media.

The Soviet Union had in place what was called the *Kazbek* communications system, with terminals in the offices of the highest military officials. This two-way link would allow receipt of a "missile-attack" signal and could transmit orders from the supreme command. Military officers on duty had access to *Krokus* terminals that could "display information about the scale of the possible attack and the projected impact area of the warheads."

A unique method of backup communication was the Soviets' *Perimetr* system, which was integrated into the battle-management system. Command rockets could be launched that would act while in flight (for as long as 20-50 minutes) as temporary communication-relay stations, transmitting retaliatory missile-launch orders and authorization codes. Although the *Perimetr* system supposedly could be activated in a mode that would require a positive signal to be annulled, it reportedly was never deployed in a configuration that would allow automatic nuclear launch.

The U.S. NAVSTAR Global Positioning System (GPS) of two dozen satellites became available for modern communications and for determining locations, including corrections for missile-guidance systems.

Submarines can communicate by either low-frequency or extremely low-frequency (ELF) radio waves. ELF is very-long-wavelength electromagnetic radiation, which can penetrate well below the surface of the ocean.

Very Low Frequency (VLF) systems were an earlier means of long-distance communication, but a submarine's receiver can only be a few feet under water. Extremely Low Frequency (ELF), instead, can transmit to a depth of about 330 feet, but ELF requires land-based transmitting antennas that are many miles long and has to be above a layer of low-conductivity rock. ELF has a very low data rate, but it can transmit terse coded emergency-action messages. It is more secure for submarines that need to remain well submerged.

"The best U.S. geology for [ELF] transmitters lay in northern Wisconsin and Michigan's Upper Peninsula," but one government official said that "the worst, most protracted, and most nonsensical environmental fight" by environmentalists and scientists opposed "funding and constructing these ELF antennae."

EMP from high-altitude nuclear detonations and from nuclear-armed antisatellite (ASAT) weapons is a potential threat to unhardened communication systems. Ground and satellite communications are highly dependent on electrical and electronic equipment that could be affected by EMP unless appropriate shielding is installed or other protective measures are practiced. All U.S. military systems are reputed to be protected against EMP

"Hot Lines," which are direct telephone links between the superpowers, permit exchange of information during critical situations. They have been in place since the Cuban missile crisis, under special agreements between the parties. A hot line would be particularly valuable if a nuclear explosion were to take place for a reason either related or unrelated to a contentious political or third-party situation.

****

The technology available or developed for support of deterrence ranged from the more specialized active design and fabrication of weapons, to basic developments that passively provided the strategic underpinnings. The production complexes of the nuclear weapons states were able to turn out in quantity many types of devices and delivery systems to guarantee nuclear incineration.

Open-air and underground test facilities bore the burden of nuclear detonations during the Cold War.

The controlled release of atomic energy was harnessed for propulsion and power of military systems, mainly surface and subsurface naval vessels.

A strategic triad of ICBMs, SLBMs, and bombers assured that retaliation after a nuclear first-strike would be assured—that the attackee would be able to avenge and destroy the attacker. Of the three legs of the triad, land-based missiles were the most vulnerable, but they had the accuracy and speed needed for a counterforce response; therefore, warning as early in the attack as possible was seen as crucial for a damage-limiting counterblow. Of course, sliding that far along the slippery slope to national suicide would have signaled failure of all aspects of deterrence, intelligence collection, diplomacy, and common sense.

To monitor potential adversaries, all available means were used to gather information from outer space, the atmosphere, the oceans, and underground (seismic waves). Earth-orbiting satellites were especially useful.

The flood of data was supposed to be integrated into a C-cubed framework that provided control of, and response with, nuclear weapons. But much of the short travel time of ballistic missiles — 30 minutes or less — would be consumed by detection, recognition, and reaction. This left only a few minutes for making momentous decisions and little time for government officials — much less civilians — to reach shelters.

When they emerged, the survivors might find themselves immersed in a radioactive darkness and a couple of thousand major cities and military installations obliterated.

D. ARMS DEVELOPMENTS AND LIMITATIONS

Creating an uncertain and frightening Cold War aura — not only the quantity of nuclear weapons, the multiplicity of delivery systems, and the expanding support infrastructure — also were the constant pressure of technical advances and the ongoing search for immunity from such weapons. All this war-focused activity made the situation look even more dangerous and unstable.

In this Section we trace technological upgrades in offensive systems — the variety, delivery and lethality of weapons — which were accompanied by attempts to implement passive civil defense and to devise active defenses against strategic bombers and missiles.

Observing the unbridled growth of armaments, the uncommitted public feared mutual annihilation, and many citizens mounted campaigns to achieve limitations through arms control negotiations.

Advances in Delivery Systems

Improvements in design, principally the development and deployment of large numbers of MIRVs — multi-warhead weapons with sufficient accuracy and yield to destroy missile silos — gave the United States a transitory military advantage. At the same time, the improvements jeopardized Soviet retaliatory capability, even though that was essential for strategic nuclear stability. Another destabilizing trend centered on possession of retaliatory weapons that could inflict unacceptable damage in response to a first strike or pre-emptive attack. The evolution and deployment of weapons with sufficient accuracy to destroy specific missile silos was a destabilizing counterforce development, as was the Stealth bomber, with its detection-evading profile.

Other technological advances bearing on the strategic balance were attempts to acquire anti-ballistic missile (ABM) systems, fractional-orbit bombardment systems (FOBS — deployed by the Soviet Union from 1969 to 1983), multiple re-entry vehicles (MRVs — precursor to MIRVs), longer-range SLBMs, armed satellites (not yet developed or deployed, to our knowledge), long-range standoff missiles from bombers, and terrain-following guidance for aircraft and cruise missiles. Often their proponents did not consider the destabilizing nature of these technologies, and — if they did consider it — welcomed the apparent, though fleeting, advantage to be gained.

Stealth. Radar is the predominant means of detecting and tracking aircraft, whose radar *profile* (signature) can be minimized by special design. The B-2A "flying wing," informally known as the *Stealth* bomber, is coated with composite materials to reduce greatly the reflected radar signal. Designed and tested initially under very secret conditions, this aircraft was the most expensive ever built ($55 billion to develop, in 1996 dollars). The Stealth bomber saw action with conventional HE weapons in Iraq during the 1989 Gulf War.

Technology Creep. In 1978, while the SALT treaty was being considered and debated, the issue of gradual, unavoidable improvements in underlying technology was coming to the forefront. The incessant creep of technology was jeopardizing stability.

Advances in MIRV technology and bomber stealth are examples we have already cited. With enough accurately delivered warheads, large numbers of hardened silos were virtually certain to be destroyed in a surprise attack. Too late to be helpful before MIRVs were developed and deployed, Henry Kissinger has admitted that

> he wished he had given more thought to the "[counterproductive] implications of a MIRVed world" at the time.
>
> Cruise missiles are another type of weapon system that gradually became of strategic importance, with contentious arms control drawbacks because of verification difficulties.

Improvements in accuracy came out of both civilian and military R&D in many fields, including electronics, materials, gravitometry, and geodesy. Computer microminiaturization added vast progresses by making guidance instrumentation more compact and by gaining reliability through redundancy. These advances were the key to MIRVs. The inertial navigation systems for missiles underwent refinements and design revisions that improved guidance and control.

Physical measurements and computer modeling of gravitational fields, coupled with satellite global positioning systems* and in-flight astronomical observations, considerably improved specification of location, time, and velocity. Infrared, radar, terrain-matching, laser, and electro-optical sensors have diverse military applications. Improved materials allowed construction of lightweight turbofan engines, propellant casings, and reentry heat shields.

Qualitative increments in weaponry ranked high on a list of factors that stimulated action and reaction in the cyclic arms race.

Civil Defense

A natural desire for defense manifested itself in two contrasting and complementary ways: passive and active. The passive approach was through after-the-fact mitigation by measures such as fallout shelters.

Sheltering, evacuating, and managing civilians in the event of nuclear war come under the rubric of "civil defense." In the 1950s state governors, such as Nelson Rockefeller of New York, spoke out in favor of bomb shelter construction.

But in the 1960s, Civil Defense became hotly debated before it became a back-burner issue until resurfacing in the 1970s. One of the reasons civil defense became controversial was that, in a age of nuclear deterrence, it was viewed by a potential adversary (and by some of its advocates) as more than a purely defensive activity: It had the appearance of preparing the population and the leadership logistically and psychologically so that forceful diplomacy, *i.e.*, "nuclear brinkmanship," would be tolerated by a populace who felt protected if war were to break out.

Information provided by Office of Civil Defense
DEPARTMENT OF DEFENSE
Washington 25, D. C.

During the early days of the Cold War, air-raid sirens would signal "drop drills" in the schools, where students were to fall to their knees under their desks. The drill played a psychological role, making people feel that they could do something in case of a nuclear attack. Stocking up on bottled water and emergency supplies might be useful for natural disasters, but would have little more than public-relations value in a nuclear war. The ritualized paramilitary practices — air-raid sirens, orderly evacuations, diving under desks, voluntary duties — helped avoid public challenge to government policies that drifted into larger-scale risk.

In military jargon, "civil defense" meant all measures to protect the civilian population; during World War II, it had consisted of air-raid shelters, voluntary observers, and post-attack cleanup. Civil defense became synonymous with the

---

*GPS has made tremendous strides since the Cold War, and now provides universal opportunities for even better navigation and targeting. This makes the satellites a more tempting target in a pre-emption scenario.

construction of fallout shelters. Driven partly by fears of a missile gap favoring the Soviet Union, mass evacuation and fallout shelters became the goal of major private-interest and government campaigns in the United States.

Although reducing the effects of fallout at a distance from ground zero was technically feasible, protection of anyone near the epicenter of a nuclear explosion was practically hopeless, in view of the intense blast, heat, and prompt radiation. But the limited value of urban shelters did not discourage their proponents, who claimed that they had worth it if missile fields were the only targets.

Fallout shelters and evacuation schemes were central to plans for population survival. Stockpiles of emergency food, medical supplies, strategic materials, and petroleum are central in preparing for shortages due to war or other disaster.

The goal of the civil-defense program was to limit extended damage from atomic explosions, particularly to civilians — or, as already mentioned, to permit a more assertive foreign policy by convincing the general public that they could be protected during nuclear war. Even if 50 million urbanites were killed, survival of the ruralites could be construed as a triumph.

After the Korean War started, on the heels of the first Soviet atomic test, President Truman established in 1951 the Federal Civil Defense Administration. The Interstate Highway Act of 1956, which created the modern U.S. interstate road system, was justified and promoted partly on the basis of its ability to allow swift exodus from cities in case of attack.

Because the Russian land mass has had a traumatic history of being invaded by armies from the west, civil-defense measures were routinely accepted in the Soviet Union. Defenses against nuclear-armed aircraft and missiles were in line with a long-standing tradition of protecting populations and facilities against foreign attack.

However, the effectiveness of fallout shelters during a nuclear war became the subject of worldwide dispute, especially from scientists.

The 1960s. In part because of the Cuban missile crisis, there was a large burst of civil-defense activity in the early 1960s. Practically overnight, a home-shelter business exploded. As memory of the Caribbean crisis subsided, interest waned and existing government programs became moribund.

Realistic protective measures to cope with nuclear blasts are limited to "distance" and "shielding." The first civil-defense principle — distance — led to procedures being drawn up for mass evacuation from cities. The second principle — shielding, a technically complex issue — led to underground shelters and other measures. Radioactive fallout and the need for decontamination were recognized post-attack predicaments, but these problems were too perplexing and complex for people to understand and prepare for realistically.

Existing buildings and basements were the first to be adapted for civil defense, with supplies stockpiled, as would be done in the case of preparing shelters against natural disasters.

The 1970s. In the 1970s the United States began planning to improve the accuracy of long-range ballistic missiles. To make credible a new strategy of planning for limited nuclear (counterforce) strikes against the opponent's offensive capability, a program of civil defense was psychologically important.

The government asserted that the war could be limited to military targets, but many citizens remained unconvinced, sensing an unquantifiable and unacceptable risk that there could be an all-out exchange. A presidential study in 1977 concluded that "140 million Americans would be killed in a nuclear war, leaving 80 million survivors." Secretary of Defense James R. Schlesinger hoped that reducing nationwide civilian fatalities to under one million would be enough to keep the public from challenging a policy that risked limited nuclear war. To implement that policy, the Carter administration rejuvenated the civil-defense program, and a new organization was put in charge, the Federal Emergency Management Agency (FEMA).

Evacuation and shelters remained the primary features of civil defense. While evacuation from large cities could not be accomplished in the half hour it takes a missile to reach a target, population exodus could be initiated when a crisis situation came up. Often the U.S. government's arguments were based on its perceptions of routine Soviet preparation for civil defense, including early evacuation and population sheltering. However, émigrés to the West denied that the Soviet Union had any widespread civil-defense rehearsals. The Chinese (who feared a nuclear attack from both the United States and Soviet Union) were said to have converted tunnels into fallout shelters.

Emergency Management. Whether the attacks were on military or civilian targets, the surviving inhabitants and industrial centers would suffer damage from fire, blast, or radioactive fallout. Measures to cope with floods, earthquakes, and tornadoes do save lives, and many of these preparations would be applicable to at least a small-scale disaster, including an accidental nuclear explosion.

Evacuation, medical care, and emergency supplies would be needed under conditions that would strain local authorities and services; nevertheless, prearranged sites for treatment, training of responders, and stockpiles of provisions to support these services are merited, especially for natural calamities.

Trained emergency management, such as that offered by FEMA or by the corresponding Soviet ministry for emergency affairs and by local agencies, was intended to organize government response to a nuclear attack, as well as to deal with natural disasters.

Under a preparatory civil-defense strategy, investment could be made in special preventive measures or redundancy for heavy industry, oil refineries, communication systems, nuclear-power reactors, and the electrical-power grid.

Fallout Shelters. Shelters could be helpful in minimizing the accumulated radiation dose from post-attack fallout far from the epicenter of an atomic blast. People would have to seek protection with relatively little warning. To be effective when close to major cities, the shelters would need to be somewhat blast resistant, accessible in minimum time, provisioned for survival, and not too costly. Studies in the 1970s estimated that an "effective" network of shelters could be built in the United States at a cost of $23-$35 billion and would reduce casualties by a factor of twenty.

If a nuclear attack on cities were configured for maximum immediate damage (*i.e.*, by blast and fire), the explosions would be at a high enough altitude that there would be little or no local radiation fallout. In that case, fallout protection is irrelevant — although shelter against blast and fire could help if far enough from ground zero. However, a counterforce attack (against missile silos) would necessarily involve ground bursts, and there would be massive fallout for hundreds of miles downwind.

A thermonuclear weapon exploded over a metropolitan area is so large that the fireball might make contact with the earth, especially if it explodes at a lower altitude than intended. That would cause considerable fallout nearby and extend a radiation pattern called a "footprint" downwind from the epicenter.

There was no significant radiation fallout from the bombs used on Japan, so the public has had little experience in dealing with it. The most relevant episode was the downwind radioactive fallout from the military test Operation Crossroads in 1946, which seriously irradiated the occupants of a Japanese fishing boat. Less relevant is the Chernobyl reactor accident of 1986, which deposited light radioactive dust over a wide area, with no identifiable health effects except a disputed increase in nearby thyroid cancers.

Blast-rated shelters protect against the direct effects of flying objects, heat, fire, and ionizing radiation. One foot of earth attenuates radiation by an order of magnitude.

People might have to stay in shelters for a long time — days or weeks — to avoid secondary contamination from the outside environment.

Those opposed to the nuclear-shelter illusion challenged both the underlying assumptions and the effectiveness of civil defense in the 1960s and 1970s. For example, as much as a month after a limited nuclear attack about a third of the population would still have to remain in their fallout shelters to avoid overdose.

Survivors would be highly dependent on weather, protection factors, and parameters of the attack; medical care would be needed. Casualty levels would probably have been an order of magnitude larger than projected by the government at the time. The problems of emergency planning, preparation, rehearsals, provisioning, training, policing, firefighting, communicating, cleaning up, and other societal tasks were likely to be much greater than assumed by shelter proponents. Massive and timely evacuation and relocation of the population would have been fraught with the complications of widespread chaos.

Other arguments raised decades ago against an extensive civil defense program included potential harm to strategic stability, waste of resources, and unwillingness to engage in arms control discussions. Overconfidence in the protective value of passive civil defense could have led to unwarranted risks, including misplaced military reliance on a nuclear-war. The mitigative effect of fallout shelters could be offset by attacking with new, faster, better, and more destructive weaponry.

In any event, less than one percent of the American population ever built shelters, partly because of their cost, partly because of their futility, and partly because of foreseeable inequities in their availability. Homeland civil defense was an elusive concept, especially without amelioration of the causes of the underlying insecurity.

The public debate about civil defense and fallout shelters also contributed to recurrent arguments over war and its consequences. Aside from its effects on life in all its forms, a nuclear war would have been extremely harmful to the structures, artifacts, and institutions of civilization. Even those nations outside the arena of conflict would be directly impacted by radioactive fallout and indirectly affected by high mortality, medical burdens, refugees, and the disruption of the combatants' economies.

Strategic Defenses

Faced with the massive destruction that can come from even a single strategic weapon, the superpowers looked for means of active, rather than passive, defense against ballistic-missiles. The problems and the solutions were, and are, mind-boggling: Should the defense be configured to reduce the threat of attack, should it be designed to protect the population against the consequences of an attack, or should it be directed at blocking every possible avenue for delivery?

The 20-century exploitation of aerial bombardment gave rise to appropriate countermeasures, such as radar and anti-aircraft gunnery. After World War II, hostile access from air or space grew to become a major vulnerability. Somewhat downplayed has been the parallel exposure to attack from the sea or from under the sea, especially with nuclear weapons.

Nuclear defense became a goal not long after the end of World War II:

> [Americans realized that] the devastation visited upon Hiroshima and Nagasaki by one B-29 bomber carrying one atomic bomb could soon occur in the United States. Initial defensive efforts focused on the skies, as for more than a decade after World War II large aircraft were the only practical means of delivering a nuclear weapon over long distances. When ballistic missiles were developed in the late 1950s, efforts turned toward countering them. Around the same time, programs were initiated to attack and

destroy satellites in orbit and to sink ships and submarines carrying nuclear weapons. While enormous sums of money were spent on all these efforts, none was able to significantly lessen the threat posed by the Soviet and later, Chinese, nuclear arsenals.

In the early 1950s, U.S. air defense consisted of a radar network for detecting and tracking nuclear-armed bombers, with SAMs to intercept the bombers. The expected long-range strategic bomber threat against the United States never really materialized.

Because of the rapid development of ICBMs, the existing air-defense system was initially upgraded as part of an attempt to devise a defense against missiles: The nuclear-tipped Nike-Hercules bomber-defense system was transformed eventually into ABM systems consisting of Nike-X and Nike-Zeus nuclear-armed interceptors.

Heavy dependence on nuclear-tipped missile interceptors is evident in the mass production of U.S. warheads. Cranked out for the air-defense mission in just a few years, were 10,900 nuclear warheads — representing a third of the 1965 stockpile, almost 16 percent of total U.S. warhead production.

Immediately after World War II, the Soviet Union had found it necessary to devise an air defense against U.S. nuclear-armed bombers; so their shift to ballistic-missile defense (BMD) naturally followed. By the mid-1950s, Soviet development of BMD technologies had started, and a test range was set up near the village of Sary Shagan in 1956. The defense-minded military had its own extensive R&D program for intercepting missiles. In 1961 they succeeded in destroying a rocket in flight. These developments prompted Khrushchev's widely quoted hyperbole that they had developed a capability to "hit a fly in outer space."

As the Soviets made progress in long-range rocketry, the U.S. Army increased its R&D on ballistic-missile and aircraft defenses, starting work in 1955 on the ground-based Nike-Zeus system, which laid the basis for the U.S. BMD program.

There was always pressure for premature deployment. One year after *Sputnik*, the U.S. Army wanted to deploy its Nike-Zeus system, which depended on mechanically steered radars to guide a nuclear-armed interceptor. It's 400-kiloton warhead was to explode in the atmosphere and destroy the incoming missile. The Secretary of Defense put the project on hold, saying "We should not spend hundreds of millions on production of this weapon pending general confirmatory indications that we know what we are doing." The more advanced Nike-X program took its place.

The history of BMD has been nicely summarized:

> For almost as long as there have been modern missiles, there have been programs for missile defense. In 1946 the Pentagon awarded the first antiballistic-missile contracts. Those earliest projects, Wizard and Thumper, are unmemorable but stand at the head of a genealogy of missile-defense efforts that have had remarkable success reproducing themselves over the years. For it might be said that Thumper and Wizard begat Nike II, which begat Nike-Zeus (also Nike-Ajax, Nike-Hercules, and Plato); and Nike-Zeus begat Nike-X, which begat Sprint and Spartan, and they were known as Sentinel; and Sentinel begat Safeguard, which begat LoAD, which begat Sentry; and all of those, along with Defender and BAMBI and numerous forgotten others, begat Ronald Reagan's Strategic Defense Initiative; and SDI was known as Star Wars, and it begat ERIS and HOE, which begat National Missile Defense.

Tallinn. In the early 1960s, Soviet homeland defenses were directed primarily against nuclear-armed bombers, culminating in an alignment of silo-based conventionally armed SA-5 anti-aircraft missiles, called the Tallinn system. Over a half-million individuals were reportedly detailed to air defenses to counter the perceived threat from a proposed U.S. B-70 long-range bomber fleet. The Tallinn air-defense system was obsolete by the time it was installed, because the United States had by then abandoned the B-70 program.

Those 3600 surface-to-air missiles that surrounded Moscow in 1955 were later replaced by more modern missiles, with more deployed in vast quantities throughout the Soviet Union. Interceptor aircraft were procured in large numbers. No expense was spared in satisfying the strong Soviet interest in defense against attack by aircraft.

In the meantime, the U.S. Army countered with its own plans to deploy a huge system of 7000 anti-ballistic missiles across the continental United States.

In any event, while the Tallinn construction was taking place, ICBMs were becoming dominant, and U.S. intelligence agencies, incorrectly interpreting reconnaissance photos, concluded that the Tallinn bomber defense was a ballistic-missile screen. Fearing that the Soviets had created a shield against the ICBM, the United States considered and eventually implemented two counter-responses: penetration aids (penaids) and MIRVs. The penaids were to confuse the assumed Soviet ABM system; the MIRVs to overwhelm all defenses.

Galosh. By 1962 the Soviets had started deploying a BMD system near Leningrad; however, despite putting on a big, successful "show and tell" act of disinformation, they had already decided to discontinue its development in favor of a system code-named by NATO as *Galosh*. Intended as a response to MIRV, the Soviets planned to defend their capital with a nuclear-tipped ABM. Special warheads with high neutron and x-ray production were developed at Chelyabinsk-70.

As a last-ditch defense, nuclear explosives were to be detonated in the atmosphere when incoming missiles (or nuclear-armed aircraft) arrived above Moscow. However, the United States had already compensated for this by targeting an overwhelming number of warheads at the Soviet seat of government. The advent of MIRVs and penaids doomed the Galosh system to impotence.

By the late 1960s the Soviets were struggling to keep up with U.S. arms deployments. While they lagged in strategic offensive systems, they continued to concentrate on defensive measures. New anti-aircraft radar systems "dotted the Russian landscape."

All U.S. and Soviet schemes for ballistic-missile defense appeared easy to penetrate or overwhelm. Some civilian scientists in the Pentagon and White House realized that the Soviets would compensate for deployment of a U.S. ABM system by increasing their offensive weapons.

The only actual Soviet ABM deployment was the Galosh system, around Moscow. Aa total of 64 missiles were installed; however, Galosh did not actually become operational until 1977.

Both the Americans and the Soviets used nuclear warheads for their defenses because of inability to destroy incoming missiles by other means. (More than three decades of R&D later the situation has not improved.)

Under the 1972 ABM Treaty, the Soviets and the Americans were allowed only one ABM system each, which could protect either a military installation or the capital city. The Soviets chose to defend their capital, while the Americans placed their systems near missile-launch sites.

The Galosh system was still in place at the end of the Cold War. The surface-to-air missiles have a range of about 210 miles, and the warhead has a reported yield of 1 megaton.

Sentinel. In 1968 the Johnson administration, under strong political pressure, especially from Republicans in Congress, announced plans for deploying newly developed Nike-X interceptors and radars as a missile defense in a "thin" nationwide system that became known as Sentinel. The decision was interpreted as follows:

> The threefold mission of the Sentinel ABM system ... was (1) to provide a thin "area defense" of the entire United States against missile attack by China, assuming that China would soon develop the capability of launching nuclear missiles against the United States; (2) to provide protection against a nuclear missile "accidentally" launched by the Soviet Union; and (3) to provide — "as a concurrent benefit" — a very limited defense of U.S. land-based Minuteman intercontinental nuclear missiles against Soviet attack.

The focus shifted to Chinese ICBMs when realization hit that defense against a massive Soviet attack was beyond reach. This was seen as the best way to relieve political pressures until an agreement could be reached with the Soviet Union to limit strategic arms.

President Johnson's Secretary of Defense, McNamara "decided in 1968 that the [ABM] technology was not yet mature enough to warrant [deployment]. Furthermore, he recognized that building [a missile defense] would likely force the Soviet Union to respond, not just by deploying a system of its own but by building even larger numbers of offensive missiles...."

Safeguard. In Richard Nixon's 1968 presidential-election campaign, the Republican party assailed the Johnson administration's BMD effort as inadequate.

When Nixon became president the following year, he had the existing program reconfigured for early deployment, reversing his predecessor's emphasis. Nixon primarily sought to protect U.S. land-based retaliatory forces against direct attack by the Soviet Union.

The new Secretary of Defense, Melvin R. Laird, warned that the Soviet Union was moving toward a "first-strike" capability. Laird also concluded that the U.S. Polaris submarine fleet was endangered by Soviet advances in antisubmarine warfare, and that the land-based bombers might be vulnerable to sneak attack by new Soviet missile submarines or by their developing Fractional Orbital Bombardment System (FOBS). Defense officials insisted that Laird's assessment was based on "new" intelligence. (None of these allegations turned out to be valid.)

To replace Sentinel, the Nixon administration proposed in 1969 the comparable yet more modest ABM program, Safeguard. This decision was based on "using

Section D: Arms Developments

political judgments and manipulated intelligence data to justify its case" All three BMD systems up to then — Nike-X, Sentinel, and Safeguard — centered around similar hardware designed to protect missile installations, not cities. National-security advisor Henry Kissinger played a major role in Nixon's decision to move in that direction.

In changing the ABM program from Sentinel to Safeguard in 1969, President Nixon had to admit:

> The heaviest defense system we considered, one designed to protect our major cities, still could not prevent a catastrophic level of U.S. fatalities from a deliberate all-out Soviet attack. And it might look to an opponent like the prelude to an offensive strategy threatening the Soviet deterrent.

Two nuclear-armed interceptor missiles for the Safeguard system were chosen: the Spartan, designed for long-range (400-mile) intercepts, and the Sprint, intended to detonate at shorter ranges (20-40 miles) from the target. Blast pressure, intense heat, and penetrating radiation generated by its nuclear explosion was intended to compensate for imprecision and lag in reaching incoming warheads. The Spartan, for its high-altitude intercepts, was to have a 3-megaton warhead, and the Sprint — because it would explode just a few miles above the city it was defending — a few kilotons.

[Nixon's ABM Decision]. As reported in a 1969 *Newsweek* article:

Last week, Richard M. Nixon, just eight weeks in office ... decided that the nation must proceed with at least a limited deployment of ABM.... The President rejected outright the proposals for a hardened ABM system designed to protect the population in the event of all-out nuclear war. "There is no way," Mr. Nixon told the nation bluntly, "even if we were to expand the limited Sentinel system ... that we can adequately defend our cities" against a determined Soviet attack.

Mr. Nixon also vetoed the compromise approved by President Johnson in late 1967. This so-called Sentinel system, already in the process of deployment ... was designed ostensibly to protect the nation's major cities from an unsophisticated Chinese attack or an accidental missile launching.

In the arcane logic of nuclear confrontation, ABM defenses deployed around cities can actually be interpreted as provocative.... Mr. Nixon went out of his way to stress that there would be no increase in the U.S. offensive weaponry despite the recent Russian buildups.... The ABM's opponents point out that any defensive system can always be neutralized if the enemy strengthens his offense.

Many of the nations now resent being asked to renounce such weapons while the nuclear powers continue to build up their own arsenals.

A independent study in *The New York Times* described the proposed system in 1978 as follows:

> The chief components of the [Safeguard] system are missiles, radars, and computers. A large, long-range radar, called the Perimeter Acquisition Radar, or PAR, is designed to detect attacking missiles while they are still some 1000 to 2000 miles away.... As the missile approaches closer, it is picked up by a second radar, known as the Missile Site Radar, or MSR, which tracks and then guides a large ABM missile, called Spartan, to an intercept point high above the atmosphere and several hundred miles away. If the

Spartan fails to destroy the enemy warhead, the MSR then sends up a fast-accelerating Sprint ABM missile to intercept the incoming warhead within the atmosphere, at a distance of 25 miles or considerably less. Both the Spartan and the Sprint ABM missiles carry nuclear warheads.

One problem with ABM, as in all defenses, is the immense difficulty of testing under operational conditions — a complete systems trial with realistic simulation of an attack. To counter criticism that the ABM system was untested and wouldn't work, the Nixon administration pressed for the Atomic Energy Commission to promptly set off nuclear warheads in evacuated compartments deep underground in Nevada. Even if such underground explosions suggested that the warheads could destroy incoming missiles in space, they would hardly verify that a full-scale system would always be prepared and ready to launch successfully in a matter of minutes of warning. This was especially so, given the limited and contradictory real-time processed information that might be all that was available.

The deployment decision was extremely controversial, and eventually Congress voted to terminate funding, marking the only time in the history of the nuclear weapons program that the Congress canceled a strategic-weapons system supported by the executive branch.

One experienced arms control specialist observed that the Safeguard system was "clearly ineffective, vulnerable and expensive and was operational for less than a year." As was the case for President Johnson's earlier BMD plans, intense public opposition arose over the Nixon ABM system. Ease of evasion, even by Chinese missiles, was a key objection.

An inherent limitation of all antiballistic-missile systems: the warning time is so short that system designers must eliminate human decision-makers; so computers must choose the precise moment for firing interceptors in a narrow time window. Moreover, the system has contradictory requirements of having a "hair trigger" to cope with surprise attacks and a "stiff trigger" to avoid going off accidentally.

Other objections were noted at the time: inability to test the system under realistic (simulated wartime) conditions; the possibility that missiles, radars, or computers would fail; errors that would be made in computer programing and attack modeling; and the impact of countermeasures — such as EMP, decoys, multiple warheads, radar neutralization, evasive maneuvers, salvage fusing, and penetration aids.

In any event, the Soviets were pushing for a treaty as early as 1969. In 1972 the two sides agreed on the ABM Treaty, which limited the Safeguard system to two sites. The United States had originally intended to defend both a city (Washington, D.C.) and an ICBM base, but in 1974 a protocol reduced the U.S. and Soviet ABM to one site each. For a brief and wasteful four months in 1975, the United States deployed 100 nuclear-tipped Spartan and Sprint interceptors at a missile-launch site in South Dakota. The Soviets continued to maintain their Galosh system around Moscow.

SDI. The *Strategic Defense Initiative* was an exorbitant missile-defense plan devised during the Reagan administration; it was an even grander scheme to defend against intercontinental ballistic missiles, but it did not address alternative means of delivery, such as off-shore submarines or cargo ships in commercial sea lanes.

President Reagan, uncomfortable with the thought of mutual assured destruction, stated that if a breakthrough in defense were to occur, "free people could live secure in the knowledge that their security did not rest upon the threat of instant U.S. retaliation."

In looking back at the 1980s, the hope for national missile defense was that

a massive and growing Soviet threat that would be rendered "impotent and obsolete" — President Reagan's words — by new and exotic technologies such as x-ray lasers, rail guns and "death ray" particle beams. [However,] these technologies, and the strategic concept underlying [SDI], proved to be impractical, infeasible, unworkable, unaffordable and unattainable, and by 1990 the program began to be phased down.

Although President Reagan directly appealed to the scientific community to turn its talents toward a technical solution to the problem of strategic defense, formidable laws of physics and technological obstacles could not be overcome at any practical cost. In addition, because there are many alternative methods for delivering nuclear weapons, all ballistic-missile defense systems constitute a porous and circumventable "Maginot Line."

From the time President Reagan unveiled SDI in a nationally televised speech in March 1983, the missile-defense program changed steadily. It had already come through three major incarnations, each marked by less ambitious technical and performance goals. The trend confirmed doubts by skeptics that anti-missile systems were needed or were feasible. Perhaps the best accomplishment of the SDI project consisted of discovering some of the technologies that would not work.

The program received significant R&D funding during the 1980s and early 1990s, with much of the effort spent on a space-based x-ray laser intended to shoot down Soviet ballistic missiles. That device was the focus of research funded by the SDI organization at weapons laboratories. It was a technically controversial and ill-fated plan to use a nuclear detonation to create narrow beams of X-ray energy directed at Soviet ballistic missiles in flight.

The multi-layered space defense envisioned by SDI proponents would have operated both within and beyond the earth's atmosphere: First would have been detection of missiles in their initial boost phase by infrared sensors, followed by laser-beam, particle-beam, or space-weapons destruction of the booster. If this did not succeed, intercept of the mid-course warhead "bus" would be attempted before the RVs deploy. If all that failed, terminal-defense monitors and weapons would have the more difficult task of distinguishing between warheads, penetration aids, decoys, and debris.

Terminal defense would try to take advantage of "atmospheric sorting" (decoys, being lighter than the warhead, are slowed and deflected upon entering the atmosphere). This last-ditch terminal-defense phase would have the urgent job of destroying warheads that escaped the earlier layered-defense phases.

"The receding dream of technological perfection led to the decision in 1987 to concentrate on those anti-missile technologies that could be deployed within about a decade." In 1990, "growing disenchantment with the technical and military prospects of a system oriented toward the declining Soviet threat" persuaded Congress to make an unprecedented call for significant reductions in the Executive Branch's SDI budget.

Directed-Energy Weapons. Because of technical limits in protecting that the outer shell of a MIRV bus or RV cone, kinetic or beamed energy could be used against missile stages during mid-course. Laser or particle beams would have to focus their energy quickly and precisely at a rapidly moving target. Atmosphere irregularity and beam attenuation would confine most directed-energy weapons to space platforms or chase rockets.

Target acquisition (which includes detecting and aiming at the object) has to be rapid and reliable for directed-energy weapons. Lasers, if promptly and accurately pointed, have the advantage of being able to deliver energy to a target at the speed of light.

If outer-space positioned, directed-energy defense against ballistic missiles and RVs would require platforms in orbit close to the potential trajectories. At high orbits fewer defense stations are needed, but the strike range is correspondingly longer; at low orbits, hundreds of platforms would be required.

A nuclear-explosive-driven x-ray laser was proposed by Lawrence Livermore Laboratory. Intended as a space-based weapon against ballistic missiles, the idea was to use collimated x-rays generated from the detonation of a nuclear device. Besides lacking technical confirmation, the nuclear-pumped x-ray would have caused diplomatic problems, being in violation of the ABM and Outer Space Treaties. Nevertheless, in 1982 Edward Teller, claiming without substantiation that the x-ray laser would be feasible, persuaded Ronald Reagan to go ahead with the program. An underground test of the laser-pumping principle was carried out in 1987. At first there were claims that the test gave favorable results, but it later became known that the report of success was highly exaggerated.

Neutral-particle beams in space seemed to hold some promise, but major technical hurdles included obtaining sufficient beam-current density and avoiding beam divergence. With charged-particle transmission, which is an alternative to neutral beams, it proved hard to obtain rapid pulsing and avoid beam divergence. Another fundamental problem is that the beam bends unpredictably in the earth's magnetic field. Also, according to a important technical study, "the beam cannot propagate stably through even the thinnest atmosphere and must wait for an attacking booster to reach very high altitude."

For the different types of directed-energy beams, not only were substantial improvements in brightness and beam management required, but the energy supply needed in space was daunting.

Unconfirmed reports of Soviet research in directed-energy weapons technology were cited as a rationale for continuing the U.S. program. In particular, a possible antisatellite laser at Sary Shagan was assumed to be capable of a future role in ballistic-missile defense. Such expectations were incorrect.

Terminal defense, for interception within the atmosphere, requires a range of only 10 or 20 kilometers. However, intervening air degrades laser and other directed-energy beams. The beams would have to be intense, bright, and instantaneously steered to any RV that escaped earlier interception.

In 1987, the American Physical Society carried out, in the words of the DOD SDI office, "an objective independent appraisal of various [ballistic-missile defense] technologies." The report's main conclusion, though, was that "development of an effective ballistic missile defense utilizing directed-energy weapons would require performance levels that vastly exceed current capabilities." The report also mentioned problems with inadequate databases, the amount of intensive research still needed, and difficulties of system integration and effectiveness.

Nor did kinetic-kill vehicles fare well either. As a matter of fact, "in 1993 four former Reagan administration officials, including two military officers, admitted to the *New York Times* that the Army Ballistic Missile Defense Command "rigged a key 1984 interceptor test and ... other test data were fabricated as part of a program to deceive the Soviet Union...." DOD confirmed in September 1993 the existence of the mid-1980s deception program The effort failed "because the interceptor vehicle never came close enough to the target to make a rigged explosion appear credible and realistic."

The Soviets were strongly opposed to SDI. Mikhail Gorbachev, denounced it as destabilizing, and one of the most famous Soviet dissident physicists, Andrei Sakharov, expressed serious doubts about its technical feasibility. Without renegotiation (or unilateral withdrawal), the ABM Treaty stood in the way of eventual SDI deployment.

Opposition by the Soviets, despite their historical inclination favoring defenses, was a serious obstacle. Even if interceptions could be reliably accomplished, missile defenses could be offset by deploying more offensive weapons or by using simple, comparatively inexpensive evasive countermeasures. A combination of those and other inherent technological and strategic weaknesses doomed the SDI program from the start. Moreover, efforts at unilateral defense are destabilizing because they stimulate the opposing side to develop offenses that offer a premium to the side that strikes first.

Directed-energy antiballistic technology has turned out to be largely unsuccessful, expensive, wasteful, and over-hyped.

Antisatellite Technology

Finding technical methods for disabling or destroying satellites was another reflexive technological competition that slowly evolved. Both earth- and space-based ASAT weapons were contemplated not long after *Sputnik* appeared. One problem was (and is) that such an skirmish would be self-defeating because both sides use satellites for critical reconnaissance, stabilizing communications, and treaty monitoring, all of which became especially important when furtive aerial-surveillance transits over the Soviet Union became too risky and controversial.

Co-orbital rendezvous (between a satellite launched from the ground with another already in orbit) were performed by the Soviet Union in the late 1960s.

While that might have had ancillary nonmilitary objectives, it had the appearance of testing unfriendly satellite interceptions.

Because satellites and their instruments are delicate and difficult to shield, their operation and function can be impaired by moderately high-intensity beams of directed light, radiation, or particles. Thus, beam technologies could also be developed for ASAT attack from space or earth.

National governments, the international community, and non-governmental organizations recognized the implications of an arms race in space. Suggestions were made for self-restraint and for international control of development and deployment of antisatellite weapons.

Overkill

Overkill is a term that applies to retaliatory destruction beyond that which would be needed to destroy a reasonable set of targets under a strategy of mutual deterrence. To gauge overkill, not only the quantity of weapons must be considered, but also qualitative attributes that contribute to their effectiveness.

A strategic doctrine that relies on assured ability to retaliate destructively should, in principle, cap the number of warheads to that needed to achieve convincing deterrence. However, superpower arsenals far exceeded the minimum. In 1966 the United States possessed 32,000 nuclear warheads, although the number of targets in the SIOP list was much smaller. Even if the target list had contained 10,000 aimpoints, the overkill would have been a factor of three. In 1970 the nation had about 20 times more weapons than targets and when MIRV came along, the overkill factor reached perhaps 50.

Some analysts considered destruction of just a few Soviet cities to be sufficient for deterrence. Allowing for redundancy and uncertainty, a few hundred warheads, distributed among the strategic retaliatory forces, would have been more than enough for assured countervalue retaliation. For instance, the U.S. Defense Department in 1969 estimated that a hundred 1-megaton warheads reaching Soviet targets would cause 37 million direct fatalities and destroy 59 percent of its industry. Doubling the number of warheads in a retaliatory strike would increase estimated fatalities to 52 million people and destroy 72 percent of industry; another doubling would bring the number of dead to 74 million but eliminate only another 4 percent of industry.

In comparing lethality of weapons, it's not enough just to consider explosive yields: various qualitative factors pertain too, such as range, accuracy, reliability, and mobility. Taking such factors into consideration, lethality of weaponry has increased almost exponentially since ancient times; a 1-megaton thermonuclear weapon is about 1000 times more lethal than a World War I 155-mm field gun; it is some 30 million times more lethal than a sword or bow and arrow. A strategic missile with a 25-megaton warhead inflicts 50 million times more harm than an assault rifle.

Section D: Arms Developments **155**

> **[Nuclear Warheads Galore].** Standing in a room with enough nuclear warheads to destroy all of civilization is an experience I'll never forget.
>
> In the mid-1980s, participating in a inter-laboratory program to devise weapon-verification procedures, I was part of a DOE team visiting practically all the major U.S. nuclear development, production, and storage sites. Our team's assignment was to figure out if on-site inspection of these sites, as proposed for the START I treaty, could inadvertently reveal government secrets to foreign inspectors. Stockpiled at various sites were hundreds of nukes.
>
> Just about any small warhead stored in this particular underground room was powerful enough to wipe out a city the size of Hiroshima; a big one would devastate the entire New York metropolis. All of them were slender enough to wrap my arms around. So much destructive power in such little packages! It took a long time for the queasy feeling in my stomach to subside.
>
> The briefings were matter-of-fact: Numbers and destructive powers, sizes and weights. Arming and transportability.
>
> This was the first time I was actually close up to a real nuclear warhead.
>
> When I emerged later into the open world above and could see sunlight, it felt like I had just seen the phantom of death. Or a lot of them.
>
> I rededicated myself to eradication of nuclear weapons.
>
> **********
>
> My dedication, though, was never enough to counterbalance that of Thomas C. Reed, a technocrat who became a nuclear weapons designer and later an advisor to Presidents. He was dazzled by the nuclear weapons he saw in 1958 at Sandia nuclear-weapon laboratory. In his 2004 autobiography, *At the Abyss,* Reed admits:
>
>> My first visit there [Sandia] was mind-bending. At our business meetings we discussed the technical details of integrating warheads into reentry vehicles, but I had glimpsed the nuclear genie. It would never let me go.
>
> Reed later served in several Republican administrations, including a stint as Secretary of the Air Force.

In spite of the capacity that each superpower possessed in strategic-delivery vehicles, including the number and capability of the warheads, many further advances in weaponry were contemplated. Some of these advances would have been (or were) strategically stabilizing, and some escalated the arms race. Worst of all, some developments tended to increase capability to launch a first strike.

Overkill was also a product of misplaced perceptions about the military utility and public acceptability of nuclear warfare: In the mid-1950s the U.S. Army requested 151,000 atomic warheads, anticipating that it would need more than 400 in a single day of "intense combat." In 1968 the Joint Chiefs recommended that a stockpile between 51,000 and 73,000 be produced.

For those analysts who abhorred and feared the effects of a single nuclear bomb, overkill was achieved very early in the Cold War; the potential for annihilating a single city was deterrence enough. For others, many of whom were victims of Nazi or Communist atrocities, there never was enough assurance: Extra nuclear weapons, better means of delivery, stronger ballistic-missile defenses — all were justified in the name of national security.

An "equivalent megaton" is a weighted value of the blast effects for various nuclear weapons — one megaton corresponding to one million tons of TNT. In 1964, Secretary of Defense Robert McNamara had concluded that 400 equivalent megatons would be sufficient to destroy the Soviet Union as a functioning society. The U.S. arsenal was then rated at 17,000 equivalent megatons — 17 billion tons of TNT. A catch phrase of the time was "a bigger bang for the buck."

Improved Weapons Safety

The nuclear weapons labs deserve credit for significant generational improvements in weapons safety. The year 1950 alone was marked by five publicized incidents involving nuclear weapons, three where aircraft crashed and two where weapons fell into the ocean. Information about incidents from 1950 through 1955 was still not publicly available more than a half-century later.

In any event, those aberrant episodes became a wake-up call about the potential for more serious weapon accidents that could lead to a highly destructive nuclear chain reaction. Many nuclear-weapon incidents and accidents have occurred, but none has resulted in a publicly acknowledged dangerous state of nuclear criticality.

John S. Foster has been credited more than anyone for improving nuclear safety. Foster started designing safety into primaries in 1958 and brought nuclear safety to fruition in 1962. (Foster later headed up the U.S. ABM-development program.)

Edward Teller's scientific trademark was optimistic new ideas, not specific technologies like safety mechanisms. Teller personally stimulated development of megaton-class thermonuclear weapons, promising U.S. Navy officials as early as 1956 that he could develop one that would be "no larger than a small trunk" and could be carried on and fired from a submerged submarine.

To help fulfill that promise of Teller, Foster became largely responsible for devising the nuclear trigger, a primary called *Birdie*. By the late 1950s he achieved higher yield, lower weight, and better materials efficiency. This met Teller's needs, but Foster also conceived of ways to add safety devices into *Birdie*. Foster favored and vigorously promoted profoundly important nuclear-safety measures, such as: (1) resisting accidental detonation — one-point safing, (2) precluding the arming sequence unless the warhead was outside U.S. boundaries, (3) introducing insensitive high-explosives, and (4) adding PALs for better control. *Birdie* technology became the basis for all Livermore-designed weapons and most Los Alamos weapons in the U.S. stockpile, and there have been "no nuclear accidents of any sort" in the inventory since 1980.

Arms Control

Universal dissatisfaction with the hazards, uncertainties, and economic burdens from a quarter-century of unbridled nuclear competition eventually pushed the great powers into substantive *arms control* negotiations. But the weapons states had difficulty agreeing on the ultimate goal of the process. The peace movement saw negotiations as a necessary first step toward total elimination of nuclear weapons, while governments did not see that result as feasible or even desirable — governmental objective usually was to define and limit the perceived military threat. In addition, the more public-relations-oriented administrations used the negotiation process to mute citizens' unrest regarding the very real prospect of an ugly nuclear war.

It was not until 1972 — halfway through the Cold War — that limitations in quantity and development of strategic weapons were settled in a series of bilateral treaties between the United States and the Soviet Union: SALT I and II, the TTBT, START I, and eventually START II. The negotiations took place against a fluid backdrop of domestic and international politics, dogma, and preoccupations.

Some of the issues that influenced East-West relations and negotiations are collected in the paragraphs below.

To guard against surprise military developments, a number of factors were taken into account in treaties and agreements, including weapons R&D, treaty verification, and escape clauses. The need to avoid technological surprise drove the topics of verification and inspection to be major subjects for public and government debate. Those who took a very conservative approach feared a sudden and disastrous treaty breakout that could not be detected ahead of time by national intelligence or agreed verification measures. Major public and intragovernmental debates took place during the 1980s and 1990s about the likelihood of treaty compliance and the effectiveness of verification.

SALT I. By the end of the 1960s the U.S. nuclear arsenal consisted of 656 submarine-launched Polaris missiles (each having three MRVs), about 1050 missiles (mostly Minuteman) in silos, 650 long-range bombers (each capable of transporting four high-yield thermonuclear bombs), and perhaps 10,000 substrategic weapons. The latter could be delivered by low-flying aircraft, planes from aircraft carriers, and land-based rockets and artillery. At the same time, the Soviets had 200–300 ICBMs, but considerably fewer (about 90) submarine-borne missiles, and only about 150 strategic bombers.

On 26 May 1972, the SALT I agreement and the ABM Treaty were signed, placing limits and controls on long-range ballistic missiles and defenses, respectively. Both sides agreed to freeze their offensive and defensive weapon systems, including submarine-launched ballistic missiles, ballistic-missile submarines, fixed land-based ICBM launchers, and ABM systems. The agreement contained no reductions in the total number of warheads or launchers.

Soviet megatonnage was greater than that of the more diversified and more accurate U.S. weapon systems, and there was considerable public and intragovernmental debate about the appropriate indicators of military strength and the meaning of "nuclear sufficiency." Although the disparity was by choice of the U.S. military, who would not have wanted to swap arsenals with the Soviets, hardliners made an issue of the difference in megatonnage.

For a realistic comparison, however, it was not enough just to add up the total deliverable megatons. The hardliners were not taking into account other factors that had to be considered, including the number of weapons, their yields, methods of delivery, "hardness" of silos, quality and reliability of command and control, and multiplicity of warheads.

As it turned out, SALT I fell well short of the goals of arms control advocates, who wanted a halt in testing of long-range missiles (especially multiple warheads) and a freeze on the quantity of delivery systems. Moreover, SALT I omitted limits on new strategic systems, notably (and deliberately) MIRVs, and was silent on Soviet deployment of newer, larger missiles, such as the SS-19. Agreement was not reached on cessation of missile testing that could have seriously impeded guidance-accuracy improvements. (Guidance accuracy, which required extensive impact testing, was crucial to developing a first-strike capability, something that hardliners wanted to preserve even at the risk of undermining long-term stability.)

Blunted as it was, SALT I was beneficial to arms control in that it formalized the use, for verification, of NTM (national technical means — satellites, seismographs, radars, and other intelligence gathering), and the treaty banned deliberate attempts to conceal tests or otherwise to interfere with agreed verification measures.

Nuclear Testing Limitations. A weapon designed for military use (in contrast to one intended for committing a terrorist act) must have extensive explosive testing and quality certification — to ensure that it functions as intended, to prove compatibility with the delivery system, and to clarify the military effects of the explosions. As a result, the weapon states carried out more than 2000 nuclear detonations before it all ended. Public sentiment slowly turned against atmospheric and underground explosions because of fears that there would be environmental consequences (and because hardliners and weaponeers created such extravagant arsenals).

The first legal restrictions were imposed by the 1963 Limited Test Ban Treaty between the United States and the Soviet Union, which caused testing to be shifted underground. It was not until the Threshold Test Ban Treaty of 1974 that underground explosions were constrained — and then only to 150 kilotons, an order of magnitude larger than the bombs used in Japan.

To supplement the TTBT, the Peaceful Nuclear Explosions (PNE) Treaty was signed in 1976, legitimizing underground explosions for peaceful purposes, prohibiting only those that exceeded 150 kilotons.

The TTBT became a bone of contention in later years, partly because it lacked agreed on-site instrumentation and inspection, and partly because hardliners needed a political football. Pending ratification, the threshold limit was observed, and no violations are likely to have occurred. The TTBT and the PNE Treaty were eventually ratified, after much debate, and entered into force in December 1990, sixteen and fourteen years, respectively, after being signed.

In 1996 the Comprehensive Test Ban Treaty (CTBT) was approved by the United Nations and signed by most of the parties. It prohibits essentially all nuclear-explosive tests, and, if ratified by designated nations, would implement extensive networks of monitors and authorize on-site inspections. Of the 44 nations whose signature and ratification are necessary for the Treaty to enter into force, 41 had signed and 28 had ratified by the end of the millennium. The Russian Duma ratified the treaty in April 2000. In October 1999 the U.S. Senate voted not to ratify the treaty. Although technically unenforceable, all signers as of 2008 had observed the CTBT terms.

SALT II. Both Soviet and American arsenals grew rapidly in the 1970s. The United States introduced new strategic weapons systems: the B-1 bomber, the Trident submarine with its Trident-II missiles, the MX land-based ICBM, the new Mk-12A warhead, and cruise missiles that could be launched from land, sea, or air. Such an unchecked arms race aroused massive public opposition, putting pressure on the governments to negotiate a treaty more meaningful than SALT I.

In 1979 the second Strategic Arms Limitation Treaty (SALT II) was signed. It is an extremely technical document, with 19 detailed Articles, accompanied by nearly 100 "Agreed Statements" and "Common Understandings." The treaty restricts the superpowers to equal numbers of strategic-delivery vehicles. Aggregate ceilings of 2400 long-range weapon delivery systems were accepted (lowered to 2250 by the end of 1981), with sub-ceilings for MIRVs and ALCMs, long-range bombers, and SLBMs. Constraints were established on deployment of new ICBMs and on the number of warheads on MIRVed missiles. In particular, the Soviet SS-18 (which has a capacity of 20–30 RVs) was limited to 10 RVs, and new missiles of this type were prohibited.

When Presidents Carter and Brezhnev signed the agreement, they intended to make up for the temporary and incomplete nature of its predecessor, which had been put in place seven years earlier. The parties issued a joint communique that addressed then-current arms control issues. Still, neither the agreement nor the communique mentioned strategic stability, and they omitted a number of other unresolved hurdles — such as compliance with the ABM Treaty, progress on a Comprehensive Test Ban Treaty, the importance of nonproliferation, continued negotiations on mutual force reductions in Europe, and the search for an ASAT agreement.

The terms of the SALT II treaty imposed restrictions on strategic bombers as well as missiles, established a ceiling on strategic-delivery vehicles, and limited the numbers of permitted MIRVed delivery vehicles. Not agreed to were Soviet proposals to ban the development of U.S. Trident missile submarines and B-1 bombers.

Without SALT II, it was estimated that the Soviets were going to build up their strategic forces to nearly 3000 launchers and 13,000 weapons, while the United States would increase to more than 2300 launchers and as many as 17,000 weapons. The Carter administration had started to move toward giving the United States the wherewithal to wage a limited conflict — a so-called "nuclear war fighting" strategy — in addition to its ability to engage in large-scale war. This ominous turn of events included bolstering the nation's civil defenses against nuclear attack, building a new generation of larger, more accurate ICBMs, and improving missile-launch command and control.

Commenting on these U.S. developments, Senator Daniel Moynihan wrote, "Herein resides the final irony of the SALT process. Not only has it failed to prevent the Soviets from developing a first-strike capability; it now leads the United States to do so. The process has produced the one outcome it was designed to forestall. And so we see a policy in ruins."

Nevertheless, President Carter called SALT II "the most detailed, far-reaching, comprehensive treaty in the history of arms control." Specifically, the Soviet inventory of missile launchers and bombers would be capped. The numbers of warheads on missiles, their throw-weight, and the qualitative development of new missiles were limited. The Soviet Union was to destroy 250 strategic missile systems.

Verification enhancements for SALT II, besides NTM, included banning concealed silos and forbidding encryption of missile-test data.

As of 2008, the SALT II treaty has never been brought up for ratification by the U.S. Senate nor ratified by the Soviet Union or the surviving Russian Federation; however, both sides have complied as though it were in effect.

Here's a 1981 perspective about the fuss over SALT I and II:

> Partial recognition that we can no longer defend ourselves [against an all-out nuclear attack] is implicit in [SALT I]. These negotiations have been going on with the Soviet Union since the late 1960s, and have so far resulted in one treaty — an interim agreement that limits certain strategic weapons.
>
> Ratification of SALT II has been stalled in the U.S. Senate, and has apparently even been repudiated ... by the incoming Reagan administration. SALT has, unfortunately, been the focus of almost all public debate on arms control, distracting public attention from questions far more fundamental than the largely issue-skirting topics considered at the SALT [II] negotiations....
>
> The SALT negotiations are evidence of lip-service to the thought that unrestrained development of certain technologies may threaten all parties, regardless of political or ideological differences.... The U.S. Senate's foot dragging over ratification of an almost do-nothing SALT II treaty is merely symptomatic of the myopia that afflicts us....
>
> The irrationality of the cultural resistance to arms control measures is clearly demonstrated in political debates over control of nuclear weapons. Complex and tedious arguments, often technically incorrect, attempt to show how the other side might manage to cheat. Claims are made that the other side is irrational and cannot be trusted.... Humanitarianism aside, nothing prevents one side from attacking except the rational understanding that the other can strike back.

INF. Formal talks on intermediate-range nuclear forces began in November 1981. Two years later, after six rounds of negotiations between the Soviet Union and the United States, the Soviets discontinued the talks when new U.S. missiles (Pershing-II and GLCM) appeared in Europe. This deployment was part of a 1979 U.S. "dual-track" decision to "modernize" NATO sub-strategic forces while engaging in INF arms control negotiations. The deployment, scheduled to begin in 1983, would place Pershings in Germany and GLCMs in Belgium, Italy, Germany, the Netherlands, and the UK.

The initial INF position of the United States was the "zero-zero" option — complete elimination of American and Soviet longer-range missiles on a global basis. The longer-range missiles, not just short-range, later did become part of the negotiations. The Soviets were facing a force imbalance because Britain and France sided with the West; nevertheless, the U.S. State Department labeled as "contrived" the Soviet view that independent national-deterrent forces warranted a countertrade. Of course Moscow viewed all of them as a combined threat.

Soviet deployment in European Russia of SS-20s with triple-warhead capability, starting in 1978, was to the West a major provocation; the missile had a 2700-mile range, enough to reach Western Europe, the Middle East, South Asia and China, but not the continental United States.

When signed in 1987, the INF Treaty eliminated Soviet SS-20s and U.S. Cruise and Pershing-II missiles; this became the first bilateral agreement to do away with an entire class of nuclear-armed missiles: The intermediate-range missiles were to be verifiably demolished. All Soviet and American missiles with approximate ranges between 300 and 3400 miles were to be scrapped, regardless of where they were deployed. Included in the U.S. total were 120 Pershing-II and 309 cruise missiles; for the Soviets there were 65 SS-4, 220 SS-12, 405 SS-20, and 33 SS-23.

Neither the sub-strategic weapons nor the 72 shorter-range Pershing-IA missiles, with their U.S.-controlled warheads, were part of the treaty. Within three years of ratification, the intermediate-range missiles, their launchers, and their support structures and equipment, were to be destroyed — while the shorter range missiles were to be eliminated in 18 months. On-site inspectors were allowed to witness the destruction process.

All terms of the treaty have since been completed to the full satisfaction of both parties.

START I. Real strategic limitations and cutbacks had to wait until the Strategic Arms Reduction Treaty (S T A R T) process matured, with the help of new political leadership — more than a decade after the SALT agreements.

The START I treaty was signed by the Soviet Union and the United States (Presidents Gorbachev and G.H. Bush) on 31 July 1991, and entered into force on 5 December 1994. Soon thereafter the five new successor states of the FSU acceded to the terms. Each party agreed to reduce or eliminate deployed strategic warheads and launchers (missile and bombers) on its territory. Ceilings were placed on launchers (ICBMs, SLBMs and bombers) at 1600 each. Separate ceilings were set for warheads at 6000 each (4900 on ballistic systems, 1540 on *heavy* — large, mostly Soviet — missiles and 1100 on mobile, land-based missiles.)

Strategic arsenals were cut overall by 30 percent; the approximate numbers are: U.S. missiles and bombs went from 12000 down to 9000, and FSU from 11000 to 7000. SLCM sublimits were set at 880. Certain categories of heavy ICBMs were reduced in number and throw-weight. In particular, the heavy SS-18 had a 50-percent reduction in allowed warheads.

The five current parties — United States, Russia, Belarus, Kazakhstan, and Ukraine — periodically exchange information on data for accountable strategic weapons. No unresolvable difficulties have arisen in connection with START I implementation.

The 30-percent reductions of the first Strategic Arms Reduction Treaty were to be carried out over an eight-year period, with various verification procedures authorized, including on-site inspection. Restrictions were placed on the transfer of weapons technology to third parties.

****

This Section has described some of the major technological developments in nuclear arms and arms control — the enhancements that threatened to jeopardize a quasi-stable balance of power.

Improvements in delivery accuracy, evasive stealth, and warhead multiplicity changed the apparent purpose of some weapon systems from deterrence (second-strike) to counterforce (first-strike). Predictably, there were two possible outcomes of for the less-advantaged party: either a calculated, resigned sense of impotence (one-sided deterrence), or (more likely) a vigorous quest for an equalizer — another stimulus ratcheting the arms race.

Reflexively carried over from the city-bombings of World War II was civil defense. Initially the military ignored its political impact, thinking strictly in terms of minimizing casualties. Before long, the public realized that ducking and taking cover was simply a game, one that should never — in the first place — be trifled with.

Taking shelter near ground zero was a government-nourished delusion: Realistically, if cities were attacked, very few civilians could get deep underground in time, and those who did could well be suffocated when the firestorm exhausted breathing oxygen — as had happened after the fire-bombing of Dresden in World War II. If missile sites were the only targets, fallout shelters could theoretically have offered a few weeks of protection for distant population centers, but survivors who emerged would find a different, difficult world awaiting them.

Civil defense was promoted in large measure because of a desire to get the public to tolerate a harder line in foreign policy. A catch phrase frequently heard was, "Better dead than red."

Also fruitless was the quest for strategic defense. Out of thousands of missiles and bombs, only a few needed to get through to kill millions of inhabitants.

Given the futility of passive or active defense, a reflexive option was to seek increased overkill. But that simply intensified the destruction implicit in MAD. Underlying the desire for overkill was a stubborn temptation in the West to seek superiority, to bring the Soviet Union to its knees, to rid the world of Communism. Once thermonuclear weapons were possessed by both sides, no such possibility existed. However, many doctrinaire hawks persisted in pushing the idea, gaining approval for far more weapons and delivery systems than needed for simple deterrence.

Another growing was that real-time reaction to ever-faster (and more-plentiful) warhead-delivery systems meant a decision was needed in a few minutes. The decision that might end civilization was to be made either by fallible humans or by preprogrammed computers (with algorithms installed by fallible humans). If the head of state were available, the judgement had to have clarity of mind and situation; if someone lower on the chain of command pushed the button, the reflexive response at the bottom could hardly take into consideration the immense consequences.

Nuclear warfighting, not just deterrence, became a deliberate political and military gambit, although its adverse implications were never fully appreciated. Because nuclear weapons and strategic-assessment information was sequestered and bewildering, it was difficult for outsiders to knowledgeably comprehend the ramifications of nuclear overkill. Fear of communism trumped self-preservation.

Dangerous shibboleths — embodied in civil- defense planning, ballistic-missile defense, and massive (rather than assured, measured) retaliation — characterized the arms race. Eventually, when leaders less committed to military solutions came to power, they realized that it was necessary to initiate negotiations for arms control, a process that went on for many years. Outsiders, meanwhile, had to vigorously push and petition for mutual restraints.

Slowly, some treaties did get negotiated, but most were windowdressing to allow leaders to claim domestic political credit. But the SALT I, START, and INF treaties did begin to reverse a confrontational trend that had been worsening.

****

Like Alice in Wonderland's *Red Queen*, it was necessary during the Cold War to run fast to remain in the same place. Every new development that gave a military advantage was inherently transitory, inviting ever more weapons research by the other side. The superpowers ended up with the explosive equivalent of billions of tons of TNT, and no more security than they had forty years before. With open treasuries and bonds sold on future generations, they bought the ability to destroy themselves many times over, to no real humane advantage.

The immense arsenals accumulated by the superpowers, not to mention the highly lethal armaments of lesser nuclear-weapon states, stirred at least one of the attitudes desired by their proponents: Don't tread on me! That posture , some argue, deterred major wars during the second half of the 20 century. One consequence was a dangerous and senseless atmosphere of mutual fear.

Some historians, physicists, and political scientists contend that nuclear weapons, at the very least, "were **mainly** responsible for having prevented direct superpower conflict during the Cold War." Be that as it may, the accessibility of nuclear weapons for war-fighting when disputes arose sometimes impelled governments dangerously close to disaster and the possible end of civilization as we know it.

In 1966, the United States alone had more than 32,000 nuclear weapons in its arsenal. Years before that (in 1960), in terms of something more ponderable, the United States had aggregated the destructive power of 1.4 million Hiroshimas. Sad to say, the other nuclear powers, especially the Soviet Union, also had built themselves highly destructive nuclear arsenals.

Except for tactical nuclear weapons with very low explosive yield, it takes only one warhead to devastate a medium-size city. In simple terms, one nuclear weapon = one city. Practically every city on earth could have been obliterated. At one time more than 50,000 nuclear weapons were poised to counter each side, with many types of delivery systems: aircraft, missiles, ships, artillery — you name it.

Even a regional battle could have become nuclear. "Tactical" weapons can be fired from cannon, howitzers, or recoilless rifles, or detonated as land mines. Fighter-bombers and naval vessels carried thousands of "theater" bombs or shells. In retrospect, another alarming factor was the extent to which control over the use of nuclear weapons had devolved down to the military field level: Authorization for use was not reserved solely for the national political authority.

The action-reaction, offense-defense sequences of missile and interceptor development programs were an example of the military "prudential rule" in operation: Axiomatic among defense planners is the principle that, because of lead times required for research and development, prudence dictates action to counter any anticipated developments by the adversary.

No defense against a nuclear broadside was (or is) truly feasible. The simple fear of being destroyed in retaliation (MAD) should have been deterrence enough against deliberate attack. Yet, at no time in the second half of the century were the nuclear powers immune to sudden disaster. Various efforts to develop strategic and civil defense faltered on the reality that there was no way to evade devastation once the missiles were on their way. Arguably, deterrence and arms control — including the ABM Treaty — worked to prevent widespread aggression and the outbreak of major wars.

Vulnerability to nuclear conflict — initiated deliberately, by crisis escalation or by accident — was a standing feature of life. Realizing that any serious conflict could be globalized, the Soviet elite conceded, perhaps more than others, that neither they nor their motherland could survive an all-out war.

Defense by either side against nuclear weapons was a chimera — a fanciful idea — intended to inure people to the existence and use of nuclear weapons as instruments of policy. Urban shelters provided little or no refuge from direct attack. If comparable retaliation were limited to military bases, timely mass evacuation would have enabled many survivors to escape more serious harm from radioactive fallout. However, any massive exchange of nuclear detonations would have either shattered the cities or created radioactive wastelands, or both. Emergency facilities would have been overwhelmed.

The mere hint of defensive strength that might undermine assured destruction was soon countered by a further buildup of offensive arms.

Slowly, through the 1970s and 1980s, arms agreements and treaties relaxed tension between the antagonists, and the demise of the Union of Soviet Socialist Republics extracted a major thorn from the side of Western society.

Nevertheless, not enough has been done to relax critical-decision reaction time — measured in minutes, rather than hours or days, or indefinite. Improvements have been mostly grudging or fortuitous. There been little lessening of the decision-makers' burden to launch a salvo that would kill millions of people in retaliation to a threat that might be spurious.

With termination of the Cold War, contemporary factors are now driving the two superpowers (1) to ease the financial burdens of maintaining arsenals and (2) to make significant reductions in the tens of thousands of nuclear warheads — which are incongruous during a time of peace and international cooperation.

A single word suggested to summarize the Cold War quest for national security is **hyperthermia***: an abnormally high institutional "temperature" — nearly self-destructive — reached by the superpowers in attempting to preserve or project their political and ideological identities.*

The next Chapter looks more closely at what we learned from the near-fratricidal Cold War experience. The residual consequences, the hangovers, passed on to future generations, are detailed in Volume 2.

Chapter III:
NUCLEAR LESSONS
(Do We Learn?)

A. MYTHS AND REALITIES 170
Who Was Ahead? 171
   Bomber Gap 171
   Missile Gap 171
      [Bang for the Buck]................................. 172
   Window of Vulnerability................................... 172
   Beam Gap 173
Initiatives for Peace and Disarmament 174
   Acheson-Lilienthal Plan 174
Arms Control Negotiations and Treaties 176
   Negotiations During the 1980s 177
Alleged Treaty Violations 179
Half-Truths and Misconceptions 180
   Half-Truths About the Arms Race 181
      ??? Political rivalries do not matter — (the problem is runaway technology) ??? 181
      ??? The arms race was accelerating ??? 181
      ??? The arms race was accelerating ??? 181
   Misconceptions About the Arms Race 181
      ??? Overkill exists in nuclear arsenals ??? 181
      ??? The Soviet Union followed the U.S. lead ??? 182
   Fallacies About Arms Control 182
      ??? Political instruments (arms control agreements) cannot constrain technology ??? 182
      ??? Arm- control agreements inevitably benefit the Soviets ??? 182
      ??? Technological innovations are always destabilizing and undesirable ??? 182
      ??? The Soviets must be trusted ??? 183
Verification as a Red Herring 184
   OSI 185
   SLCMs 185
   ASAT 186
   MIRV 186
   Warheads 187
Perspective About Alleged Violations 188
   Treaty Experience 188
   Symmetry 188
   Vagueness... 188
   Military Significance 188
   Balance of Terror 189
   Constraints on Treaties 189
   Verification Improvements 190
   Intangibles 190
Alternative Courses of Action 191
   NATO Nuclear Issues 191
   Offensive Strategic Forces 192
   Command & Control 193
   Nuclear Defense 193
The Role of Espionage 194
   Atomic Spies 195
   Thermonuclear Espionage 197

B. THE UNTHINKABLE 199
Strategic Assessments 199
   [Nostril to Nostril] 201
War Planning 202
   Second Thoughts 205
Security Paranoia 205
   McCloy-Zorin Agreement 206
   Scowcroft Commission 206
Worst-Case Scenarios 208

Nuclear Bravado 209
   Nuclear Threats 210
      [Nuclear Weapons Use Considered by Eisenhower] 210
      [Lessons From the Cuban Missile Crisis] 211
      [DEFCON: Defense Condition] 212
   Nuclear Illusions 214
      [Doomsday Machine] 215
National Security at Any Cost 216
   Radiation Effects of Nuclear War 216
   Denver's Close Call 217
   K-19: The Widowmaker 218
   Deterrence Theory 218
Doomsday 219
NATO's Nuclear Deterrent 222
Broken Arrows 225
Accidental or Unauthorized Launch Risks 227
   Nuclear Safety and Authorization 228
Radiological Warfare 229

C. THE NUCLEAR PRIESTHOOD 232
Think Tanks 233
   RAND 234
Military-Industrial Complexes 237
   Beltway Bandits 238
   Armaments Lobby 239
   Soviet Defense Establishment 240
   Technological Innovation 240
The Scientific-Technological Elite 241
   [Livermore] 243
   The Consummate Role of Secrecy 243
JCAE 247
Reagan's Intractability 248
   The "Evil Empire." 249
   The Revelation 251
Hardliners 252
   Committee on the Present Danger 253
   Perle 256
   Pipes 257
   Nitze 257
   Teller 259

D. COST OF THE COLD WAR 263
Dollars for Defense 263
   Foreign Aid 263
Distorted Statistics 265
Economic Impact 266
   Stimulating the Economy 267
   Infrastructure 267
   Productivity 267
   Jobs 268
   Inflation and Taxes 268
   Research and Development 269
Post-Cold-War Impact 269
Nuclear Worthiness 270
Rubles for Defense 271
Vietnam War 273
Indirect Costs 273
   Psychological Costs 274
   Radical Compromises 275
   [The Enemy of My Enemy] 275

Chapter III: Nuclear Lessons

    Nuclear Weapons and Information Proliferation 275
        [The Enemy of My Enemy (Getting to America)] 276
    Nuclear Institutionalization 276
        [The Enemy of My Enemy (In America)] 277
Who (What) Won the Cold War? 277
    The Cold War as History 277
        [CIA papers show agency knew where hunted Nazi was hiding] 278
    The Cold War Was Won Before Reagan 278
    Fallacies of Monopoly and Secrecy 280
    Did Nuclear Deterrence Prevent War? 281
    We All Lost the Cold War 282
    Peace Works 283
        [Personal Recollection About the Cold War *(Vadim A. Simonenko)*] 284
    CIA Findings 285
Benefits of the Cold War 285
    Who, Where, and When 287
    What 288
    Why 292
    How 294

"Can we still remember that to destroy hundreds of millions of human beings is an atrocity beyond all history?"

How did we get into a situation where such a question could even be asked? And how did we avoid the ultimate barbarity?

Answers to these and other queries about the Cold War in the perilous half century require that we retrace the path of tenuous coexistence. What did it mean, after all, to be ahead or behind in the nuclear arms race? How close to the precipice did we come, how much did it cost, and who — if anyone — "won"?

What were the driving forces, and who were the drivers on this unmarked route, where one wrong turn could have resulted in mutual devastation?

A. MYTHS AND REALITIES

Claims and counterclaims — about nuclear superiority and inferiority — were often voiced during the Cold War.

While at the time some of the assumptions were pertinent, others had mythic underpinnings. In this Section, the real from the unreal is winnowed down, aided by the passage of time and the subsidence of passions. In later Sections, some Cold War driving forces and the ultimate costs — monetary and otherwise—are examined.

Section A: Myths and Realities   171

Who Was Ahead?

"Ahead" was difficult to gauge at the time. Did it mean more, or did it mean better? Partisans of big numbers sought to impress with their analyses. Frequently, in the absence of hard facts, estimates were off the mark. Deliberate secrecy by all parties added to the difficulties; intelligence sources were of limited value and often politicized; strategic thinking was sometimes outmoded. For example, while emphasis on the quantity of weapons is perhaps appropriate for conventional warfare, its relevance is questionable when one nuclear weapon can destroy an entire city.

The so-called bomber and missile "gaps" of the 1950s and early 1960s were two politicized sound-bites whereby Soviet military strength was overestimated for partisan political advantage. Another trumped-up travesty was the "window of vulnerability" in the 1980s. Even a so-called "beam gap" was projected.

Let's take a retrospective look at these controversies.

Bomber Gap. Having started many years behind the United States in deliverable strategic weaponry, the Soviet Union attempted in the 1950s to build a long-range force by developing two bombers with intercontinental range, the *Bison* jet and the *Bear* turboprop. U.S. Air Force Intelligence persuaded the CIA that a "bomber gap" loomed because *Bisons* were being produced faster than American bombers. To counter this feared disparity, the Air Force in the mid-1950s began to procure in large numbers more advanced B-47 bombers, and later B-52s. While U.S. intelligence agencies estimated that as many as 500 Soviet *Bisons* were to be produced by the early 1960s, only 200 were ever built.

As it turned out, technical deficiencies had rendered that generation of Soviet bombers to be an unsatisfactory means of delivering nuclear weapons; so the bombers were never produced in large numbers. Instead, the Soviets put considerable effort into anti-aircraft defenses to protect their cities from the growing American bomber threat. In reality, the bomber gap that finally emerged left the Soviet Union with numerical *inferiority*. A retrospective assessment of intelligence data found that the United States was favored in 1958 by a ratio of nearly 21 to 1 (1769 vs. 85 *total* bombers).

The USSR had quietly capped their long-range bomber production because they were turning their attention to ballistic missiles.

Missile Gap. As the 1950s ended, a new claim of a developing disparity in intercontinental missiles drew attention away from the nonexistent bomber gap. While the initial liquid-fueled SS-6 Soviet ICBM, was large and cumbersome, it was thoroughly tested and was probably operational by 1960. Since U.S. coverage of the USSR by U-2 reconnaissance planes was constrained, actual Soviet ICBM deployment was difficult to ascertain at the time. Air Force Intelligence postulated widespread deployment, the Navy came up with a very low number, and CIA estimates were in between.

After *Sputnik* was launched in 1957, the CIA estimated that the Soviet Union could have 500 ICBMs as early as 1961. The issue became highly politicized, with both Senators Stuart Symington and John F. Kennedy exploiting the alleged imbalance during their respective presidential campaigns in 1960. Having made the status of U.S. strategic missile forces an election issue, Kennedy gained office promising new vigor in national security. Although the missile-procurement program he inherited was already extensive, Kennedy accelerated it in his first six months in office. But after U.S. reconnaissance satellites came into operation in August 1960, it was discovered that the Soviet Union had not gone forward with extensive deployment of the SS-6: the "missile gap" of the 1960 election never materialized. Just the opposite: by 1964 the United States had deployed 834 missiles, the Soviet Union only 120.

> [Bang for the Buck]. Military slang believed to have first been used when the Air Force was making the case for government funding of ballistic missiles, which, it was argued, could do more damage than a Navy aircraft carrier and therefore represented a better investment.
>
> The phrase was later used to express the (questionable) idea that nuclear weapons were far cheaper than conventional forces for projecting military power.

Not long after the 1961 Bay of Pigs debacle, the Soviets began a series of nuclear-weapon tests in the atmosphere, which served to reinforce Kennedy's decision to expand nuclear forces. This was carried out even though underlying intelligence information was indicating that U.S. ICBM production was substantially ahead and would be for the foreseeable future. Because deployment information was secret, little informed public discussion could take place. As time passed, the Soviet Union fell farther behind in the arms race.

Missiles were not the only measure of comparative strength. In 1974 President Carter's defense secretary, James Schlesinger, was quoted as saying that "by 1980 the Soviet Union could have 7000 one-megaton warheads capable of attacking our [missile] silos." As a result, the administration asked Congress for major increases in research and development of more accurate guidance, larger land-based missiles, and possibly a mobile missile. The Soviets did have perhaps 7000 warheads by 1980, but many had yields much smaller than one megaton and were delivered by rockets not accurate enough to destroy missile silos with those warheads. Their most powerful warheads were bombs that had to be delivered by aircraft, which could not make a surprise attack against missile silos.

In 1981 President Reagan inherited an ICBM buildup program that fitted his political commitment to expand U.S. forces. Even so, it wasn't enough for hardliners, who had helped Reagan get elected by proclaiming a "window of vulnerability" to pre-emptive Soviet missile attack.

Window of Vulnerability. In the early 1980s, the U.S. defense establishment proclaimed a "window of vulnerability, contending that its Titan and Minuteman ICBMs were vulnerable to a disarming first strike by large and "highly accurate" Soviet ICBMs." Even if it had been true, the proposed remedy that Reagan promoted — the multi-warhead MX — would itself have been a prime target for pre-emptive strike.

Here is how Senator Daniel Moynihan understood the events leading up to the alleged window of vulnerability:

> In 1976 a group of outside hardline nuclear analysts was invited by CIA director George [H.W.] Bush (over the objections of his top analysts) to review the data used to develop the national intelligence estimate and issue its own findings. The "Team B" report concluded that the CIA had seriously underestimated the Soviet threat, and was used by conservatives throughout the 1970s and 1980s to support large increases in defense spending.
>
> The B Team made a powerful case — more so than had been anticipated. In October, word of the exercise leaked out; in December the *Times* reported the results. The B Team, headed by Richard Pipes, of Harvard, had come to the conclusion that the Russians were seeking strategic superiority.
>
> In throw-weight — the pounds of "payload" that can be sent aloft — [Paul] Nitze estimates that the Soviets by 1977 had an advantage of 10.3 million pounds to the United States' 7.6 million, this being the effect of the Soviet heavy missiles. By 1985, he projects a widened gap: 14.5 million [pounds] for the Soviets, eight million for the United States. The gap is even more dramatic in the critical category of explosive power — in what is called "equivalent megatonnage." Nitze gives the Soviets a nearly three-to-one advantage for 1977: 9319 equivalent megatons for the Soviets, 3256 for the United States.

At the time of the B-team report, a first strike could not have deprived the United States of its ability to deliver devastating retaliation with its strategic triad of ICBMs, SLBMs, and bombers.

DOE has since acknowledged that as early as 1960 the total megatonnage of U.S. weapons in the active stockpile was about 20,500 megatons, and the Soviets did not catch up until the mid-1970s. The B-team had chosen a misleading parameter to optimize their arguments; the United States had deliberately moved in the direction of lightweight warheads with more accurate delivery capability.

Meanwhile, the MX mobility plan was abandoned, largely because of publicized objections, especially from people and officials in the western deployment states. The MX missile eventually replaced some silo-based Minuteman strategic rockets. The alleged window of missile vulnerability never materialized.

Beam Gap. Allegations were made during the 1970s that there was a "beam gap" too, referring to a supposed U.S. shortfall in ability to deploy laser- or particle-ray systems that could destroy ballistic missiles from ground- or space-based platforms. These ideas were pushed by Edward Teller and his allies from the Committee on the Present Danger and the High Frontier.

Lasers were then in their infancy regarding beam intensity, atmospheric penetration ability, and target-destruction capability. Also, speed and reliability of target-tracking technology were not yet up to the job. Even greater technical challenges faced beam targeting for neutral and charged particles.

However, intelligence sources discerned that the Soviets had initiated some research and development; so it was not long before this was leaked, without appropriate qualifications, to eager hardline ears. While the Soviets did begin some

R&D on lasers, they never did achieve any practical success in applying them to ballistic-missile defense.

Of the four alleged "gaps" and "windows" in national defense described in this subsection, none was supported by reliable intelligence data or analysis. A bomber gap, indeed existed, and a missile gap did develop—but both strongly favored the United States. The window of vulnerability never opened, and the beam gap was fantasy.

Meanwhile, spurious claims had been made to gain political advantage, resulting in policy decisions that raised the ante in armaments and expenditures.

Initiatives for Peace and Disarmament

The very first resolution of the U.N. General Assembly in January 1946 called for "the elimination from national arsenals of atomic weapons and all other major weapons adaptable to mass destruction."

Soon after World War II, specific initiatives to prevent or curb a nuclear arms race were undertaken by the superpower adversaries and by hopeful intermediaries. Fearing a dangerous outcome, private parties or groups also advanced some initiatives, one of which emerged with government backing and named after two of its foremost backers, U.S. diplomats Dean Acheson and David Lilienthal.

This earliest concept for nuclear disarmament was introduced by Robert Oppenheimer and backed by the Chicago-based atomic scientists' movement. The impetus for banning atomic weapons came from scientists who conceived and built the bomb, such as Niels Bohr, Leo Szilard, and James Franck.

Acheson-Lilienthal Plan. Proposed in 1946 to implement the first UN resolution, the plan was to put atomic energy under international control. The hope was that national secrecy and suspicions that originated in wartime projects would be overcome by creating an international agency (dominated by the great powers) that would make atomic energy available worldwide, while retaining control of all aspects of development and implementation, from mines to power plants. But the emerging icy relationship between East and West dampened enthusiasm for the initiative.

The Acheson-Lilienthal plan was to share technology before nuclear capabilities gave rise to an arms race. However, Bernard Baruch revised the concept, adding preconditions that undermined chances for agreement. He called for inspection and control mechanisms applied to all facilities, including those within the Soviet Union, before the United States would agree to turn over its weapons and nuclear materials to international management. But the Soviets, who had yet to test their first nuclear weapon, viewed the agreement as lopsided: It would allow the United States to dominate the international agency that carried out the inspections and had the power to impose sanctions. The Baruch revisions were "obviously designed [by the United States] to be unacceptable to the Soviet Union."

A Soviet counterproposal for international control was also unrealistic, calling for dismantling atomic weapons immediately and giving the international agency only limited powers of inspection and no authority to impose sanctions.

Soviet nuclear initiatives, often dismissed in the West as propaganda, began in 1946 too with a proposal for a "complete and unconditional ban on manufacture and use of atomic weapons." A year later the Soviet Union suggested "strict international control of atomic energy." International inspections were to be limited to non-weapons activities (ultimately adopted). In subsequent years, the Soviets frequently proposed banning atomic bombs, including destruction of then-existing weapons. In 1958 the Supreme Soviet ordered "a unilateral halt to tests on Soviet territory of nuclear and thermonuclear weapons." The USSR in 1963 agreed to an international ban on nuclear tests in the atmosphere, outer space and under water. In the ensuing two decades the Soviet Union and the United States negotiated and signed several arms control treaties and agreements.

Back in April 1946, some American scientists had issued a statement on "denaturing" weapons-grade nuclear materials, the idea being to make the materials inherently much less useful for weapons. The proposal never made much headway, and controversy on denaturing plutonium has persisted to this day. At the end of 1946 a UN commission on atomic energy rendered its report to the Security Council, advising on measures for taming the new capabilities of nuclear physics so they could be applied in controlled peaceful international uses.

After Josef Stalin's death, the term "peaceful coexistence" began to appear. In 1953 the Eisenhower administration introduced the "Atoms For Peace" program, placing a greater effort on limiting the spread of nuclear weapons while assuring peaceful uses of atomic energy. In 1954 Britain and France presented a comprehensive blueprint for general and complete disarmament — a plan that induced the Soviets to accept key provisions, including the principle of inspection. Soon afterward, however, the United States disowned proposals it had previously backed, on the grounds that "low levels and drastic reductions in armaments would ... increase the danger of the outbreak of war."

When President Kennedy came into office in 1961, he authorized presidential advisor John J. McCloy to discuss the possibility of nuclear disarmament with his Soviet counterpart, Valerian Zorin; their meeting and correspondence resulted in an eight-point proposal for general and complete disarmament, a proposal endorsed by the UN General Assembly. This was the most detailed and comprehensive disarmament plan ever developed by the U.S. government, and the 1962 draft specified a three-stage process of arms reduction.

The Soviets submitted their own plan, which would have led to complete elimination of nuclear weapons, deep cuts in conventional forces, a halt to weapons production, international inspection, and strengthened UN forces.

Timetables for both plans were so optimistic that they seemed to be closer to propaganda than realism (the McCloy-Zorin plan became a target for rabid hardliners). Meanwhile, the arms race was being pushed into higher gear despite the talk of disarmament. In any event, the Cuban missile crisis of October 1962 eclipsed the McCloy-Zorin plan, ending for rest of the century any suspicion that U.S. policy favored nuclear abolition.

The negotiations from 1958 through 1989 have been thoroughly chronicled by historians. While disarmament never came about, and peace existed only as a state of mind (the absence of nuclear conflict), a little progress was made on agreed restraints for some types of weapons, aimed at throttling their unfettered development.

Arms Control Negotiations and Treaties

Despite an atmosphere of hostility between the superpowers, many treaties and agreements were put in place. They dealt with commerce, diplomacy, international conventions, conventional arms control, and nuclear arms control. The latter included treaties and agreements to prevent the spread of nuclear weapons, reduce the risk of nuclear war, limit nuclear weapons testing, and reduce the number of deployed nuclear weapons.

The high points of arms negotiations in the 1960s were SALT I/II and the test bans. None of those treaties actually reduced nuclear arms, as peace groups would have liked; they tended to accomplish what governments wanted. But, by putting a numerical cap on the arms race, the military threat was defined and limited.

Another highlight of the 1960s was formal recognition of the dangers of proliferation, resulting in a series of treaties attempting to limit the spread of nuclear weapons beyond the acknowledged states.

Arms control negotiations during the 1970s floundered for years over disputes about weapon capabilities. The Soviet Union favored agreements that would result in equal capabilities, rather than equal numbers; the United States emphasized quantitative restrictions, especially on the number and throw-weight of missiles. For instance, the alleged Soviet "advantage" in throw-weight had the potential for allowing more warheads to be deployed in the future. Moreover, the Soviets insisted that (NATO) forward-based systems located in Europe should count in the overall strategic balance.

At the 1974 Vladivostok summit, the two superpowers reached agreement on the outlines of what turned out to be the SALT II Treaty. The total number of strategic launchers was to be limited; the United States agreed to include strategic bombers in the limits and to refrain from redefining the interpretation of heavy missiles so as to include smaller missiles.

But after the summit, differences in interpretation continued, such as questions about the range of the Backfire bomber and problems with counting rules for cruise missiles. Although limits were set on strategic armaments, and the provisions of the unratified treaty were largely complied with, SALT II did little to affect strategic modernization programs, most of which were already under way. In fact, according to a Russian post-Cold War survey, the "renewed U.S. modernization program caused serious concern among the Soviet leadership, in part because it could undermine Soviet efforts to achieve parity ... and provide the United States with the capability of launching a disarming strike."

Negotiations During the 1980s. As soon as President Reagan came into office (1981), his administration made vocal claims that arms control treaties were unverifiable. Somewhat of a contradiction was his declared "vision" of nuclear arms abolition, which implicitly would be dependent on successful verification by expanded (and intrusive) means. While incoming administration spokespersons were charging Soviet violations, government agencies were verifying continued adherence to existing treaties. In addition, international and non-government organizations (NGOs) were collecting data relevant to and supportive of treaty compliance.

In the United States, intelligence agencies (CIA, DIA, NSA) had a major role in collecting human and technical intelligence, especially by satellite reconnaissance and electronic surveillance. Separately, military services participated in operations that produced information related to treaty compliance. The Department of Defense — including its advanced research arm, DARPA, and its nuclear-custodial agency, DNA — had major roles in information collection and in verification improvements.

At the same time, depending on funding allocated by Congress, other government agencies began to support verification research and development. These included the Department of Energy (through its national laboratories) and the Arms Control and Disarmament Agency.

Various analytical arms of Congress, including the Office of Technology Assessment, Library of Congress, General Accounting Office, and the National Academy of Sciences contributed surveys of verification capabilities. Universities, private organizations, and science foundations carried out their own evaluations.

A favorite argument put forth by right-wing treaty opponents was that the other side was not to be "trusted" to abide by agreements. The flaw in that logic was, of course, that trust never was and never will be a central ingredient in a realistic treaty. What the parties assume is they all will adhere to a treaty as long as they think it suits their national interest, and not a day longer.

The trust problem became prominent during the Reagan era (1981-1989), when it was officially declared that treaties should be designed to be verifiable, without relying on trust. "Trust but Verify" became Reagan's slogan. While this was fundamentally a sound approach to providing confidence in treaty compliance, the slogan unfortunately came to symbolize an unspoken ***distrust*** of arms control.

Despite the belligerent administration rhetoric, a major arms control agreement was signed by President Reagan and Soviet General Secretary Gorbachev at a Washington Summit in December 1987. The INF Treaty eliminated all nuclear-armed ground-launched ballistic and cruise missiles with ranges between 500 and 5500 kilometers (about 300 to 3400 miles), along with their infrastructure. This treaty greatly eased the nuclear confrontation in Europe. The INF Treaty remains a landmark standard for mutual concessions and intrusive inspection to reduce the nuclear confrontation.

Reagan left office two years later, just before a series of agreements were reached to phase out the Cold War.

S T A R T Negotiations. By 1983 it was increasingly realized that MIRVed ballistic-missile systems were more destabilizing than single-warhead missiles. Under intensified pressure from the public and from U.S. allies, it became necessary for the superpowers to give more than lip-service to strategic-arms reduction treaty (S T A R T) negotiations.

Incorporated in the U.S. S T A R T I negotiating position in the early 1980s was the "build-down" proposal — called a "mutual, guaranteed reduction" of strategic forces — endorsed in principle by President Reagan. The controversial proposal was stimulated by the 1983 Presidential Commission on Strategic Forces (named the "Scowcroft Commission" after its chairman, Brent Scowcroft). It recommended *inter alia* that modernization of strategic forces be channeled toward less-destabilizing, single-warhead systems.

However, implementation of the Scowcroft single-warhead suggestion was undercut by U.S. determination to proceed simultaneously with deployment of the highly destabilizing 10-warhead MX system. The build-down rhetoric was considered a "copout" by those who favored realistic, stabilizing reductions. Under build-down, both sides were to reduce the number of delivery systems while permitting "modernization" of the remaining forces, effectively increasing the number of deliverable warheads. The critics pointed out that, under the U.S. formula, stabilizing programs (such as hardened land-based missiles) would be penalized, and the deployment of a vulnerable and destabilizing MX would be rewarded.

Another cause of bilateral friction in the 1980s was the SDI missile-defense program, which the Soviets considered to be an attempt by the United States to alter the strategic balance that had prevailed during the 1970s. Thus, Soviet concessions for S T A R T were conditional on restricting the SDI program, a condition the Americans would not accept.

The Reagan administration also shifted the judgement standard for verification from "adequate" to "effective." This transition in terminology was characterized by Edward L. Rowny, the chief U.S. S T A R T negotiator, in the following way:

> Inadequate verification was one of the major flaws of the SALT II Treaty. This has made it clear that we will insist that any S T A R T agreement be effectively verifiable. [The Soviets] in the S T A R T talks have indicated a willingness to consider cooperative measures [and an on-site presence to witness weapons destruction] to supplement so-called national technical means. [The] Soviets must understand that we will settle for nothing less than a fully and effectively verifiable agreement.

Even if the Soviets did understand U.S. insistence on comprehensive verification, the process of reaching agreement was bound to be difficult when Rowny maintained this entrenched opinion:

> ...the Soviets refer to their principle of "equality and equal security"' [One] sees that it is an Orwellian formulation denoting the Soviet right to superiority.

As the two-term Reagan administration came to and end, rumblings of public dissatisfaction were getting louder both in the West (about the lack of arms control) and the East (about communist control). Partly because of the difficulties in reaching agreement on S T A R T, the INF treaty was suddenly and unexpectedly resolved, with quick implementation and with highly intrusive verification for both parties.

Shortly after, as the Soviet Union was visibly falling apart, accommodation on the START I Treaty was quickly reached, its signing taking place in mid-1991 — with much unfinished business: without settling the ABM compliance issue, without resolving the SDI problem, and without placing legal restrictions on sea-launched cruise missiles.

Alleged Treaty Violations

Charges and countercharges of treaty violations were frequently made by governments and by hardliners during the 1980s. Rarely did the charges have substantive backing; most often the underlying purpose was to scuttle arms control negotiations and agreements.

Legitimate allegations were made about the Soviet Krasnoyarsk Radar, and in 1989 the Soviets agreed to cease construction and dismantle it. The operation of some U.S. radars was also open to question.

Other Soviet practices came under suspicion: the testing of two new types of Soviet ICBMs, encoding of missile data, and deployment of SS-16 at Plesetsk. However, the United States never ratified SALT II, a treaty that would have created mechanisms for resolving compliance issues. Regarding the dispute over whether the Soviets exceeded the limits of the TTBT, the United States did not ratify the treaty that would have put the proper verification in place.

Lack of verification measures in the treaties dealing with chemical and biological weapons left those issues without a means of resolution. Both of the superpower adversaries (as well as other nations) were producing lethal agents in very large quantities. (The real excuse for opposing verification was probably to avoid disclosure of secret chem-bio warfare programs that might have stirred public revulsion.)

Other possible violations were cited by the Reagan administration, some so minor that they could have no strategic significance. At the time, Congressional conservatives were determined to undermine arms control in order to justify a major nuclear buildup; so they created a list of charges — now known to be spurious — in an effort to convince the public and Congress that a pattern of unilateral Soviet violations was occurring. Unsubstantiated, legalistic charges could be made because no mechanism existed for resolution or adjudication, and the public found it difficult to understand the complexities.

In 1979, during debate over ratification of SALT II, the Department of State had released its own official summary of U.S. experience with the Soviet Union's treaty compliance:

> We have had more than six years experience in monitoring Soviet compliance with [SALT I], and we have demonstrated our ability to verify Soviet compliance ... with these agreements....

> We have observed activity which we believed raised questions with respect to Soviet compliance.... **In every case the activity has ceased, or subsequent information has clarified the situation....**
>
> Despite the closed nature of Soviet society, **we are confident that no significant violation** of [SALT II] could take place without the United States detecting it [emphasis added].

In addition, there already had been considerable experience with successful implementation of other treaties — pertaining to the Antarctic, the seabed, Latin America, and outer space. Even so, this positive foundation of treaty experience, extending through decades of Cold War discord, could not withstand the politicized assault that came from the Reagan administration.

The accusations of the mid-1980s induced considerable public debate. A close look reveals that the disputes centered on ambiguities in the agreements. Were it not for the hidden hardline political agenda, the disputes could have been resolved by mechanisms that were in place: ratifying treaties already negotiated, utilizing the consultation processes set up by existing treaties, improving methods of verification, and — if needed — clarifying ambiguities through renegotiation.

Many of the U.S. hardliner allegations were aimed at the SALT II and Threshold Test Ban agreements, which had been ratified by the Soviet Union but not by the United States. If treaties had been in full force, verification mechanisms would have been available to clarify ambiguities.

A publicized outbreak of anthrax near Sverdlovsk was cited as a violation of the BWC ban on producing and stockpiling biological weapons. Because outside inspectors were not invited (or authorized under the BWC), it was not possible to verify the allegations. It has since been confirmed that the Soviet Union did, in fact, secretly violate the treaty.

The United States was not without inadvertent or deliberate violations: in 1984, Congress required that the Pentagon report "on the implications of deploying a new nuclear submarine, the U.S.S. Alaska, in apparent breach of SALT II limits on nuclear launchers."

While both sides sometimes fell short of rigorous compliance with their negotiated obligations, the balance of terror was never in jeopardy. Even if all the above allegations had been valid, no significant impairment of anyone's national security would have been threatened — there would have been no consequences of strategic importance.

Half-Truths and Misconceptions

In 1982, an independent coalition of knowledgeable individuals, known as the Harvard Study Group, searched for lessons that could be gleaned from more than three decades of confrontation. Now, in retrospect, one can look at the Group's conclusions regarding the arms race, arms control, and the "inevitability" of nuclear arms competition. Since their study made such a big splash, their analysis and recommendations will be discussed extensively in the remainder of this Chapter.

Section A: Myths and Realities 181

Half-Truths About the Arms Race. In the boxes below are statements (surrounded by ???) about the arms race, arms control, and verification that the Study Group addressed, along with a condensed verdict as to whether each was a half-truth, misleading, or wrong.

The Harvard team concluded that several common beliefs about the nuclear arms race were only half true:

> ??? **Political rivalries do not matter — (the problem is runaway technology) ???**
> **Half-truth.** Technology is advanced only by choice, as reflected in funding decisions: it is subject to political control. However, domestic political rivalries hindered resolution of long-standing disputes about Soviet hegemony and the export of militaristic communism.

> ??? **The arms race was accelerating ???**
> **Half-truth.** Increased technical sophistication, accuracy, and diversification of delivery systems made part of this assertion true. On the other hand, comparing 1964 with 1982, the number of missiles, bombers, and weapons had stabilized, and strategic forces consumed less than about 15 percent of the total U.S. defense budget.

> ??? **The arms race was accelerating ???**
> **Half-truth.** Increased technical sophistication, accuracy, and diversification of delivery systems made part of this assertion true. On the other hand, comparing 1964 with 1982, the number of missiles, bombers, and weapons had stabilized, and strategic forces consumed less than about 15 percent of the total U.S. defense budget.

Atomic historian David Holloway concludes that even with a nuclear monopoly (1945-1953), the United States could not force the Soviet Union "to do things it did not want to do." Once it developed its own nuclear weapons, the Soviet Union actively participated in the arms race by developing an arsenal to fit its own technical, geographical, historical, and political conditions. Coupled with the dominance of "worst-case analysis" as an analytical tool, secrecy on both sides fueled the arms competition by making it easy to sell the idea that the worst had to be assumed in the absence of validated information.

Misconceptions About the Arms Race. Some facets of the arms competition, according to the Study Group, were not so much half-truths as they were pervasive misconceptions.

> ??? **Overkill exists in nuclear arsenals ???**
> **Misleading.** While the capacity of nuclear arsenals to inflict huge casualties and devastation existed, the weapons were not aimed directly at people. And, concluded the Harvard Study Group, redundancy was needed in weaponry.

> **??? The Soviet Union followed the U.S. lead ???**
> **Half-truth.** It was true that the United States usually was first to introduce new weapons, its technology being stimulated by the unknown — an uncertainty enhanced by Soviet secrecy and obfuscation. But the Soviet Union had a number of firsts: the deployment of ICBMs, an initial ABM system, ASATs, heavy ICBMs, reloadable ICBM silos, a super H-bomb (tested at 59 megatons in 1961), and FOBS in 1967.

Here, the Harvard Group erred in labeling as "misleading" the prevailing impression of nuclear overkill; the Group was too deferential to assurances from the military and the government: A longer-term retrospective makes it clear that there were more nuclear weapons than needed for assured deterrence. The Group made itself an apologist for the policy of keeping more than one form of invulnerable retaliatory force. An even more significant issue, however, was (and remains) not the form but the number of powerful weapons. The Group's arguments tended to discourage significant reductions in levels of armaments.

Fallacies About Arms Control. Conclusions reached by the 1982 Harvard Study Group about some other then-raging issues in arms control and verification were as follows:

> **??? Political instruments (arms control agreements) cannot constrain technology ???**
> **Half-truth.** To the contrary, some treaties have successfully created limits or a total ban on nuclear weapons in certain domains: The Outer Space Treaty prohibits nuclear weapons in space, the Seabed Arms Control Treaty limits nuclear weapons employment, and the ABM treaty constrains testing of technologies.

> **??? Arm-control agreements inevitably benefit the Soviets ???**
> **Misleading.** Treaties are necessarily crafted by a process that benefits both sides. Without the mutual perception of benefit, the treaties would never have been ratified and placed in force.

While treaties, such as the three mentioned in the box above, ban deployments, they do not constrain any technologies. But nuclear testing is an important exception; treaty restrictions do help keep a technological lid on both horizontal and vertical proliferation.

> **??? Technological innovations are always destabilizing and undesirable ???**
> **Wrong.** Many technological innovations can be adapted for warfare; e.g., aircraft, missiles, nuclear reactions, explosives. However, a few examples indicate how wrong the premise was: Satellites became an important part of direct or indirect arms control enforcement, naval nuclear propulsion resulted in a capability for invulnerable retaliation, and PALs minimized the possibility of accidental or unauthorized firing of nuclear weapons.

Section A: Myths and Realities

> **??? The Soviets must be trusted ???**
>
> **Half-truth.** Lack of confidence in treaty compliance was a major Cold War arms control issue. Yet, already on hand were many encouraging examples: the Antarctic treaty, the NPT, the verification provisions of SALT II, and the ever-readiness of NTM. The old Russian proverb "trust, but verify" (*doveryai, no proveryai*) was used by President Reagan a cautious philosophy.

Overall, the Harvard Study Group likened the nuclear arms competition to a *track meet* rather than a single race: There were many challenging "events," and each side's score was tallied from its performance in events such as the evolution of nuclear forces, declared policies, targeting policy, and arms control policies and agreements.

Was the nuclear arms competition inevitable? The Harvard Group drew the following conclusions:

1. *Genuine security requirements* provided an incentive for an arms buildup, with nuclear arms compensating for relative weakness in conventional military power.

2. *Uncertainty and misperceptions* between superpowers were the main causes of arms buildup, resulting in an action-reaction cycle. This cycle was based on "illusions, mistakes, or inadequate military intelligence." Officials on both sides tended to focus on the worst case, a situation aggravated by excessive official secrecy.

3. *Pressure from politicians, buttonholing by military services, and lobbying by industrialists*, i.e., the military-industrial-laboratory complex (MILC in the United States, "metal-eaters" in the Soviet Union), who promoted their various domestic causes, were a major influence. Sometimes the U.S. *ad hoc* alliance of Congress, the Pentagon, and the defense industries was referred to as the "iron triangle." (Because their role was important, if not well recognized, we consider the nuclear weapons laboratories to be worthy of explicit mention as part of the lobbying structure. Their interests tended to coincide with those of the military and industrial factions.)

4. *Technological determinism* was a factor. (The "sweetness" of atomic-bomb physics and other weapons technology intrigued many scientists.)

5. *International political rivalry* in the quest for global power provided motivation both to continue the arms race and to control or manage the arms competition.

According to Thomas Naylor, considering the human dimension to the Cold War,

> Exaggerated fear of an opponent is one of two forms of delusion often associated with paranoid personalities.... Macho pride ... often accompanies an overstated fear of an opponent's strength and hostility....
>
> To support the claim that the [USSR was] an evil empire, the Reagan administration embarked on a broad-based campaign of accusations of Soviet adventurism, Soviet arms-treaty violations, Soviet war plans, Soviet use of chemical warfare, Soviet human-rights violations, and Soviet support of international terrorism.

(Ignored, typically, was the U.S. record of having systematically violated more than 300 treaties with Native Americans.)

Naylor was convinced that "With the exception of the 1957 launching of Sputnik, the Soviets did not initiate any other nuclear-weapon development." (Notably, Sputnik was not a nuclear weapons development, *per se*, but was part of a program that led to a weapon-delivery system — the ICBM.)

"Very often," says Naylor, "when [the United States] accused the Soviets of developing excessive military capabilities, they were simply trying to reach a state of parity with the United States." For instance, the Warsaw Pact was not formed until six years after NATO. Naylor notes that the Reagan administration

> created a myth of Soviet noncompliance by repeatedly representing, as established facts, allegations about Soviet behavior that represented "worst case" interpretations based on ambiguous evidence.

Verification as a Red Herring

To divert attention from its own freewheeling arms buildup, the Reagan administration often used verification issues as a red herring: Paul Nitze (well known as a hardliner), when reviewing developments in arms control negotiations that he was conducting in the mid 1980s, frequently cited verification obstacles.

For INF, for example, he declared — without specificity — what had become a superficially self-evident, but nevertheless disingenuous mantra: "We require that any agreement be verifiable." Despite Nitze's opposition to arms control, or perhaps helped by his insistence on verification, years later the formative mutual verification provisions in the INF Treaty broke all precedent.

Referring to the Reykjavik talks, Nitze saw "movement" on verification, meaning Soviet "agreement to the U.S. proposal that an effective verification package must include [data exchanges and] onsite inspection...." (It was always the Soviets, according to Nitze, who had to be brought along.) Nitze added that the United States must "insist on adequate verification measures" (whatever that might mean). Because of its closed society, the Soviet Union was inherently resistant to any intrusion that did not appear to be warranted, limited, and reciprocal.

For S T A R T negotiations, Nitze identified several issues that separated the United States and Soviet Union, including "inadequate progress on verification, especially of SLCMs and mobiles. " Nitze concluded that "We must work on identifying adequate verification measures." For "defense and space" (i.e., SDI), the United States did not make an issue of verification (because it wanted no treaty); thus Nitze saw no problem in going without verification.

In 1984 the following official Soviet understanding was reported about U.S. verification capability: "Reagan's national security advisor ... says that the United States cannot agree to limit certain nuclear weapons [e.g., cruise missiles and ASAT weapons] because it does not have the technology to ensure Soviet compliance."

How robust were the tools for treaty verification? Was it possible to meet Nitze's elusive standards of "adequate verification"? In essence, yes. Several separate and distinct methods existed: national technical means, intelligence techniques, on-site inspection (when authorized), cooperative measures — all assisted by advances in verification technology.

NTM included land, sea, air, and space surveillance. Monitoring of nuclear tests through national and international seismic networks was well established. Human intelligence (spying and collection of public information) was commonplace. Formal cooperative data exchanges existed in several treaties. Besides the agreed cooperative measures implemented, on-site inspections were in effect and in use for three treaties and the Antarctic Treaty.

One important U.S. verification project was Project VELA, a multifaceted NTM program created to help monitor compliance with the Partial Test Ban Treaty of 1963, and later the Threshold Test Ban Treaty of 1974, by detecting nuclear explosions in the atmosphere and in space. A surveillance system called VELA Hotel used pairs of satellites to recognize nuclear explosions in space or on the earth's surface. A ground-based network designated VELA Uniform operated seismic and other equipment to detect underground and underwater explosions, and another named VELA Sierra applied earth-based equipment to sense explosions in the atmosphere or in space. To assess and improve VELA Sierra's detection capabilities between October 1963 and July 1971, the Atomic Energy Commission conducted seven nuclear tests (four in Nevada, two in Mississippi, and one in Alaska).

In order to verify compliance, the U.S. Air Force Technical Applications Center operated the Atomic Energy Detection System, a network of some 100 sites in more than thirty-five countries capable of detecting nuclear detonations underground, underwater, in the atmosphere, and in space. Data was collected 24 hours a day from monitoring sites.

OSI. One of the biggest stumbling blocks was accessibility to sensitive controlled facilities and treaty-limited weapons. The two sides were highly distrustful of each other, and wary of inspections either by the opposing side or by international inspectors. On-site inspection (OSI) was, during much of the Cold War, anathema to the military and the weapons labs, who viewed the visitation of inspectors as a form of espionage. The labs feared their "crown jewels" — their weapons secrets — would be disclosed to the "enemy."

The ambit of OSI access had to be mutually accepted and limited by agreement. Physical presence of inspectors was necessary because many objects that might have to be verified were not visible to satellites — they were in buildings or underground — and thus needed visual confirmation. For example, under the proposed S T A R T Treaty, rocket motors and possibly RVs had to be counted.

Analysis showed that it would be possible to regulate the inspection process so as to avoid disclosing sensitive information, using a concept called "controlled access"— later adopted in the Chemical Weapons Convention as "managed access."

One of the biggest political problems that evolved was when the Soviets agreed to the pioneering inspection process: The surprised U.S. Executive Branch had to accept its own proposals and convince its agencies to do so too.

SLCMs. Cruise missiles that could be launched from naval vessels (SLCMs) were a difficult challenge for arms control because of their relatively small size and their ambiguity (they could be armed with nuclear or conventional warheads — that is, strategic or tactical weapons).

In an effort to overcome an impasse in negotiations, NGOs combined their resources and their technical advisors in a search for methods of verification. The

U.S. government resisted the concept, although the Soviets were cooperative and offered access to one of their nuclear-armed cruise missiles on the naval cruiser *Slava,* which patrolled the Black Sea. In what became an unprecedented demonstration, known as the Black Sea Experiment, a team of United States and Soviet non-government scientists was allowed to make external measurements of the mechanical and radiation characteristics of a designated cruise missile aboard the *Slava*.

The experiment, as a pilot program, was a success: Measurements showed that on-site inspection with the proper instruments did indeed determine non-intrusively that the SLCM was emitting radiation that would be expected from a nuclear weapon of a particular fissionable composition. For a real-life inspection, that would be sufficient evidence to classify the SLCM as nuclear. But before such a procedure could be built into a treaty, to make sure that there would always be observable radiation if a nuclear warhead were present, the procedure would have to be replicated under carefully controlled experimental conditions, covering the range of possible warhead types.

ASAT. The Soviet Union had undertaken development of an anti-satellite capability in the 1960s, co-orbiting a space target in 1968. By 1971, they were able to demonstrate an ability to destroy satellites orbiting "at altitudes ranging from 250 to 1000 kilometers." A ground-launched system was commissioned and put on combat duty in 1979; Soviet ASAT tests were unilaterally suspended in 1983.

The United States concentrated on land-based laser ASAT systems, using high-power carbon-dioxide and chemical lasers developed rapidly from the 1970s into the 1980s. Even today, ground-based lasers apparently have strictly limited ability to damage military satellites.

Meanwhile, the U.S. Department of Defense claimed that laser development at Sary Shagan and Dushanbe meant that the Soviets could damage satellites in low-earth orbit. In any event, because their space-surveillance radars could not detect satellites in geosynchronous orbits, the Soviet Union had been working on optical and laser systems to detect and track space vehicles.

Despite the importance of satellites for strategic reconnaissance, no mutually agreed controls were ever reached over weapons that could be used to destroy satellites in the early stages of an all-out nuclear war. Without reconnaissance information from satellites, it could become extremely difficult to descend from a ladder of nuclear escalation. Recognizing this, the FAS embarked on a project to show that technology existed to verify an ASAT arms control agreement.

A cooperative FAS ASAT project produced detailed analysis confirming that a combination of instruments and inspections could forestall development of ground-based lasers which had sufficient power to damage satellites. It was ascertained that in-country monitoring with ground-based equipment could determine laser brightness, thus verifying that the energy density of a laser that could be aimed at a satellite did not exceed an agreed upper bound.

MIRV. Whether a missile was armed with one or a dozen nuclear warheads made a big difference in arms control equations. Because it was not possible to determine warhead multiplicity from external appearance, technical means were needed to count the number of warheads from external observations. Argonne National Laboratory scientists demonstrated that, given adequate access to the vicinity of the nose cone, the number (multiplicity) of warheads could be determined with relatively simple instruments without exposing sensitive design information. Notwithstanding various official claims that such measurements would be inadequate or that design information would be revealed, it was shown that, technically, MIRV verification was clearly feasible and had little downside risk.

Warheads. In the long run, arms control and disarmament must involve keeping track of nuclear warheads, which come in many sizes and shapes. Because both superpowers had sophisticated packages for their warheads, there was little sensitive information that would be revealed if access were limited to the outer container of the nuclear explosive device. Thus, a warhead's outer casing could be uniquely marked or tagged to identify it for the purposes of tracking the number of warheads in inventory and the number of genuine warheads dismantled. One way proposed was to make a three-dimensional "fingerprint" of a distinctive surface feature, such as an inscribed identifier normally used to inventory the warhead. A simple process for making a plastic cast of the inscription demonstrated that this three-dimensional fingerprint was a durable and reliable identifier. The technique is potentially a good way to track warheads being dismantled according to post-Cold War agreements.

The intrusiveness of verification was and remains a thorny issue, one that often sidetracked negotiations. Intrusiveness is highly subjective and dependent on what is already known, what knowledge can be protected, and who are the inspecting parties. The risks of intrusiveness can be and have been overstated in order to avoid compromise.

As a result of NGO arms control verification initiatives, alleged obstacles to arms control agreements were one-by-one shown to be political, not technical. With minimal risk of disclosure of truly sensitive information, every object of arms control interest could be verified by a combination of procedural and technological methods.

The Soviet stance on verification was evolving, which could be seen from their modified positions on the CW and CFE treaties. Before the Soviet Union broke up, they agreed to data declarations and intrusive access for monitoring chemical weapons. They accepted instruments and other technical devices for verification. Some difficulties surfaced in CW Treaty verification and monitoring of the civilian chemical industry, stemming in part from differences between private and state forms of ownership; nevertheless, "anywhere, anytime" on-site challenge inspections were agreed to for unresolved crucial situations. Alternative measures to verify compliance were also accepted. In addition, on-site inspection and open-skies monitoring of conventional-force reductions were negotiated.

However, during the height of the Cold War, it is unlikely that the Soviets would have been able to acquiesce to their own offer of on-site inspection for chemical-weapons facilities. Now that wholesale violation of the BWC has been revealed, the Soviet Union would have had to find a way to avoid inspection of biological-agent facilities, which had external characteristics that would have triggered inspection for CW-Treaty-related activities. Evidence of residual biological-warfare agents would have been pervasive, readily detected by intelligence agents accompanying inspectors in the vicinity of or entering biological facilities as part of OSI.

Perspective About Alleged Violations

Most allegations of treaty noncompliance were trivial or invalid.

Treaty Experience. By the early 1980 there was a considerable history of undisputed adherence to bilateral, multilateral, and international treaties by the superpowers. Relevant examples included the following treaties: Antarctic (1961), Limited Test Ban (1963), Latin America (Tlatelolco, 1967), Outer Space (1967), Seabed (1972), and Non-Proliferation (1968). It was misleading of the U.S. government to levy blanket charges that the Soviet Union was not a dependable treaty adherent.

Symmetry. The charges and countercharges by each side (examined by the Harvard Study Group) were notably symmetrical, and often it was very difficult to sort out the validity of each one. The charges by the United States drew responses from the Soviet Union, either in public or in governmental exchanges, some through procedures established formally by the treaties as forums for clarification. When the charges became public, they were often matched by countercharges. What one nation considered to be compliance was interpreted by the other as a violation. Hardliners on each side pushed their claims to the limit, often using them as excuses to justify another cycle of arms buildup. To the bystanding public, the parties sounded like a conclave of lawyers incessantly pleading their own cases.

Vagueness. Many alleged violations (and standards of verification) were couched in deliberately vague terms such as "possible," "likely," and "probable." Much was made of what were described simply as "possible" violations, without quantification. The mechanisms for independently and dispassionately assessing compliance were in need of major improvement. However, it was not clear that the antagonists wanted improvement, because that might have led to resolution — which would have been incompatible with their barely hidden agenda for increased armaments.

Military Significance. The military consequences of the alleged treaty violations were difficult to quantify in terms such as balance of power or mutual assured destruction. While some of the allegations, if true, might have been technical violations of the letter of the relevant treaty (as interpreted unilaterally), none alone gave military advantage to either side. And, most important, despite the uncertainties in compliance, the underlying objectives of the treaties were upheld, and joint consultative commissions resolved many disputes. Some commentators observed that each charge of violation had a political implication that transcended its military significance.

Even hardliner Kenneth L. Adelman, a weapons-lab scientist who was appointed President Reagan's director of ACDA, admitted at one point, "What is its military significance? You could say that none of these violations in and of themselves have immediate and profound military significance."

This is especially true of the now-revealed extensive biological-weapons production program of the Soviet Union which was (probably known to other nations who chose not to divulge the information). While the program represented a brash violation of the BW convention, it never had any military significance other than deterring nations from engaging in biological warfare with the Soviet Union.

Balance of Terror. In 1984, the nuclear balance was clearly stalemated with deterrence by pure terror. About 6000 forward-deployed U.S. tactical nuclear weapons were stationed in Europe, and others were on ships and submarines. Also, France and Great Britain had about 1000 nuclear weapons of their own. In all, nearly 15,000 nuclear weapons were available to NATO to fight a war in Europe; the Soviet Union had far fewer.

The United States by itself at the time had 13,000 nuclear warheads it could explode on the Soviet Union; the latter could retaliate or attack with about 8500. The United States had advanced bases from which it could strike the Soviet Union, and about half of the weapons were on invulnerable strategic submarines; the Soviets had an edge in aggregate explosive power. One way or another, deterrence or catastrophe was assured.

But numbers of warheads and their explosive power were not the only measures. The United States had a qualitative edge in terms of modern missile systems with greater accuracy. Many American missiles were in hardened silos and were propelled with solid fuel, advantages not then available to the Soviet Union. American strategic bombers were high-speed jets that could be refueled in flight, while many Soviet bombers were still propellor driven. The United States was adding air-launched nuclear cruise missiles to the bombers' destructive power. Many Soviet strategic nuclear submarines were kept in port, making them sitting ducks in a pre-emptive strike. The United States was then beginning to deploy the more advanced, more secure, and more deadly Trident missiles.

Plenty of weapons and durable delivery systems were available to assure the incapacitation of all fixed military targets. But whenever the human and societal cost were tallied for even a single nuclear explosion, the overabundance of weapons and secure delivery systems invoked such frightening possibilities that atomic clashes were effectively self-deterred.

Constraints on Treaties. Some of the treaties for which violations were alleged had not even been ratified (e.g., SALT II); others were "agreements," rather than carefully worded treaties fortified with provisions for proper verification and dispute resolution. Thus, it was disingenuous to raise a ruckus about them. Another factor was that many weapons and systems in dispute were mobile, small, or ambiguous; this would make verification difficult, even by means that had already been agreed. Clearly, irrespective of claim validity, additional means for assuring compliance were needed — but not vigorously sought. This is particularly the case for the BWC, where President Richard Nixon's 1969 unilateral renunciation of biological agents and weapons went unreciprocated, given the absence of a protocol establishing verification procedures.

(A post-Cold War wrinkle is U.S. unilateral renunciation of the ABM treaty, which became null and void on 13 June 2002.)

Verification Improvements. One logical outcome of the 1980s compliance disputes could have been improvements in verification procedures and technologies (which, to some extent, occurred). Particularly relevant would have been to seek verification methods that were more accurate, innovative, and informative — without leading to undue revelation of design details or other sensitive information.

In fact, specious claims of inadequate technical means for verification were a particular incentive for scientists to become directly, personally, and professionally involved in improving verification technologies. Areas of investigation included unique identification of fissile material, remote detection of missile launches, and verification of several arms control objectives: multiplicity of warheads, cruise-missile armaments, chemical-munitions content, and explosive yields of underground nuclear tests. While considerable funding was allocated to verification-technology research and development, very little new technology was ever implemented.

While SALT I negotiations were proceeding (late 1960s, early 1970s), the means existed for verifying the number of nuclear warheads mounted on MRV/MIRV missiles.

Eventually the principle of short-notice on-site inspections was accepted, reducing the need for intrusive verification technologies. So-called "cooperative" measures of verification gained ascendency, especially for multilateral treaties; these allowed more exacting procedures, such as calibration of detectors or confirmation of data declarations, to be followed by reciprocal inspections at military sites.

National technical means (primarily satellites) benefitted from improved capabilities for visual and electronic surveillance.

Intangibles. Motives were often difficult to ascertain. Many individuals who exacerbated Cold War rancor had undisclosed agendas — including development of personal power bases, economic incentives, and psychological insecurities left over from World War II. While development and use of atomic explosives had radically changed the calculus of war and peace, many of the older generation still sought solutions based on European history in the first half of the 20 century. Some individuals sought to achieve short-term military or political advantage, without seeing the long-term downside of an unbounded arms race.

No treaty was ever simply based on trust; the catch phrase "trust, but verify" need not have been co-opted as a political slogan: It should have been a fundamental basis for treaty architecture from day one, especially when the stakes were high. In retrospect, one conspicuous solution to the 1980s debate about treaty violations would have been to more rapidly establish an institutionalized verification structure.

Alternative Courses of Action

Another solution, dealing specifically with each nuclear weapons-issue on an *ad-hoc* basis, would have been to introduce constructive alternative verification measures, such as those described earlier, that might have resolved the disputes. But technological solutions during the hyped-up period of tension were not always welcome if they got in the way of political dockets. As a result, many alternatives were driven by military and arms control policy.

The aforementioned Harvard Study Group concentrated on negotiated arms control; rarely did they bring up unilateral self-restraint. During the 1980s unilateral measures were derided as though they represented capitulation. (Later, as the Cold War was winding down, unnegotiated initiatives became a major means for disarming tactical nuclear forces.) Of course, in the early stages of the Cold War there was ample evidence that nuclear arms momentum was not conducive to self-denial; e.g., as Holloway has remarked

> there was no plausible missed opportunity to avoid the development of thermonuclear weapons, because Stalin would not have reciprocated American restraint.

Morton H. Halperin, a civilian national-policy strategist, evaluated three models of nuclear policy available to the United States during the Reagan years: treating nuclear weapons (1) as "regular weapons" to be integrated into all military planning, (2) as "special weapons" to be carefully considered for first use, and (3) as "explosive devices" for demonstrating national resolve — not using them to fight wars. Taking the best and most stabilizing features of these models, Halperin advocated emphasis on survivability, separation from conventional forces, and a no-first-use policy. As for NATO, Halperin was concerned that its nuclear posture resembled Herman Kahn's doomsday machine — a process of uncontrolled escalation that could be set off by any of several triggering events: "Soviet pre-emption, the use-them-or-lose-them syndrome, [or] unauthorized use by battlefield commanders...."

Having shaken many shibboleths, the Harvard Study Group considered alternative ways to slow down the unbridled arms race, based on several criteria: the likelihood and consequences of war, the economic costs, and political implications. They advocated options like negotiated arms control and mutual restraint. The Study Group looked at ten alternatives to specific contemporary nuclear issues, lumped under four categories: NATO Nuclear Issues, Offensive Strategic Forces, Command & Control, and Nuclear Defense.

NATO Nuclear Issues. Three major decisions by NATO (regarding deployment, negotiation, and general policy) impacted nuclear forces assigned to European defense. The first decision, in 1955, was American deployment of nuclear weapons in western Europe to counterbalance Soviet superiority in conventional forces. For storage in 70 European locations by 1983 were 6000 U.S. nuclear weapons — warheads deliverable by short-range ballistic missiles, aircraft, artillery, and anti-aircraft missiles, as well as landmines. The Federal Republic of Germany operated nuclear-capable delivery systems, while the United States maintained custody of the nuclear warheads. Some of the older weapons did not have PALs. Although NATO forces had newer and better weapons, the Soviet Union had more of them.

By 1977, the Soviet had produced their SS-20 missile and the Backfire bomber, which were perceived by NATO to tip the balance in favor of the Warsaw Pact.

In response to Soviet modernization, the second major NATO decision was to take a "two-track" approach: deployment + negotiation. The first track consisted of deploying 108 Pershing-II ballistic missiles and 464 ground-launched cruise missiles (GLCMs); the second track was "a [simultaneous commitment to pursue] negotiations aimed at limiting theater nuclear forces." Because the Soviets figured that NATO missiles could reach Moscow, they feared their command and control structure would become vulnerable to a surprise attack. With neither Pershings nor GLCMs limited by SALT II, the Soviets viewed these deployments in Europe as an American effort to circumvent that unratified treaty.

The promised second-track arms control negotiations, which were the INF talks, did open in October 1980, but they dragged on for seven of the eight Reagan years. In the early 1980s many questions arose about what should be included in the negotiations and what should be the outcome. Unfortunately, NATO modernization proceeded on schedule, while ongoing negotiations — getting nowhere — legitimized the buildup.

The third of the three major decisions by NATO dealt with its policy on exploitation of nuclear weapons. NATO's policy was to use them if conventional forces were unable to repulse a Soviet attack. Many individuals in and out of government urged a "no-first-use" declaration; but both the arms control value and the operational effectiveness of such a policy were challenged. Disappointing arms control advocates, the Harvard Study Group was "not yet persuaded" of the value of a no-first-use policy.

Despite having been armed to the teeth with nuclear weapons — and ready to use them — General Lee Butler, former commander-in-chief of the U.S. Strategic Command, agreed with a post-Cold War Russian scholar's assessment:

> The much-vaunted nuclear capability of NATO turns out, as a practical matter, to have been far less important to the eventual outcome than its conventional forces. But above all, it was NATO's soft [non-nuclear] power that bested its adversary.

General Butler acknowledged that "the presence of nuclear weapons played a significant role in the policies and risk calculus of the antagonists. They made the United States and the Soviet Union exceedingly cautious when one another's vital interests were at stake." But to Butler,

> it is equally clear that the presence of these weapons inspired the United States and the Soviet Union to take risks — especially in launch-on-warning postures — that brought the world to the brink of nuclear holocaust and left it there.

As a chief strategist, General Butler was in a position to observe that "senior leaders on both sides consistently misread each other's intentions, motivations, and activities." He subsequently added, "Their successors still do so."

Offensive Strategic Forces. Three problems involving offensive strategic forces were prominent during the 1980s: ICBM vulnerability, strategic-delivery modernization, and cruise missiles.

Back in the late 1950s, in anticipation of Soviet deployment of ICBMs, various practices were devised to keep U.S. bombers from being susceptible to a sudden attack; later, U.S. missiles were based in concrete silos. But improving missile accuracy and deployment of multiple warheads jeopardized the ability of strategic forces to ride out a first strike. (Because so many means of delivering nuclear weapons existed for both superpowers, deterrence was never truly undermined; instead, what was impaired was "political leverage," that is, being able to dominate or prevail on the political landscape.)

To offset perceived vulnerability, launch-on-warning strategies were improvised. New, often awkward, basing schemes such as "dense pack" were considered. The Harvard Group reasoned that "no politically acceptable land-based force can remain invulnerable to attack by a large number of accurate nuclear warheads." (It would be another decade before vulnerability would be reduced.) Although favoring retention of a land-based missile force, the Harvard Group advocated greater reliance on bombers, cruise missiles, and SLBMs — asking for high priority given to a more reliable and secure SLBM communication system.

Continuing modernization of U.S. strategic delivery was a thorn in the side of arms control proponents. The MX (advertised to have potential value as a "bargaining chip" in arms control negotiations) and the D-5 missile for the Trident were two troubling upgrades. Bomber modernization via the B-1 "Stealth" aircraft was viewed by arms control advocates as an unnecessary expenditure because the older B-52s remained effective and cruise missiles were being improved.

However, the introduction of advanced cruise missiles tended to upset the balance of forces, especially when the Reagan administration had more cruise missiles assembled while simultaneously resuming production of the B-1 bomber. Up to 1980, arms control had relied on quantitative limits, such as counting the number of launch vehicles and launchers; however, the relatively small and indistinguishable features of the cruise missile made it difficult to verify. Even so, the Harvard Group approved the deployment of air- and sea-launched versions.

Command & Control. In the early 1980s a U.S. concern was the growing vulnerability of networks for nuclear command and control, including satellites for intelligence missions. The components were always extremely difficult to protect because of their known locations; at the same time, an attack on these systems would invite massive nuclear retaliation. If indispensable information and control were lost, capability to ride out a nuclear attack might not survive. Such inherent weakness inadvertently encouraged the early and massive use of nuclear forces. Modernization and diversification of the command and control network therefore gained almost unanimous support.

Anti-satellite (ASAT) warfare had loomed as a destabilizing prospect since the 1960s. Both superpowers were developing operational capabilities for attacking satellites either with space-borne or earth-based weapons. An agreement to forgo ASAT weapons would reduce threats to valuable military satellites that provide both sides with vital reconnaissance, communications, and treaty monitoring. While the Harvard Group encouraged negotiations to inhibit ASAT capability, it also incongruously supported the U.S. ASAT program.

Nuclear Defense. The 1972 ABM Treaty imposed severe constraints on Soviet and American development, testing, and deployment of missile defenses. A decade later Washington had shut down its single ABM site at Grand Forks, North Dakota, but Moscow was still surrounded by nuclear-armed interceptors (of dubious capability). By then, both nations had abandoned serious efforts to install passive civil-defense measures.

Because offensive capabilities had outpaced defensive measures, the Harvard Group recognized that ABM systems alone could not solve ICBM vulnerability questions nor protect other valuable targets against a nuclear salvo; so the Group recommended that "no action should be taken to modify or withdraw" from the ABM Treaty. A "vigorous" compliant BMD R&D program was encouraged.

"Forswearing defense of one's country in the interest of national security," the Group said, "is a counterintuitive idea and one that many people still find difficult to accept." This has been a dilemma with which advocates and protestors have struggled. On one hand, ballistic-missile defenses offer "crisis stability" by diminishing the prospect of missiles hitting their targets; on the other hand, unilateral deployment of an ABM system threatens "arms-race stability." ABM development and deployment were a "major factor contributing to an arms race" in offensive as well as defensive weapons.

However, ABM systems are not considered "cost effective at the margin": They are not only high-priced, but they are more expensive for once side to build than the other side to defeat (by deploying additional offensive forces). In 1987, a Pentagon official responsible for strategic defense told the House Armed Service Committee:

> The Soviets have been deploying their Moscow [Galosh ABM] defenses for over ten years at a cost of billions of dollars. For much less expense, we believe we can still penetrate these defenses.... [If] the Soviets should deploy more advanced or proliferated defenses, we have new penetration aids ... under development [and] we are developing a new maneuvering re-entry vehicle that could evade interceptor missiles.

Regarding civil defense, the Harvard Study concluded that it would be "unlikely to have a significant effect on the probability of nuclear war," although it could "mitigate the terrible consequences if deterrence fails." But the divided study group could not reach agreement on justifying the economics and social costs of renewed civil defense.

Although prestigious and influential, the Harvard Study Group could do no better than reflect the arms vs. arms control discord so prevalent in the Reagan years.

The Role of Espionage

With the end of chronic tensions and suspicions have come cathartic revelations about the role of espionage in the development of the Soviet atomic and thermonuclear weapons. Two post-Cold War technical-historical symposia have been held, one at Kurchatov, primarily focused on the first Soviet atom bomb, and the other at Dubna, concentrating on the Soviet thermonuclear weapon. Both symposia revealed a high level of indigenous scientific competence that overcame very difficult resource and political problems not faced in the West. Fittingly, the espionage history of World War II and the Cold War offers many lessons.

Section A: Myths and Realities

Knowledge about German wartime nuclear research was sought by the Allies. In particular, shipment of heavy water from Norway to Germany led the British to destroy storage and production facilities. The Soviets learned of these undisclosed Allied operations and about secret nuclear research in Germany.

In the context of espionage, it is important to recall the U.S. publication of the Smyth Report immediately after World War II. Not only did the report reveal unpublished historical and technical information about the development of the atomic bomb, it also had a significant influence on Soviet officials who for the first time began to appreciate the effort and planning needed to make nuclear weapons.

Thirty-thousand copies of a Russian translation of the Smyth Report were published early in 1946 and distributed widely to scientists and engineers in the Soviet atomic project. Along with information obtained from spies, it helped guide the technical choices made in the Soviet program. For example, the British government ascertained — from espionage and from their own research — that Soviet scientists adopted a U.S. method partially described in the Smyth report for sealing (canning) plutonium-production reactor fuel rods.

Atomic Spies. Soviet physicists were engaged in nuclear studies before 1941; so — despite an Allied wartime embargo on sharing nuclear information — they understood the basic science and technology needed to develop atomic bombs. Because Stalin required the scientists to realize success as quickly as possible, they had to base their design on a detailed description of the American atomic bomb received from Klaus Fuchs, who was a scientific insider at the Los Alamos Manhattan District. In parallel, the Soviet scientists undertook (successful) development of a smaller, more efficient weapon of their own design.

Apparently, extensive leaks of secret information reached the Soviets through British spies and sympathizers. "According to [Vladimir] Barkovski [of the 1940s Soviet intelligence services], there were at least ten British experts supplying the Soviet Union with information on the atomic bomb."

When the feasibility of an atomic bomb was originally recognized, that revelation was included in secret study commissioned by the British government, the 1941 "Maud report," which advised proceeding with the bomb's development. Several authors of the Maud report (Frisch, Rotblat, Peierls, and — most significantly — Fuchs, a GRU* agent) later became members of the British mission to Los Alamos.

Ted Hall, an American who worked at Los Alamos, is considered to have been another important source [Mlad] of nuclear information for the Soviets; he evidently disclosed the key fact that plutonium could be brought to a supercritical condition by surrounding it with a spherically shaped high-explosive "lens," which rapidly and uniformly imploded the plutonium core.

Barkovski, who was part of the espionage chain that collected the Maud report for the Soviets, concurs that intelligence data saved development time; indeed, its collection accelerated the first Soviet atomic test. However, the information was

---

*GRU was the Soviet Army's intelligence service.

only useful when fed to technically competent scientists who could correctly interpret the clues being provided. Academician Roald Sagdeev quotes Andrei Sakharov, who said there was only one secret: that it was possible to make the nuclear bomb. Lev Artsimovich says that espionage saved the Soviets about a year, and Yuri Khariton estimated a speedup of about two years.

Only Stalin, Lavrenti Beria, and Igor Kurchatov had access to all the stolen secrets. The intelligence windfall — even its existence — was known to just a few people; so only at critical junctures was the stolen technical information parceled out (in order to influence or confirm technical programs).

The first Soviet fission-bomb test, *Joe-1* in August 1949, was detected by the U.S. Naval Research Laboratory from atmospheric radioactive particulates. Monitoring for radiation carried aloft by winds in the atmosphere had been organized by Lewis Strauss of the AEC. The Soviet Union made no announcement of its success until U.S. President Harry Truman declared, three weeks after the test, that there had been a Soviet atomic explosion.

Certain crucial technologies are more art than science, which makes the details virtually impossible to steal; examples of this are fuel-rod reprocessing, materials metallurgy, and machine-fabrication skills. These techniques and skills can be indigenously developed or circumvented, though at the expense of time and other costs.

One important Soviet time-saver derived from espionage occurred when Igor Khurchatov was apprised by his intelligence sources about an initial major American setback: Reactor-criticality poisoning by fission-product xenon had been discovered and overcome for the first U.S. plutonium-production reactor design. Khurchatov was thus able to avoid a serious delay with weapons plutonium from their first production reactor built at Mayak.

Lack of expertise has not been the main impediment to rapid progress in the development of atomic weapons: Insufficient uranium supply, partly due to inadequate knowledge of deposits, was a greater limitation on headway during the Cold War.

Former Soviet intelligence officers have claimed that they were able to obtain details of all aspects of the development of the atomic bomb from sources working within the American, British, and Canadian projects:

> Almost all of the sources working for the KGB and GRU were volunteers, often with communist backgrounds, who believed that information about such a destructive weapon should be shared with the Soviet Union.... The most serious and damaging were Klaus Fuchs and the American, Theodore Hall [Mlad], who between them passed information about the design of the first U.S. Plutonium bomb which was sufficiently detailed to provide Soviet scientists with a blueprint for replicating its manufacture.

Long after both nations had developed mature arsenals, Cold War espionage continued. Its more productive results were information about strategic plans and operations, although details about each other's weapons were also sought. A classic case was the CIA-supported Glomar Explorer apparent retrieval of nuclear weapons from a sunken Soviet submarine. Evidently the salvage ship was able to recover some part of the Soviet missile submarine K-129, which sank in the North Pacific in 1968. The official story is that the bow section of the submarine was recovered in 1974, along with two nuclear-tipped torpedoes, but that three SLBMs were lost. Details remain classified.

Thermonuclear Espionage. As was the case for fission explosives, espionage played at least some role in the Soviet development of hydrogen bombs. According to a British investigation, Klaus Fuchs handed over "an important collection of materials related to the superbomb" in March 1948, but later told the FBI that he had not done so. The Teller "superbomb" thermonuclear concept, however, was a dead end, and so Fuchs' information was not particularly helpful to Soviet scientists.

Although initial Soviet interest in thermonuclear weapons was stirred by intelligence data, former Soviet scientists say (justly) that the early U.S. concepts did not lead them to a workable design for a weapon. The Soviets earned bragging rights from their own endeavors; they trace success to their August 1953 test, which they distinguish as a "true" thermonuclear bomb, not a boosted fission weapon. Nor did Soviet physicists acquire crucial intelligence data from the 1952 U.S. *Mike* test because, according to Holloway's findings, their radiochemistry was not far enough advanced and too poorly organized to do intensive analysis of the airborne products from the thermonuclear explosion.

The 1952 American thermonuclear device exploded was huge (it weighed 50 tons), immobile (land-based — the size of a two-story house), and experimental (cryogenically condensed thermonuclear fuel); it was a trial step on the way to creation of a true hydrogen bomb. The indigenous and successful Soviet design was managed without this very complicated and costly experiment and without the benefit of espionage.

****

The myth-permeated rhetoric of the Cold War heavily influenced political decisions about nuclear armaments. Usually the rhetoric rationalized a massive weapon's buildup; sometimes it ominously positioned the superpowers at the edge of nuclear annihilation. Most accusations levied by American hardliners about arms buildups, treaty violations, and verification inadequacies fail to withstand present-day scrutiny. Even though vigorous (usually futile) objections were raised, many unsubstantiated accusations were given an aura of credibility by half-truths, foisted upon policymakers and on a public that lacked sophistication to sort out competing technical and legalistic claims.

The upshot of the dramatic but essentially baseless claims of weakness throughout the 1950s through the 1980s — the "bomber gap," the "missile gap," and the "window of vulnerability" — was a series of U.S. political decisions to ratchet the arms race with more or better nuclear weapons. Now we know that these claims were highly exaggerated excuses to further the goals of American hardliners. If anyone was "ahead," it was the United States. In any event, the forces of deterrence were at such high levels that additional weapons added nothing to national security, instead increasing the risk of mutual nuclear annihilation.

Highly significant is strategist General Lee Butler's assessment that both sides consistently misread "each other's intentions, motivations, and activities," to which he added, forebodingly, that "Their successors still do so."

Claims that arms control treaties were unverifiable turned out to be red herrings. This was evident to careful observers at the time, but outsiders' efforts to explain this to policymakers were futile. During the Reagan years, technical initiatives to improve verification technology and analysis were often sidelined, and tirades against Soviet untrustworthiness continued unabated.

Strategic treaties could be, were, and still are verified well enough to prevent technological surprise; none of the doomsayers' destabilizing shocks has ever occurred. For treaties under negotiation, a panoply of instruments and procedures were available and could have been implemented if the will to do so had existed.

Another lesson to draw from that experience is the need to look behind publicly stated objections to binding treaties that have stringent verification. Although the Soviets reflexively were suspicious of ideological interference, the underlying resistance of both superpowers to intrusive verification turned out to have had more to do with unwillingness to genuinely reduce the weapons arsenal. An additional unpublicized factor for both sides was the fear that strategically unrelated sensitive activities might be discovered.

The original U.S. nuclear monopoly, the attempts of both sides to maintain secrecy, the unchecked development of new and more destructive weapons and delivery systems — none of these had much impact once each side had counterbalanced the other with at least a few nuclear weapons. Mutual deterrence became irrevocably intrinsic, and it turned out that a half-century nuclear standoff, a metastable status quo, *had already commenced at the very beginning of the Cold War, back in the mid-40s.*

The equilibrium of this East-West standoff was reinforced by military, political, and economic alliances among nations; multilateralism has to be credited with tempering the emotional crises of the Cold War.

With regard to the advantages and shortcomings of deterrence, compellence, or avoidance as theoretical algorithms to apply in judging the standoff and outcome, more will be forthcoming in this and subsequent Chapters.

****

Even though this book delves primarily into nuclear history, it's difficult not to notice similarities with post-Cold War institutional "wars" against poverty, crime, drugs, hunger, terrorism, AIDS, energy shortage, and global warming. Too often, balanced and expert judgment seems to get elbowed aside by more powerful influences.

B. THE UNTHINKABLE

Often termed "unthinkable," nuclear conflict could have occurred during the Cold War at any time. A weapon could have been exploded by accident or as a deliberate, unauthorized act. War could have broken out between the superpowers, either deliberately actuated or through a chain of events that got out of control — a precipitating incident escalating into an unstoppable frenzy of nuclear annihilation.

What national policies and military technologies placed civilization in this precarious position?

Nuclear weapons — their number, destructive capabilities (explosive yield), and means of delivery to distant targets — were very significant parameters of national policy during the Cold War. We now examine the decisions and underlying rationales that made it so.

In the mid-1950s, an atmosphere of crisis prevailed (within governments, mostly in secret). The (ballistic-missile) arms race was on in earnest because the United States was regarded itself as "behind." It was feared that the Strategic Air Command deterrent would be "negated," thus emboldening

> the Red Army, or its proxies to move as it wished. It was thought [within the U.S. government] that a forced reunification of Germany on Soviet terms would be next.

So reports Thomas C. Reed, a former Secretary of the Air Force, who seems to have discounted, as others did, the 15-to-1 U.S. lead in nuclear warheads in 1950: approximately 3000, compared to about 200 for the USSR.

One might argue that U.S. intelligence would not be aware of that disparity, but that was unlikely. Certainly the Soviets knew (through their spy network) of the inequity. Even in the absence of inside information, it was not that difficult to estimate the comparative number of warheads simply from mirror-imaged production data.

Strategic Assessments

By 1966, the Soviet Union had accumulated 1000 strategic nuclear warheads, and the United States 8000 (numbers and dates approximate). By 1974 the two arsenals had grown to 3000 and 13,000, respectively. It wasn't until 1990 or so that numerical parity was reached, with about 11,000 strategic nuclear warheads on each side.

The U.S. armed services requested many more strategic delivery vehicles than they received. In 1956 and 1957 the total stated requirement for the U.S. Army alone was 151,000 nuclear weapons, mostly tactical. Some U.S. naval officers in early 1958 envisioned a Polaris fleet of 100 SSBNs. In the late 1950s the Air Force fancied several-hundreds to many-thousands of Minuteman ICBMs.

As for strategic-delivery vehicles, the United States, without reaching numbers as large as some wished, had attained an immense superiority by 1960 — a 15-to-1 ratio in long-range bombers. It wasn't until thirty years later that the Soviet Union had attained a 25-percent margin in the number of delivery systems.

The total number of nuclear warheads that the United States cumulatively manufactured has been more than 70,000.

Initially, the weapons designed by the United States were large in size and payoff. In 1960 the U.S. arsenal totaled around 20,000 megatons in explosive yield, but thereafter the aggregate dropped as big warheads were traded for a larger number of compact ones to arm smaller ICBMs. Meanwhile, unable to match the accuracy of U.S. delivery systems, the Soviets compensated by building bigger warheads that required large rockets. This gave them greater throw-weight and megatonnage — but not a nuclear arsenal that the U.S. Joint Chiefs of Staff wanted to emulate.

A question that pops up is, what were the rationalizations that determined the sizes of the two arsenals? And what other factors — tactical nuclear weapons, delivery systems, conventional forces, geography, political stability, economics — had a major impact on containment and national security?

> Available evidence suggests that the production and deployment goals proposed and realized for nuclear weapons were essentially arbitrary.... The reality is that the eventual size of a weapon program was arrived at through a number of interconnected factors and influences, including budget trade-offs, the perceived Soviet threat, interservice and intraservice rivalry, the promotion of jobs in states and districts by elected officials, corporate lobbying, cycles of technological obsolescence and novelty, and political charges and countercharges, to name a few....
>
> It is important to remember that although the "triad" of nuclear bombers, land-based missiles, and sea-based missiles was and is often touted as a pragmatic and time-tested means both of ensuring that an adversary cannot, in a single blow, destroy the entire U.S. nuclear arsenal and insuring against the unforeseen catastrophic failures of any one leg, the triad itself developed not by design but in large measure because the Air Force and Navy each wanted "a piece of the action"....

The chronic lack of systematic strategic assessment is corroborated by insiders such as Herbert York, once head of the Livermore nuclear weapons laboratory. With reference to the arbitrary nature by which warhead outputs were often determined, he disclosed that the yield for the Atlas ICBM was one megaton simply "because one million is a particularly round number in our culture."

Government-paid consultants — closely allied with military officers who fancied career and patriotic benefits derived from a "strong" defense — were (and are) usually making a case for more or better weapons, almost always resulting in the continuous flow of funds for research, development, and production. For instance, during 1979 and 1980, nuclear strategists argued that 200 MX missiles — to be shuttled around and hidden in eastern Nevada and western Utah — were "absolutely essential to the national security of the United States." Somehow, national security was preserved without shuttling missiles.

The rationalization for boosting armaments was usually based on "worst-case analysis." In 1988, former Undersecretary of the Navy R. James Woolsey wrote:

"In strategic modernization, we are dealing, in a sense, with a major insurance policy against an admittedly unlikely eventuality. But it is insurance against the most catastrophic of imaginable losses."

Meager intelligence data underlay these strategic "assessments," whether arrived at by formal analysis or political pressure. Lacking direct access to the closed, secretive Soviet society, U.S. intelligence agencies relied on their own espionage and technical sources. However, to protect "sources and methods" — or to influence the national security debate — the information they gleaned was often withheld from policy makers and the public. In fact, American and Soviet intelligence created

> a unique symbiosis between some purposive Soviet deception and some equally purposive U.S. gullibility.... Increasingly deluged by the products of our technological marvels, the [U.S.] intelligence community was disinclined to listen to those still small voices suggesting that the exertions of the Cold War and the deformities and contradictions of the Soviet system were extracting a greater toll than perhaps anyone realized.
>
> ...politicians expected [intelligence] assessments to conform to their view of the Soviet threat, and the military services naturally valued estimates that helped justify their programs and budgets. Hence Soviet military strength was frequently overestimated and conclusions about Soviet doctrine and strategy were skewed.

[Nostril to Nostril]. Roald Sagdeev recalls an "impressive feature" of Gorbachev, which was "his ability to learn and, subsequently, to make up his own mind about complicated issues of strategic stability and arms control." Gorbachev is quoted as having said,

> I bet there are as many definitions of strategic parity as we have people sitting in this room. I am ready to defend my own. Real strategic stability does not necessarily require that both sides follow each other, nostril to nostril.

Further confirmation of unduly alarmist presumptions about strategic intentions has emerged from previously classified Soviet internal documents. Two examples are (1) Khrushchev's 1956 repudiation of the inevitability of war between capitalist and communist countries, and (2) Brezhnev's 1977 denial that the Soviets were seeking a first-strike capability. These were valid Soviet policy declarations that were ignored or discounted in the West.

Emerging now is a picture of a typical bureaucratic process involving the Soviet defense industry. A well-informed émigré, Nikolai Sokov, portrays a "defense industry that enjoyed the same power and status as the Ministry of Defense." Thus, weapons designers were able to push through their favored proposals, and the Strategic Rocket Force was able to gain priority. This explains why production of MIRVed ICBMs gained precedence in the Soviet Union over submarine- and air-launched missiles, both of which were more stabilizing.

War Planning

Although deterrence strategy differs from war planning, the distinction was blurred. War-fighting, a strategy that transcended policies of assured or measured retaliation, was seen to require a capability that was mainly offensive. Rather than simply equipping and training military services to retain tactical nuclear weapons in a state of readiness, with strategic forces kept in the background, war-fighting strategies spelled out specific escalatory steps for using them against conventional attacks, provocative moves, or policy failures.

> U.S. policy ... has rested on the principle that there are interests for which the United States would make the first use of nuclear weapons rather than concede. Although not illegal, all of this occurred without informed public consent. Provocative and risky deterrent practices ranging from strategic targeting to launch-on-warning to pre-delegation were developed under minimal democratic oversight and influence. Civilian masters of the military establishment failed to exercise vigilant oversight of strategic planning and failed to deal decisively with deficiencies in the safety and performance of nuclear command, control, communications, and intelligence systems. All the while, the nuclear planners misrepresented U.S. strategic and command and control capabilities as sophisticated, flexible, and resilient.

Senator Daniel Moynihan has written:

> As no other subject, strategic-arms doctrine has been the realm of the intellectual and the academic. This is military doctrine, to be sure, but it has never, in this nation, been formulated by military men. It began with the physicists who created the weapons — men such as J. Robert Oppenheimer, Hans Bethe, and Leo Szilard — who were then joined by other physicists and scientists, and also by social scientists.

The social scientists that Moynihan refers to happened to be mostly RAND Corporation alumni — men such as Albert Wohlstetter, Herman Kahn, Fred C. Iklé, Alain C. Enthoven, Henry Rowen, and Henry Kissinger — who came to be known collectively as "defense intellectuals." They moved in and out of Washington, but in the main they kept to their campuses and think tanks, where their task, in Kahn's phrase, was "thinking about the unthinkable."

As early as 1951, Project Charles, a civil defense study at MIT, warned that "there was a serious question of the propriety of scientists trying to settle such grave national issues [as war planning] alone, inasmuch as they bear no responsibility for the successful execution...." In that report, Robert Oppenheimer agreed: "Such follies can occur only when even the men who know the facts can find no one to talk to about them, when the facts are too secret for discussion and thus for thought."

Gradually, social/political scientists and "operations analysts" took over government strategic-policy planning projects.

The U.S. national planning document embodying deterrence and war strategy was (and still is) the SIOP (Single Integrated Operational Plan), signed off by the chiefs of the armed services. The details of the SIOP were perhaps the most important secrets of all, because it contained the plans for waging war, the list of designated targets, and the justification for nuclear arsenals.

Before the SIOP was in place, it was the President who designated an "enemy" and authorized weapons release. The military services took care of planning the actual placement of weapons onto targets. The first SIOP was a joint effort of a JCS planning staff and a Scientific Advisory Group, which included nuclear specialists from the weapons laboratories.

Inasmuch as Thomas C. Reed was assigned to the Scientific Advisory Group during the Reagan administration, he has been able to accurately outline how the system worked: After the President, through the Secretary of Defense, decided who the "bad guys" were, the JCS planning staff built up a data base of potential targets that included options for attacking the opponent's leadership, industrial base, or hardened targets. "Aim points" (Designated Ground Zeros) are then picked. Depending on target hardness, proximity, and other factors, the weapon yield, accuracy, and other options are chosen. Timing of sorties is based on defense suppression, avoidance of radioactive clouds, and air defenses.

Although for years Reed was part of this "arcane targeting process," it was not until 1996 before his "mind turned to how awful it might have been if something had gone wrong." He rehashes the dilemma that

> we Cold War targeteers had tried to spare the [Leningrad museum] Hermitage, [but] with dozens of weapons raining down all around, it would not have made much difference.

In the 1960s, NATO's war plans were based on early first-use of some 6000 tactical nuclear weapons in response to a presumed Soviet conventional attack. Both NATO and WTO countries routinely trained their forces for nuclear operations. The SIOP-dictated war-fighting doctrine and preparation thus created a desire for increasingly sophisticated nuclear weapons, and both sides became committed to programs that threatened their adversaries' vital military assets with increasingly swift destruction.

Out of this uneasy situation grew the fear of a "first strike." If Soviet missiles could, at one stroke, eliminate the land-based Minuteman ICBMs, it would be left to the surviving submarines to retaliate against cities of the Soviet Union. But that would be too risky because the Soviet Union could counterattack against America's urban populations, and thus, this type of reasoning went, the United States would have to yield to Soviet demands. (All this assumed the SAC bomber force was neutralized too.)

Henry A. Kissinger, a leading architect of confrontational engagement, was instrumental in President Nixon's 1974 national security decision memorandum NSDM-242, "Planning the Employment of Nuclear Weapons," which prescribed a flexible response to a nuclear attack. In his first published book, Kissinger had argued that, because of the devastation that could be wrought by all-out nuclear war, only limited objectives could be achieved. (Subject to debate, though, was just what limits were realistic.)

When the Ford administration came into office in 1974, intermediate-range ballistic missiles (SS-20s) were deployed by the Soviets throughout the Eastern Bloc. This caused the Ford administration to adopt nuclear war-fighting plans. The Reagan administration countered in 1983 by deploying intermediate-range nuclear forces in Western Europe. The stage was set for a nuclear confrontation.

The subsequent Carter administration continued a "war-fighting" strategy, but his staff drew up a wider choice of preplanned options (PD-59).

President Reagan's initial directive to his staff in 1981 was "to seek the dissolution of the Soviet empire." Accordingly, NSDD-32 was drawn up under the coordination of Reed, who summarized the policy directive as follows:

> NSDD-32 listed five integrated strategies to achieve this result: economic, political (at times to include covert action), diplomatic, information (both the promotion of unfettered communication and the use of propaganda), and military (to include arms control).

It is curious to an outsider that covert action was placed in the category of "political" rather than "military" strategy. One explanation for this curiosity is a remark by Reed, who says that after the President signed the NSDD-32 on 5 May 1982, Deputy Secretary of State Eagleburger undertook "initiatives to counter the nuclear freeze movement and to take the political war to the Soviet Union." President Reagan's ensuing speech to the British Parliament on 8 June 1982 was the basis of which became the "Reagan Doctrine."

Leadership "decapitation" was of constant legitimate concern. Without survival of the leadership, it would be even more difficult to stifle a nuclear exchange. After being briefed on the nuclear-options notebook (the "football"), the President learned of the procedures for releasing an Emergency Action Message (EAM — the presidential order to attack with nuclear weapons). The President was also informed of forces at disposal, time available for response, and options of counterattack and withhold.*

President Reagan's "resolve to do something about a [ballistic-missile] shield" was stiffened when he sat through his first crisis exercise and saw "the United States of America disappear" in less than an hour of simulated Soviet attack. A SIOP briefing that came later is described by Thomas Reed as "scary" too, in that the President had to quickly choose from a "menu of plans," in which "tens of millions of women and children who had done nothing to harm American citizens would be burned to a crisp."

Reed became alarmed when he found out that the Command and Control system President Reagan inherited "would have been headless within minutes of an attack." He also realized that "in reality" the nuclear button would be in the hands of the Joint Chiefs.

Moreover, Reed credits then-Secretary-of-Defense Dick Cheney for asking, in 1989, "as the Soviet satellite states in Eastern Europe were spinning out of Moscow's control," why the United States was "targeting the military infrastructure of nations no longer embracing communist doctrine or tied to the collapsing Soviet empire."

At the time "over 10,000 nuclear weapons were targeted on the disintegrating Soviet Union ... with hundreds ... targeted at the greater Moscow area alone." Reed's explanation for the latter is as follows: The Soviet antiballistic missile system around Moscow "had to be defeated one launcher at a time." (That questionable rationale lacks corroboration; most evidence points to the ineffective antiballistic missiles having little impact on strategic targeting.)

---

*Nuclear forces withheld from an initial strike.

When later asked by the G.H. Bush administration to rejoin the SIOP planning group, Thomas Reed makes clear that "for the first time we were considering a *reduction* in the strategic nuclear forces of the United States." Reed was impressed by how Cheney overcame bureaucratic resistance and "was able to drive down the nuclear weapons required in the SIOP by over 10 percent every year." (Since that was on the eve of the formal disestablishment of the USSR, the United States evidently could afford the risk to reduce the 10,000 doomed aim-points to only 9,000!)

Second Thoughts. Lee Butler — a retired general who had been director of the Joint Strategic Planning Staff during President G.H. Bush's term in office, and thus responsible for the SIOP on his watch — eventually changed his mind too about nuclear weapons and their utility in national defense. After examining the course of his career in the military, he came to these "unsettling judgments":

▸ From the earliest days of the nuclear era, the risks and consequences of nuclear war have never been properly weighed by those who brandish them.

▸ The stakes of nuclear war engage not just the survival of the antagonists but the fate of mankind.

▸ The likely consequences of nuclear war have no political, military, or moral justification.

▸ The threat to use nuclear weapons is indefensible.

General Butler admits that "For much of my life ... I saw the Soviet Union and its allies as a demonic threat, an evil empire bent on global domination." Reflecting on "a deterrence strategy that required near-perfect understanding of an enemy from whom we were deeply alienated and largely isolated," Butler realized that "While we clung to the notion that nuclear war could be reliably deterred, Soviet leaders became convinced that such a war might be thrust upon them, and, if so, it must not be lost."

General Butler eventually concluded,

> Deterrence was [so] flawed that the consequences of its failure were intolerable. Deterrence failed completely as a guide for setting rational limits on the size and composition of military forces.... Deterrence is a slippery conceptual slope. It is neither stable nor static.

Thomas Reed praised General Butler, who had assumed command of the Strategic Air Command (SAC) in January 1991, as one of the "new thinkers" in the bunkers of Omaha, in contrast to General Curtis LeMay, whom he described as "the first and most truculent Commander in Chief of SAC." Butler, on the other hand,

> did not think that nuking any nation "back into the stone age" was a great idea. In time, Butler decided that he did not believe in nukes at all, although he would enunciate those views only after his retirement.

Security Paranoia

Bordering on paranoia, fear of military, economic, psychological, or political conquest was a significant factor that exacerbated and extended the Cold War. While this book is not the place to examine all dimensions of this national-security emphasis, one cannot avoid delving into anxieties that sustained the nuclear arms race.

The West had been antagonistic to communism from its birth, and, conversely, the Soviets perceived the American nuclear monopoly as detrimental to their national survival. Realizing that the Soviet Union would move toward its own nuclear capability, some U.S. military leaders (including Air Force Generals Spaatz, LeMay, and Norstad) advocated "preventive war" against the Soviets.

These palpable fears of nuclear war lead to governmental policies that seesawed between accommodation and bellicosity.

McCloy-Zorin Agreement. In 1961 Chairman Nikita Khrushchev and President John F. Kennedy authorized senior negotiators to have private talks about a framework for comprehensive disarmament, resulting in a joint statement of principles. The U.S. negotiators were John J. McCloy and Arthur H. Dean, both senior members of the American establishment and both Republicans; the chief Soviet negotiator was Valerian A. Zorin, Deputy Soviet Foreign Minister and Ambassador to the United Nations.

The negotiators reached broad agreement about steps toward "general and complete disarmament," leaving details for future resolution. One fuzzy point was the nature of inspection and verification: Judging from written accounts, "There was no provision for any further inspection should suspicion of violations be aroused, nor any provision for inspection to make sure that the level of armaments retained was what had been designated as the legal maximum." Zorin declared that "the Soviet Union is resolutely opposed to the establishment of control ... over armed forces and armaments ... during any state of disarmament," labeling it "an international system of legalized espionage." Lost in the rhetoric of the day was the expectation, realized decades later, that inspections could be strictly limited by agreement. In no way would inspection or verification lead to loss of control over armed forces and armaments.

One prominent organization to embrace the McCloy-Zorin agreement was the World Federalist Association, which enthusiastically promoted the accord as a basis for peace and disarmament, describing it as "a safe and sound roadmap toward a disarmed world with adequate controls, adequate verification, and adequate provision for enforcement and for peaceful settlement of international disputes." But because the McCloy-Zorin principles appeared unrealistic and involved some loss of national autonomy, hawkish conservatives were in fervent and unceasing opposition. Many decades later, hardliners would still decry the McCoy-Zorin episode as a near "sellout" of U.S. national interests.

Scowcroft Commission. In an April 1983 report to President Reagan, a Commission on Strategic Forces that he had appointed reviewed the nuclear modernization program of the United States. The conclusions and recommendations were unanimous, according to the chairman, General Brent Scowcroft. On the Commission were well-known individuals such as John Deutch, Alexander Haig, Jr., Richard Helms, William Perry, and R. James Woolsey; senior counselors were Harold Brown, Henry Kissinger, Melvin Laird, Donald Rumsfeld, and James Schlesinger. The results of this study were as a typical product of formalized inquiries driven by anxiety about national security. (Rumsfeld, later to become Secretary of Defense under President George W. Bush, headed in the late 1990s an alarmist study about third-world ballistic missiles; another member of that study, Woolsey, served later as director of the CIA).

The Scowcroft Commission judged that "Effective deterrence and effective arms control have both been made significantly more difficult by Soviet conduct and Soviet weapons programs in recent years.... The Soviets have shown by word and deed that they regard military power, including nuclear weapons, as a useful tool in the projection of their national influence." (This also mirrored U.S. policy at that time.)

The report continued with the judgment that "it is ... clear that the Soviet leaders have embarked upon a determined, steady increase in nuclear (and conventional) weapons programs over the last two decades — a buildup well in excess of any military requirement for defense." In support of this determination the Commission cited Soviet ICBM growth, giving emphasis to throw-weight, while ignoring strong qualitative advantages of U.S. forces; in addition, they attributed to the Soviet SS-18/19 missiles "excellent accuracy" (without actual knowledge). The Commission stressed a "narrowing" gap in the U.S. advantage in SLBMs, and asserted that the Soviets were giving "high priority" to antisubmarine warfare and submarine quietness (U.S. proficiencies that the Soviet Union never matched). While acknowledging that the Soviet heavy strategic bomber force "is considerably less capable than the total active U.S. bomber force," the Commission pointed to a "new Soviet intercontinental bomber (the Blackjack)" that was then being flight-tested (but barely made it into production before the Cold War ended).

To the Commission members, there was another frightening element:

> Soviet strategic defenses are extensive, consisting of a dense nationwide air defense network and a limited ballistic missile defense at Moscow, [and] both are undergoing modernization. Their vigorous research and development programs on ballistic missile defense provide a potential ... for a rapid expansion of Soviet ABM defenses, should they choose to withdraw from or violate the ABM treaty.

(How times have changed! The United States has since withdrawn from the treaty.) In any event, American strategic military commanders, when asked, never offered to trade their offensive capability for any of the purported Soviet defensive potentials.

The Scowcroft Commission assessed strategic trends according to five criteria: accuracy, superhardening, mobility, ASW, and BMD. Noting that the accuracy of strategic weapons (of both superpowers) would continue to improve, they realized that fixed launchers (especially MIRVed ICBMs) were increasingly at risk. Superhardening of ICBM silos was considered "quite promising." Mobile strategic systems were deemed "more feasible" because of improvements in guidance, miniaturization, electronic hardening, and solid fuels. The Commission recognized that ballistic-missile submarines would be able to evade ASW measures "for a long time" (which is still the case). Regarding BMD, the Commission thought that "substantial progress had been made in the last decade in development of both endo-atmospheric and exo-atmospheric ABM defenses" (In actuality, the only progress was with new ideas for research, rather than substantive development).

One of the Commission's more sensible recommendations was overdue: a "single-warhead missile ... may offer greater flexibility ... to obtain an ICBM force that is highly survivable." (Both the United States and Russia have been moving in that direction.) Other recommendations included putting the MX in Minuteman silos, implicating accepting that other possible MX basing modes were not promising.

A Section of the report was devoted to arms control, which the Commission considered to be a "legitimate, ambitious, and realistic objective ... to channel the modernization of strategic forces, over the long term, in more stable directions...." In contrast, they acknowledged that counterproductive SALT II and the ongoing S T A R T proposal "led to an incentive to build launchers and missiles as large as possible and to put as many warheads as possible into each missile." The Commission advocated

> an evolution toward forces in which — with an equal number of warheads — each side is encouraged to see to the survivability of its own forces in a way that does not threaten the other.

In summarizing that portion of the report, the Commission urged "the continuation of vigorous pursuit of arms control [and] innovation in verification techniques."

Though not to be implemented, that foresight could have help tame the most extensive and dangerous arms buildup in history.

Worst-Case Scenarios

A highly abused methodological tool is "worst-case analysis." The most extreme, unfavorable outcome, however unlikely, is postulated, and then this "worst case" boundary condition is used as the basis of a strategic plan. (It is a fundamental analytical tool in the mathematical and physical sciences, but not well established for societal-behavior predictions. Human activities have too many variables.)

Cold War policy planners, it seems, have tended to be oblivious to the possibility that unintended consequences of ensuring against the worst can worsen security rather than improve it. Moreover, the comparatively small probabilities that the best and worst cases might never really happen are often forgotten or deliberately ignored later on.

In order to maintain the arms buildup, great efforts were made in responding to worst-case scenarios that were "largely imaginary." This lead the United States to spend huge sums in developing weapons that reacted to "bogus" threats from the Soviet Union. The billions spent by the United States to forestall imaginary weapon systems almost suggests that disinformation was successfully fabricated by Soviet double agents.

A 1994 headline in the Chicago *Tribune* is illustrative:

CIA viewed arms race as mostly folly
Agency knew Cold War buildup was irrational

According to the article that followed the headline, records declassified after two decades indicated that "the nuclear arms competition had gone beyond anything that could reasonably be explained by the legitimate security needs of either Moscow or Washington." CIA analysts wrote in a top secret report in 1974 that "The weapons competition nowadays is largely a technological race. Each side is impelled to press forward ... lest it be left behind." Nuclear-production programs attained "a momentum of their own," pushed by an "immense apparatus" of "government and military organizations, installations, employees, and vested interests."

Much of this competition can be attributed to the pervasiveness of worst-case analysis and its inherent misrepresentations — a situation that continues to haunt long-range planning, as in the antiballistic-missile program pushed by President Bill Clinton at the turn of the millennium and continued by President G.W. Bush.

Although it is prudent to avoid underestimating a potential opponent, often the conservative assumptions entering into an analysis are not forwarded along with the results or are frequently ignored, deliberately or carelessly. The "better-safe-than-sorry" approach contributed to the arms race because it led to decisions that stimulated worst-case analysis on the other side. Compounding this problem is the distinction, or often lack of distinction, between capabilities and intentions. While actual capabilities can sometimes be assessed from technical intelligence, intentions are often concocted by worst-case analysis. Intentions are also subject to partisanship, each side seeing its own as peaceful and the other's as aggressive.

Nuclear Bravado

The nuclear establishment's willingness to use nuclear weapons should not be understated.

First and foremost, of course, is the reality that atomic bombs were dropped on Japan without fully exhausting timely alternatives.

Second, one must consider the extensive investment that nations have made in nuclear weapons development; in fact, validation of this huge World War II venture was one of the reasons given for using the bombs against Japan.

Third, because of hype about national security, thousands of nuclear weapons were tested despite environmental consequences. The process of developing and testing nuclear weapons had acquired an irresistible constituency and momentum.

Anyone who still believes that nuclear weapons are simply for deterrence should read the following closely.

Nuclear Threats. The threat — explicit or implied — to use nuclear weapons has often been an instrument of national policy. A 1998 chronology listed more than a dozen nuclear threats made. Some were implicit, delivered by elevating nuclear alerts or deploying nuclear forces to a crisis area. Stationing aircraft or naval carriers known to be nuclear-capable near an area under controversy has been a form of implied threat used fairly frequently.

A more subtle form of intimidation by a nuclear weapons state is simply refusing to adopt a stated policy of no-first-use.

An early example of nuclear hectoring occurred in 1946 when President Truman explicitly pressured the Soviet Union, which had not yet developed atomic weapons, to withdraw from northern Iran. Needless to say, this reinforced Stalin's determination to develop nuclear weapons.

During the post-World-War-II conflict in Korea, indications occurred several times that nuclear weapons might be used. In 1950, Truman announced that he was considering nuclear weapons; in fact, he ordered the transfer of non-nuclear components to Guam, Canada, and an aircraft carrier. During 1951, the Joint Chiefs are said to have authorized nuclear retaliation in northern Korea in case of a major attack against UN forces, and Truman is reported to have authorized the use of nuclear weapons against China.

[Nuclear Weapons Use Considered by Eisenhower]. The use of nuclear weapons was evidently considered at the beginning and end of President Eisenhower's two terms. The first occasion was in regard to the Korean War, if the Chinese had not agreed to a truce.

The second episode was in 1959, when Chairman Khrushchev was "provoking a new Berlin crisis." A secret NATO war plan known as *Live Oak* was created to maintain Western access to Berlin. One option was to force access to Berlin via East German highways, "to be accomplished by an armored division using tactical [nuclear] weapons if need be."

The secret plan, though, was found out by the Soviets. Evidently Khrushchev blinked, because a Soviet nuclear brigade was pulled back into Soviet territory.

President Dwight Eisenhower in 1952 is said to have tried *sub-rosa* nuclear threats, and he publicly hinted at the use of nuclear weapons against the Beijing government. Backing this up somewhat was the implicit impact of ten atmospheric nuclear explosions conducted by the United States, including the 10-megaton *Mike* thermonuclear test that vaporized an island in the Pacific.

In 1953 the White House acknowledged that nuclear-armed missiles had been placed in Okinawa; the top-secret 1953 U.S. "Basic National Security Policy"

(NSC 162/2) had asserted: "In the event of hostilities, the United States will consider nuclear weapons to be as available for use as other munitions."

Interestingly, as Chinese nuclear weapons officials point out, U.S. nuclear threats in the mid-1950s stimulated China's leaders to develop nuclear weapons. According to Chinese officials, "having experienced humiliation and calamity caused by foreign aggressors in more than one century and still facing military encirclement and war," China decided in 1956 to get atomic bombs. (Is there any wonder that China, the object of many threats, embarked on a program to develop its own nuclear arsenal and a variety of short- and long-range delivery systems?)

In 1954, when French troops were surrounded at Dien Bien Phu in Vietnam, Secretary of State Dulles reputedly offered two tactical nuclear weapons. Fourteen years later, the United States considered using nuclear means to rescue its surrounded Marines at Khe San. Later, President Nixon was more explicit about the possibility of using nuclear weapons in Vietnam. In fact, in order to indirectly prod the Soviet Union to force concessions from Vietnam, he ordered a secret nuclear alert, a bluff which he called a "madman strategy" — to no avail, as the Soviets took no visible notice.

During the Suez crisis, President Eisenhower is said to have made a nuclear threat to the Soviet Union. In any event, in 1958 he apparently twice authorized the use of nuclear weapons: in the Middle East and against China.

[**Lessons From the Cuban Missile Crisis**]. Perhaps a central lesson was the politicized nature of the episode. Seemingly the American response had more to do concern over public perceptions of Soviet nuclear missiles stationed so close to the Florida than it had to do with strategic relevance. After all, the Soviets already had submarines capable of launching nuclear missiles near American shores and ICBMs that could hit cities in the United States. And *vice-versa*.

Particularly noteworthy is that the outcome narrowed down to a few individuals, who — on a given day — made critical decisions that averted escalation or nuclearization of the conflict. Some of the challenged individuals were government leaders; others were simply submarine commanders and officers cut off from communication with their superiors.

Nothing in the episode changed the underlying validity of nuclear deterrence: Each side had more than adequate and dispersed capabilities to assuredly retaliate, irrespective of warheads positioned in Cuba.

Some lessons applicable to present-day events (40 years later), are (1) don't exaggerate the danger of nuclear weapons in the hands of hostile regimes; (2) governments want nuclear armaments for self-preservation; and (3) don't go to war to address a danger that nuclear deterrence already suppresses.

Two other lessons pointed out in Chapter I are (1) that the Soviets — as a result of the public indignity that they suffered — greatly expanded their nuclear arsenal afterwards and (2) that the two superpowers — because of the sobering realization of nuclear obliteration — began to focus on negotiations for a nuclear test ban and a direct hotline between Moscow and Washington.

Acting as an American proxy, Taiwan had a plan in 1958 to retake mainland China using nuclear weapons supplied by the United States. Secret plans declassified in 2002 indicate that:

> The plan was scuttled when the United States pulled back, fearing that a nuclear attack would cause a heavy death toll in the People's Republic of China and could prompt China to get nuclear technology from the Soviet Union.

The Cuban missile crisis in 1962 is a classic example of the heightened DEFCON alerts and brandishings of nuclear weapons. Robert McNamara, who instituted reliance on assured nuclear retaliation as the declared policy of the United States, suggests that the Cuban episode was an event with low probability but high risk of catastrophe. Retrospective reports indicate that "We came within a hairbreadth of nuclear war without realizing it."

[DEFCON: Defense Condition]. In the event of a national emergency, a series of seven different alert Conditions (LERTCONs) can be called. The 7 LERTCONs are broken down into 5 Defense Conditions (DEFCONs) and 2 Emergency Conditions (EMERGCONs).

Defense readiness conditions (DEFCONs) describe progressive alert postures primarily for use between the Joint Chiefs of Staff and the commanders of unified commands. DEFCONs are graduated to match situations of varying military severity, and are numbered 5, 4, 3, 2, and 1 as appropriate. DEFCONs are phased increases in combat readiness. In general terms, these are descriptions of DEFCONs:

DEFCON 5 Normal peacetime readiness
DEFCON 4 Normal, increased intelligence and strengthened security measures
DEFCON 3 Increase in force readiness above normal readiness
DEFCON 2 Further Increase in force readiness, but less than maximum readiness
DEFCON 1 Maximum force readiness.

EMERGCONs are national-level reactions in response to ICBM (missiles in the air) attack. By definition, other forces go to DEFCON 1 during an EMERGCON.

DEFENSE EMERGENCY: Major attack upon U.S. forces overseas, or allied forces in any area, and is confirmed either by the commander of a unified or specified command or higher authority or an overt attack of any type is made upon the United States and is confirmed by the commander of a unified or specified command or higher authority.

AIR DEFENSE EMERGENCY: Air defense emergency is an emergency condition, declared by the Commander in Chief, North American Aerospace Defense Command. It indicates that attack upon the continental United States, Canada, or US installations in Greenland by hostile aircraft or missiles is considered probable, is imminent, or is taking place.

During the Cuban Missile Crisis, the U.S. Strategic Air Command was placed on DEFCON 2 for the first time in history, while the rest of U.S. military commands (with the exception of the U.S. Air Forces in Europe) went on DEFCON 3. On 22 October 1962 SAC responded by establishing Defense Condition Three (DEFCON III), and ordered B-52s on airborne alert. Tension grew and the next day SAC declared DEFCON II, a heightened state of alert, ready to strike targets within the Soviet Union. 15 November 1965 was the day Strategic Air Command (SAC) postured down to defense condition (DEFCON) III.

On 6 October 1973 Egyptian and Syrian forces launched a surprise attack on Israel. On 25 October U.S. forces went on DEFCON III alert status, as possible intervention by the Soviet Union was feared. On the next day, normal DEFCON status was reestablished except for the Sixth Fleet. which t resumed its normal DEFCON status on 17 November 1973.

In addition to the confrontation over the nuclear-armed missiles, Soviet submarines at sea off Cuba were engaged by American surface ships in a dangerous

Section B: The Unthinkable 213

effort to force them to surface; one submarine commander allegedly ordered the arming of a nuclear-tipped torpedo.

It has since been learned that the Soviet general in charge on mainland Cuba had been given authority to launch his 98 (tactical rocket, aircraft bomb, and cruise-missile) nuclear weapons if he could not contact Moscow.

Notwithstanding the sobering experience in the Caribbean, nuclear threats — implied or specific — continued sporadically in international affairs.

The Soviet Union, when it had an ideological dispute with China in 1969, hinted at nuclear attacks against them.

After an Egyptian blitzkrieg against Israeli troops in 1973 ("the Yom Kipper war"), Israeli forces reversed the initial strike, chased the attackers back into Egypt, encircled their forces and drove toward Cairo. The Soviet Union, taking Egypt's side, interceded with a cease-and-desist ultimatum to the Israelis. At that pivotal point, the United States and the Soviet Union briefly put their strategic missile systems on high alert.

In 1974, concern about nuclear security sprang up when Turkey invaded Cyprus, resulting in an armed conflict that involved Greece, which was another nuclear partner of the United States. Both nations harbored U.S. nuclear weapons that could have been forcefully confiscated from the American custodians.

During the Reagan administration, a beefed-up war-fighting contingent of tactical and intermediate-range missiles and artillery was deployed in Europe under NATO auspices. Implicit in the assignment was not just a defensive stance, but a slightly veiled threat of invoking the nuclear option as a pre-emptive measure. This U.S.-NATO atomic "coupling" potential was recognized not only by the Soviets, but also by allies who worried about being drawn involuntarily into a nuclear exchange that would take place on their homelands.

As late as 1991, in the first Gulf War, the United States threatened to use nuclear weapons under certain circumstances. During the post-9/11 retaliation against Al Qaeda and the Taliban in Afghanistan, nuclear weapons were said to have been readied at island bases in the Arabian Sea.

Before being appointed to President Clinton's national security staff, Morton Halperin reported — from his studies of nuclear strategy — that "the world of nuclear threat has been a secret one, hidden from review or assessment by the American public." In the nineteen crises he carefully examined, the value of nuclear intimidation was often either "exaggerated" or "entirely wrong." In each instance, Halperin concludes, the critical variable has not been the nuclear threat, but "the nature of the security interests at stake, the balance of conventional weapons, and skillful diplomacy accompanied by a willingness to compromise." In many situations, the public's perception of diplomacy was cloudy, partly because of secrecy behind undertakings that led to crisis resolution. Halperin stresses that when the possibility of using nuclear devices was considered, "the intention was not to win a battlefield victory with them but to coerce the political leadership of the enemy": Presidents who gave serious consideration to using nuclear arms were trying to persuade an adversary to back down.

Former President G.H. Bush subsequently wrote:

> The Cold War was a struggle for the very soul of mankind. It was a struggle for a way of life defined by freedom on one side and repression on the other. Already I think we have forgotten what a long and arduous struggle it was, and how close to nuclear disaster we came a number of times. The fact that it did not happen is a testimony to the honorable men and women on both sides who kept their cool and did what was right — as they saw it — in times of crisis.

As reassuring and upbeat as Bush's message ends up, it also confirms some frightful facts and illusions: that maintaining the boundary between humankind and disaster depended heavily on the coolness and honor of men and women. (Many people then, even more now, preferred that much less be at stake and that much stronger institutional and technological barriers could help humans keep "cool" all the time.)

Nuclear Illusions. In a Cold War peace-movement article entitled "Nuclear Illusions," psychologist Robert J. Lifton dissected a number of misconceptions about nuclear weapons — misconceptions stemming from fear caused by "Hiroshima-like images of extraordinary destruction." That fear is heightened by the invisibility of what was then a comparatively new phenomenon, radiation assault. Inadequate understanding and a sense of infinite destruction combined to induce a "doomsday" impression. These fears and illusions resulted in feelings of helplessness.

Lifton identified a set of "mental formulizations" (illusions) about nuclear weapons. One is the *illusion of limit and control*, which he believed underlies virtually all of the false assumptions about atomic warfare. As an example, he cited the concept of "tactical" or "limited" nuclear war, which was strongly promoted by Herman Kahn and Edward Teller. This concept was based on the theory that

> limited nuclear action and equally bold, more or less *unlimited* nuclear threat can enable to *control* the situation and keep it *limited*.

According to Lifton,

> That assumption defies virtually all psychological experience. Controlled nuclear war ... is an illusion.

"The second great nuclear deception is the *illusion of foreknowledge*," the idea that people can survive a nuclear attack if they are taught what to expect. This leads to Lifton's third case, the *illusion of preparation*: that evacuation plans, shelter construction, etc., based on an assumption of a limited nuclear war, will help manage nuclear destruction. Closely related is the *illusion of protection*, that shelters will protect against nuclear war. Failures of the *illusion of stoic behavior* and the *illusion of recovery* were well depicted in the movie "Fail Safe."

Lifton surmises that after a nuclear bombing, survivors will have "psychic numbing so extreme that the mind would be shut down altogether." This is connected to the *illusion of recovery*, that the outside world will come in and help. He anticipates "the most extreme form of collective trauma stemming from a rupture of patterns of social existence, with no possibility of outside help" (also chillingly pictured in "Fail Safe").

The final one of eight self-deceptions identified by Lifton is the *illusion of rationality*, which is usually couched in fashionable terminology like "reasonable individuals," "rational nuclear war," and "systems rationality" (a construct developed by systems and operations analysts, such as those working at RAND. Herman Kahn and Edward Teller symbolize the illusions of recovery and rationality when they assert that "rational behavior" consists of "courage," "readiness," and being "prepared to survive an all-out nuclear attack." Lifton says we are dealing here with "nothing less than the logic of madness — of social madness and collective 'mad fantasy'."

> [Doomsday Machine]. As envisioned by Herman Kahn, the Doomsday Machine was a vast computer network wired to a stockpile of H-bombs. In case of nuclear attack, it would be automatically detonated, creating worldwide fallout that would kill billions of people on earth—a catastrophe so horrendous that supposedly nobody would risk it by setting off a nuclear weapon. "Doomsday Machine" later became a generic, less-explicit term for any arrangement that endangered all of civilization.

He warned of the need to be on guard against spurious appeals to reason and rationality. The builders of such "rational systems" of weapons and strategic concepts are "confronted with an image they [like the rest of us] do not know how to cope with, and seek desperately to call forth, however erroneously, the modern virtue of reason."

Another misconception was, and is, that nuclear explosives should not be feared as "the absolute weapon." (The term is from RAND analyst Bernard Brodie's book with that title. Brodie's original conclusion was that nuclear weapons were so destructive that deterrence was their main function, but he later espoused the possibility of limited nuclear war.)

Michael May, former director of the Livermore weapons laboratory, also seems to think that nuclear war can be limited: he objects to what he labels as a "myth" that nuclear weapons are absolute weapons. May, for instance, takes issue with the estimate, attributed to Lord Solly Zuckerman, that six nuclear explosions would destroy Great Britain. To Michael May, a nuclear weapon does not differ qualitatively from any other weapon, although he admits that one or two hundred could destroy the heartland of the United States or Russia. Clearly May was opportunistically defining his scenario (using Hiroshima-size weapons) so as to support his conclusion.

Consider a more realistic scenario, in which six megaton-size weapons were exploded at ground-level, or even as air bursts, in the six largest metropolitan areas of Britain; it would surely mean near-total devastation of the island, not at all mythical. That's about as absolute as a weapon can get. Worse yet would be six of the largest thermonuclear weapons ever designed — Soviet behemoths of 100 megatons or more.

Some nuclear analysts believe that a large metropolitan area could rebound after nuclear demolition if proper advance preparations — shelters, hospitals, and supplies — had been made. However, those analysts tend to overlook the complexity of societal chaos and human psychology in reacting to such an assault.

Remedial efforts would be hampered by radiation phobia and by horrendous environmental consequences.

Another popular illusion is the assumed coercive utility of nuclear weapons. In actuality, the major nuclear weapons states have either lost wars or found their massive nuclear arsenals useless for winning wars — in Korea, Vietnam, Afghanistan, Suez, and the Falklands. While the history of nuclear self-denial reflects sound policy on the part of governments, that decision has been easier because experienced military leaders tend to recognize the inherent disutility of nuclear weapons in contemporary small-scale conflicts and policing actions. Unfortunately, that same recognition does not carry over to civilian politicians, candidates, or columnists.

National Security at Any Cost

The design, manufacture, and delivery of nuclear weapons became an end in itself, rationalized as essential for "national security," whatever the price. After all, the thinking went, the very survival of Americanism or Communism was at risk. Yet the consequences of nuclear warfare — including the physical, psychological, medical, and societal damage — were usually understated.

Many instances during the Cold War can be cited in which the governments abused their mandate and public trust. Most were camouflaged by secrecy. Here are a few prominent examples:

Radiation Effects of Nuclear War. Nuclear weapons create destruction, injury, and havoc via direct blast, heat, and radiation. An air-detonated blast, though, would produce very much less airborne radioactive fallout than a ground-level burst. Thus, air bursts were touted as a way of significantly reducing the secondary fallout effects at a distance from the target. The stratagem was thus embedded in the concept of limited "surgical strikes" that would destroy certain military targets, such as troops, ships, airfields, or missile silos without doing harm to the civilian population.

But few members of the public realized that hardened targets, such as missile silos and command centers, can be destroyed only by ground bursts. If it happened to be raining during such an attack, the precipitation-induced radioactive fallout ("rainout") could spread radiation levels as high as those from fallout after a surface blast. Describing what he called "dangerous illusions," U.S. Senator John C. Culver wrote in 1979 that "the 2300 or so warheads in the theoretical limited strike would kill between 3 and 22 million Americans, depending on wind and fallout protection."

According to Nobelist Henry W. Kendall, "12 to 18 million people would be killed from the collateral (or unintended) damage resulting from a Soviet attack intended merely to wipe out America's land-based missile force." Most of those casualties would be due to intense fallout extending hundreds of miles downwind from ground bursts at missile installations.

"The central fact of the nuclear age," concluded Henry Kendall, "is that nuclear arms are too powerful and numerous to be used to gain a nation's political or military objectives." In fact,

> the cumulative effects of an all-out nuclear war would be so catastrophic that they render any notion of "victory" meaningless.

Denver's Close Call. The primary weapons-plutonium fabrication plant of DOE was installed at Rocky Flats, Colorado, 30 miles upwind of Denver (upwind by mistake — the planners misjudged prevailing wind directions). In 45 years it produced some 70,000 plutonium "pits," the cores of nuclear weapons. Rocky Flats suffered at least two major fires which released radioactivity that drifted toward Denver. Exacerbating the potential hazard was the need to use water to extinguish plutonium fires: Water was the worst possible fire suppressant for the purpose, because it could have led to accidental nuclear criticality or to a steam explosion, followed by release of a great deal more radioactivity.

One fire at Rocky Flats, in September 1957, was in a building where there was enough plutonium to constitute 14 critical masses (slightly more than one critical mass is enough for an uncontrolled nuclear excursion — a serious accident). The second major fire, twelve years later, was in a building containing more than 750 critical masses of weapons-grade plutonium. About 10 percent of the plutonium was actually damaged or burned to oxide. The potential for major off-site contamination was exacerbated by the fact that plutonium-oxide residue created in the fire became a powdery ash — readily dispersed and moderately radioactive.

For this type of situation, firefighters are taught and ordered not to use water (because water can help a mass of fissile material go critical), but they had to do so in both fires because their carbon-dioxide extinguishers were inadequate. They were simply lucky, and so were the inhabitants of Denver: Subsequent investigations, which largely went unpublicized, confirmed the close brush with an accidental chain reaction.

The price tag of the 1969 fire was given as $70.7 million, which broke all previous records for U.S. industrial accidents. But little information was released by the public about how near Denver was to a minor disaster (there was little or no possibility of a nuclear explosion, but there could have been serious irradiation of the firefighters and the spread of radioactive contamination). "Secrecy's thick walls," says one reporter, "prevented full disclosure about the Rocky Flats fire." "Had they known the full truth about the risks to which they and their families were being subjected, [the public] would have been outraged."

The prevailing illusion that national security depended on having lots of nuclear weapons compounded the problems at Rocky Flats by leading to emphasis on throughput — production of plutonium pits — rather than safety.

In November 1989, with increasingly evident environmental problems, and with the Cold War winding down, the Rocky Flats plant ceased operations. The cleanup process began in February 1992.

(Horrifying tales about accidents and incidents in the Soviet weapons complex are slowing becoming public, a few being cited herein.)

K-19: The Widowmaker. A movie titled *The Widow Maker* has been made about a nuclear-submarine accident kept secret under the Soviet regime. On 4 July 1961, their first nuclear-powered ballistic-missile boomer on its maiden mission suffered a rupture of its starboard-engine reactor pipe, causing it to loose cooling water. The crew averted disaster only after rigging a makeshift cooling system using the submarine's drinking water. That action prevented a meltdown that would have killed the entire crew, destroyed the submarine, and added somewhat to the natural radioactivity of the ocean. Contrary to the movie's dramatization, it is very unlikely that complete loss of coolant or a reactor meltdown would have created a nuclear chain-reaction explosion.

Before that reactor accident, the boomer had successfully launched a ballistic-missile from under the Arctic ice pack. This was intended to send a then-urgent message that the Soviet Union had the capability for assured nuclear retaliation. The incident reflects the insecurity felt by Soviet officials because the United States had many more deliverable nuclear warheads, and American hardliners were uttering bellicose statements.

After its successful missile test, the submarine had taken its station off the Atlantic coast of the United States; it was during this routine cruise that the accident occurred.

Taking turns welding new piping on top of the reactor in order to provide emergency cooling, eight heroic crew members received lethal radiation doses. While the crew was evacuated 13 hours after the rupture, it is reported that seven or eight died within a week or so and fourteen others within two years. Surviving crew members suffered lingering consequences of excessive radiation exposure.

Actually, the K-19 had several serious accidents, the first being an in-port loss-of-pressure in the cooling circuit of the reactor, half-a-year prior to its maiden 1961 voyage. Eight years later, the sub collided with an American submarine that was shadowing it. In 1972 a fire broke out; although a 30-ship rescue operation saved the sub, the lives of 28 crew-members were lost. Widowmaker was an apt epithet for the K-19.

Its 1972 collision with an American submarine is a hint of what might be many untold tales of gamesmanship at sea, on land, and in air during times of extended political tension. Another underwater collision, near a Soviet naval station in Petropavlovsk, took place in 1974 between the Soviet nuclear submarine K-108 and the American attack submarine *Pintado* (SSN-672).

As time goes by, additional accidents — some more fatal — during the Soviet era are coming to light. Keeping the K-19 episode and its aftermath secret, from both their people and from the West was of paramount political importance to both governments.

Deterrence Theory. Risks from domestic production of nuclear weapons were inherently far less than the dangers of their deliberate use. One lesson learned belatedly by Robert McNamara, Defense Secretary during the Cuban nuclear missile crisis, is that

> [there are many] who still don't understand the degree of destruction that would occur if nations used these weapons [nor do they] understand the risk that, quite

unintentionally, we can maneuver ourselves into a position where these things would be used. They don't understand the fog of war.... It has become very clear to me as a result of the Cuban missile crisis that the indefinite combination of human fallibility (which we can never get rid of) and nuclear weapons carries the very high probability of the destruction of nations.

To Jonathan Schell, a well-known writer for *The New Yorker*, this disclosure revealed a dilemma:

> At the heart of deterrence theory, it had seemed to me, lay a basic contradiction: According to the theory, the way to prevent the foe from launching a first strike was to threaten to launch an annihilating second strike. But if a first strike was actually launched, ensuring that the United States would [promptly] become a wasteland, the retaliatory strike would have lost its purpose.

While McNamara, as Secretary of Defense, once praised "controlled escalation," as an outsider he has since become "quite confident" that a first strike against the United States would have unleashed "a very, very substantial [retaliatory] force."

Nobel Laureate Joseph Rotblat is another who derided the idea that nuclear deterrence prevented wars. He was one of the first to act on the concept of deterrence, personally helping develop an atomic bomb to stop Hitler from using one against the Allies. Later, Rotblat quit the Manhattan Project when it was clear that Germany had failed in its nuclear quest (despite efforts by General Groves to withhold that information from the Project staff).

Rotblat came to believe that his original decision to participate in the development was "mistaken" in several respects: First, he did not know that the Germans had given up work on the bomb long before he quit; second, he did not realize that "so-called nuclear deterrence doesn't work with people who are irrational, as Hitler was"; and, third, he was "naive in believing that once we scientists had produced a weapon, the military and civilian leaders would listen to us regarding how it should be used."

Some war planner developed scenarios for "graduated" response. Their thinking went as follows: After one violent exchange, both sides — appalled by what happened — would pull back from total nuclear exchange. (A big problem with this type of analysis was that it assumed good intelligence, communications, control, and calm thinking in a time of crisis — none of which can be taken for granted. Pressures would be immense on both sides for quick reaction or revenge.)

A combination of war planning, security paranoia, worst-case analysis, and nuclear bravado led a journey toward doomsday — the dreaded end of civilization.

Doomsday

"We all muddled into war," said Lloyd George in 1914 about World War I. "How [it] began, and why it occasioned surprise, are among the biggest puzzles of the century."

The Great Powers wanted neither war at all nor the type of conflict they got. World War I began because events got out of hand, and hostilities in the east of Europe (Austria *vs.* Serbia) meant hostilities in the west of Europe, because diplomats had spent more than 30 years tying alliances together. Russia, Germany, France, Britain, Italy, and eventually the United States were drawn in. The timing of each European country's entry into the war was highly dependent on capability for troop mobilization, which at the time was tied to railroad timetables. An all-or-nothing situation arose because of the extent to which mobilization was bound to the railroads and their availability ("use it or lose it!").

Analogously in the Cold War, U.S. strategic forces — armed with about 10,000 nuclear weapons — were "wired [together by] systems of incomprehensible complexity [under the control of] fantastically complex nuclear command organizations [that would] find it technically difficult if not impossible to limit a nuclear conflict — or even to limit a nuclear alert — once begun." Making the situation more perilous by 1985 was a "shift from loose to tight coupling," referring to the command and control links between mechanisms of warning and response:

> The possibility exists that each side's warning and intelligence systems could interact with the other's in unusual or complicated ways that are unanticipated, to produce a mutually reinforcing alert. Unfortunately, this last possibility is not a totally new phenomenon; it is precisely what happened in Europe in 1914. What is new is the technology, and the speed with which it could happen.

During the Carter administration, a move toward war-fighting as a national strategy took place with PD-59, a policy directive intended to offer a wider choice of preplanned options. Even if the plan set up a system that would "fail safe" in peacetime, it could "fail deadly" in war. The latter scenario resulted from pre-delegation of authority to release nuclear weapons. Pre-delegation made it extremely unlikely that the Soviets could immobilize the U.S. strategic triad through a decapitating attack on command and communication centers. At the same time, pre-delegation made nuclear exchanges very difficult to stop. Although the submarine leg of the triad was relatively secure, a "use it or lose it" fear could develop for the bomber and missile legs.

While some analysts argued that "no sane leader would deliberately take the fatal step in starting a nuclear war," the tight coupling of the "fail deadly" response to the early-warning and command-and-control systems could have let escalation take place without time or means to exercise deliberate restraint. Here are some questions that needed careful attention: Should every indication of an incoming attack trigger a convulsive retaliation? Should the entire nuclear retaliation force be launched if unidentified space objects are detected on the horizon? Where is the threshold set? What constitutes an "attack?" (These lingering questions led some strategists to support the quest for antiballistic missile defense.)

The preceding uncertainties indicate that there are two significant problems with the unstable nature of a hair-trigger response. First, a minor incident can escalate so rapidly that control cannot be exercised, and second, the response could end civilization.

The term *"mutual* assured destruction" (MAD) was coined in 1969 by Donald Brennan, of the Hudson Institute. Even earlier, in making the 1964 film, *Dr. Strangelove*, director Stanley Kubrick had caricatured the proposal of RAND analyst Herman Kahn for a "the doomsday machine."

Section B: The Unthinkable 221

George Kennan is often credited as the architect of the policy of Soviet "containment." He came to the same conclusion as Manhattan Project senior scientists who "realized from the start that, since there was no theoretical limit to the destructive power of nuclear warheads, the latter could not be regarded as just a new form of armament."

Jonathan Schell is quoted as explaining that the history of massive nuclear deterrence has revealed

> the unresolvable contradiction of "defending" one's country by threatening to use weapons whose actual use would bring on the annihilation of one's country and possibly of the world as well. And the emergence of the contradictions was in turn propelled ... by a recognition — this time on the part of nuclear strategists rather than citizens at large — of what a doomsday machine really is, and what it means to intend, in certain circumstances, to use one.

A profound observation on the "taste of blood" comes from Schell:

> Yet the bombing of Nagasaki is in certain respects a fitter symbol than Hiroshima of the nuclear danger that still hangs over us. It is proof that, having once used nuclear weapons, we can use them again.

Mikhail Gorbachev once remarked to Jonathan Schell,

> More likely [than giving the order to use nuclear weapons in retaliation for a nuclear attack] was that nuclear weapons might be used without the political leadership actually wanting this, or deciding on it, owing to some failure in the command and control systems. [My "new thinking" was] that the presence of nuclear weapons, a colossal threat to humankind, could run out of the control of politicians.

Another analyst who has studied the risks of international war, wrote:

> [The] possibility of a nuclear holocaust being triggered unintentionally is a matter of growing concern.... It is quite conceivable that an acute international crisis may act as a catalyst to trigger a nuclear war not in fact intended by the Governments concerned. What is being envisaged here is not accidental nuclear war, but rather nuclear war based on false assumptions, i.e., on misjudgement or miscalculation by the persons legitimately authorized to decide on the use of nuclear weapons....
>
> There is every reason to assume that strategic doctrines in East and West are mismatched. This mismatch has its origin in deep-rooted difference in philosophical and national traditions.

On more than one occasion East-West antagonism threatened to escalate to nuclear war. At the height of the Reagan administration's confrontational approach to the Soviet Union (November 1983), a NATO military exercise (*Able Archer*) was launched to practice procedures for release of nuclear weapons. This induced "the largest intelligence operation in Soviet history" in an effort to discover U.S. plans for what Soviet leaders suspected was a surprise nuclear first strike: "Some Soviet intelligence officials feared that [the NATO exercise] would turn into a real attack...." Moscow sent emergency telegrams to its intelligence officers abroad reporting an alert at the U.S. bases involved in the exercise. Had the Soviet Union already deployed a "doomsday machine," considerable danger would have existed.

In the aftermath, Soviet Defense Minister Dmitri Ustinov remarked that "it became more and more difficult to distinguish [U.S./NATO military exercises] from the real deployment of armed forces for aggression." After being briefed by the CIA director on the Kremlin's concern, President Reagan "recorded in his memoirs that he was surprised to learn that the Soviet leaders were genuinely afraid of an American attack."

Three years earlier there had been another false alarm:

> As reported by former CIA director Robert M. Gates, [President] Carter's assistant for national security, Zbigniew Brzezinski, received a call from a staff officer at 2:26 A.M. on 3 June 1980. The officer reported that the U.S. warning system was predicting a nuclear attack of 220 missiles on the United States. Brzezinski was convinced we had to retaliate, but knew that the president's decision time for ordering retaliation was from three to seven minutes after a Soviet launch, and wanted a confirmation call from the staff officer.
>
> Shortly after, Brzezinski received a second call and was told that the warning indicators were now pointing to an all-out attack of 2220 missiles. U.S. bomber crews were now manning their aircraft, and the Pacific Command's airborne command post had taken off.
>
> One minute before his call to the president, Brzezinski received a third call that informed him that only one of our warning stations had reported the impending attack, suggesting that someone had mistakenly put military exercise tapes into the computer system. The crisis came and went in a matter of minutes, but the implementation of a U.S. launch-on-warning doctrine would have brought [mutual] nuclear devastation.

With the bellicosity of the Reagan administration in mind, a strategic-policy analyst warned against brinkmanship:

> Despite the threat of nuclear disaster, the frequency of international crises has not declined in the nuclear age. "Coercive bargaining" is and continues to be a dominant factor in international politics.... The use of force for the purpose of conveying signals to the opponent in a situation of coercive bargaining may easily produce incidents of all kinds which have a propensity to escalate. This is even more serious if nuclear weapons serve as demonstrations of resolve ... [because] the two major powers have more opportunities to "test" each other by means of crisis confrontation and coercive bargaining. Hence the risks involved in unintended consequences of such behaviour become proportionately greater.

NATO's Nuclear Deterrent

One way of escalating a confrontation into a nuclear conflict was through the East-West arms competition in Europe. Western military planners, in fact, assumed Europe to be the main theater of skirmish. WTO conventional forces were alleged to be superior to NATO's, partly because of size and partly because of proximity. The Soviets had nuclear weapons for theater use, but they relegated them to a secondary, defensive role.

The role of NATO's nuclear weapons was not quite as secondary as it appeared. As presented to the public, NATO was compensating for its asserted conventional inferiority by requesting nuclear weapons from the United States, planning to use them to stop a conventional Soviet attack in Europe. However, that claimed inferiority did not withstand scrutiny: Realistic assessments, regularly done in the Pentagon, of Soviet conventional forces *versus* NATO's, showed no credible scenario for a successful Soviet incursion; for one thing, Soviet forces were not deployed in a configuration that could provide logistical support for a sustained conventional invasion into western Europe.

Perhaps the NATO nations, particularly West Germany, felt the need of a "nuclear umbrella" to counter a potential threat of nuclear blackmail, or perhaps they did not believe the strategic analysts. In any event, nuclear weapons were deliberately deployed with NATO to *couple* U.S. strategic forces — a doomsday tripwire to convince the Soviets that any attempted takeover of West Germany would not be localized to Western Europe but would put the heartland of the Soviet Union at risk.

Hence escalation of the nuclear arms race during the 1970s and 1980s brought perhaps one of its most dangerous dimensions: a massive buildup of medium-range nuclear rockets in Europe. According to a high-level Soviet official (Roald Sagdeev):

> In the Soviet arsenal, the deadliest of rockets were known in the West by the name SS-20. They were carried by trucks, which made them mobile, thus practically invulnerable. They were counterbalanced on the Western side by Pershings.... At that time the Kremlin elite was paranoid about the technical capability attributed to Pershing rockets.... According to leaks in the press, allegedly the accuracy of these rockets was enough to "kill the Soviet leaders in their bunkers." At that time the Politburo took the possibility seriously.

No wonder: NATO in 1971 had 11 types of U.S. nuclear systems deployed in Europe, including nuclear landmines, surface-to-air missiles, intermediate-range missiles, ground-launched cruise missiles, howitzers, antisubmarine missiles and depth bombs, and aerial gravity bombs. This was besides British and French mobile and home-based nuclear components.

The viability of NATO's reliance on nuclear weapons has often been challenged. Lord Solly Zuckerman, an advisor to the British government, put it this way:

> For the sake of argument, take the case in which the Soviets are stupid enough to invade Western Europe, NATO forces are being overrun, and we get to the point when the supreme allied commander asks permission to fire a few tactical nuclear weapons. What the hell happens next? We fire a couple. The Soviets respond with four, maybe bigger ones. We respond again, and so on.

Lord Zuckerman stressed that he, for one, did not accept the "worst-case scenario of Warsaw Pact forces slicing across West Germany and defeating NATO land forces within a couple of days unless stopped by use of tactical nuclear weapons."

Zuckerman also rejected the argument that the neutron warhead would be particularly effective in stopping an invading army. He described the enhanced-radiation device as "the most ludicrous weapon that has ever been designed."

As for the deployment of almost 600 GLCMs and Pershing-IIs in Europe, Lord Zuckerman said it would be an illusory addition to security because nuclear-weapon delivery systems already in service would continue to hold the Soviet homeland at risk.

One analyst's point of view was that

> NATO's strategy creates a twofold credibility problem: first, the Soviet Union firmly rejects the idea that a nuclear war, once started by NATO's first use of nuclear weapons, could be kept limited or controlled; secondly, Soviet sources threaten to launch a "full-scale devastating and annihilating" retaliation strike against the homeland of the United States, thus casting doubts on the credibility of extended deterrence, implying nothing

less than the incineration of Chicago and New York for the sake of Hamburg and Hanover.

In the words of Jonathan Schell, who reported on and analyzed the overseas reaction,

> During the Cold War, Western Europe was beset by two opposite anxieties in relation to nuclear arms. One was that the United States would not come to Europe's defense in a crisis. The other was that it would.... The second anxiety became more acute in the mid-sixties, when the United States adopted the doctrine of "flexible response" — an option that supposedly made limited nuclear wars possible. The Europeans feared that such wars would be limited to Europe.

A military historian has pointed out that "Both NATO and the Warsaw Pact regularly rehearsed the use of tactical nuclear weapons in their exercises, treating them as some superior form of artillery." This did little to diminish the fact that "The greatest single spectre haunting the political and military leaders throughout the Cold War was that of nuclear war, and it was a threat that influenced every decision of any significance." "Nuclear warfighting" was a school of thought, particularly in the United States, that considered a protracted conflict with nuclear weapons to be possible. However, it is more likely that, after a nuclear exchange, surviving military forces would have ceased to function, and further fighting would have become impossible. Such an dismal outcome as the aftermath of a climatic nuclear exchange was consistent with NATO mock field exercises during the 1970s and 1980s.

Two major difficulties were presented by tactical nuclear weapons: The use of only one would have crossed the nuclear threshold, and, further, it would have been extremely difficult to differentiate between tactical and strategic weapons. As a case in point, the U.S. Pershing-II stationed in Western Europe had a range of 1,800 km, sufficient to reach targets inside the Soviet Union, and the Soviet SS-12, with its range of 900 km, could hit aim points in England or France. All of these target nations would have regarded the use of "tactical" weapons as "strategic" attacks.

Regarding the use of enhanced-radiation weapons ("neutron bombs"), it is difficult to predict if any American leader would have authorized nuclear release. President Jimmy Carter is quoted as having said, "A decision to cross the nuclear threshold would be the most agonizing decision to be made by any President. I can assure you that [low-yield, enhanced radiation weapons] would not make that decision any easier."

Whether NATO or the Warsaw Pact would have actually used nuclear weapons remains debatable. The Soviets might have avoided them, thinking that they could succeed with conventional forces without reducing Europe to a nuclear wasteland. NATO's leaders would have been in a quandary, especially if they had to use them first. Both sides realized that any conflict could escalate from conventional to nuclear, from tactical to strategic.

Although it is quite possible that war in central Europe was avoided because of the existence of nuclear weapons, either type of war — conventional or nuclear — would have been extremely bloody for both sides. In any event, the eventual closure of the East-West confrontation without direct military conflict was very fortunate.

Broken Arrows

As described, nuclear wars could have been started either intentionally or by miscalculation. A third way to trigger a nuclear war would be by accidental nuclear detonation at the wrong place or the wrong time (begging the question of whether there is a right place or a right time). The unforeseen (or unauthorized) detonation of a nuclear weapon has inevitably been credible; in fact, the U.S. military has adopted the term *Broken Arrow* to refer to nuclear-weapon incidents that have the potential for unanticipated detonation.

While no unplanned full-scale nuclear detonation has occurred, there have been many incidents of less consequence that involve explosion of the non-nuclear detonators or burning of a nuclear weapon. According to the U.S. government, "...there is always a possibility that, as a result of accidental circumstances, [a nuclear] explosion will take place inadvertently."

In addition to the potential for triggering a nuclear exchange, the accidental detonation of a nuclear weapon can lead to release of plutonium — an event that might cause public panic and a perceived need for cleanup action, but most of the actual damage would be done by the explosion of the conventional high-explosive, not by the plutonium. U.S. ground-transported nuclear shipments were said in 1990 to be roughly "more than one thousand vehicle trips and one million miles per year."

Some Broken Arrow incidents are outlined below. The Department of Defense has reported thirty-two potentially serious occurrences involving U.S. nuclear weapons from 1950 through 1980.

Although it was not revealed until 1986, one of the first major incidents took place in the outskirts of Albuquerque, NM, in 1957. According to a 1986 news story in the *Chicago Tribune*, "Newly released documents disclose that a 42,000-pound hydrogen bomb, one of the most powerful ever made, accidentally fell from a B-36 bomber near Albuquerque 29 years ago.... Most researchers believe [its design yield] was more than 10 megatons."

Also in 1956, a U.S. bomber crashed into a storage site in the United Kingdom that contained three nuclear weapons. Although a fire damaged the bombs, their conventional explosives were not ignited.

The Center for Defense Information, investigating further, found that, on 22 May 1957, a weapon dropped out of the bomb bay of a bomber that had taken off from Kirtland Air Force Base. The nuclear-fission capsule was not in the bomb: With some early models of nuclear weapons, it was standard procedure during most operations to keep the essential capsule of nuclear material separate from the weapon as a safety measure to preclude a fission chain reaction. However, the high explosive of the bomb did detonate, presumably dispersing some low-level radioactivity in the vicinity.

In June 1960 a nuclear tipped Bomarc air-defense missile at McGuire AFB near Trenton, New Jersey, was incinerated in a fire energized by its liquid-fuel first-stage rocket. The intensity of the fire was so great that the 10-kiloton nuclear warhead fell from the upper stage into the molten mass. Warhead-safety systems functioned so as to avoid nuclear criticality, and the nuclear core simply melted into a radioactive slag. Some plutonium was injected into the atmosphere by the fire. After cleanup, about an acre of land surrounding the site was sealed in concrete.

On 24 January 1961, two nuclear bombs fell from a bomber flying near Goldsboro, NC. The parachute deployed for one of the bombs, so it received little impact damage. The other bomb went into free fall and broke apart on impact; there was no explosion, but it is reported that the uranium portion of the weapon was not recovered. Despite the benign outcome, it had the potential to be a very serious nuclear-explosive accident because the bombs went through their arming sequence in preparation for detonation. "Five of the six safety devices failed.... Only a single switch," said nuclear physicist Ralph E. Lapp, "prevented the bomb from detonating and spreading fire and destruction over a wide area."

One of the worst and most embarrassing incidents took place on 17 January 1966 over Palomares, Spain. A collision in air between a B-52G and its refueling tanker caused four unarmed (Mk-28) hydrogen bombs to fall, three of them on the ground and one at sea. One was found relatively intact on the ground, and one was recovered three months later from the sea. Fortunately the high explosive in the hydrogen bomb that fell into the Mediterranean sea did not detonate. The high-explosives of the other two bombs detonated on the ground, spreading slightly radioactive fissile materials. Approximately 1400 tons of plutonium-contaminated soil and vegetation were removed to the United States. This was the most publicized of the Broken-Arrow incidents; it resulted in what has been described as "the most expensive, intensive, harrowing and feverish underwater search for a man-made object in world history."

Another incident on foreign territory took place on 21 January 1968 at Thule, Greenland, at a U.S. Air Force base. A nuclear-armed B52-G aircraft crashed on the ice near the end of the runway with four nuclear weapons. All were destroyed by fire, resulting in extensive low-level radioactive contamination because the high-explosive charges detonated. About 237,000 cubic feet of contaminated ice, snow, and water, with crash debris, were removed to a U.S. storage site.

According to Newsweek, President Kennedy was informed after the Goldsboro accident that "there had been more than 60 accidents involving nuclear weapons ... since World War II ... including two cases in which nuclear-tipped anti-aircraft missiles were actually launched by inadvertence."

Accidents involving the military occur frequently, but we usually don't find out whether nuclear weapons were involved until years later. Pentagon policy has been to neither confirm nor deny the presence of nuclear weapons; for example, nuclear components are known to have been shipped more often than reported. At the time that DOD had conceded 11 Broken Arrows, there had been at least 4 AEC weapons accidents that were not acknowledged.

The implications of these accidents were frequently noted by critics of unlimited arms buildup: Admiral Gene LaRocque, head of the Center for Defense Information, commended the armed services in 1981 for their safety record, but said: "Before we build 17,000 new nuclear weapons for the MX, cruise, Trident, Pershing-II and other planned weapons systems, we should give very careful consideration to how we are going to deal with the dangers inherent in multiplying the numbers and complexity of our weapon systems."

The risk of additional accidents remains, for several reasons: a large number of nuclear weapons are still in inventory and being transported, many weapons-delivery systems are deployed in active standby, and stressful conditions result from rapid reaction to strategic alerts.

Information about Soviet accidents is still sparse. However, there was a policy of not airlifting nuclear bombs outside of Soviet airspace, so international incidents like Thule and Palomares would not have occurred with Soviet weapons. In view of the history of known accidents at Soviet military-production facilities like Kyshtym and Sverdlovsk, it was a wise policy.

Accidental or Unauthorized Launch Risks

Another way nuclear war could start is through accidental or unauthorized launch of a nuclear missile. The risk of accidental missile release depends on the effectiveness of safeguards, both intrinsic and procedural.

The consequences of an unintentional launching of a nuclear-armed missile could be:

- merely a bad public-relations day,
- or dispersal of slightly radioactive fissile material,
- or (unlikely but possible) a nuclear explosion,
- or — most worrisome, although (one hopes) most unlikely — an automatic launch-on-warning response from the inadvertently targeted nation, without waiting to see whether the warhead would detonate.

Missiles and bombers have many built-in and procedural safeguards to reduce the risk of accidental or unauthorized nuclear release, as outlined in the following two paragraphs.

At least three classes of safety devices or controls hinder unintentional missile firing and subsequent detonation of its warhead: (1) mechanical or electronic devices inhibit improper or unauthorized launch; (2) various safety interlocks keep the warheads from detonating unless arming is enabled by the highest national authority; and (3) a Permissive Action Link (PAL) for many — but not all — missiles is an important end-of-the-line safety feature.

Nuclear-delivery aircraft have inherently more safeguards: They need not be armed with nuclear weapons unless under a high state of alert; the nuclear weapons can be "safed" by having separate nuclear capsules; airplanes can be recalled after takeoff since they are piloted by humans; and, moreover, defense against aircraft — especially a single errant bomber — is feasible, far more so than defense against ballistic missiles.

Even if a missile were launched without national-command authorization, many of the arming mechanisms of the warhead would automatically be activated in the correct sequence. Several nuclear-armed missiles have indeed come very close to inadvertent launch: Some nuclear warheads are said to have been on the verge of triggering, and the PAL did not become a standard safety-backup feature until the second-half of the Cold War.

The circumstances under which an unauthorized or accidental missile launching might occur depend in part on the national state of military readiness. A high state of readiness reduces safety margins, even though nuclear release requirements are coupled to another key policy objective: ensuring that a war does not start by accident. An order to fire a missile is more likely to be carried out without hesitation in time of crisis. Now, in the post-Cold War environment, a stand-down in readiness reduces hair-trigger risks.

As mentioned above, accidents with potentially serious consequences were almost routine during at least the early phases of the Cold War: Between 1950 and 1968, eleven nuclear weapons that were never recovered fell out of or crashed with U.S. aircraft — in addition to the ones discussed above that were recovered.

Because false alarms of an unidentified foreign missile launch were not uncommon, the C^3 networks and hair-trigger launch policies adopted by both superpowers heightened the prospects of accidental or unintentional nuclear war. That was an unadvertised downside to the so-called "nuclear insurance" policy of assured retaliation.

For example, only the good judgment of Soviet Lieutenant Colonel Stanislav Petrov prevented a missile-launch crisis on 26 September, 1983. A missile-attack warning alarm was triggered unexpectedly when he had the late-night duty watch in the officer's early-warning bunker south of Moscow. Petrov was aware that the United States and NATO had been organizing a military exercise that could be a cover for an actual invasion; so he had to decide within a few minutes on what action to take based on the indications he was receiving that U.S. ICBMs were being launched: Five rapid rocket firings were being reported by his space-based early warning system. Despite the political tensions of the time, Petrov decided that "nobody starts a war with just five missiles." He was right, as it turned out; an infrared satellite-based signal from the Soviet Cosmos 1382 was evidently triggered not by rockets but by light reflections from clouds over the U.S. missile base at Malmstrom, Montana.

> Despite the enormous investment ... the systems to control the use of nuclear weapons have always been vulnerable to attack, set on a hair trigger, susceptible to false alarms, and dependent on dubious measures of control such as predelegating nuclear launch authority to military officials other than the president.

Nuclear Safety and Authorization. Although the prevention of nuclear accidents has always been a priority, there was an inherent contradiction between nuclear safety and the weapon-authorization procedures. "The vulnerability of command and control systems to nuclear attack created strong pressures to buttress the credibility of the retaliatory threat by pre-delegating alert and launch authority and shortening the reaction time of nuclear forces." Thus, launch discretion was granted to U.S. military commanders who were outside the chain of presidential succession, and authorization and unlock codes used to unleash the strategic forces were even more widely distributed.

Section B: The Unthinkable

Cold War military decisions often pre-empted public safety, and there were delays in implementing safety features such as PALs and insensitive high explosives (IHE).

> The delay in introducing PALs for bombers, ICBMs, and SLBMs, and the decision by the DOD, DOE, and the Navy in 1983 to forgo the use of insensitive high explosives (IHE) and other safety features on the W88 warhead ... are further evidence that the perceived urgency of deterrence overrode safety considerations.

Another example is the choice of an extremely volatile propellant in combination with a high-yield (475-kiloton) warhead to maximize the range and effectiveness of Trident-II missiles at a time when safer alternatives existed.

At the time of the 1962 Cuban missile crisis, a particularly dangerous confrontation took place in another part of the world when a U-2 aircraft flew deep into Soviet airspace. Two U.S. F-102s, armed with Falcon air-to-air nuclear missiles, had to be scrambled to protect the U-2 from MiGs that were overtaking the slower-flying reconnaissance plane as it returned to U.S. airspace near Alaska. The possibility of an accidental, unauthorized, or mistaken use of a U.S. nuclear weapon during this incident was not inconceivable. Furthermore, given the already tense international climate, the risk of escalation after a conventional clash would not have been negligible.

As recently as 1995, the world came within perhaps a few minutes of experiencing a Russian retaliatory nuclear attack on the United States. Russian radars detected a rocket launch they interpreted as a Trident missile from a U.S. submarine. President Boris Yeltsin convened a threat-assessment conference and deliberated for about eight minutes on "whether to launch a counterattack before the incoming missile arrives." With only two or three minutes to spare, Russian military officers determined that the rocket was heading away from Russian territory. Actually, the rocket was a routine scientific launch from Norway, which had previously notified the proper Russian authorities.

We have cited some publicized unpremeditated incidents that caused the C^3 system to respond — sometimes taking the issue all the way to the nation's chief executive. Fortunately, all the human decisions were correct; the erroneous information led to no counterattacks. But we do not have public information on how many unpublicized incidents really took place, nor do we know how close to nuclear Armageddon these incidents really brought the world. Much about these incidents still remains secret.

Radiological Warfare

A feared potential form of nuclear warfare has been the use of radioactivity to contaminate people and terrain. Radiological weapons are intended to disperse highly radioactive material (without a nuclear explosion) so as to cause injury or paralyze facilities. While the potential for such warfare was recognized during World War II (Germany planned to spread radioactivity on American cities), the amount of radioactive material in existence was small. During and since the Cold War, the quantity of radioactive materials increased significantly and governments have ready access to the materials. For instance, radioactive substances could be extracted from spent nuclear fuel, inserted in canisters, and scattered by bombs or missiles.

Even Robert Oppenheimer once favored the spraying of the radioactive fission-product strontium-90 on German soil during World War II. He is reported to have written a letter to Enrico Fermi "saying that we should not begin the [nuclear reactor] project unless we could produce enough [strontium-90] to [poison the food] and kill half a million men."

The Soviets engaged in research and testing of R-2 (SS-2) missiles with a warhead filled with radioactive liquid that was to be dispersed over a target.

Radiological warfare, however, would have little military value, because the effects of dispersed radioactivity are likely to incapacitate only slowly, and are less likely to impact military personnel as much as unprotected civilians. The harmful effects of nuclear weapons are more immediate through blast, fire, and prompt radiation. Although radiological warfare would be far less directly destructive to life and property than nuclear explosions, the direct and long-term effects in heavily contaminated areas could be damaging to human health and the food chain. Radioactively contaminated farms, water, and livestock would pose long-term cleanup problems. Contemporary experience indicates that the psychological outburst would be severe.

The greatest impact from radiological weapons would occur via dispersion in the atmosphere as an aerosol small enough to be inhaled. Aside from a nuclear explosion, such pulverization is very difficult to accomplish, with either radioactive materials or with a fissionable substance like plutonium. The more likely outcome of a radiological attack would be psychological trauma, with few near-term or long-term casualties.

In 1979 the superpowers together proposed a multilateral treaty banning radiological weapons. After successful bilateral talks on the subject, the two parties agreed that the United Nations should follow through in making an international prohibition. The proposed treaty would have forbidden the deliberate development, stockpiling, and possession of radiological weapons. Not covered in the treaty would be nuclear weapons, which are in a category of their own. An agreement on radiological weapons, however, never came to fruition.

****

During the political ice age, situations and policies that are now widely seen to be unacceptable were seriously considered, experienced, or tolerated: mutual assured destruction; risky, unsafe nuclear weapons; shaky, weak chains of nuclear command and control; and announced or implicit policies for first-use of nuclear weapons to frustrate conventional attacks and to retaliate by destroying cities. As long as large arsenals of nuclear weapons remain in existence, some of these conditions persist as facts of daily life — despite the nominal warm up in relations.

The quantities of weapons acquired were in large measure the result of special-interest pressures fundamentally unrelated to strategic considerations, with a rationalizing veneer of judgements about military strategy and national security.

Very dangerous conditions resulted from a process of assessment and planning that was war-oriented, with little precedent and few checks or bounds. Nuclear policy treated nuclear warfare as though it was a game between reasoned and informed adversaries, both of whom understood the rules and agreed to play by them. Escalation control was presumed, despite possible failure of communication links and despite nearly instantaneous strategic weapon delivery and unprecedented firepower.

During moments of grave decision, the ultimate weapon gave U.S. presidents difficult choices — especially Truman, Eisenhower, Kennedy, and Johnson. Relying on a deteriorating value system for dealing with perceived national threats, each president slid into, subscribed to, or exploited the exaggerated rhetoric, worst-case analysis, policy oversimplification, and excessive secrecy that were inherent to the times. President Franklin D. Roosevelt's heavy hand against those who opposed intervention before and during World War II contributed to erosion of the presidential value system. President Nixon pushed the envelope so far that he got caught lying about his arrogant and extensive abuses of public trust.

Paranoia over national security was pervasive and tended to dominate the public and intra-government debate (when open discussion was permitted) and much of the discourse that did take place was kept secret. Paranoia was subtly incorporated in the worst-case-analysis methodology that was used to justify the expansion of nuclear arsenals. Whether through ignorance or error, the extreme fears about the adversary were usually imaginary.

We are beginning to read memoirs of American Cold Warriors who still rationalize the huge weaponry buildups. Their own writings betray the ascendency of fear. Even with 10-to-1 or 15-to-1 domination in warheads and delivery systems, U.S. insiders trembled.

Little of this buildup would have passed open scrutiny had the numbers been made public. Eventually the Soviet Union caught up and passed the United States in some of the deadly statistics, but never did that lead to any of the feared dominance. What it brought about was a stalemate.

It is not unusual to find unreconciled hawks (hardliners) who insist that the United States might have surrendered to the Soviets or to the inroads of Communism were it not for their hardline resistance. (Ignored often is the one major contribution of nuclear weapons: deterring direct Soviet incursion against the West.)

Illusions about the utility of nuclear war and expectations of a favorable outcome were widespread. The planners kidded themselves and others that nuclear wars could be won, glossing over the irreparable consequences of mutual destruction. Even doomsday concepts were tolerated — witness the slogan "better dead than red," once popular with a large segment of the U.S. population. Though a powerful and motivating motto for hawks, it was contrary to the ideals of peace, justice, and accommodation that guided public-minded protesters.

A widely syndicated columnist, William Pfaff, wrote a commentary about World War I and its lesson for the 21st Century. He believed the effects of 1914-1918 have yet to subside, noting that the first war produced yet another. Pfaff deemed the most important aspect of World War I to be "that it was unexpected, unpredicted, and one can even say uncaused."

> What really happened in 1914, and what followed during the rest of the 20th Century was ... literally unimaginable, and hence it was not imagined.
>
> That is the lesson. You can make forecasts, write scenarios, project trends, but the future in its real dimensions is unimaginable. You can really only look in the past, and that demonstrates ... tragedy is a fundamental principle....

He wrote that in 1998, fully aware of the tempestuous Cold War, its placid culmination, and its hopeful aftermath.

Nuclear weapons are meticulously designed to explode in a devastating chain reaction. As long as the design-condition control and authorization systems prevail, the weapons — with or without their delivery systems — are safe and stable. But accidents have happened, and the risk of unauthorized launch of a nuclear missile is real. Simply an errant missile or an accidental explosion could conceivably trigger a hasty, irreversible nuclear retaliation.

Now nearly everyone realizes that nuclear war would have been catastrophic for both superpowers, destroying the warring nations in their misguided attempt to be secure — and that bystanders like those in Western Europe would have fared no better. National security — everyone's — would have vanished in a nuclear war.

C. THE NUCLEAR PRIESTHOOD

Dubbed the nuclear "priesthood," a faction of influential scientists, technologists, analysts, and policymakers maintained the momentum of the Cold War. They were a driving force that brought humanity — in the name of national security — to the brink of disaster. They were the purveyors of myths that undermined arms control. Who were they, and what interests drove them?

After interviewing influential protagonists, Jonathan Schell arrived at the following explanation for excesses of the arms race: "A nuclear priesthood taught that to threaten genocide, even to carry it out, was not only justifiable but [an] inescapable duty."

Schell was referring primarily to U.S. hawks, but we should keep in mind that American allies, especially members of NATO, were like-minded. Great Britain and France had their own nuclear priesthoods, and those two anti-Soviet allies had developed and tested atomic weapons.

U.S. strategy was largely in the hands of a self-nurturing conflict-inclined *ad-hoc* clique. First, it influenced or orchestrated strategic assessments used by the military; second, by intellectualizing the war-planning process, it made it appear more like science than speculation; and, third, by working with like-minded individuals and organizations, the priesthood dominated and bureaucratized nuclear decisions.

Morton Halperin characterizes this inner-circle group as a "national security bureaucracy," in which he includes career civil servants, military officers and politicians who believe that first-use threats of nuclear weapons are credible and have played a critical role in resolving crises.

In the United States, such vested interests that produced their own momentum included defense intellectuals, military officers (active, retired, and assigned to bureaucratic posts), scientists and technologists, congressional members and staff, and hard-line hawks. They frequently shuttled between offices in and around Washington, D.C. Organized labor and executive management in the defense industries constituted two other important hawkish pressure groups.

Think Tanks

Being responsible at the beginning of the Cold War for strategic planning, the U.S. Air Force turned to civilian or quasi-military consultants for advice. Many former World War II military officers nurtured their influential connections. These officers joined or formed consulting firms, often hiring other officers who were buddies or had good connections. In addition,

> The Air Force found further institutional support for its [nuclear strategy] at the Atomic Energy Commission's weapons laboratories, where many scientists were more interested in developing the large hydrogen bombs carried by strategic bombers ... than the small nuclear weapons that might be needed for strategic air defense (a choice that figured prominently in the increasing conflict between Edward Teller and Oppenheimer over the Super program).

Producing tomes that justified a massive build-up of the military were some right-wing policy-analysis associations: the Hoover Institution, American Enterprise Institute, Center for Strategic and International Studies, Heritage Foundation, Hudson Institute, Cato Institute, Freedom House, and American Security Council.

Specialists were hired by defense-consulting firms like BDM and SAIC to deal with the more complex issues that arose. They concentrated on military technologies, on intelligence assistance, and on support for specific government agencies. One think-tank that flourished (and still prospers) is RAND, with its main office in Santa Monica, California.

RAND. The development of various American deterrence theories and policies originated at the RAND Corporation — an Air Force creation that started in the late 1940s. Having the "big bomb," the Air Force controlled a generous slice of defense funding well into the 1950s. As it turns out, by 1983, "Almost all of the major government defense planners have at one time worked in the think-tank's offices in Santa Monica, California."

RAND was given its intellectual start by John von Neuman, the inventor of mathematical game theory. The "prisoner's dilemma" became a strong paradigm at RAND, leading to academic, unsubstantiated arguments in favor of military expansion, despite consequences of a nuclear arms race.

Drawing from wartime experience with "operations research," RAND developed "systems analysis" as a tool that was widely exploited although there was very little substantiated evidence of its validity.

Among the earliest RAND analysts who gained significant influence was Bernard Brodie, whose 1946 book *The Absolute Weapon* asserted that deterring — not fighting — was the main function of the military. This led him to conclude that the United States should keep its atomic arsenal invulnerable to attack. It was also important, he thought, to generate a universal fear of nuclear warfare. But Brodie later retreated from that position and embraced the idea of limited nuclear war, saying "we can hardly afford to abjure tactical use of such weapons without dooming ourselves and allies to a permanent inferiority to the Soviet and satellite armies in Europe."

Brodie slowly became disillusioned by the war in Vietnam and rejected what had become the underlying RAND "ethos" of *flexibility* and *options* that made nuclear strategy even possible. Faced with potential massive destruction by nuclear weapons and fundamental uncontrollability of strategic warfare, Brodie ultimately acknowledged that these RAND beliefs were "illusions."

The "Doomsday Machine" concept was the creation of Herman Kahn, a RAND nuclear physicist. In order to] enable the U.S. to take a much firmer position, Kahn advocated a first-use policy for nuclear weapons and the construction of a massive civil defense. Kahn was sometimes likened to Dr. Strangelove.

In his treatise, *On Thermonuclear War*, Kahn argued that American security depended on "a willingness to incur casualties in limited war just to improve our bargaining position moderately," his illustration being the ill-fated war in Vietnam. The second Chapter of Kahn's book is entitled "Will the Survivors Envy the Dead?" Kahn then believed that a thermonuclear war could be won. He summarized his views as follows:

> If our nation can survive the actual attack and has made some minimum preparation, then in all probability, "the survivors will *not* envy the dead"; there will be more than a mere technical or academic difference between victory, stalemate or defeat. Under these circumstances, in addition to having a deterrent capability, we might want an ability actually to fight and survive a [nuclear] war.

Besides Brodie, another RAND analyst disillusioned by Vietnam was Daniel Ellsberg, who leaked the so-called *Pentagon Papers* to the press in 1969. Comprising a 7000-page top-secret history of U.S. decision-making about

Vietnam from 1945 through 1968, the documents were copied by Ellsberg from his safe at RAND. Their publication by the *New York Times* came about only after civil libertarians won a landmark legal appeal. The Supreme Court supported constitutional First Amendment protection, overturning prior-publication restraint by the executive branch which claimed "grave and immediate danger to the security of the United States."

Ellsberg later became a vocal anti-nuclear-weapon activist. In explaining why he released the Pentagon Papers to the press in 1971, Ellsberg reveals how he had "learned that President Truman, followed by President Eisenhower, had fully financed the French colonial war from 1945 through 1954." He also found out "that President Nixon was as determined as Lyndon Johnson had been to avoid U.S. failure in Vietnam; however, his secret plan to achieve this was to threaten to enlarge the war dramatically if North Vietnam did not withdraw its troops from the South (as we withdraw ours)."

Ellsberg explained:

> I believed that the threats would fail, so I gave the [Pentagon Papers] to the Senate Foreign Relations Committee. Almost two years later, defying four court injunctions granted at the President's request, I gave the document to *The New York Times, The Washington Post,* and 15 other newspapers.

RAND was closely associated with the development of U.S. strategic concepts. In 1951 RAND physicists learned of the "Super," Edward Teller's concept of a hydrogen bomb. (At the time, Oppenheimer objected to its development, partly because he believed the U.S. H-bomb project "might prompt the Soviets to build one too." He was right.)

It was a RAND analysis that led to John Foster Dulles' famous 1954 "massive-retaliation speech." Secretary of State Dulles was "an unbending opponent of the Soviet Union and a firm believer in the existence of a monolithic communist conspiracy to overthrow the Western political system." Dulles declared that the United States would thenceforth feel free to meet any threat, anywhere in the world, through "a great capacity to retaliate instantly [with nuclear weapons] by means and at places of our own choosing."

Albert Wohlstetter, a mathematical logician, was another RAND analyst who had considerable influence on U.S. strategic policy. Wohlstetter is credited with conceiving of the "second strike" as the key concept of deterrence. His primary concern then was his calculated vulnerability of U.S. strategic forces, and he went public on the issue, claiming that SAC was terribly exposed to a Soviet first strike. One outcome of this analysis was the development of hardened ICBM silos.

Another of Wohlstetter's causes was ballistic missile defense. In the 1960s he advocated defending Minuteman missile silos because the Soviets had begun to deploy their own missiles in hardened silos. During the 1969 Safeguard ABM debate, Wohlstetter testified before the Senate Armed Services Committee that "the percentage of the Minuteman force that would be destroyed, if undefended, comes to about 95 percent."

Wohlstetter was eventually fired, for undisclosed reasons, from RAND. After joining the faculty of the University of Chicago, he continued to deal in ABM hyperbole. He denounced "minimum-defense theorists" who "call for no defense of our civilians and nearly total reliance on [retaliatory] threat to bombard enemy civilians."

RAND, meanwhile, prospered on controversies such as the "bomber gap" and the "missile gap," which were projected estimates of imbalances that never materialized in favor of the Soviet Union. When the alleged missile gap was forecast, there was no hard supporting evidence. Although the CIA picked low numbers in its estimate, the RAND-advised Air Force prevailed with its higher predictions. One reason for the high projections is that the Kremlin successfully undertook a systematic campaign of deception, with widely reported exaggerations regarding long-range rockets entering service; in effect, with their own quest for deterrence, the Soviets unknowingly nourished the exaggerated American rhetoric.

Robert McNamara, in his early exposure to strategic policy, was impressed most by RAND's graphs of costs that the Soviets would incur in offsetting the U.S. mix of damage-limiting measures — fallout shelters, ABMs, bomber defense, and counterforce capability. McNamara began to argue that an era of mutual assured destruction (MAD) was arriving; thus, it would be cost-effective to counter a Soviet buildup of offensive and defensive forces, to retain superiority in strategic forces, and to deploy an ABM system. However, it turned out that the Soviets did not have a nationwide ABM after all, and their feared SS-9 was not MIRVed and was not accurate. Much, much later McNamara realized this and changed his mind.

James R. Schlesinger, who became quite influential in government, had been director of RAND's strategic studies division. He devised a rationale for nuclear operations: a nuclear war-fighting force with MIRV as the nucleus (MIRV was predominantly a RAND conception). According to Thomas C. Reed, who served in the U.S. government at the same time, Schlesinger viewed communism as "a treacherous, cruel, and implacable world movement that needed to be defeated, not accommodated."

Another RAND analyst who gained prominence was Fred Iklé (member of the Committee on the Present Danger). Iklé is credited as the intellectual "father" of the Permissive Action Link (PAL), by virtue of a paper he wrote in 1957-1958 on accidental nuclear war, in which he suggested a mechanical lock for weapons; however, nothing has surfaced to show that Iklé or RAND promoted this concept or pushed for its implementation.

Andrew Marshall, less well known publicly, spent 21 years at RAND and became the Pentagon's most influential insider, especially with his advocacy of military-strategy "revolutions" in technologies and operational concepts.

RAND was deeply infiltrated and cozy with neoconservatives — "neocons" — best known for advocating aggressive foreign and military policies. A neoconservative label is generally associated with formerly moderate (or Democrat) policy advocates, who came to favor "a hawkish or assertive foreign policy to implant democracy and American values abroad."

In the final analysis, the central concept that the RAND operations analysts developed for nuclear strategy was that "nuclear war could be calculated with precision." It took many years of much increased mutual danger, before it was realized how unrealistic this conjecture was.

Military-Industrial Complexes

One prominent Cold War scholar, David Holloway, has concluded, "The causes of the Soviet Union's collapse were many, but one of them was surely the economic and political burden of [their] military-industrial complex."

President Eisenhower, during his oft-quoted farewell address in January 1961, presciently warned about the U.S. burden and its consequences:

> In the councils of government, we must guard against the acquisition of unwarranted influence, whether sought or unsought, by the military-industrial complex. The potential for disastrous rise of misplaced power exists and will persist.
>
> We must never let the weight of this combination endanger our liberties or democratic processes. We should take nothing for granted. Only an alert and knowledgeable citizenry can compel the proper meshing of the huge industrial and military machinery of defense with our peaceful methods and goals so that security and liberty may prosper together.

The hybrid complexes that emerged East and West from World War II initially consisted of (1) the military services, with officers eager for retention and promotion, and (2) vested defense industries, with a network of suppliers and research facilities in search of profits or patronage. What had been an important patriotic contribution to the Allied cause in World War II became a form of government largesse to be continued indefinitely in the name of national security. Similarly in the East, under a controlled government and managed economy, a defense establishment became embedded in the Soviet Union.

Senator Barry Goldwater of Arizona in 1969 extolled the value of the American way: "Thank heavens for the military-industrial complex. Its ultimate aim is peace in our time, regardless of the aggressive, militaristic image that the left wing is attempting to give it." By that year, the juggernaut was consuming $80 billion annually — a tenth of the nation's gross national product, and the largest single activity. On the payroll were 120,000 suppliers, hundreds of universities, and contractors in every state in the Union. Many key civilians in government came from defense-oriented industries. Military officers who retired often took jobs with firms doing defense work. Cost-plus-fixed-fee contracts helped induce reluctant contractors to take the risk; after all, they got paid for their billed hours and expenses, regardless of the outcome.

Overruns on military contracts sometimes reached 200 percent before Defense Secretary McNamara reduced it to "only" 100 percent. Questionable contracting practices, such as "low-balling," gave the military-industrial-laboratory complex its unique economic status. Nevertheless, industry was not getting high profits from its relationship with the Pentagon. Not until June 1969 did defense stocks on Wall Street reach their all-time high.

One of the greatest mistakes blamed on the military was America's involvement in Vietnam. In 1965, with many generals optimistic about the outcome, the Joint Chiefs of Staff recommended putting massive forces into Vietnam, and approval was given by President Johnson and his top civilian advisors. Senator William Fulbright spoke of the military-industrial complex in 1969 as "a direct threat to American democracy" because Congress had only perfunctory misgivings, because it spawned global military involvements, and because it "has become a powerful force for perpetuation of those involvements."

In *Fortress America*, a thoughtful 1998 study, William Greider does not fault private industry nor other elements of the complex, all of which exploited a policy vacuum. He recognizes continuance of the problem into the 1990s as "a failure to confront the implications of the end of the Cold War." Why was it, he asked thoughtfully, that "we should be prepared to fight two simultaneous full-scale wars.... Why ... not three or one? No one seems to know."

Given an opportunity in the late 1960s to comment on U.S. missile-defense policies, Georgi A. Arbatov, director of a Soviet Academy of Sciences institute, attributed much of the ongoing arms "intensification" [such as the RAND-supported Safeguard ABM system] to "increased pressure from [U.S.] internal problems whose solution requires large funds...." Referring to Herman Kahn's work, Arbatov noted that the lack of an external threat deepens a nation's internal problems and prevents it from consolidating itself, and he added that Americans were again being frightened by the myth of the "Soviet threat." He blamed "Kremlinologists" like Kahn for inventing argument that the Soviets were developing weapons systems to strike a "preventive blow."

Beltway Bandits. As East-West relations chilled, clusters of think-tanks (and foundation-supported organizations) formed around the nation's capitol and eventually became known as "beltway bandits." Firms such as SAIC, SPC, Titan, BDM, etc., became contractors to government agencies. Most of these companies provided very competent, essential, and timely services to agencies that did not have suitable staffing for policy "fire drills" — urgent demands from higher echelons to respond with qualified personnel and thought-out policy positions.

SAIC has been one of the largest providers of programmatic, logistical and technical support, for example operating a verification-technology center for a Defense Department agency.

Lured to high paying employment, many ex-military became "double-dippers" in the form of beltway-bandits, civil-service qualifiers, or political appointees. Most attractive to private companies were former civil-service employees who had fresh knowledge of intra governmental workings and still had contacts with key decision-making or contractor personnel. The new employees were often able to channel a contract to consulting firms, or even bring the tasking over when they departed government. To their credit, experienced personnel were often able to rapidly provide competent assistance and also had inside knowledge on forthcoming governmental problems that needed resolution.

Deriving their funds from the Pentagon, consulting organizations, such as RAND and the Institute for Defense Analyses, originally focused on technical problems, but later evolved into advisors on strategic issues. The "defense intellectuals" that emerged were primarily mathematicians and economists who lacked the technical expertise of physical scientists.

Section C: The Nuclear Priesthood

Because of changes in presidential administrations, surges in priorities, and sudden shifts in policy or treaty negotiations, the beltway bandits fulfilled an essential role in providing qualified and expedient response that could not have been expected from civil servants, who had many other duties. Until some more independent-minded think-tanks were formed in the 1980s, much of the advice was part of a positive feedback loop — like-minded people reinforcing each other — with little nonconformist thinking.

The pressure for armaments exerted by the combination of the DOE nuclear weapons laboratories and the beltway bandits made the military-industrial complex into an "Iron quadrangle" — military, industrial, laboratory, and congressional — or, more aptly, an "atomic quadrangle." Nuclear weapons laboratories became eminently successful in making the most of their congressional representatives and of personnel assignments to critical offices at DOE and DOD, and on the Hill.* As mentioned, there is a scientific component of the iron (atomic) quadrangle that extended beyond the nuclear weapons laboratories.

Despite the widely heralded claims of the weapons complex, the Soviet Union never had a margin of strategic superiority over the United States. General Lee Butler, former Commander-in-Chief of the U.S. Strategic Command, who acknowledged that he "was present at the creation of many of [the Cold War nuclear] systems," has observed that

> a succession of leaders on both sides of the East-West divide directed a reckless proliferation of nuclear devices tailored for delivery by a vast array of vehicles to a stupefying array of targets. They nurtured, richly rewarded, even reveled in the industrial base required to support production at such levels.

Armaments Lobby. When President Kennedy sought to reduce armaments, he ran into protests from politically powerful representatives of states with defense industries, especially California and Texas, who feared any disarmament measure which might have a deleterious effect on the economies of their states.

Helmut Schmidt, former Chancellor of West Germany, a nation that was more than a bystander, told Jonathan Schell:

> You know, within the nuclear-weapon states — particularly in the United States and Russia — there is an enormous body of vested interests. And they are influential in many ways — not only through lobbying in Washington and Moscow but through influence on intellectuals, on people who write books and articles in newspapers or do features on television. It's very difficult as a reader or as a consumer of television to distinguish by one's own judgment what is led by these interests, and what is led by rational conclusion.
>
> Twenty-five years ago, I met Edward Teller. He was an irate man, full of argumentative capacity and power. But he had vested so much of his personal prestige in his position that he was unable to rethink anything. There are thousands of people who are unable to rethink. In addition, there are the military and civilian bureaucracies. These influence the thinking of members of Congress, senators, members of the legislature in Washington and Moscow.
>
> Then, on the other hand, you have the extreme, psychotic enemies of nuclear things, whether weapons or reactors, and they also try to influence people. They also invested their prestige to a degree and cannot rethink.

---

* "The Hill": Capitol Hill, the site of the U.S. Capitol in Washington, D.C., and of the Senate and House staff.

So it's going to be a very, very long process of struggling [for nuclear abolition] within the nuclear-weapon states.

Soviet Defense Establishment. The official organization in the Soviet Union for development and production of new weapons systems was based on a number of defense ministries, each of which had responsibility for a particular type of weapon or military equipment. Nuclear weapons were the primary responsibility of the Ministry of Medium Machine Building, and delivery systems were developed by the Ministries of General Machine Building (missiles and space vehicles), Shipbuilding Industry, and Aviation Industry. Ministries that supported nuclear and conventional development and production included Defense Industry, Radio Industry, Electronics Industry, Communication Equipment Industry, and Machine Building (explosives and propellants).

Each of the industries had its own scientific research institutes, design bureaus, test ranges, and production plants. The Military-Industrial Commission was responsible for coordinating the activities of all industries.

Academician Roald Sagdeev, director during the height of the Cold War of the Soviet Space Research Institute, wrote a book largely about their military-industrial complex. Although Soviet propaganda denied the existence of such a complex in their nation, Sagdeev narrates his many encounters with Soviet reality, even though "for many decades we had even been unable to find out if we had a military budget at all." Because their system was organized on principles different from Western nations, the defense establishment could not be separated from the political environment of the country.

Before becoming director of his Institute, Sagdeev "had no knowledge about this enormously influential hidden part of the military-industrial iceberg." As soon as he took office, he began to get phone calls and visitors from "the mysterious Commission on Military-Industrial Issues [VPK]," which consisted of the most important influential people, mostly ministers of industries classified as an "indispensable part of the defense establishment." Although the existence of the Commission was a state secret, "The government, even the ideologues, admitted quietly among themselves that our country also had something similar to the military-industrial complex as defined by President Eisenhower."

Sagdeev relates a story about a conversation with Gorbachev, who was amused that the U.S. military thought its share was only about 3 percent of the total American budget.

Georgi Arbatov observed that "you have to take into account all the industries, the infrastructures, and services that are related in one way or another to the military-industrial establishment. On top of it, the [U.S.] military-industrial complex is very well organized in the political sense, with a strong lobby on the Hill."

Technological Innovation. The nuclear arms race, like most historic developments in military science and technology, had and has its own internal spawning mechanism. Much of the arms race might be attributable to a momentum sustained by scientific and technological advances. Rather than strategic doctrine determining the weapons to be fielded, it seems that more often the doctrine is shaped by improvements in weapons systems and the advent of new kinds of weapons. At the same time, scientific research and discoveries are made possible by national policies and budgets. A certain sense of manifest destiny — a feeling of "divine will," particularly in the United States, that the nation is bound to be the overwhelming influence in the world — seems to have pervaded funding and direction of research and development in military technology.

While governmental proclivity to establish and nurture large projects accounts for the actual development of the atomic bomb and many major military developments, technological determinism still offers politicians high-stakes playthings to champion for war or peace. Submarine-launched ballistic missiles came about because the military planners wanted them. On the other hand, scientists like Edward Teller promoted the development of thermonuclear weapons, not because they were needed but because they were possible and technically "sweet."

As long as governments provide funds for R&D, especially for military applications, creativity and technological advances are likely to engender new weapons. That kind of innovation, if so channeled, could also be applied to safer, more stable weapons, and to improved technological means of treaty verification.

The seemingly inexorable forward-going process in armaments is sometimes a result of technological creep, a result of government funding and technical ingenuity than can slowly lead to new weapons not originally intended, and no longer needed (if they ever were), in the decade or so that it takes to develop them. In short, we get weapons looking for applications. And there is no shortage of proponents who equate technological possibility with manifest destiny.

The Scientific-Technological Elite

Less quoted from President Eisenhower's farewell address is his warning of the undue influence and synergism of the military-industrial and scientific-technological sectors:

> In holding scientific research and discovery in respect, as we should, we must also be alert to the equal and opposite danger that public policy could itself become the captive of a scientific-technological elite.

Lord Solly Zuckerman, who was in charge of many nuclear development projects of the United Kingdom, observed that those working on nuclear weapons were really "not contributing to greater security for their nations." Instead, they acquired inordinate influence to protect their own interests. One consequence identified by Zuckerman was of their major contribution to the failure of the United States, Soviet Union, and United Kingdom to arrive at a comprehensive test ban treaty in 1963. To appease the weapons labs, the nations settled for a treaty that banned nuclear tests in the atmosphere, but permitted them underground.

The labs' role was especially self-serving and effective in promoting their agenda. Because three of the U.S. weapons laboratories were situated in two low-population states, New Mexico and Tennessee, the labs through their senators were able to wield a disproportionate influence within their states, and nationally.

In particular, the weapons labs used their influence to affect national policy on the development and testing of atomic weapons. Their efforts to emasculate or bypass the CTBT have been described by Jack Evernden, who worked on nuclear-testing verification in the U.S. government:

> The weapons labs tried everything possible to make senators nervous about the treaty. It has always been so. In earlier years, the weapons labs raised any number of false bugaboos calculated to make a test ban treaty seem dangerous.

Evernden cited two examples: (1) questions raised about stability of warheads in storage; "there is not now nor has there ever been any evidence of failure of warheads placed in storage after full testing," and (2) the fact that Livermore in the late 1980s "declassified a highly classified document that purported to present an effective evasion scenario largely based on seismology." (Evernden, who had written part of the document, had long since shown that the evasion scenario would fail — that the seismic evidence would easily point to a nuclear explosion.)

A rationalized but shrewd means of achieving influence has been the practice of assigning weapons-lab staffers to key positions in government. This sanctioned infiltration was funded by finding government assignments that pay salaries or by using money furnished by government agencies — usually Energy, Defense, or Intelligence.

The U.S. weapons labs have had considerable latitude in spending, because they were able to allocate a significant portion of their funding to discretionary purposes. While the non-weapons labs were limited to spending not more than 20 percent of their budget at management's discretion, the weapons labs were allowed a much larger fraction, close to one-third. Although this money was intended to allow leeway for innovative research and development, it was often allocated to special assignments in Washington.

Although not obvious, many personnel in federal administrative positions were thus seconded from weapons labs or military services. They found assignments especially in the Department of Energy, the Arms Control and Disarmament Agency, the Department of Defense, the CIA, and the Department of State. Not only were they in a position to affect government policy, they also kept the weapons labs promptly informed of changes in policy or funding. This is not to say that they were other than conscientious, in the main, but the fact that they usually returned to their laboratories or military posts after the assignment was a factor in fostering their allegiance.

The weapons-lab managements also had an effective strategy for urging Congress, the Executive Branch, and the public to support their operations. By proclaiming special expertise — insider technical knowledge about nuclear weapons — they were able to ensure that their labs had a seat on every deliberation that affected them. In DOE, for instance, they used this claim or their connections to have personnel named to influential positions, such as chairing conferences, working groups, and committees.

Many Livermore staffers were placed in key nuclear-strategy and arms control billets at the Pentagon. One Livermore individual who secured a powerful appointment was Bob Barker, becoming a key deputy at ACDA; his like-minded

Section C: The Nuclear Priesthood   243

wife, Kathleen Bailey, had published extensively in the public literature, pushing a hard line on arms control issues.

Another Livermore individual who sided with Edward Teller in his quest for missile defense was John S. ("Johnny") Foster, Jr., who was director of Livermore from 1961 to 1965, and Director, DOD Research and Engineering from 1965 to 1973. In the latter post, he reported directly to the Secretary of Defense, and was in a key position to influence the evolution of the Sentinel ABM system that Secretary of Defense Robert S. McNamara was directed to approve in September 1967, against his better judgement. Foster provided technical continuity in DOD, spanning the transitions from McNamara (who left office February 29, 1968) to Clark Clifford and then to Melvin Laird (who took office with President Nixon in January 1969). It was Johnny Foster who made the 1968 decision to deploy ABMs near major cities (trying to intercept missiles aimed at them). The ensuing public backlash against "missiles in the back yard" led eventually to the ABM treaty.

> [Livermore]. Mot as scary or mysterious as Los Alamos, Livermore was surrounded by vineyards and situated an hour or two driving distance from San Francisco. While visiting many times for symposia and business meetings about arms control and treaty verification, I found their scientists and supervisors to know a little more about published scientific information that than weaponeers at Los Alamos, who kept themselves more isolated. I also found the Livermore staff to be just as partisan as the other weapons labs, not only aggressively against their rivals at Los Alamos, but also viewing themselves as in a competition with other DOE laboratories.
>
> Aside from its weapons-related programs, Livermore had an intelligence-analysis division and a well-funded verification program. Its scientific staff has been somewhat more inclined to support arms control, perhaps a heritage of former director, Herbert York. Livermore is also noteworthy for the constructive role that dissident physicists Hugh DeWitt and Ray Kidder publicly played in numerous technical issues otherwise burdened with unnecessary secrecy.
>
> I remember once (and only once) walking out with a classified document, which the guard also missed in his search of my briefcase. I didn't realize I had it until home in Chicago; I promptly turned it over to Argonne security and got a deserved rap on the knuckles.

Los Alamos lab personnel who took prominent positions in the Washington area include Theodore B. ("Ted") Taylor. He had managed the Defense Nuclear Agency when it was responsible for taking custody of nuclear weapons on behalf of the DOD, and for carrying out the U.S. test program to study the effects of nuclear explosions. Taylor ultimately became an ardent supporter of arms control, contributing much time and expertise to non-governmental working groups.

From Sandia Lab there was C. Paul Robinson, a manager converted to treaty negotiator, who ultimately earned the title of "Ambassador" before returning to the leadership of the laboratory. Many DOE laboratory personnel — especially from the weapons labs — were assigned as technical experts to treaty negotiation teams. In some cases nuclear weapons expertise was useful.

The Consummate Role of Secrecy. Authorized secrecy is, of course, essential to military security and to the protection of sensitive technical information, such as that related to nuclear weapons design. Laws such as the U.S. Atomic Energy Act and the British Official Secrets Act codified the protection of classified information.

Because of a perceived threat to national security, the weapons states practiced a high degree of secrecy, espionage, and disinformation. Information "warfare" was routinely conducted to mislead other governments or to deceive the public.

Secrecy within the Soviet Union was endemic, maintained on practically every sensitive aspect of government information. They attempted to keep a lid on their first nuclear-explosive test (called *First Lightning* or "Joe-1" [RDS-1]) until they had time to build a significant inventory of bombs. Its radiation fallout was detected by the United States. Former intelligence official Thomas C. Reed contends that Stalin publically issued a "lie," denying the test because he was "justifiably" afraid of a U.S. pre-emptive attack.

Some attention-getting misrepresentations made bad press for the USSR. They stonewalled the international community and the Russian people about a large and lethal leak of anthrax; in fact, they hid from their own public and from the West massive a bioweapons-production program that was the source of the leak. Years later, the Soviet government, despite the advent of *glasnost*, tried for many days to downplay the extent of damage and consequences of the Chernobyl reactor accident.

Government secrecy kept public constituencies in the dark about radiation exposures and experiments that affected them; exigencies of the moment were used as the justification. Testing nuclear weapons on the surface created radiation fallout that spread with the wind to large public areas, and testing in the atmosphere resulted in global fallout; information about this radiation was universally kept secret as long as possible. In some cases, radiation-release experiments involving ill-informed military personnel and un-informed members of the public were conducted. The unexpected exposure of foreign nationals by American, British, and French thermonuclear tests in the Pacific could not be kept secret, although every effort was made by the governments to understate the effects. When testing was forced underground by treaty, only minimal information about subsurface effects was disclosed to the public.

One of the most egregious excesses of government secrecy by the five nuclear weapons states was the coverup of exposures to workers who were involved in producing the nuclear materials needed for weapons; many were often exposed to harmful levels of radiation, especially in the early 1950s. Radiation accidents were concealed from the public in the name of national security. As the veil of secrecy is slowly being removed, the public is becoming increasing aware of these rationalized exposures to radiation and to hazardous materials, such as beryllium and uranium dust.

It seems that in the former Soviet Union, numerous sacrifices were made by dedicated, but unaware workers.

Employees of private firms under contract to the U.S. government were frequently exposed to hazardous and toxic particulates. Not all unwitting exposures were from nuclear materials: Hazardous testing was also performed with chemical and biological agents.

Peaceful applications of nuclear power have suffered because of the backlash against government secrecy resulting from distrust generated by misleading or deficient official assertions. Conducting Cold War medical experiments on humans without informed consent was a major adverse factor; these and other failures of disclosure have helped to create a radiophobic society.

Indiscriminate dumping of nuclear waste, underground, in ponds and rivers, and elsewhere, without safeguards, was done in secret under the cover of a wartime-type exigency. Contamination surveys conducted by government officials were classified as secret. The locations and quantities of fissile nuclear materials were also kept from the public; nearby populations did not realize how much added risk they were exposed to. Large expanses of the United States and the former Soviet Union are still contaminated with radioactive materials, and some locations and hazards were yet to be declassified.

The most significant and legitimate role of secrecy was to protect military activities and strategic policy. This practice began with the wartime decision to develop nuclear weapons and carry out the atomic bombing of Japan; secrecy was subsequently maintained in order to deal with various East-West threat scenarios. Just how close the world came to reflexive mutual destruction was a closely guarded secret. Of course, all specific operations and details about delivery systems and weapons were carefully safeguarded.

During some administrations, government officials frequently used secrecy provisions to reinforce their own agenda, which might have included the sabotaging of arms control agreements. Often independent scientists and members of the public were able to contradict government assertions by making use of deductive reasoning or scientific principles. Secrecy was also used to hide bureaucratic ineptitude and errors.

Sometimes overzealous efforts to enforce secrecy have been counterproductive. The *Progressive* case is a well-documented example of federal officials failing to stand by the established adage, "neither confirm nor deny": The events resulted in unintended confirmation of specific sensitive technical information.

To push the funding of favored programs, defense contractors and other proponents used the mantle of secrecy, often liberally invoking the label "national-security information." Compartmentalization through "need-to-know" became a license to protect not only legitimate intelligence and military information, but also to hide funding and policy decisions.

By manipulating these restrictive provisions of regulations and law, crafty practitioners could propose projects, get approval, inveigle funding, and carry out major spending without external scrutiny. Audit and review processes were often stymied by astute application of confidentiality barriers.

Having access to classified government information was another way to either get a timely heads-up or to gain useful competitive insight. Access begins with security clearances, which were given out almost wholesale, sometimes with minimal background investigation. Temporary, *ad-hoc* clearances could be readily acquired with the right connections.

At one time, though, you could not even do innocuous work in the fields of atomic energy or nuclear weapons without high-level clearance. In the hierarchy of access clearances, compartmentalization resulted in a vertical and horizontal process of protecting both legitimate and convenient secrets. Clearance badges were the emblems of admission to the inner circles.

The other side of the coin was the use of security clearances and secrecy restrictions to constrain government or laboratory experts who wanted to speak out on public issues. Sometimes conscientious individuals censored themselves. More often, managers or government officials applied pressure; this could be either indirectly or by threat of sanctions against the individual or the organization for which the outspoken individual worked. Either implied or expressed, threats could be made to withhold an employee's access to classified information or to sensitive facilities, or even go so far as to revoke security clearances.

Through an extensive network of contacts, assignments, and lobbying, camouflaged by restrictive provisions of the Atomic Energy Act, the U.S. weapons labs were able to exert a disproportionate influence on government policy and spending. Influence preserved and expanded their role, while ensuring that their views on arms control and verification were given strong recognition, stronger than accorded to other competent voices outside that scientific-technological elite. In particular, the weaponeers often brandished their high-level "sigma" clearances required for certain types of restricted information; those who lacked the clearances were kept out of the discussion; those within the circle became privy to inane details that often had little to do with productive policy. In the Soviet Union, government ministries used secrecy to promote their own bureaucratic aspirations.

A. David Rossin, a former DOE official has looked into the use of secrecy to promote commercial or political aims. He mentions an example: "the expert who explains that there is a real problem but the scientific information is classified. [The expert] knows the answer but ... can't reveal it, so others should believe [him or her]." Rossin called it "marketing secrecy."

Having made a thorough study of former President Carter's decisions on nonproliferation, Rossin concluded that if the information used by Carter to reach his decision to ban nuclear-fuel reprocessing had been accessible to the public, a different policy might well have emerged. Carter's nonproliferation study, called PRM-15 — thought to be the basis of his reprocessing ban — was classified Secret and remained so. After the President issued his nuclear-policy statement, the White House ordered the destruction of a draft final report on safeguards for reprocessing and separated plutonium, as well as an AEC pamphlet that described reprocessing and the nuclear fuel cycle.

Also obscured by government secrecy have been several related matters of fact that could have clarified the value or limitations of an artificial ban on reprocessing. These factual matters, which remain contentious, have to do with the usability of low-grade plutonium in military-quality weapons. Several bomb designers from Los Alamos — Robert Selden, J. Carson Mark, and Ted Taylor — were delegated by the U.S. government to preach that any grade of plutonium could explode. (While narrowly true, only weapons-grade plutonium has actually been used in military quality weapons, and — with the exception of one U.S. oft-cited experiment in 1962 — no nation has been known to use or try anything except high-quality fissile materials for weapons.) Dissident physicists endeavored to counter this misleading impression and to put the matter into a more meaningful and practical perspective on plutonium safeguards. The dissidents also sought, unsuccessfully, to have more technical information about the dispute and the 1962 test placed in the public domain.

Another case of what Rossin calls "selective marketing" to support a "political policy" is that some results of the aforementioned 1962 nuclear explosion were leaked so as to bolster statements being made by the administration at the time. While the test has been frequently cited by the U.S. government to support claims about the usability of low-grade plutonium in weapons, Rossin says its yield "was minuscule by nuclear-weapon standards — of the order of a large chemical ordnance."

Many past misuses of secrecy in the name of national security have since come to light. But some continue to this day. Additional publicized infringements have included one agency of the government deliberately keeping things secret from another agency or body of the government, such as the U.S. Executive Branch failing to disclose to Congress information pertinent to its nuclear decision-making.

The scientific-technological elite inside and outside of government have been most adept at using national-security fears and government-secrecy restrictions to serve their objectives. Appropriate secrecy certainly has a legitimate function in protecting national security — but manipulation of the system usually gives someone an entrepreneurial edge.

JCAE

One study of the American military redefined the old military-industrial complex as the "iron triangle" — the third leg being Congress, which has historically been eager to push for pork-barrel programs. Reportedly, when President Eisenhower drafted his famous farewell address, he originally wanted to warn against the "military-industrial-congressional complex," but removed the third part reportedly because "it was unfitting for a president to criticize Congress."

Congress's Joint Committee on Atomic Energy (JCAE, 1947–1977) was a fixture that had extraordinary power. The Atomic Energy Act of 1946 created the Atomic Energy Commission to oversee the entire atomic power and weapons program, along with a Military Liaison Committee and a General Advisory Commission. "To ensure a strong congressional role in overseeing the program, the act created the JCAE, the only committee to be brought in being by legislation (and thus held to have rights under that law) and the only joint committee to have full legislative powers."

The Atomic Energy Act stipulated that the JCAE consist of eighteen members, equally divided between the House and Senate. The committee was to be vested with full jurisdiction over "all bills, resolutions, and other matters in the Senate or the House of Representatives relating primarily to the [Atomic Energy] Commission or to the development, use and control of atomic energy."

The JCAE and its members gained extraordinary and unprecedented power through their jurisdiction to hold hearings and make recommendations that led to authorization and appropriation of funds for the nuclear complex. Because of intense secrecy about nuclear matters, the JCAE (and also the intelligence committees) had a special role and responsibility in oversight of the weapons complex. Although hampered by lack of scientific or engineering understanding, Committee members and staff were granted special security clearances for access to technical programs and details.

Oversight, it was noted, consisted of

> [periods of] intense scrutiny ... interspersed with periods of relatively little attention.... Consistent annual scrutiny of budgets, plans, and operations — when practiced at all — rarely proceeds beyond the staff level. This and the relatively high levels of secrecy surrounding most nuclear programs have allowed the nuclear weapons program to oversee itself to a large extent, a practice that unfortunately is not conducive to efficient government.... Congress must devise and implement mechanisms to improve its vital oversight functions. Failure to do so will inevitably lead to the wasteful expenditure of billions of dollars and the perpetuation of unnecessary or unwise programs.

A very worthwhile program that the Committee pushed was the use of mechanical locks on nuclear weapons. The JCAE took credit for creating the permissive action link (PAL) and claimed that its executive director originated the idea, but several people (including Iklé at RAND) had suggested the concept (and later the physical apparatus) of "locking" nuclear weapons to prevent unauthorized use. Early PALs were little more than mechanical locks that could have been defeated, given enough time and a hacksaw. In 1961 the Committee sent a summary of its overseas nuclear-weapon inspection report to President Kennedy, describing "the fictional weapons-custody system now in use." By September of 1962, five-digit mechanical lock PALs had been installed on many weapons.

Reagan's Intractability

Ronald Reagan was elected president in November 1980, after an election campaign in which he scorned détente and disarmament. Although lacking experience in foreign and military affairs, he immediately authorized a massive expansion of U.S. nuclear and conventional forces, causing widespread public fear. Reagan was strongly influenced by a circle of supporters, advisors, and cabinet members who implemented this expansion.

His presidential platform explicitly pledged to achieve what already existed: "technological and military superiority" over the USSR.

In his 1990 autobiography, Reagan explained the reason for his presidential challenge to the Soviet Union:

> During the late seventies, I felt our country had beg1un to abdicate its historical role as the spiritual leader of the Free World and its foremost defender of democracy. Some of our resolve was gone, along with a part of our commitment to uphold the values we cherished. Predictably, the Soviets had interpreted our hesitation and reluctance to act [as lack of resolve] ... and had tried to exploit it to the fullest, moving ahead with their agenda to achieve a Communist-dominated world.... The Soviets were more dedicated than ever to achieving Lenin's goal of a communist world.... I deliberately set out to say some frank things about the Russians, to let them know there were some new fellows in Washington.

Since 1963, Reagan had publicly opposed every major nuclear arms control treaty. In his first term as president, none of the 14 negotiations previously under way was completed nor even continued. In 1981, he defined communist philosophy as the "right to commit any crime, to lie, to cheat." Soon after that, Reagan approved a 53-percent increase in U.S. defense spending, the biggest military buildup in U.S. peacetime history. He restored the previously canceled B1-bomber and neutron-bomb programs, authorized development of the Stealth aircraft, pushed for preparation of the mobile MX missile, and confirmed imminent deployment of cruise and Pershing-II missiles in Europe.

In 1982, to put a positive face on U.S. arms control policy, the State Department took credit for accomplishments of prior presidents. In order to excuse the long delay in achievement, the Department explained that:

> Arms control agreements with a highly secretive adversary like the USSR cannot be based simply on trust.... Agreements that cannot be effectively verified are not acceptable.... It will be essential to supplement NTM with some form of "cooperative" verification measures.

The "Evil Empire." In March 1983 Reagan made a speech that became a hallmark of his administration, referring to the Soviet Union as "that totalitarian darkness." He asserted it was "the focus of evil in the modern world," urging awareness of "the aggressive impulses of an evil empire." Thomas C. Reed, who was an insider at the time, confirms that "Reagan meant every word."

That speech made public the "thrust" of a national-security directive, NSDD-75, described by Reed as "the blueprint for the end game." The then-secret Reagan directive focused on "three fronts":

> the reversal of Soviet expansionism abroad, the promotion of a more pluralistic Soviet society at home, and the engagement of the new Soviet leadership [Yuri Andropov, after the death of Leonid Brezhnev] in negotiations to protect and enhance U.S. interests.

Reed labels NSDD-75 "a confidential declaration of economic and political war." It did not include, he adds, "any plan for initiating a nuclear attack on the USSR." Reed reports that while hardliners like Henry Rowen, Richard Pipes, and William Casey were involved in its codification, "détente-minded members of the administration fought against the plan." No wonder Reed boasts that the tone of Reagan's "evil empire" speech based on NSDD-75 "drew gasps from those who wished to accommodate the sinking USSR."

At about the same time, Yuri Andropov, general secretary of the Communist Party, justified the Soviet Union's "strengthened" defense capability on the grounds that the United States was involved in a "feverish" effort to establish military bases near Soviet territory and to develop "ever new types of nuclear and other weapons." A few days earlier, Reagan had announced the Strategic Defense Initiative (later derisively dubbed "Star Wars") for ballistic-missile defense.

Reed, who was responsible for defense issues on the National Security Council, gives us a good view of the inner workings and thoughts of the Reagan administration. He acknowledges that "pushing the Soviet Union to the brink of economic collapse," which became a goal of the administration, "was a gamble not without risk." He credits President Reagan with putting "the pieces in place to end the Cold War." One piece was to install "a team of hawkish policy advisors" in the Pentagon.

Although Reagan wanted to peacefully end the tension, he was "not sure" how to do it, only that "there's got to be a way." Reed asserts that "Reagan was not a hawk," and "He was worried about the nation's security and the possibility of nuclear war."

In a 1984 *New York Times* opinion piece, W. Averell Harriman, a former Ambassador to the Soviet Union, reluctantly went public with his concerns about "not the risk but the reality of nuclear war." Harriman spoke up because "the behavior and the proposals of the [Reagan administration] in both the strategic and European nuclear discussions have raised serious doubts ... about whether there ever was any intention to reach any reasonable agreement."

Harriman objected that "Long-time opponents of arms restraint have been put in charge of policy making. American delegations have arrived at the Geneva negotiations empty-handed." Regarding verification, he noted that "Three years after taking office, the [Reagan administration] still does not have a policy on verification." Moreover, "The issue of verification — so central to arms control — has been blurred by the administration."

President Reagan's anti- arms control policies became a catalyst for vigorous domestic and international public opposition, just like President Johnson's expansion of the war in Vietnam. His "negotiate through strength" stance led Reagan to revive and accelerate production of the MX, widely viewed as a first-strike weapon. He ordered nuclear-armed Pershing-II and cruise missiles positioned in Europe (so close to Russia that they could reach Moscow with a flight time of only a few minutes), and he initiated the "Star Wars" program, all of which helped rekindle the arms race.

In connection with the so-called "modernization" of NATO INF forces, the United States planned to deploy 108 Pershing-IIs and 464 GLCMs (cruise missiles), all nuclear armed, beginning December 1983. This decision was justified by a self-fulfilling contention of the Reagan administration about the "absence of full arms control agreement arising out of U.S.-Soviet INF negotiations...." The argument was put forth that the modernization stemmed from Soviet deployment of triple-warhead SS-20s: Pershings and GLCMs were part of a "bargaining chip" negotiating strategy. Discounting the UK and French nuclear forces aimed at Russia's heartland, the State Department simply dismissed the Soviet claim that a nuclear balance already existed. President Reagan's Chairman of the Joint Chiefs was General John Vessey, who felt strongly that U.S. INF forces "had to be deployed if [all nuclear missiles] were ever to be removed from both sides of the iron curtain." That was an expensive and risky bargaining chip. Reed admits that if a fight started, it would take some "luck" to keep if rom going nuclear.

Ill-informed about nuclear issues, Reagan at one time or another asserted that nuclear missiles could be recalled after launching, that B-52's don't carry nuclear weapons, and that a nuclear war could be limited and contained. He also misled the public by proclaiming a nonexistent "window of vulnerability."

Perhaps President Reagan's grandest folly was his hyperbole about the prospects of ballistic-missile defense. Based on overly enthusiastic advice from hardliners, he embraced the porous and discredited idea of a defensive shield against nuclear-armed missiles. Aside from the false hopes and dead-end investments, this ballyhooed ABM mission had another downside: It predictably expedited Soviet countermeasures, one of which was deployment of the Topol road-mobile missile (said to have a low-trajectory and to be equipped with penetration aids). The cold-launched Topol was the first Soviet solid-fueled missile. Just a little afterwards, the Soviets commissioned their 10-warhead solid-fueled SS-24, which could be rail- or silo-based.

The Revelation. Despite his original intractability, President Reagan in his second term belatedly noticed and began to understand the Soviet point of view. Acknowledged in his own memoirs, Reagan "began to realize that many Soviet officials feared us not only as adversaries but as potential aggressors who might hurl nuclear weapons at them in a first strike; because of this, and perhaps because of a sense of insecurity and paranoia with roots reaching back to the invasions of Russia by Napoleon and Hitler, they had aimed a huge arsenal of nuclear weapons at us." Reagan's change of heart, rather than his earlier bellicose policies, opened a "window for Soviet reformers and their transnational allies" to bring in "the new thinking that brought down the Cold War."

Through the four years of his first presidential administration, Reagan had to endure the constant challenge of political pressure brought about by peace movements in the United States and Europe. Peace activism grew, significantly influencing public opinion and creating a political climate in Washington for arms control and restraint. His administrations saw the largest demonstrations in history against the arms race. It cannot be ascribed simply to coincidence or backroom machinations that "the Reagan administration quietly dropped the concept of superiority, abandoned its initial harshly anti-Soviet rhetoric, and within a year of taking office was forced to begin negotiations with the Soviets."

Reagan also had to back away from the promotion of civil-defense shelters and federally mandated emergency-evacuation plans; he had to adopt the peace movement's proposal for elimination of INF missiles in the East and West; he was forced to accept MX-missile deployment at one-fourth the original proposal, and confined to silos, rather than mobile; he had to tolerate budget cuts and legislative restrictions on the SDI with which he was so enamored; and he suffered his worst political crisis when the Iran-Contra scandal surfaced to stymie his overthrow of Sandinistas in Central America.

Reed, who had left the White House in 1983, blames the Iran-Contra events on "the absence of graybeards [like himself]" and on "a succession of staff chiefs who had no insight into President Reagan's mind."

Despite these setbacks, and irrespective of pressure from the peace movement, Reagan was able to reward the internal structures of militarism in the United States, particularly by enlarging the military budget, strengthening intelligence agencies, and reinforcing military institutions — legacies that have endured.

Hardliners

People who took a "hard line" on defense fell into a camp ideologically opposite to those who promoted "liberal" national-security policies. As with all simplistic nomenclature, this terminology does not faithfully represent the opposing camps, but it was typical of the language used at that time. Basically, the hardliners, who for the most part were politically ultraconservative, stood for a "strong" defense establishment. Contrary to the contentions of the hardliners, the liberals rarely were "soft" on defense; they primarily wanted a better deal and less risk from the massive expenditures on nuclear weapons; they also saw value in reaching formal agreements with the Soviet Union. The generalizations — that hardliners tend to be (right-wing) Republicans and liberals to be (left-wing) Democrats — have had some validity, but there are notable exceptions. The espousal of excessive defense spending and war-fighting has transcended party affiliation.

Just as a single mind set cannot be assigned to liberals, neither can a simplistic label describe hardliners nor conservatives. With that caveat in mind, we recognize the dichotomy as a convenient artificiality of the times. Discussions among like-minded people often lapsed into those terms, which are still somewhat in use.

Thomas H. Naylor, the economist who analyzed arms control and pressure groups, says the support of the U.S. defense establishment came from "a coalition of right-wing American anticommunists consisting of émigrés from the [Soviet Union], Eastern Europe, and Cuba; the Israel lobby; ultraconservative political foundations and think tanks; a group of scientific technocrats; and a number of high priests of the military-industrial complex." Specifically, Naylor mentions émigrés Zbigniew Brzezinski (Poland and Canada), Richard Pipes (Poland), Alexander Solzhenitsyn (Russia), and Edward Teller (Hungary). Richard Pipes was one of 60 members of the Committee on the Present Danger who eventually served in the Reagan administration.

"Technocrats," says Naylor, are "scientists and engineers who design, develop, and implement weapons" and who have a "psychological need to justify their decision to spend their lives producing instruments of death." Edward Teller was a "guru" of the technocrats who made the most of the Soviet Union's role as a "formidable enemy." Robert McNamara and his team also conformed to the technocrat mold, although mostly as financial specialists. Thomas C. Reed, the Reagan-Nixon-Ford administration official quoted throughout this book, also fit Naylor's definition of a technocrat.

Naylor included Radio USA and the Voice of America as part of the nuclear establishment. Propaganda was rampant on both sides. A slick and controversial brochure, *Soviet Military Power,* initiated by Secretary of Defense Weinberger in 1981, was answered in 1982 by a comparable Soviet publication, *Whence the Threat to Peace.* Ironically, after Weinberger stepped down, there were revelations of "massive fraud" in his Department's military spending.

Some "highly opinionated, anti-Soviet intellectual hardhats" that the right-wing think tanks relied on included Jeane Kirkpatrick, Richard Perle, A.M. Rosenthal, William Safire, George Will, William F. Buckley, Jr., Patrick Buchanan (who has since run for President), Rowland Evans, and Robert Novak. Many were strident columnists for newspapers and magazines.

Naylor designated as "the high priests of the temple of doom" ten very powerful conservatives who had profound influence on policy. Some were protégées of Senator Jackson — Richard Perle, Frank Gaffney, Elliot Abrams, and Ben J. Wattenberg. Other prominent hardliners mentioned by Naylor were Zbigniew Brzezinski (national security advisor to President Carter), Eugene Rostow (ACDA Director under President Reagan), and Caspar Weinberger (Secretary of Defense under Reagan).

Based on such interlocking relationships, Naylor concluded that "Since the late 1940s the American public [had] been the victim of faulty scholarship, unrestrained prejudice, and unbalanced financial support only for right-wing, anti-Soviet views."

Committee on the Present Danger. Strongly influencing policies, according to Naylor, were "right-wing" groups such as the Committee on the Present Danger (CPD).

> The CPD became the political catalyst behind a plethora of hard-line philanthropic foundations, think tanks, and lobbying groups whose stated objectives included putting an end to détente, electing Ronald Reagan as president, and bringing the Soviet Union to its knees. It successfully achieved most of its objectives, and in so doing prolonged the Cold War for at least another fifteen years at enormous cost to the American taxpayers and the rest of the world.

The Committee on the Present Danger strongly opposed the SALT II treaty and promoted increases in military spending.

Many Directors of the CPD held important positions in the Reagan administration. Besides President Reagan himself, the list includes Director of the Arms Control and Disarmament Agency Kenneth Adelman, CIA Director Bill Casey, Under Secretary of Defense for Policy Fred Iklé, Secretary of State George Schultz, White House Science Council member Edward Teller, and many others. Also becoming officials of the Reagan administration were CPD members Eugene Rostow, Richard Pipes, and Richard Perle.

Financial resources for the CPD and similar anticommunist institutes were provided by "a network of interlocking conservative foundations," such as Carthage, Scaife, Olin, Hertz, Pew, and the National Endowment for Democracy. Naylor, who traced sources of right-wing funding, found that "anticommunist think tanks supported by Sara Scaife Foundation in 1988" included Cato, CPD, Freedom House, Heritage Foundation, Hoover Institution, etc. The Hertz

Foundation is located in Livermore, California, near the Lawrence Livermore National Laboratory, which has employed scientists educated with Hertz grants. Edward Teller was a director of the Foundation. Many SDI concepts "originated with Hertz Fellows working at Livermore lab."

Directors of the Hoover Institution have included Committee on the Present Danger members Jeane Kirkpatrick, David Packard, Donald Rumsfeld, and Richard Scaife, as well as former Reagan-administration officials Richard Allen, Edwin Meese, and Reagan himself. Richard Perle also was a fellow of the American Enterprise Institute. The Center for Security Policy, an anti-Soviet think tank that opened in 1988, headed by Frank J. Gaffney, Jr., was "one of the most militantly anti-Soviet policy groups." Gaffney and his Center remained vocal on defense issues, promoting, particularly, ballistic-missile defense; his sponsors include many members of the CPD and recognizable organizations of the military-industrial complex.

Richard Allen was an influential foreign-policy advisor to both Nixon and Reagan, and for a short time Allen was Reagan's national-security advisor. Active in the right wing of the Republican National Committee and a founding member of the CPD, Allen had responsibility for analyzing the Soviet threat. He ran into trouble for obtaining lucrative business contracts for himself and his friends, and for other business dealings that damaged his reputation.

A *Chicago Tribune* reporter wrote the following:

> The recent report [1984] on Soviet arms control violations by a presidential advisory committee ... is a truculent polemic, not a calm evaluation of the evidence.... No fewer than 9 of the committee's 12 members are present or former members of the board of directors of the Committee on the Present Danger, which counts the demise of SALT II among its noblest achievements.
>
> Looking at some of the arcane and antique charges in the report, it becomes plain that the point is not to advance arms control by enforcing its provisions, but to destroy it by holding it to an absurd standard.... An administration less hostile to arms control could most likely resolve these [alleged breaches in the Standing Consultative Commission, set up to settle treaty problems]. Unfortunately, the Reagan administration is hamstrung by its refusal to seek ratification of the SALT II treaty, under which the first two violations fall, and by its eagerness to scrap the ABM treaty, which is involved in the third.
>
> [The CPD is] sowing popular distrust of any negotiations with the Soviets.

Lord Solly Zuckerman, former principal scientific adviser to the British Ministry of Defence, alluded to opinions of the CPD as "the proclamations of a band of military camp followers who pretend to provide intellectual backing for the controversial defense policies of Ronald Reagan and Caspar Weinberger...." He further described them as "a handful of civilians who offer guidelines for all-out nuclear war, as though its consequences would be little worse than a succession of severe droughts...."

The Reagan people, suggested Zuckerman, assumed

> that the Politburo would deliberately risk intruding into NATO territory, in the near certainty that such aggression would be likely to entail nuclear war.... They also now say that America has to plan for a "protracted" war because the Russians believe that they could fight and win such a war. [Yet] Soviet leaders, while deploying nuclear arms with

their forces, publicly declare that a nuclear exchange could never be contained and that, once started, the result would be scores of millions of deaths on both sides.

During the alleged window of vulnerability for U.S. ICBMs to a Soviet first strike, the U.S. capability was indeed insufficient for the "counterforce" attack desired by the CPD, but it was sufficient for retaliatory destruction of the Soviet Union. One analyst concluded that the fundamental error of the neocons was unreasonable hostility to the concept of mutual assured destruction (the profound reality of nuclear weaponry's absolute destructiveness).

A thread of neoconservatism stitches many of the CPD members together. Names like William Kristol, Kenneth Adelman, and Dick Cheney are often included as prominent neocons.

The false note, a fundamental error of the hardliners, in the neoconservative alarm was that there was always residual uncertainty whether an American president would actually use nuclear weapons in war, or even in retaliation. While it was certainly possible, maybe likely, it was hardly assured. In fact, the only case where a president did use nuclear weapons was with the intent of quickly ending the war with Japan; there was no prospect then of retaliation in kind. The post-war photos of Hiroshima and Nagasaki have never been forgotten; the absolute nature of nuclear destruction remained visible, and the fear of unleashing a nuclear holocaust has nagged every president, including Reagan.

Although it could be perceived that U.S. interests in Europe were less than vital (if it meant exposing American cities to nuclear retaliation), Soviet interests in attacking Europe also were short of being vital (if it risked destruction of the motherland).

The anticipated "breakdown" of containment (as prophesied by the Reagan-era neocons) was based on "a sophisticated intellectual construct: the impending 'Finlandization' of Western Europe." Historian D.H. Allin detected neoconservative "pessimism" about the likely political stability and moral resolution of the European allies if they were to be faced with Soviet pressure (e.g., Finlandization). He mentions neocon "illusions of nuclear blackmail." One part of the debate between the neocons and the non-ideologues was a "subjective argument" about the reliability of West European allies; the other part was an "objective argument" about the effectiveness of nuclear weapons as instruments of political coercion. Allin observes that neoconservatives mixed these two arguments together to raise the alarm about the imagined failure of containment.

Neoconservative rejection of containment was connected to one of their more dubious assumptions: the irreversibility of Communism. One example is the a tortured distinction made by Jeane Kirkpatrick (American ambassador to the UN) between "totalitarian" and "authoritarian" regimes: Totalitarian regimes, she argued, "exercised total control over every aspect of society," using it to ensure permanence in power — while authoritarian regimes, which "exercised more limited control over society," were "capable of peaceful evolution to democracy." Although taken very seriously then, the Kirkpatrick thesis that totalitarian regimes would never evolve to democracy, proved within a few years to have been "spectacularly wrong." According to Allin, their pessimistic view was not only "exaggerated" but "the reverse of reality:"

> The essential weakness of the Soviet position was acknowledged by some neoconservative writers [but they made it] indistinguishable from unlimited danger.

In any event, these neoconservative voices were not stilled by the peaceful demise of the Cold War. Their reincarnation later came about through the Project for the New American Century, launched by many carry-overs from the Committee on the Present Danger.

Some prominent Cold Warriors are singled out below.

Perle. A highly influential hardliner, Richard Perle, was at one time a staffer for the hawkish Senator Henry Jackson. Naylor comments that Perle was "known best for the vehemence of his anti-Soviet views." D. H. Allin classified him as an "ideologue."

As a senatorial staff member during the Nixon and Carter administrations, Perle was an unabashed opponent of arms control agreements with the Soviet Union. When the Nixon administration agreed to reduce the number of American ABM sites from four to two, Perle and Senator Jackson became angry. They launched an attack on the corollary SALT I agreement and organized a purge of ACDA officials who conducted the negotiations.

As an assistant secretary of defense during President Reagan's administration, he was one of its leading hardliners. Perle was responsible for shifting the debate over arms control to the question of Soviet trustworthiness and the need for strict treaty verification. His hard line was especially unpopular with America's European allies, one British spokesman labeling him "The Prince of Darkness." Perle reciprocated by accusing West European leaders of being "mealy mouthed," adding that they resorted to "misty blandishments" instead of speaking outright of (alleged) Soviet violations and cheating.

Perle resigned as assistant secretary in 1987, after charges (which never went to trial) that he had passed classified information to Israel and, later, that he had accepted payments from an Israeli arms manufacturer.

Perle had little tolerance for opposition to Cold War military strategy, belittling public views as a "moral argument." He wrote:

> The period of the Cold War was marked by a most extraordinary inversion of moral argument applied to complex questions of military strategy. We had the bizarre spectacle of clergy marching against ballistic missile defenses in the late 1960s and early 1970s as the nation debated whether to proceed with the Safeguard antiballistic-missile defense system — a system that would have protected our retaliatory forces by shooting down missiles aimed at destroying them. This "moral" opposition to ballistic-missile defense, which resurfaced in the 1980s after the SDI was proposed, would have had the effect of leaving the United States with only offensive weapons with which to deter attack. It was never clear by what moral principle offensive weapons were to be preferred to defensive ones; one's intuition would seem to lead in the opposite direction.

In that part of his diatribe, Perle avoided mentioning the still-unsurmounted technical difficulties that plague missile-defense systems: He was assuming without foundation that such missile defenses would have protected U.S. retaliatory forces, and that was the only way to do it. Nor did he give much credence to the strategic consequences of unilateral deployment of a ballistic-missile defense, as his following statement indicates:

> The most likely explanation is that the judgment of the moralists was, not a moral, but a strategic one, one that followed from the belief that the deployment of strategic defenses by the United States would lead the Soviet Union to deploy additional offensive weapons with which to attack the U.S. deterrent. This was a doubtful conclusion.

Perle expanded his tirade against public opposition, ridiculing those who objected to the introduction of neutron weapons into Europe by the United States for NATO forces:

> But ... the moralists soon lined up again to oppose the "neutron bomb" in the mid-1970s. This time, the moral authorities who took to the streets to protest were demanding that the United States not replace its tactical nuclear weapons with weapons whose effects would be confined to the battlefield — or at least to an area substantially smaller than the weapons they would have replaced. Thus, the moral position seemed to be on the side of a greater radius of destruction.

Here again, Perle claims the high ground, ignoring political and strategic implications of expanding the arms race and of crossing the nuclear threshold. As we now know, neutron weapons were not much different from ordinary nuclear weapons — just another nuclear-explosive device, still with horror, symbolism, and risks.

Perle subsequently aligned himself with Frank Gaffney's Center for National Security Studies.

Pipes. An historian at Harvard University, Richard Pipes, whom Solly Zuckerman characterized as a "notorious anti-Soviet hardliner," had declared that keeping Soviet fatalities in a nuclear war to the level of 20 million would be a "tolerable figure."

> He also believes that if all Soviet cities with a population of a million or so "could be destroyed without trace of survivors, and, provided its essential cadres had been saved, it [the USSR] would emerge less hurt in terms of casualties than it was in 1945."

Pipes headed the a non-government group, Team B, that second-guessed the CIA in the 1970s. He was closely associated with another hardliner, Richard Allen.

Pipes gave an interview with Reuters in 1981 in which he said that war with the Soviet Union was inevitable if it did not abandon communism. He also accused West Germany of being too susceptible to Soviet persuasion.

Nitze. One of the most articulate, enduring, and flexible hardliners was Paul H. Nitze. His career as a defense advisor stretched from Truman's administration to Reagan's. He had a major role in the formulation of NSC-68, which became the foundation of America's post-World-War-II strategic policy. Nitze favored a mix of conventional and nuclear weaponry, at increased levels, to deter Soviet aggression. He also wanted NATO to develop its own nuclear force. Nitze is a study in contrasts.

As a deputy secretary of defense under Lyndon Johnson, he advocated the Sentinel ABM system to defend against Chinese nuclear missiles. His assistant, T.K. Jones, is quoted as nonsensically advising, "In the event of nuclear attack, dig a hole, cover it with a couple of doors and throw three feet of dirt on top."

Because of Nitze's testimony in 1977 against the appointment of Paul Warnke as director of ACDA, he was not offered a government position during the Carter administration. He became one of the strongest critics of the SALT II negotiations, arguing that the treaty would give the Soviet Union a 10-to-1 advantage in land-based ICBMs by 1985. He also worried that it would leave the Soviets with superiority in megatonnage and throw-weight. Although Nitze shared President Reagan's opposition to SALT II, by 1987 he had become isolated within that administration because of his belief that the SDI program was a breach of the ABM Treaty, which he had helped negotiate.

Despite Nitze's hard-line views, in 1985, speaking as a special advisor to the President and the Secretary of State, he publicly recognized that "a relationship of offsetting deterrent capabilities can be made more secure, stable, and reliable — and perhaps less costly — if we and the Soviets can agree on effective, equal, and verifiable arms control constraints."

It is instructive to take a new look at Nitze's criteria for balancing deterrents. While arms constraints were eventually achieved, many of them were unequal and not directly verifiable. Nevertheless, tensions and the level of deterrence diminished significantly. And there is no "perhaps" about the lower levels being less costly to maintain. The evolved situation is what Nitze sought: deterrence is more secure, stable, and reliable. However, by the standards of the 1980s, the hardliners would not have accepted *implicit* verification as "effective." (So much for one of the primary political obstacles raised at the time to stymie progress in arms control!)

Nitze defined two alternative strategic approaches to deterrence: (1) the "very existence" of nuclear weapons would "in itself prevent war" due to the horrendous nature of nuclear weapons, or (2) deterrence would be greatly strengthened by "military capabilities" and a "war-winning" strategy. With the alternatives so framed, Nitze proclaimed his adherence to the second option, "manifest military capability" — i.e. strong "defense," meaning strong nuclear offensive capability.

To reinforce the validity of this hardline strategy, neoconservatives had set out a red herring: that those who feared the dangers of nuclear warfare (Nitze's first alternative) would give way to unilateral disarmament or insufficient conventional defense. Neither of those dire alternatives was close to the truth.

According to Nitze,

> In the late 1960s, we were completing our [ICBM and SLBM] deployment programs and were pursuing an active [ABM] program. The Soviets also had vigorous — and growing — programs in both the offensive and defensive fields.

(As dissenters figured and as later confirmed, the Soviets were at that time actually far behind the United States.)

In the mid-70s, according to Nitze, "the Soviets passed the United States in the number, size, and throw-weight of offensive missile systems." They "proceeded to develop and deploy one generation after another of more modern systems. Meanwhile, we had frozen the number of our weapon systems and restrained our modernization programs." (So, an outsider might now ask, why should it have been a surprise that U.S. opposition to arms control agreements in years past finally came back to haunt?)

It was Nitze who quantified and demonized the throw-weight "advantage" of the Soviets — he chose to concentrate on a yardstick that gave the Soviets the biggest advantage, and refused to acknowledge the importance of the U.S. advantage in missile accuracy. The U.S. Defense Department had made a deliberate military tradeoff, choosing small, precisely delivered weapons instead of large, imprecisely delivered weapons. Nitze persistently drummed up fears of "gaps" that most people found difficult to understand.

Even as late as May 1989, an "unreconstructed" Paul Nitze continued to defend the "neoconservative alarm." He credited "various steps" taken by the West with preventing "an unambiguous [detrimental] change in the strategic nuclear balance of power."

A studied repudiation of Nitze asserts that the "fundamental neoconservative error" was "an unreasonable hostility to [the mutuality of] assured destruction: The hardliners transformed the actuality of deterrence into a more difficult burden of trying to win nuclear wars."

Nitze, in retrospect, contended that "In the 1950s, total nuclear disarmament was the declared objective of both sides, but it was wholly impractical." Setting up such a straw man — total nuclear disarmament — in place of stepwise restraints and agreements, is typical of hardliners. Although such detractors of arms control as Nitze experienced decades of glory, it should be no surprise that they added little in progress toward peace.

Subsequently, Nitze seems to have moderated his position. General Lee Butler and the State of the World Forum made public on 2 February 1998, a "Nuclear Abolition Statement by International Civilian Leaders," which says, in part,

> Leaders of the nuclear weapons States, and of the de facto nuclear nations, must keep the promise of nuclear disarmament enshrined in the Non--Proliferation Treaty of 1970 and clarified and reaffirmed in 1995 in the language codifying its indefinite extension. They must do so by commencing the systematic and progressive reduction and marginalization of nuclear weapons, and by declaring unambiguously that their goal is ultimate abolition.

As of 4 March 1998, the statement had been subscribed to by 128 leaders from 48 nations. Among the signer was Paul H. Nitze.

Teller. One of the inventors of the H-bomb, Edward Teller, vigorously associated himself with "strong defense" positions. His personal role in developing nuclear weapons is described elsewhere in this book, along with his resentful verbal mugging of Robert Oppenheimer during the latter's security hearings.

Teller, during the protracted Cold War, used all his influence to push for antiballistic missile systems and to oppose strategic arms limitation treaties. As a vocal member of the Committee on the Present Danger and the White House Science Council during the Reagan administration, he was able to draw the President's personal attention to the grand vision of a strategic defense, which materialized as the controversial SDI program.

In 1979, Teller contested the SALT negotiation process, invoking his European roots and memories of Hitler, whom he described as "a single fanatic [who] had a chance to initiate Armageddon." Teller openly feared that "our superiority [has] disappeared and even turned into a position of inferiority vis-à-vis the Soviet Union" and that "hardly anybody will deny that the continuing trend is shifting the balance of power in favor of the Soviet Union." He also believed that "Today [1979] the United States negotiates [arms limitations] from a position of inferiority." Teller wanted to scrap SALT II, expand civil defense and evacuation programs, deploy cruise missiles, and modernize and make land-based missiles mobile.

In mustering arguments against arms control treaties, Teller asked, rhetorically, "How shall we check that an SS-18 carries fewer than ten warheads? ... Could the Russians not fire a missile carrying 40 warheads, release only seven?" (This turned out to be another nonissue, since the number of warheads is easily subject to verification.) Noted physicist Wolfgang Panofsky responded to Teller in part by noting that

> *None* of the programs Teller strongly advocates for the United States — expanded civil defenses, tightened alliances, increased emphasis on electronic warfare, cruise-missile development, strengthening of the strategic Triad — are inhibited by SALT.... Past arms control agreements with the Soviets have not caused us to ignore our defense. The Limited Test Ban Treaty has actually accelerated our rate of testing nuclear weapons underground, as permitted by that Treaty.... It was the United States that refused to include the 'Forward Based Systems' ... in SALT I and II.

Teller's fears emanated from his perception of the Soviet Union: "Let us remember," he said, "that in 1961 the test moratorium was broken by Russia on a 48-hour notice" — not mentioning that the voluntary moratorium did not have the legal commitment of a treaty, nor acknowledging Eisenhower's announcement that the United States would stop observing the moratorium.

According to Panofsky, Teller strongly believed

> there does not seem to be any method by which the proliferation of nuclear weapons could be controlled in a society such as Russia's.... The U.S. is an open society, while the USSR is not; and the U.S. has no plans to dominate the world, while the USSR has a clearly stated program to extend their philosophy and their rule around the globe.

Teller, along with Leo Szilard, Eugene Wigner, and John von Neuman, were clever scientists driven from Hungary by Nazi anti-Semitism; they all contributed significantly to the Manhattan Project. Teller drove Szilard to visit Einstein at Long Island to compose the famous letter to Roosevelt, and in 1942 helped Robert Oppenheimer organize Los Alamos.

Teller's most significant technical and promotional contribution was in the development of thermonuclear weapons, which he had pursued since as early as 1941. Controversy remains about Teller's claim for a breakthrough on a militarily practical fusion explosive; others, including Stanislaw Ulam, have also been credited for major insights.

During many years of debate over enhanced radiation weapons — neutron bombs — Teller was their strong advocate; in fact, it was his laboratory, Livermore, that developed them. In 1981 Teller warned that "Those who are against the neutron bomb are proposing the Finlandization [and neutralization] of [Western] Europe.... To simply build up conventional armaments means going down the road of desperation." He noted that neutron warheads could be as effective against aircraft as against mass tank assaults, implicitly endorsing uninhibited use on the battlefield.

Teller once told a gathering of scientists that

> the beginning of World War III could be signaled by a radio broadcast from Moscow calling on Soviet citizens to take refuge in bomb shelters.... The current Soviet superiority in nuclear weapons and bomb shelters could tempt Moscow to tell the West to surrender or face World War III. The Soviets have 80 per cent of all the atomic bombs in the world.... Moreover, they possess a bomb shelter system that guarantees the survival of 95 per cent of their population.

How those fallacious and outrageous assertions — more than hyperbole — could be made by a person of Teller's intelligence and expertise remains a mystery.

Contemporary historian Richard Rhodes wrote that Teller, in his *Memoirs* "frequently denies or distorts [the historical record]." Rhodes further characterizes him as "a divisive force in American science and politics, someone who lent his scientific reputation to dangerous belligerencies that cost the nation lives and treasure but left it shockingly unprepared."

These are legacies of a brilliant scientist who championed nuclear superiority by developing new and more atrocious nuclear weaponry, by endless nuclear testing, and by deploying illusionary missile defenses — convinced to the end, like many neoconservatives, that hydrogen bombs played a significant role in ending the Cold War with "an American victory."

****

Nearly half a century of virtual nuclear conflict kept civilization at the brink of self-destruction. In learning what went wrong so we can keep it from happening again, it has been instructive to scrutinize the individuals in the United States who promoted this brinkmanship, who caused the nuclear arsenals to be expanded, who were willing to risk national devastation in the name of national security. Dubbed the "nuclear priesthood," they were a composite of self-styled defense intellectuals, who banded together in "think tanks," raising primal fears of national vulnerability to the perceived threat of world-gobbling communism.

So much of the national discourse about foreign policy was not about reality but about domestic politics, history, and mythology.

One of the most fecund of the think tanks was the RAND Corporation. By tracking the flow of its personnel into government positions, RAND's pervasive influence on U.S. national policy has been illuminated. RAND was part of — a foundation of — the military-industrial-laboratory-congressional atomic quadrangle that became the backbone of Cold War nuclear weapons promotion. Beltway bandits aggregated around the capital, and the armaments lobby vigorously exerted its influence on Congress. The weapons labs found ways to solicit endless support, and were able to carry out their schemes without public or independent oversight.

The intractability of President Reagan and his advisors can be traced to a cabal of hardliners — very conservative, mostly Republicans, who pushed and shoved for constant buildups. On the basis of feeble evidence, they pressed for remedies to perceived or pending military imbalances, almost all of which failed to materialize. The prominent, highly vocal hardliners who were effective in swaying decision-making processes of the time included Richard Perle, Edward Teller, Paul Nitze, Richard Pipes, Jeane Kirkpatrick, Fred Iklé, and Frank Gaffney. With camouflaged funding from groups with far-right ideology, the particularly belligerent Committee for the Present Danger ingratiated itself into the government, almost swallowing the executive branch whole during the Reagan administration.

It should not be a big surprise that most Reagan hardliners have persisted, to re-emerge with major influence in post-Cold War administrations. Not only are the neocons retaining their convictions, they took sustenance from others — such as like-minded former UK prime minister Margaret Thatcher — who has eulogized Reagan's role.

Needless to say, there were hardliners in the Democrat party, especially those that gained a foothold during the Carter administration. And, in fact, many former Democrats became neoconservatives who helped propel the Reagan administration in its hardline direction.

Yet the hardliners had challengers, such as Senator Daniel Patrick Moynihan, who understood the fundamental muddle of the Reaganites:

> The enterprise [to convince the Soviets of American views about nuclear doctrine] failed. And why? Because the Russian situation is not our situation, the Russian experience not our experience.

As far as the American governmental system of check-and-balances goes, there was little sign during the Cold War of institutional restraint. The Congress generally approved warlike resolutions requested by the Executive Branch, and by the time a related issue reached the Supreme Court, the meter had expired.

Because any military posture becomes vulnerable over time, especially if an adversary is working hard to make it so, there can be no ultimate deterrent. As Jonathan Schell noted,

> [While] nuclear weapons are the first means man has developed for destroying his own species, they will not be the last. Scientists can invent other means for the full destruction of the species.

D. COST OF THE COLD WAR

While no fixed monetary price can be placed on national security, it's a cost that must be eventually be paid in currency. A minority of outsiders vainly argued that security could be achieved at much less cost and with much less risk to humanity. Right or not, it's time for a retrospective assessment. What was the total expenditure for the Cold War? And what sacrifices besides financial were made? If less money had been spent, how much of the warlike confrontation could have been avoided? What are the residual costs to present-day and future generations for ongoing environmental remediation and weapons dismantlement?

Monetary cost is but one measure of the price paid: There were other human and societal consequences. While much of the sacrifice was shouldered by the peoples of the United States and Soviet Union, repercussions were felt in other parts of the world, to be sure. The financial and psychological burdens experienced elsewhere are not examined in this book.

Dollars for Defense

According to an independent organization directed by retired senior military officers, the U.S. net cost was $12.8 trillion (normalized to 1995 dollars). They added up essentially all military spending from 1948 to 1991, averaging almost $300 billion per year. That included only direct outlays for national defense, and excluded military-related national defense costs such as foreign military aid, veterans' benefits, and interest on past military spending. (Interest accrued from national-security spending added about 36 percent to the fiscal-year 1996 military budget.)

Based on a detailed audit, U.S. military spending from before World War II through 1996, reached $18.7 trillion — from government expenditures of $51.6 trillion. Allocations for natural resources, the environment, administration of justice, and energy were each under $1 billion.

The audit tallied $5.8 trillion (1996 dollars) to build the entire arsenal of 70,000 nuclear warheads deployed on 75,000 missiles, 8600 bombers, and other means of delivery. This included 14 types of heavy bombers, 47 kinds of heavy bombers and canceled missiles, and 91 categories of nuclear warheads. Evidently nuclear weapons were much more expensive than the publicly announced "few percent": more like one-third of total military spending, taking delivery systems into account.

Foreign Aid. Although some foreign aid was unrelated to the superpower conflict, much of it was rationalized in terms of benefit to the ideological or military struggle.

Both the United States and the Soviet Union engaged in swapping guns for favors — giving conventional military equipment in exchange for surrogate support, overseas bases, missile sites, or overt alliance.

$1.2 trillion was spent on international affairs (of the $13.2 trillion on national defense and $5.5 trillion on nuclear weapons). Both national-defense and nuclear weapons spending included defense-related foreign aid.

Foreign aid "was one of the chief economic weapons of the Cold War, as both power blocs sought to buy influence in Third World countries through extensive aid programs." The Truman Doctrine extended military and economic assistance to Greece and Turkey, to reduce the prospects of a communist takeover. The Marshall Plan, unveiled in 1947, addressed the need for a massive aid program to save war-ravaged Western Europe from economic and political collapse. Both the Western European recovery and the American economy were boosted by $12.5 billion funneled into that reconstruction. Political interests in fostering democratic institutions and reducing threats to U.S. national security were also served by the Marshall Plan. The Soviet Union countered by forming the COMECON (Council for Mutual Economic Assistance: 1949–1991) to encourage trade within the Eastern Bloc.

Two international organizations heavily subsidized by the United States were the International Monetary Fund and the World Bank. Soviet bloc countries boycotted them because of their emphasis on private-sector investment and free-market economic policies. President Truman, by setting up his Point Four Program of technical assistance in 1949, recognized the need for continued support of the United Nations and the strengthening of "freedom-loving countries." Regionally based programs also ensued, especially for Latin America. In 1961, President Kennedy's Alliance for Progress took effect.

The U.S. Agency for International Development was launched "to coordinate and rationalize foreign aid programs," but it came under criticism as much money was shifted from economic development aid to military assistance.

Major recipients of American military assistance were Greece, Turkey, South Korea, Egypt, Israel, and Pakistan, irrespective of their form of government — as long as it wasn't communist.

The Soviet Union and the United States often competed for influence, one case being Egypt's acceptance of funding from the Soviet Union to build the Aswan High Dam.

The Soviet Union demanded and received World-War-II reparations from Germany and confiscated German assets in Eastern Europe. After the death of Stalin, the Soviets extended aid to Eastern Europe "to ensure Soviet control." They also granted assistance to some Third-World nations, including India, Indonesia, and Ghana. Bartering agreements were reached with Cuba, Vietnam, North Korea, Nicaragua, Angola, and Ethiopia. The Soviets directed aid at political objectives in Mongolia, North Korea, North Vietnam.

Both superpowers had to forgive debts and trade deficits during the Cold War, especially with Third-World countries. While America was able to absorb the losses, the Soviet economy was ultimately unable to bear the burden, leading Secretary Gorbachev to restructure the Soviet economy and cut foreign and military aid to its allies.

Distorted Statistics

To make extraordinary treasury expenses seem somewhat smaller than they were, governments often were less than candid. Sometimes statistics were distorted so as to obscure the purpose of expenditures.

Public-interest groups repeatedly questioned official reports of the military burden. In 1979 the Center for Defense Information contradicted government claims that military spending was decreasing. Although CDI is an organization primarily composed of former military officers who favor a strong defense, they described the Carter administration's analysis as

> a carefully designed public relations effort ... despite these easily verifiable facts: (1) Military spending has increased every year since 1976 in real, inflation-adjusted dollars, and (2) The new 1980 budget calls for the largest real peacetime military outlays in more than a decade.

CDI regarded the Pentagon's techniques as effective in misleading laymen, but well known to clever practitioners of the statistical arts:

> Develop new concepts that obscure the facts by adding together items whose separate magnitudes are more informative than the combined total; for example, [the] unified federal budget itself, which mixes together programs like military spending [that] are supported by general taxes with those like social security that are entirely self-funded through trust fund contributions.

That deception — the unified budget — was a fixture for many years. CDI emphasizes that, during the half-century of ideological tension, military spending had been increasing steadily, "tricky" statistics were being used, "inflated" estimates of Soviet military expenditures have been widely publicized, the total federal budget was presented in a form that gave a "false" impression, and high military spending was the "major" cause of the huge federal deficit and a significant factor in fueling inflation.

One prominent and abiding deception, perpetrated by Pentagon-generated charts of government spending, gave the misleading impression that, from a 50-50 split in the late 1960s, the military portion was reduced to 25 percent in 1980. Actually, the military fraction omitted veterans' benefits and interest on the national debt, both of which had substantial portions derived from Cold War military spending.

Another form of deception identified by public-interest groups was the use of "rubber yardsticks" for converting the Russian ruble, to give the appearance of a military spending "race." The Pentagon asserted that the Soviet Union was lavishing money on its armed forces. Because comprehensive, reliable data were not available on what it cost the Soviets to buy, staff, operate, and maintain their armaments, the Pentagon developed "Estimated Dollar Cost of Soviet Defense Programs," a form of "creative accounting" based on what it would cost in dollars to perform the same activities in the United States. As one commentator put it, this was largely "a hand-drawn illustration of a foregone conclusion ... unsupported statistically and meaningless practically." In other words, the comparisons were of dubious value, accuracy, and significance.

Though the confrontational period has long been over, budgets for U.S. intelligence authorities continue to be hidden within appropriations of defense and other government agencies.

Economic Impact

The U.S. economy declined noticeably in the early 1980s when the Reagan administration embarked on its extensive arms buildup. Military spending did little to help the economy. The Defense Department created fewer jobs per dollar spent than most other industries, and that it employed highly skilled people who would have had little trouble finding civilian jobs.

Military expenditures were concentrated in only a few industries, but distributed geographically rather widely, which gave the Pentagon's budget an important political base in a substantial number of congressional districts.

The annual federal budget before Reagan came into office averaged a deficit of about $50 billion. During his two terms it was roughly three times higher, peaking in 1988 with a shortage of $155 billion. The budget shortfall worsened for several years, not getting better until several years after Cold War spending came to an end. President Clinton can be credited with the rare presidential achievement of reducing the deficit to the point that it became a surplus in his second term.

In 1981, at the beginning of Reagan's tenure, the Pentagon projected the cost of 132 B-2 bombers to be about $167 million each. The Reagan administration immediately classified the cost, and at the end of the president's second term, it was disclosed that the cost had gone up to $530 million each. Eventually the Air Force acquired 21 of them at $2.2 billion per bomber, more than a ten-fold escalation in cost.

The Reagan-era B-1 bomber had an even more tortured development, ending up as a gross technical and financial failure.

SDI was often considered to be one of President Reagan's "bargaining chips" (though not by him). Because the Soviets never followed suit, SDI was viewed by others variously as a shallow chip, brave gambit, or expensive dream. Although proposals were made within the Soviet military-industrial complex to match the missile-defense program, wiser heads realized that SDI was not worth the competition. Given a genuine concern for public safety, the President was insufficiently informed and poorly advised to sign on for SDI and its over-hyped goals. Thomas C. Reed, though, one of Reagan's close advisors, still thinks that "SDI was neither a staff study run wild nor a Cold War bluff without substance."

Defense expenditures correlate with poor U.S. economic performance in the 1970s and 1980s. Although the United States had been able to build an industrial base while increasing its arms expenditures, the nations devastated by World War II were concentrating on expanding their industrial strength.

Stimulating the Economy. Often the claim was made that defense spending stimulated the economy. However, purchasing one B-1B bomber at a cost of $333 million denoted a choice to forego an alternative: building new schools in more than twenty cities; two electric power plants, each serving a town of 60,000 people; two fully equipped hospitals; or 30 miles (48 kilometers) of interstate highway.

The decision to spend heavily on defense had long-term economic-security implications, because federal money directed toward building infrastructure or promoting economic activity was capable of creating more economic growth than defense-related spending; in the long run, this could have increased national security. The ultimate breakdown of the former Soviet Union testifies to the consequences of excessive spending on defense at the expense of domestic needs.

Commercial products boost the economy more effectively than defense products: A new civilian tractor can be used to build roads, raise crops, or engage in other activities that make a positive contribution to the economy.

> Defense spending to support economic rather than security objectives is generally an inefficient way to accomplish national fiscal objectives or to generate "spin-off" civilian products.
>
> And when compared with most civilian products, the military products associated with strategic nuclear weapons have a smaller multiplier effect on the overall economy. After an ICBM is constructed, for example, it sits in its silo consuming additional resources as it is operated, manned, and supported throughout its general life. It cannot, however, be used to stimulate directly additional economic growth.
>
> Among the most important costs of nuclear weapons are opportunity costs ... what one gives up by making an economic choice. [Hidden government subsidies] kept the opportunity costs incurred in waste generation from being fully expressed and artificially depressed the stated costs of the weapons involved.

Infrastructure. The large American Cold War military burden seems to have stifled investment in mass transit, electronic products, and environmentally benign energy sources, reducing growth for several decades.

Some major infrastructure improvements did take place, such as to the national interstate highway system. But other centralized improvements aimed at better efficiency and lower pollution lagged in sectors for transportation, energy supply, water resources, education, medical care, air quality, and overland shipping.

Productivity. Purchases of nuclear weapons do not enhance industrial-production capacity or increase the quantity of consumer goods. They do create a demand for additional resources to support themselves, but nuclear weapons do not generate new goods that can be used to offset their operating costs. Once acquired, nuclear weapons do not make additional jobs; they are end items, representing a non-recoverable sunk cost.

The economic penalty for spending on nuclear weapons does not mean that no such spending should occur. It simply means that, for national-security purposes, no more than truly necessary should have been spent on nuclear forces. Basing national security on the indeterminate and largely unmeasurable concept of nuclear dominance led to huge expenses and reduced national productivity.

Military spending had a depressing effect on investment, trade, and productivity that becomes evident when economic-feedback effects are taken into account.

From 1949 to 1971, as a result of two wartime buildups with industry close to full capacity, U.S. military spending was inversely linked to economically productive investment (as military spending increased, investment decreased). After 1971, the downturn in civilian consumption caused declines in living standards and key capital expenditures.

Jobs. When the 1981 Reagan-administration military buildup began, $1 billion in annual defense spending supported about 35,000 jobs; by 1985 this number had dropped to about 25,000.

> A major difference between military and civilian projects is that military projects are, by necessity, concerned principally with performance, not economic efficiency. This is particularly true of nuclear weapons. Design follows philosophy, and this key difference creates production methodologies that often lead to higher costs.... However, these higher costs do not translate to more jobs. Manufacturing nuclear weapons is a capital-intensive activity, and since nuclear jobs tend to be more highly paid, few direct jobs are created. For the defense industry as a whole, a study by the Congressional Research Service estimates that non-defense spending produces 16 percent more jobs than defense spending.

The Bureau of the Budget recognized limitations of job-boosting through military spending in a 1950 critique:

> It is neither necessary nor desirable to regard military expenditures *per se* as a method of maintaining high employment. Large and growing military expenditures not only would divert resources from the civilian purposes to which they should be put but also would have more subtle effects on our economic system. Higher taxes, if necessary, would have had a proportionately dampening effect on incentives and on the dynamic nature of the economy, without any offsetting productive impact from the expenditures. The rate of private investment might be slowed down unless special measures or controls were undertaken. There would be a continuing tendency to reduce public expenditures for developmental purposes which are highly desirable for the continual strengthening of our economy.

Inflation and Taxes. Military spending tends to cause inflation.

> Government borrowing to finance a deficit tends to raise the general level of interest rates and "crowd out" other potential investments. This condition is further exacerbated when interest rates are kept high to entice foreign lenders to invest in U.S. debt. To the extent that demand for investment in nuclear defense has crowded out other, valid investment requirements in the economy in general, the share of gross domestic product (GDP) allocated to nuclear weapons has potentially lowered both the productivity and competitiveness of the U.S. economy (the same situation attains in Russia, although deeper structural problems exacerbate the problem there).
>
> Because nuclear weapons are not available for sale to absorb the extra income generated by their production, the government may have to increase taxes to absorb the added purchasing power it has created. If such measures fail to absorb the added purchasing power on a large scale and the large-scale budget deficits this implies, the effect will usually be inflationary.

In order to fund military expenses, the top personal-income tax rate was set at 91 percent; however, tax breaks and shelters proliferated to the extent that, in the late 1960s, 155 of America's wealthiest citizens paid no income tax, and nobody at all was paying taxes in the top bracket. This loophole led to the alternative minium tax; with gradual inflation it has been having an increasing impact on the middle class.

Research and Development. While much basic and some applied R&D carried out with defense funding has contributed significant benefits to modern society, the money has not necessarily been allocated through the most productive channels.

> Research in the United States and Britain on the general tendency of military spending to consume resources required for other, nonmilitary manufacturing (called the depletionary effect) has suggested that the demand for human and physical resources to develop and build nuclear weapons crowds out other potential users of these resources. This and other work depicts these capital outlays as unproductive investments.
>
> Because of the length of time over which Cold War funding for R&D occurred, it is now possible to view the results and to evaluate their effect on the U.S. economy. If benefits from government-funded R&D outweighed costs, one could expect growing technological superiority in the United States. However, American firms experienced their greatest losses to foreign industries during this period in areas such as aircraft, electronics, and machine tools, where military R&D predominated. There is little evidence that the United States received any benefits that could not have been achieved more cheaply through normal, nonmilitary research and development.

Post-Cold War Impact

One prominent financial commentator, Robert J. Samuelson, in the beginning of 2000 wrote a column about the previous decade of U.S. prosperity without assigning any significant economic benefit from the passing of the expensive bilateral confrontation. He specifically attributed the economic revival and boom to the taming of inflation, to better corporate management, and to the growth of technology. Yet surely it is not a coincidence that an economic uplift would occur after a long period of expensive hostilities.

Samuelson tracked the boom era back to 1991 — just after the Cold War became history. However, Samuelson might have underplayed factors that must have contributed to the economy in the closing decade of the century and the beginning of the new millennium.

Two beneficial factors that Samuelson did not mention explicitly were (1) the significant reduction in U.S. military spending, and (2) the reopening of international markets. When military research and development spending was prevalent, U.S. industry lagged behind foreign industries in such areas as aircraft, electronics, and machine tools.

A third factor surely must have been more productive governmental spending of tax revenues. Defense spending as an economic stimulus has been an inefficient policy: Military projects focus on performance, not economic efficiency. The *Atomic Audit* points out that large military expenditures diverted resources from civilian purposes and slowed down private investment. Other deleterious effects on the economy were higher cost of capital, depletion of resources, and exacerbation of inflation.

During the 1970s and 1980s, major public dissent occurred because national spending priorities emphasized tanks, guns, missiles, and nuclear weapons — rather than human needs. The economy started to recover without excessive defense spending and a fortified infrastructure. Violent crime rates were dropping too.

A fourth positive factor derived from termination of the Cold War has been a psychological lift — relief from tensions that must have stifled economic investment and consumer spending. Peace, security, and prosperity seemingly can better coexist without being under the threat of mutually assured nuclear annihilation.

The temporal correlation — of reduced military budgets, expanded international markets, more productive spending priorities, and improved psychological public temperament — seems strong.

Samuelson subsequently reconsidered his views; in an article nearly a year later, he attributed the elimination of huge budget deficits specifically to two events: "The end of the Cold War, which reduced military spending; and the economic boom, which resulted in an unexpected surge of tax revenues."

Nuclear Worthiness

Because a half-century of conflict concluded without incurring mutual destruction, we survivors can ask if the risk was worth the cost. It is not a question of whether the belligerency was tolerable: rather, an important question is whether national security could have been achieved at considerably less expense. As Eisenhower said in 1961: "We have been compelled to create a permanent arms industry of vast proportions [with] an immense military establishment."

A "central finding" by the Brookings *Atomic Audit* was that

> government officials made little effort to ensure that limited economic resources were used as efficiently as possible so that nuclear deterrence could be achieved at the least cost to taxpayers.... Many nuclear weapons programs [were justified] on grounds that often had little to do with clearly defined military requirements or objectives.
>
> [Other costs and benefits associated with employing nuclear weapons] emanated in part from the concept of deterrence, whose rationale was partly economic. Basing the nation's defenses on nuclear weapons appeared to be less expensive because it was considered easier to implement than alternative strategies for deterring the Soviet threat, principally the perceived imbalance of conventional forces in Europe....
>
> During the Cold War ... U.S. defense policy was complicated by great uncertainty regarding both Soviet intentions and the mechanics of waging nuclear war....
>
> The nuclear forces amassed by the United States during the Cold War were not necessarily the sole or even the major reason for the absence of direct conflict with the Soviet Union or China.... Far greater sums were spent on non-nuclear forces, and these forces certainly played some part in the avoidance of nuclear war.

One estimate of U.S. Cold War *nuclear* spending is that it averaged out to roughly $1/day/person, a price that Michael May, former director of the Livermore weapons laboratory, considers to have been acceptable. Others feel that it was too much, taking into account humanitarian and societal needs of the time, even though the nuclear portion was a relatively small fraction of total defense spending.

Thomas C. Reed, who got his nuclear baptism at Livermore, substantiates in his book *At the Abyss* that a primary criterion at the labs was getting more bang for the buck, rather than improvement of global stability.

Rubles for Defense

Because the Soviet Union was a centrally managed economic system and closed society, it is difficult to compare superpower defense expenditures in absolute terms. Some estimates are that the Soviets spent about one-third more than the United States on the Cold War, but that is hard to gauge because their GDP was about 60 percent less. In any event, the Soviet Union allocated a greater portion of their comparative GDP — about four times more.

Reed, whose professional career spanned the Reagan, Nixon, and Ford administrations, took a keen interest in Soviet spending (as he did in American politics). In his book, it is clear that he at least understood from access to National Intelligence Estimates that the "Soviets were going broke." Reed discounted the "intelligence graybeards" who estimated Soviet defense during the Reagan years at 15 percent of GDP; instead, he favored a Pentagon reckoning of 35 to 50 percent.

Those who estimated the larger GDP fraction took into consideration indirect defense costs: the size of the Soviet naval support fleet, over-designed transport ferries, massive reinforced concrete highways, excessive war reserve stocks, client-state support, and incredible production inefficiency. The National Intelligence Council concluded that in the early 1980s "the economic system of the USSR was broken and on the verge of collapse."

Additional evidence was available: the Soviets were selling off their assets, dumping gold on a declining market and taking shortcuts in crude-oil production. The Reagan administration made a conscious effort to suffocate the Soviet economy by trying to push down the world price of crude oil (always politically advantageous) and blocking their sale of natural gas to Western Europe. Inefficiency and mismanagement extracted a heavy toll from the communist economic system.

With perhaps as much as half of USSR national income going into defense in 1969 — and almost three-quarters in 1989 — it was only a matter of time before the unbearable shortfall in human needs would engender internal reform or societal disintegration. The Soviet military-industrial complex was bankrupting the state budget.

By comparison, U.S. defense expenditures in the early1960s were about 11 percent of national income, and 9 percent 1966.

Although it remains difficult to sort out costs borne by the Soviet Union, some analysts have examined the reasons that the United States spent so much in response to perceived Soviet expenditures. Anne Cahn, in her book *Killing Détente*, describes the history behind Team B's estimates of the Soviet nuclear triad in the 1970s. Team B was a panel of extreme hardliners outside the government, led by Prof. Richard Pipes, who developed their own estimates of Soviet strategic objectives, ICBM accuracy, and low-altitude air defense. Their estimates were intended to refute CIA analyses, despite the obvious disparity in information resources and analytical capability. Years later, former Secretary of Defense Casper Weinberger admitted that worst-case analysis was the basis for Reagan administration estimates of dangers.

Team B's assessment clearly overstated the capabilities of Soviet technology and intentions. The Backfire long-range bomber buildup was projected to be more than twice as large as occurred. The Team assumed a Soviet antisubmarine warfare capability that never blossomed and an ABM capability that made little progress and was never significantly deployed. They argued, against all logic, that Soviet leadership thought a nuclear war could be fought and won despite a highly invulnerable U.S. triad of deterrents.

Team B assumed Soviet ICBM accuracy to be equal to that of the United States, although it was widely known that the Soviet Union lagged in technological capability. Regarding low-altitude air defense, they credited the Soviet Union with capabilities they suspected, rather than actually existed. Even today, Russia has yet to be successful in developing the ability to prevent U.S. bombers and cruise missiles from reaching Soviet targets.

Post-Cold War economic analysis leads to the conclusion that, despite large expenditures by both superpowers, "There is no evidence that Soviet defense spending [roughly constant through the Cold War at 25 percent of disposable income] rose or fell in response to American defense spending."

On the other hand, it is clear that specific Soviet military programs often responded to perceived Western threats, notably the development of MIRVed ICBMs. Because the Soviets had always relied on heavy boosters, they had no insurmountable problems in designing multiple-warhead systems. They quickly followed, reacting to U.S. initiatives in ABM defense and in the ability to deliver several warheads with one rocket.

An irony in this action-reaction cycle was that this U.S. MIRV initiative created a competition between Soviet ministries that resulted in both the SS-18 and SS-19 missiles materializing. Because of vested interests in the Ministry of General Machine Building, which had the largest military budget, and in the Ministry of Defense, which was very powerful, they could not resolve their turf dispute; so both missile projects were undertaken in parallel. The competition ended in a draw, with both projects succeeding. Discovery from reconnaissance of the missile duplication created a puzzle for American strategists.

In the case of ballistic-missile defenses, the Soviets did less damage to their economy than the United States did to its own. The Soviets were able to deploy the Galosh system around Moscow and never tried to match the deep and unrewarded spending of the American SDI program.

What was the ultimate impact of these costly endeavors?

> The reforms and military concessions of the Gorbachev era resulted primarily from the internal contradictions of the Soviet system. The strongest of these domestic factors were economic. Nearly every commentator, from right to left, has pointed to the chronic economic inefficiency of the Soviet Union as the paramount consideration in the decision to begin the [*perestroika*] reform process.

Vietnam War

The costly Vietnam War had a heavy human toll too,

> in suffering, sorrow, in rancorous national turmoil... Fifty-eight thousand Americans lost their lives. The losses to the Vietnamese people were appalling. The financial cost to the United States comes to something over $150 billion.
>
> Vietnam was not just a war, it was a crusade to defend those who chose freedom from the Viet Cong terrorists. But it incurred a terrible cost. Some of those costs were tangible and obvious, but others were not. Domestic needs went unmet, the dangers of inflation were ignored, and the strategic forces of the United States fell into disrepair. By 1970 the Soviet Union had achieved parity with the United States in many measures of strategic nuclear firepower. If those trends continued unchecked, the Soviets could achieve the power to blackmail, to operate below the nuclear threshold, with a free hand....
>
> As Vietnam began to devour more of the defense budget, ICBM modernization slowed and the construction of new silos stopped. In the 1970s the new Nixon administration tried to stem the tide of Soviet advances with ABM and SALT treaties, but once those agreements were in hand, the Soviets unleashed a new series of missile tests that made clear their goal of ascension to a position of strategic nuclear superiority.
>
> ...we set about doubling the effectiveness of Minuteman III, with improved warhead yields and guidance upgrades. Then we pursued the full-scale development of an MX missile that was to replace the 1960s' weapons with a system that would redress the Soviet's advances [and] move into space....
>
> The technocrats who managed the recovery at home during those post-Vietnam years did not have an easy time....
>
> Every president is dependent on the technology and production inserted into the pipeline by his predecessor(s). A disruption of this recovery by the 1974-78 [Democratic] Congresses would have precluded the Reagan buildup in the 1980s and could have stalled the Cold War victory that followed.

Linking the word "victory" with a mention of the "Reagan buildup" is a partisan stance. The technocrat who wrote that last quote refers only to "military" recovery; omitting mention of a serious domestic economic decline during the Reagan administration.

Indirect Costs

Non-financial costs and consequences can be identified:

> The United States paid a heavy economic, diplomatic, and moral price for the long and bitter Cold War. The growing national debt, decaying infrastructure, and large trade imbalance are all attributable in part to decades of excessive military spending.

The Soviet satellite nations suffered substantially in living and in deaths. America lost more than 55,000 dead in Viet Nam alone, while the French, Vietnamese (north and south), Laotians, and Cambodians endured heavy casualties in the war over their territories — estimated close to one million military and four million civilians.

Human losses for both sides in the war on the Korean peninsula are estimated to be up to three million. The Soviets and Afghans left many injured and dead behind in the mountains.

Two abiding human and domestic costs are the disproportionate number of prisoners in American jails and the lapses in health care for the poor. The Los Angeles riots in 1992 were the result of "domestic failure at the cost of global success.... The Superpowers had become superlosers."

Economists highlighted inconsistencies in President Reagan calling for reduced government regulation while he was creating the largest U.S. budget deficit and the largest U.S. foreign trade deficit. A price tag of $11 trillion has been placed on U.S. spending, and the Reagan defense program cost $20,000 for each American household. Two-thirds of the worldwide military annual spending of $1 trillion was by the superpowers. Nations that choose military over economic superiority

> suffer eventual loss of power, influence, and standard of living.... The ravenous appetite of the military-industrial complex has not only contributed significantly to the United States' huge national debt but also to its serious production problems and record trade deficit.

In 1988, Japan was spending about 4.5 percent of its R&D for defense, and Germany about 12.5 percent, while 69 percent of U.S. federal R&D was military (it had been about 50 percent in 1973).

While Gorbachev was just beginning to reduce expenses by implementing strategic policies of "reasonable sufficiency" and "defensive doctrine," 60 percent of the U.S. defense budget was going into defense of Western Europe, resulting in a huge trade deficit with Japan and South Korea. Clearly the superpowers were neglecting their domestic priorities.

This is poignantly reflected in a bitter complaint by Mikhail Gorbachev — stated just before he was to visit the Baikonur launch site. He was there to see "the nobility of the military-industrial complex" paraded before him, displaying the "grandiosity" of the Soviet shuttle, *Buran*, which was ready for launching: "It is a paradox that we can successfully launch VEGA spacecraft to encounter Halley's comet, but we are unable to produce decent washing machines or vacuum cleaners."

Psychological Costs. The heightened period of tension represented a pervasive threat to individual and societal survival. This threat affected human behavior during that period and has ever since. Psychological experiences of the *hibakusha*, survivors of the atomic bombings of Hiroshima and Nagasaki, offer not only guidance on the impact of actual nuclear war, but they also stand as reminders of the risks that everyone one else has tolerated.

Viet Nam left unhealed scars on American society. Some nations in Europe and elsewhere in the world suffered under various forms of politically or ideologically motivated repression.

Psychological factors are considered to have been "a paramount feature of the nuclear arms race." Huge levels of nuclear weapons were amassed in an effort to influence perceptions. In looking at the psychological effects of the arms race, "psychic numbing" reflected denial of the continuous shadow that the threat of nuclear war cast on humanity. Also, a feeling of helplessness crept in, especially for children, some of whom had a hard time understanding why so many preparations were being made for nuclear war.

Nuclear illusions remain part of the legacy. Fading only slowly are the deceptions about rationality in the conduct of nuclear conflict and the likelihood of survival. The toll on present-day human behavior is difficult to assess, but it will be a long time before it is purged.

Other aftermaths that lack a scientific foundation include unfounded phobias about low-level radiation and nuclear power. These phobias persist in part due to excessive secrecy — the lack of candid acknowledgment and public education — surrounding the Cold War development of nuclear technology and impact of radiation effects.

Radical Compromises. Illustrative of moral lapses to expediency and national security would be the rapid recruitment and assimilation of former enemies into competing Cold War camps. A prominent example is Wernher von Braun, who became a major contributor to the American ballistic-missile program. Similarly, the Soviets succeeded in recruiting or mustering German engineers and scientists.

An additional illustration of expediency has since surfaced, as many more will, from documents being gradually (grudgingly) declassified: how the CIA kept quiet about the whereabouts of Adolf Eichmann in the 1950s.

[**The Enemy of My Enemy**]. As World War II was coming to a close, German specialists were widely sought in anticipation of their potential post-War contributions. Here's a summary for rocket-scientist Wernher von Braun:

Von Braun ... was one of the leading figures in the development of rocket technology in Nazi Germany and the United States. The German scientist who led Germany's rocket development program (V-2) before and during World War II, entered the United States at the end of the war through the then-secret Operation Paperclip. He became a naturalized U.S. citizen and worked on the American ICBM program before joining NASA, where he served as director of NASA's Marshall Space Flight Center and the chief architect of the Saturn V launch vehicle, the superbooster that propelled the United States to the Moon.

He is generally regarded as the father of the United States space program while also remembered as head of the team that designed the Nazi V-2 rockets that killed more than 7,000 people in Britain in 1944 and 1945, in addition to a heavy death toll of slave workers involved in the construction of the rockets.

Nuclear Weapons and Information Proliferation. The cat's out of the bag, as far as *how* a nation can make nuclear weapons.

During the Cold War, the "secret" of the bomb was held by a few governments. Now it is published in magazines... Back then, the nuclear club was exclusive. Now just about anyone can join. Then, the necessary scientific knowledge was centralized in a few places. Now it is ubiquitous, protean, osmotic — fully up- and down-loadable, just like any other information in the information age.

> **[The Enemy of My Enemy (Getting to America)].** [When the] Soviet Army was about 160 km from Peenemünde in the spring of 1945, ... von Braun assembled his planning staff and asked them to decide how and to whom they should surrender. Afraid of Soviet cruelty to prisoners of war, von Braun and his staff decided to try to surrender to the Americans. Forging a set of orders on SS stationery, von Braun authorized a convoy to move 5,000 personnel south through war-torn Germany toward the American lines. The SS had meanwhile been ordered to kill the German engineers and destroy their records. The engineers, however, had hidden these in a mineshaft and continued to evade their own troops....
>
> The American high command realized the importance of the engineers and immediately sent soldiers to Peenemünde and Nordhausen to capture the remaining V-2s and their parts before destroying both sites with explosives. Over 300 train-car loads of spare V-2 parts ultimately found their way to America. Many members of von Braun's production team, however, were captured by the Russians. The V-2 rocket plans that had been hidden near Bad Sachsa in Germany were later recovered by the US....
>
> On June 20, 1945, U.S. Secretary of State Cordell Hull approved the transfer of von Braun and his specialists to America. Since the paperwork of those Germans selected for transfer to the United States was indicated by paperclips, von Braun and his colleagues became part of the mission known as Operation Paperclip, an operation that resulted in the employment of many German scientists who were formerly considered war criminals or security threats (like von Braun) by the U.S. Army.

Because nuclear weapons became attractive as the ultimate guarantor of national security, the temptation was and is for other nations to make their own. Certainly the technical capability, especially with the rapid advancements in computational power, is available to many nations. With so much information disseminated over the decades, it was unavoidable that some crucial concepts and data would leak out; these days it is much easier to sort through extensive computerized information databases and find shortcuts for implementing proliferation-related technology.

The control of further proliferation will have to be based on limiting special-materials access, reducing incentives, and reenforcing disincentives. More than anything, nations with nuclear weapons will have to lead by the example they wish to promote.

Nuclear Institutionalization. Nuclear weapons became the Cold War norm for the *haves*. Deterrence became a living strategy. Civil defense and ballistic-missile defense were exhorted, but unattainable goals. The planned use of nuclear weapons, either in pre-emption or retaliation, became embedded in strategic policy and SIOPs.

Holding hostage a city, a nation, or all of civilization to ensure a nuclear standoff became a callous orthodoxy of life — as well as a theoretical boundary condition for mathematical modeling of nuclear operations.

Equally fundamental was the institutionalization of nuclear war in national economies; the continued diversion of effort and resources away from elementary human needs became an embedded pattern. Pressure from vested interests, the taste of power, and fear ushered in a seemingly endless quest for greater national security and national superiority under the nuclear umbrella. Meanwhile, the human condition suffered poverty, hunger, malnutrition, famine, illness, disability, trauma, displacement, drought, hatred, psychosis, and death.

[The Enemy of My Enemy (In America)]. In 1950, von Braun and his team were transferred to Huntsville, Alabama, his home for the next twenty years. Between 1950 and 1956, von Braun led the Army's rocket development team at Redstone Arsenal, resulting in the Redstone rocket. In 1955, von Braun became a naturalized citizen of the United States.

Still dreaming of a world in which rockets would be used for space exploration, in 1952, von Braun published his concept of a space station in a Collier's Weekly magazine series of articles entitled Man Will Conquer Space Soon! These articles ...were influential in spreading his ideas....

As Director of the Development Operations Division of the Army Ballistic Missile Agency (ABMA), von Braun's team then developed the Jupiter-C, a modified Redstone rocket. The Jupiter-C successfully launched the West's first satellite, Explorer 1, on January 31, 1958. This event signaled the birth of America's space program.

Despite the work on the Redstone rocket, the twelve years from 1945 to 1957 were probably some of the most frustrating for von Braun and his colleagues. In the Soviet Union, Sergei Korolev and his team plowed ahead with several new rocket designs and the Sputnik program, while the American government was not very interested in von Braun's work or views and only embarked on a very modest rocket-building program. In the meantime, the press tended to dwell on von Braun's past as a member of the SS and the slave labor used to build his V-2 rockets. It was not until 1957 and the launch of Sputnik 1 that America realized how far it lagged behind the Soviet Union in the emerging Space Race. After the U.S. Navy's attempt at building a rocket to lift satellites into orbit resulted in the very unreliable Vanguard rocket, American authorities recognized they needed von Braun and his team's experience, so they were quickly transferred to NASA.

Not all Cold War outcomes are seen as negative. Some individuals — crediting nuclear weaponry — rejoice in the endurance of democracy and capitalism, while celebrating the crash of the Soviet Union and its form of communism. Peaceful applications of nuclear energy and radiation sources have also become beneficially institutionalized for society.

Who (What) Won the Cold War?

A decade or so later, the debate — about who or what won — continues. Was there a "winner" at all? Here's a sampling of views.

The Cold War as History. In Great Britain, memory of Nazi German aggression in World War II, intensified the feeling of urgency to halt the westward progression of Soviet Communism toward the English Channel.

> The reason for the Cold War was the sudden projection of Stalin's empire three quarters of the way across Europe toward the Atlantic, under circumstances in which it seemed that it might go all the way.... Moscow... exercised a tyrannical rule over the nominally sovereign states of Eastern Europe....

> **[CIA papers show agency knew where hunted Nazi was hiding].**
> WASHINGTON -- Determined to win the Cold War, the CIA kept quiet about the whereabouts of Nazi war criminal Adolf Eichmann in the 1950s for fear he might expose undercover anti-communist efforts in West Germany, according to documents released Tuesday.
>
> The 27,000 pages released by the National Archives are among the CIA's largest post-World War II declassifications. They offer a window onto efforts to use former Nazi war criminals as spies.
>
> The war criminals "peddled hearsay and gossip, whether to escape retribution... or for mercenary gain, or for political agendas not necessarily compatible with American national interests," Robert Wolfe, an expert on German history and ex-archivist at the National Archives, said at a briefing announcing the document release.
>
> In a March 1958 memo to the CIA, West German intelligence officials wrote of Eichmann, who had a key role in transporting Jews to death camps during the war: "He is reported to have lived in Argentina under the alias 'Clemens' since 1952."
>
> But neither side acted on that information because of what he might say about Hans Globke, a former Nazi and a chief adviser in West Germany helping the U.S. coordinate anti-communist initiatives in that country.
>
> Two years later, when Israeli agents captured Eichmann, the CIA pressured journalists to delete references to Globke.

When Soviet hegemony folded, the outcome was clearly a victory for united resistance. While that phase of history was basically a struggle over balance of power, with elements of ideological conflict (liberal democracy versus Marxism), a more decisive factor in its demise was probably Soviet economic stagnation. When Gorbachev became their leader, he recognized that people in the West were living a better life, and he acted to loosen the social, political, and military burdens that then pressed upon the Soviet economy.

The Cold War Was Won Before Reagan. According to the Western historian, Dana H. Allin,

> During the second half of the Cold War, Moscow's global power, like its power at home, was vastly overrated. By the 1970s, the Soviet Union had become a vast Potemkin village, not only in the stagnation and rot of its domestic political economy, but also in its ability to maintain its power and influence in the world at large. Such a conclusion cast startling light on the American foreign-policy debates of the 1970s and early 1980s, for it suggests that the prevailing pessimism about Soviet power fed mostly on illusions.

Allin reasons that the Cold War was "won" before Reagan. Acknowledging that the Soviet Union was a "malicious antagonist," "The notion of an expanding Soviet threat was largely an illusion." To give "some measure" of validity to the Reagan arms buildup, the historian concluded that an "overblown" enemy was needed to prop up the U.S. military-industrial complex.

Heartily challenging such "opinion" are diehard conservative Reagan supporters, convinced that this view not only underplays his impact, but it fails to credit his policies — especially SDI — for collapse of the "evil empire."

The Soviet hegemonic demise — the end of European Communism — is traced back to the rise of resistance in Poland, notably in 1979 when Pope John Paul II conducted open-air masses for millions of people. That was followed by sporadic strikes over meat prices, as well as demands — led by Lech Walesa — for free trade unions and the freeing of political prisoners. Through the early 1980s the Polish Solidarity movement struggled for religious and political freedom. However, "the Soviet rulers could not tolerate free societies on their peripheries, because freedom itself constituted a moral ideological challenge to their rule at home."

The quality and durability of Soviet society had started deteriorating in the 1970s, with governmental and industrial undertakings poisoning the air, lakes, rivers, forests, and soil beyond Western imagination. Average life expectancy had severely declined. Alcoholism was rampant. Crime and corruption "drained the system's lifeblood." Communication with — and travel to and from — the West was increasingly difficult to suppress.

When Gorbachev came to power, the Soviet system was ripe for reform. It was not external pressures of capitalism, but "internal contradictions" that brought about the "global, systematic collapse" of Communism.

Allin's analysis contrasts with Thomas C. Reed's dismissive and unsupported conclusion that "Most observers now think the SDI competition was a key contributor to the Soviet collapse." His evidence to support that position is quite weak.

Reed has his own iconoclastic interpretation of the penultimate events:

> Some historians say the Cold War ended when Gorbachev came to power in 1985, when the Berlin wall came down in 1989, or when the four allied powers signed the German reunification treaty in 1990. Some say it came with the events of 1991, when Boris Yeltsin was elected, or when the coup failed or when the white blue, and red flag of Russia replaced the red hammer and sickle over the Kremlin. Those events were the death throes of the Soviet empire, the war truly ended on May 31, 1992. On that rainy night in Omaha, Nebraska, the Strategic Air Command furled its colors, and that supreme symbol of the Cold War ceased to exist.

Ever the consummate Reagan insider, Reed thus suggests that the Cold War, at least symbolically, did not cease until 1992. But, his remark above about "death throes" implicitly acknowledges that the Soviet empire (and with it, the Cold War's *raisons d'être*) had been crumbling sometime before. Reed thus inadvertently confirms his own, and therefore the Reagan administration's awareness at the time, of the Soviet empire's ongoing decline. More explicitly to the point, Reed mentions in his book that Leonid Brezhnev, who held power for 18 years, was "oblivious to the disintegration taking place around him."

Meanwhile, billions were spent by the Reagan administration for arming militant Islamists in Afghanistan and indigenous opposition forces in Central America. Both of these could be claimed to be more or less "successful" insurgencies. But the USSR was not acutely wounded by these challenges to its hegemony. Compounding chronic, long-term internal bleeding, the ultimate hemorrhage was apparently boosted more by activities of nonviolent militants in Eastern Europe plus the influence of Western peace groups on Gorbachev and his circle of advisors.

Once Reagan and his administration came into power, they thought they could apply the screws to the Soviet Union and accelerate its economic crash. Implicitly Reagan (and Reed) expected to succeed before leaving office in toppling the USSR. Since that didn't occur until a decade after Reagan's election, neither of them could have been pleased with their inability to force the breakdown on their watch.

To the credit of President Reagan, whom Allin describes as "a fairly ignorant optimist surrounded by knowledgeable and ideologically sophisticated pessimists," he shed "a great deal of conservative and neoconservative baggage" and embraced with "speed and ease" the reforms of Gorbachev. Reagan "had a better grasp of reality" than his hard-line cohorts.

Reed, though, explains that one group of "White House players" that were known as the "Pragmatists" lacked confidence in "Ronald Reagan's ability to think through and give direction to his government." The Pragmatists (particularly James Baker and Michael Deaver, supported by Nancy Reagan*) viewed Reagan as "the Great Communicator."

> To them, Reagan was a President of limited intellectual power whose role was to approve predigested position papers, documents that deliver a consensus solution to any given problem.

In any event, the very policy of détente, so maligned by the hardliners as a "snare and delusion" according to Allin, proved to be responsible for encouraging "liberal reforms and political dissent" in the Soviet bloc. Also deprecated and reviled by hardliners were the Helsinki agreements, with their solemn human-rights pledges — an evident contradiction with Communist daily life — that were publicized throughout the Eastern world. Allin concludes that, while "European Communism *did* expire on the Reagan-Bush watch," the economic dysfunction that led to *perestroika* and *glasnost* had a degree of inevitability by then.

In the course of time, however, former Soviet government officials have contended that the various Carter-Reagan arms buildups and SDI made it more difficult to pursue arms control and general reform, confirming that American hardliners indeed complicated Gorbachev's early attempts to end the superpower confrontation.

Allin's final words on the subject are a warning not to take analyses of historians too literally: "We may be prepared for present and future danger in some proportion to our exaggeration of past ones."

Fallacies of Monopoly and Secrecy. Two popular misconceptions noted by atomic-history scholar Greg Herken are a fallacy of raw-materials monopoly and a fallacy of secrecy about basic scientific facts. (A third fallacy is the futility of trying to stop the spread of technology by restricting the movement or hiring of scientists — a frequent post-Cold War occurrence.)

Herken has observed that a more-significant role of secrecy and information-compartmentalization was as an excuse for ignoring protests from the public and others in or out of government. In any event, Western policies based on these

---

*Reed asserts that "Nancy Reagan was indispensable to the ending of the Cold War [because she got her husband elected President]."

fallacious misconceptions were not instrumental in the defeat of Soviet Communism.

Vain efforts after World War II by the United States to gain a "preclusive monopoly" of nuclear source materials aided what Herken calls the evolving "myth of enduring [A-bomb] monopoly."

That the subsequent H-bomb was valued for "bargaining purposes" is evidence that the arms race had created its own momentum. Nuclear weapons became an "incitement" to war rather than a deterrent, which depended on the attitudes of prevailing governments. In Herken's mind, U.S. nuclear policy "neither guaranteed American security nor perceptibly advanced American interests." Instead, "it may have spurred parallel Soviet developments in weaponry."

Quoting psychologist Robert J. Lifton, Herken remarks that "nuclearism" became a "secular religion" which depended on nuclear weapons to prevent disruption of the global social order. Their alleged capacity to avert military confrontation is a myth that persists to the present day. Nuclear weapons intensified the Cold War and gave impetus to the arms race. There was a "deadly illusion" of enduring nuclear supremacy, and a "persistent myth" that the atomic secret was a tangible commodity stolen by communist spies. More significant was the hubristic American "false assumption" of a primitive Soviet technology and a Western monopoly in uranium. All of the above led to the "illusion of security," but did not bring an end to the Soviet threat.

(This illusion has been further shattered by the post-Cold War success of nations like India, Pakistan, and North Korea to indigenously develop nuclear weapons.)

Did Nuclear Deterrence Prevent War? Research into the question of whether building up large arsenals prevents war has provided some useful insight into arms races. Nearly 100 "serious disputes" between great powers since 1815 have been studied: More than a fourth of them resulted in full-scale war, and only about 5 percent of such wars were not preceded by an arms race. Conversely, "arms races tend to be followed by war": Arms races before 1914 "played a significant role in bringing on [World War I]." Increased risk of war has historically been associated with periodic instabilities caused by actual or perceived force imbalances.

War itself has sometimes been touted as a means of achieving peace, but the claim is unconvincing. Recalling World War I as the "war to end war," Robert McNamara observes, "...in the wars since then, we lost 160 million people, depending on how you count those dead from conflict. [The 20th century] was the bloodiest century in all human history."

General Lee Butler, who headed up Pentagon strategic targeting before he retired, admitted

> I was responsible for nuclear war plans with more than 12,000 targets, many of which would have been struck with repeated nuclear blows. [In fact] the history of the Cold War suggests that nuclear deterrence should be viewed as a powerful but very dangerous medicine.... The superpowers "overdosed" on deterrence.

While there was no actual outbreak of a third world war, the risk of mutual destruction was ever present. (Hardliners have a different perspective: They relish this as a proof of their successful "poker-game" strategy because they were "right" and "won" the "pot"—notwithstanding the fact that the collapse of the USSR did not completely end communism nor despotism.)

Despite dangers, the American nuclear umbrella, coupled with NATO's conventional forces, bought time for the devastated nations of Western Europe (and elsewhere) to rebuild after World War II. In the opinion of Richard Rhodes, "Deterrence bought us time, but people desiring freedom and opportunity (in Poland and Eastern Europe and then in the Soviet Union itself) tore down the walls."

We All Lost the Cold War. A book titled *We All Lost the Cold War*, offers the following outlook on the utility of nuclear deterrence and compellence[*]:

> Deterrence and compellence have been given credit for restraining Soviet aggression.... This book challenges all these claims. We contend that the strategies of deterrence and compellence were generally more provocative than restraining and that they prolonged rather than ended the Cold War.... The book portrays the Cold War as a contest between insecure, competitive, and domestically driven leaders with competing conceptions of security.

Lebow and Stein explain that "the deployment of ever more sophisticated weapons of destruction convinced each superpower of the other's hostile intentions and sometimes provoked the kind of aggressive behavior deterrence was intended to prevent." As examples, they analyzed specific crises, concluding that "The action and reaction that linked Berlin and Cuba were part of a larger cycle of insecurity and escalation that reached well into the 1950s, if not to the beginning of the Cold War." "For much of the Cold War, Soviet and American policymakers doubted that their opposites were deterred by the prospect of nuclear war."

The driving force for these cycles was "worst-case analysis," a paradigm that "remained the conventional wisdom for many years among militants in both the United States and Soviet Union."

In a series of generalized judgments about nuclear threats and nuclear weapons convey the following thoughts:

> 1. Leaders who try to exploit real or imagined nuclear advantage for political gain are not likely to succeed.
>
> 2. Credible nuclear threats are very difficult to make.
>
> 3. Nuclear threats are fraught with risk.
>
> 4. Strategic buildups are more likely to provoke than to restrain adversaries because of their impact on the domestic balance of political power in the target state.
>
> 5. Nuclear deterrence is robust when leaders on both sides fear war and are aware of each other's fears.

"The Cold War ended when Soviet leaders became committed to domestic reform and to a concept of common security that was built on the reality of nuclear deterrence, and when Western leaders reassured [the Soviets] and reciprocated.... Soviet officials insist that Gorbachev's withdrawal of Soviet forces from Afghanistan, proposals for arms control, and domestic reforms took place *despite* the Reagan buildup."

---

[*]*Deterrence* and *compellence*: "While deterrence threatens punishment if the enemy acts in an undesired manner, compellence threatens punishment until the enemy acts in a desired manner."

Peace Works. Activist insider David Cortright has described his experience in *Peace Works*, which details the history-making impact of the public peace movement through the tumultuous 1980s. He chose the 1987 INF Treaty as the "beginning of the end of the Cold War," because it "heralded a dramatic improvement in East-West relations and a new era of demilitarization and political change in Eastern Europe."

Based on first-hand experience, personal interviews, and extensive research, Cortright articulates his conclusions "that citizen peace activists played a significant role in ending the Cold War and that grass-roots social movements have the power to shape history."

Cortright explains why world populist and democratic action "won": The public-activist movement — which asserted "the road to peace is through military restraint and international cooperation" — prevailed over the confrontational "peace through strength" approach — which brandished nuclear weapons and sustained an expensive arms buildup. Although he found that "leading decisionmakers [of the era] generally deny [which they still do] that they were influenced by the peace movement," hardliners have acknowledged that outsider presence and pressures caused them to rein in their goals.

Some recalcitrants remain convinced that military pressure from the West brought about the changes in the Soviet Union. General Colin Powell bragged that "We are witnessing today [1990] the long-term success of [military] policies we put in motion 40 years ago." Liberal columnist Tom Wicker wrote, "The Reagan arms buildup no doubt speeded reform within the Soviet Union." Cortright challenges as too simplistic the notion that "peace through strength" explains *perestroika* and Soviet reform, contending that "the Reagan buildup had little or nothing to do with the transformations in Soviet policy and may even have retarded the process."

To strengthen that contention, Cortright quotes several sources. Georgi Arbatov, a key Soviet foreign-policy advisor, labeled it *"utter nonsense"* that Western military pressures caused *perestroika*:

> Those [reforms not] only ripened inside the [Soviet Union] but originated within it. The hostility and militarism of American policy did nothing but create further obstacles on the road to reform and heaped troubles on the heads of reformers.

Cortright quotes a Soviet scholar:

> There is some precedent, dating back to the 1920s, for Soviet leaders to use "war scares" for internal political causes.... [And typically] the Russian response [to military threats] was a mixture of authoritarian state-building at home and imperial expansion abroad.

To bolster his case about the "deeper roots" of *perestroika*, Cortright cites reports showing that

> By the 1980s, the Soviet Union was entering the "scientific-industrial-information age" with an increasing number of workers in service, information, and semi-skilled professions. An important part of this process has been the rise of the intelligentsia — scientific, political, and administrative "specialists."
>
> This group ... chafed most under the restrictions of the old system and ... became a primary constituency for *perestroika* and the process of political renewal.

In analyzing the effect of Western hardline policies, Cortright took an in-depth look at Jewish emigration from the Soviet Union. Many Americans tried to persuade Moscow to permit freer emigration. But the usual confrontational approach was followed by more repression of their Jewish citizens; this counterproductive result is illustrated by the consequences of the Jackson-Vanick amendment (to the Trade Act of 1974). The amendment made

> economic cooperation contingent on freer emigration for Soviet Jews. The tactic backfired, however, and the Brezhnev regime imposed tighter restrictions.... Emigration declined after the law's passage, and U.S.-Soviet relations became strained.

Cortright, very much a peace activist during those distressing Reagan years, created a "scorecard" of the administration's strident aspirations, and Cortright added his assessment of their impact and achievements.

Cortright notes that the "central tenet" of the Reagan administration's peace-through-strength mantra was ostensibly to bring the Soviets to the bargaining table: Supporters have argued that it forced the Kremlin to negotiate seriously for arms control. But General Alexander Haig, an administration insider, reported in his memoirs:

> [Soviet Ambassador] Dobrynin raised the subject [of arms control negotiations] in his first talk with me and never failed to mention it in subsequent encounters. [But because the White House spurned the Soviet overtures] there was nothing substantive to talk about, nothing to negotiate.

Cortright adds that "it was only when political pressures [against Reagan's policies] in the United States began to build that the White House became serious about negotiations." Cortright argues that "it was not necessary to deploy INF weapons and other missiles to get an arms treaty." Another lost arms control opportunity cited was the Soviet unilateral nuclear-test moratorium, which — by being rejected by the Reagan administration — did not help the internal Soviet reform process. Nor did it help when Reagan's administration dismissed the ambitious Gorbachev proposal at Reykjavik for complete elimination of all nuclear weapons. Only when faced with accumulating political crises at home did the White House say *yes* to its own offer, unexpectedly accepted, for elimination of INF missiles by both sides.

[Personal Recollection About the Cold War *(Vadim A. Simonenko)*]. The matter under discussion is the origin of the nuclear buildup. I have the impression that the usual Western view is that it was a Soviet initiative. Being on this [Soviet] side, I remember that, at each stage, the USSR tried to match U.S. weapons developments. Yes, sometime there was in the Soviet Union something invented one half-step ahead, but each time this was just a small spark in comparison with a tremendous technological development in the United States.

On the Soviet side we were mainly boxing a real American shadow, with multiple desperate attempts to reach the United States on each such step. Each time it was usually clear that it was technologically impossible to reach the U.S. level. For this reason there were attempts to invent so called asymmetric answers (like the Khrushchev-Sakharov superbomb [photo, Figure 1, Introduction], which was totally inconsistent with military usefulness). Even in the case of the original Soviet thermonuclear breakthrough (at least, it was so presented by our domestic propaganda for a long time), it is clear now that it was just blunt step, or more correctly, a step aside.

The Reagan administration was faced with a sharp rise in federal deficits, which created incentive to limit military expenditures. The peace movement pushed for superpower and NATO cutbacks in arms and spending.

"The ironic fact" says Cortright, "is that military pressures from the West provided political justification for the system of Soviet repression and militarism."

Perhaps no one was more entitled to have the final word on the Reagan administration than George Kennan, the architect of containment:

> The general effect of Cold War extremism was to delay rather than hasten the great change that overtook the Soviet Union at the end of the 1980s.

CIA Findings. As confirmed in documents eventually made public, the CIA had determined that it was the Soviet failure to solve their own "domestic economic and social problems — rather than inability of the Soviet military to keep up with the U.S. armed-service increases — that led most directly to the Soviet Union's collapse." In fact, from the mid-70s — long before Reagan's ascendency — "the CIA was aware that the Soviet Union was on the verge of economic and political collapse and had warned U.S. government leaders." And for the United States, ongoing periods of recession or inflation correlated with the periods of heightened conflict (like the war in Vietnam and the Reagan buildup).

Even the very partisan Thomas C. Reed, who extolls the Reagan era in his book, acknowledges that as far back as Eisenhower's presidential term

> Soviet leaders failed to understand that economic fact of life [America's powerful economy] would ultimately tip the balance at the end of the containment game.... With that mistake, they condemned the USSR to a forty-year death spiral into bankruptcy and defeat.

Affirmation by Reed that he understood the USSR was in a "death spiral" which had started 40-years earlier is difficult to reconcile with assigning the credit uniquely to Reagan. Reed also acknowledges in his book that the information revolution "destabilized the Soviet system. Copying machines, facsimiles, personal computers, and the Internet incapacitated the iron hand of the communist state.... The Soviet government simply drowned in a flood of uncontrolled information."

Benefits of the Cold War

To be balanced against the cost, both financial and societal, are palpable "benefits" derived from the long-term confrontation. The West was able to withstand further incursion of stifling communism and Soviet tyranny. The Soviet hegemony was able to sustain its power over people and lands for many decades.

But another question properly comes up. Given that the West "won" and that "freedom" was protected for those outside the iron curtain, what was the role of nuclear weapons in bringing this about?

To get some answers, we must examine what is meant by "war" (and by "winning a war").

In one dictionary of modern war, the term is described as "a form of international relations in which organized violence is used in addition to other instruments of policy." That classical, Von Clausewitz, view of compelling violence implies traditionally accepted limits that would be transcended in the case of nuclear war. A crusade leading to unconditional surrender or annihilation would fall in the category of an idealistic war, an example being a *Jihad*.

A rival conception of war was the Marxist-Leninist doctrine of class struggle between capitalist and socialist states, where it was taught that the state-controlling capitalists used labor, raw materials, and markets against the working class.

The non-violent-resistance school considers war not to be a necessary instrument of policy. For advocates of peace and conflict-resolution, war is avoidable, rather than an instrument of policy.

The term "war" has more recently been applied to low-intensity national campaigns against drugs, crime, and terrorism.

The Cold War, on the other hand, was a sustained contest with low-scale conflict and violence between rivals; it falls somewhere between the classic international-scale contest and national campaigns of a domestic (and increasingly transnational) scope.

The utility of outright war between nations seems to have been declining because the cost and risks have increased and the benefits of invasion and occupation have decreased. Thus, "winning" a war, although usually defined in terms of nationhood, is a mix of offsetting cost and benefits. Even low-scale wars and societal uprisings often go largely unresolved.

The definition of a hot "war" has sometimes been stretched to go beyond the classical definition, but its consequences in the past have included systematic and organized violence, physical casualties, and the shedding of blood. Thus, it would be fair to note that nuclear weapons have won no war of classical description (the allied campaign against Japan in World War II having reached an accelerated surrender as a result of the atomic bombings). While nuclear weapons had a significant influence during the "Cold" War, they and their instruments of delivery were brandished but not used.

That leads to another relevant observation: Much smaller nuclear inventories might have had the same deterrent outcome as did the massive, expensive, and dangerous arsenals.

In fact, after World War II, nuclear weapons did not win, decide, or terminate any conflict. Although they helped stymie further expansionism through mutual deterrence, nuclear weapons were never used during that extended period, nor were they the only forceful option. The same nugatory impact could be attributed to the superpower chemical and biological arsenals.

It is relevant to acknowledge that mere possession of nuclear weapons was a powerful instrument of policy and of defense, but not of violent war. In fact, deterrence was so neutralizing that Eastern Europe could not be freed even when the United States had dominant nuclear weapons.

Moreover, none of the hallmark proxy wars — in Korea, Vietnam, Cuba, Berlin, Afghanistan — was decided by nuclear weapons. Even though nuclear threats were evoked, and the unfathomable abyss was sometimes approached, weapons of last resort were kept on their leashes. The war in Korea played out as a conventional war between nuclear powers. The war in Vietnam was largely fought in jungles and against low-value targets. The Carribean showdown ended in a dead heat, with nuclear weapons being withdrawn from Cuba and Turkey. The Berlin wall stood and fell without an atomic puncture. Afghanistan was caught in an ill-fated, fruitless Cold War struggle that ended without rattling the nuclear saber.

Nuclear Shadowboxing

This book's thematic highlight appropriately comes at the closing of this Chapter on nuclear lessons. While the remainder of the book concentrates on post-Cold War legacies, enough about the superpower confrontation has been identified so as to spotlight policy patterns that can be emulated or avoided in the future. The term *nuclear shadowboxing* is adopted because so much of the struggle reflects the futility of boxing with one's own shadow. (*Nuclear Shadowboxing* is the title of the parent book to *Nuclear Insights*.)

Journalists are traditionally schooled to report the "who, what, when, where, why, and how" of events; their method serves well in outlining salient features of the nuclear shadowboxing theme.

Who, Where, and When. The "who," "where," and "when" of nuclear shadowboxing can be addressed more succinctly than the other queries:

The cast centers around the superpowers, whose confrontation eventually reached into nearly every corner of the earth during the final half-century of the past millennium.

A subset of the "who" includes military-industrial-laboratory-legislative complexes (MILLC) on both sides of the iron curtain that helped sustain the arms race. It is their self-perpetuating interest that dominated.

Organizations favoring an exclusively militaristic approach to national security generated an influential constituency of fervent supporters favoring unbounded weapons development. The inter-linked, inter-bred complexes nourished themselves and each other, and the synergy of their implicit collaboration reinforced symbiotic momentum and collective influence.

The armed services and the nuclear weapons labs were big players. In the United States, the Army, Navy, and Air Force raced against each other "to acquire new missions and develop new weapons" that would place them "at the forefront of U.S. military power." Nuclear weapons were considered "free goods" by the military services; that is, the cost of developing and building the warheads was in someone else's budget — at the time, the Atomic Energy Commission's. Although the armed services had budgetary allocations for nuclear-delivery systems, they did not get charged for the warheads.

Included among the nuclear-shadowboxing "who" list are the weapons labs, such as those at Los Alamos and Livermore, which

> also ended up competing with each other in the quest to develop newer and better nuclear weapons, with each coming to view the other as the "enemy." Interestingly,

despite a very different political and economic system, the same dynamics took hold in the Soviet Union.

Neoconservative zealots who earned leadership positions in the Reagan administration thought that they, more than the policies of containment and deterrence, were responsible for "winning" the Cold War. With their writings, attitudes and appointments, the faction espoused the efficacy of nuclear force, notwithstanding the risk of Armageddon.

Pork-barrel politics were an important underlying factor as well. A time-honored custom for elected representatives is to use government programs not only for legitimate appropriations, but also to enhance political position and enrich constituents and cronies. The extended East-West confrontation gave an opportunity for "pork" to be buried in a huge budget that was nearly sacrosanct and relatively immune to close scrutiny, if for no reason other than the large amount of detail engulfed in the authorizations and appropriations.

What. The first three Chapters of this book described "what" happened during the Cold War: *e.g.*, the nuclear crises, from Berlin to Cuba to Vietnam; the massive arsenals of death; as well as the myths of nuclear security and insecurity, of MAD, and of surgical strikes.

Political scientist Francis Fukuyama used the term "nuclear shadowboxing" in the title of an article he wrote for Rand Corporation in 1980 about Soviet threats of intervention in the Middle East. This book uses it in a broader sense, to delineate an underlying bilateral dynamic during the entire Cold War.

In the 1960s, political leaders of the Soviet Union decided that they would achieve quantitative and qualitative weapons parity with the United States. Meanwhile, hardliners in the United States relentlessly asserted, without foundation, that the Soviets were already ahead — a convenient assumption proved untrue long after it was too late. A one-upmanship sequence thus ignited and sustained the process of recursive nuclear shadowboxing.

Various individuals, groups, and historians have decried the cyclic phenomena. Although once a contributor to the problem, former Secretary of Defense Robert McNamara has belatedly remarked on the stupidity of an arms-race action-reaction cycle that spurred the growth of nuclear arsenals.

The University of Cambridge faculty asked in 1983, somewhat rhetorically, "Is it necessary for Russia to follow automatically — with a lag of 3–5 years — every American step?"

In his book, *The Cold War and the Making of the Modern World*, historian Martin Walker states, "An awful symbiosis emerged between the main actors of the Cold War, a rhythm of escalation between the Pentagon and the Soviet strategic rocket forces." He illustrated this escalation by noting that "The missile gap implied a technology gap, which in turn implied a research gap, and thus an education gap and so on."

David Cortright notes that

> Throughout the Cold War era, buildups by the West led not to concessions and peace on the other side but to similar buildups and hardline positions.

The Kennedy administration's Minuteman missile program was followed by a massive ICBM buildup by the Soviets. The MIRVing of U.S. missiles in the 1970s was followed by Soviet MIRVing in the 1980s. The two giants were locked in a constant cycle of action and reaction, with each side responding in kind to challenges from the other and neither gaining a commanding lead. Military confrontation was a no-win strategy creating ever greater burdens and risks for both sides.

Looking at the nuclear-armed NATO in this context, peace-activist Mary Kaldor is quoted by Cortright: "Far from countering a Soviet military threat to Western Europe, NATO legitimized the Soviet presence in Eastern Europe." Cortright adds,

> In this sense, the military establishments of the two sides were symbiotic, feeding off each other and sustaining a right-wing political climate that reinforced the war system on both sides. In the Soviet Union, the vast military-industrial complex and repressive apparatus inherited from the Stalin era were justified as a necessary response to the Western threat.

From an unexpected source is yet another confirmation of this tendency. Thomas C. Reed, who worked at the top of the Reagan administration, relates what happened when the French *Farewell* files were shared with the Americans: It was discovered that, as a result of many years of successful Soviet espionage,

> the Soviets had been running their R&D on the back of the West for years. Given the massive transfer [via espionage] of technology in radars, computers, machine tools, and semiconductors from [the United States to the Soviet Union], the Pentagon had been in an arms race with itself.

But it started somewhere. Jonathan Schell summarizes his interviews of the antagonists with this observation:

> Except for the intercontinental ballistic missile, every major innovation in nuclear arms — the intercontinental-range bomber; the hydrogen bomb; the solid-fuel rocket; the nuclear-powered submarine with its submarine-launched, nuclear-armed ballistic missiles; the multiple independently targetable re-entry vehicle; and, finally, the techniques for strategic nuclear defense — was pioneered by the United States.

Even in the absence of direct conflict, American A-bomb policy at the beginning failed to guarantee security, nor did it ultimately advance American political interests. No outright superpower war took place, but the tenuous peace was punctuated with a nervous uncertainty. Despite failed efforts by the United States to gain a nuclear monopoly after World War II, a myth about its achievability persisted. Not only was the United States boxing against its own shadow, it had to compensate for the myths it had created.

Atomic Audit concludes that American officials "systematically failed to anticipate how the Soviet Union would perceive the U.S. buildup and how it would drive the Soviets to respond with their own provocative programs."

According to Herbert York, former director of the Livermore weapons laboratory,

> The Manhattan Project, then, was based on the first of a long series of mistaken beliefs in our being "raced." Because this was the first such case, our mistake in this regard was, in my view, entirely justified, but our subsequent failure to learn anything from repeatedly making this same mistaken judgment about the existence of a "race" is less so.

The initial, excusable mistake that York refers to is the unsustainable assumption about Nazi efforts to make an atomic bomb. Although the Germans had indeed embarked on that course, insufficient intelligence data was available to the Allies. After the war it found that the Nazi project (1) did not have a high priority, (2) did not use the best German scientists, (3) had made a measurement error that diverted them,* and (4) was hampered by the Allied destruction of a heavy-water production plant in occupied Norway.** Although Nazi Germany had embarked on the right track, it dropped out of the race.

Even before the Soviet atomic test of August 1949, the decision to increase reliance on nuclear weapons and to use them against a postulated conventional attack in Europe "launched the United States on a one-sided strategic arms race...." according to historian David Alan Rosenberg.

In a somewhat similar vein, the authors of *Atomic Audit* noticed that "Official statements about deterrence during the nuclear era tended to reflect this action-reaction view of deterrence, rather than a view that derived from any historical analysis of what had deterred various states." For example, "[The Kennedy administration] failed to recognize that the large American deployment under way ... would contribute to decisions by the Soviet Union to increase its own capabilities."

Analysis by T. H. Naylor comes to the conclusion that "Relations between the two superpowers were based on a never-ending cycle of fear and distrust of each other, combined with a heavy dose of tough talk and macho pride."

Later presidential administrations did no better: "Certain defense intellectuals of the Johnson era," says Daniel Moynihan, "began to assert that Soviet strategic behavior was basically imitative of ours — two apes on a treadmill, as the image went — overlooking, presumably, that the fondest hope of the [defense community in the early 1960s] was that Soviet behavior *would* become imitative." And, from *Atomic Audit*, we find "there is no evidence ... that the Soviet Union increased its military spending to match the Reagan administration's program." They add:

> The early years of the program set the parameters for everything that followed, and during these years the chief influence was the desire to acquire as many nuclear weapons as possible. What this would cost, how these weapons would increase U.S. security, and what would happen if they were ever used were questions that were seldom raised. Although the actions of the Soviet Union during this period were of great importance in rationalizing the program, giving it a focus and propelling it forward in several great leaps, in many ways the United States was racing against itself, as became evident when the bomber and missile "gaps," among others, turned out to be illusory, the product of limited intelligence data and a propensity to characterize the Soviet Union as the literal opposite of the United States. Future scholars would do well to study these factors and assess their importance on the programs of the other nuclear powers.

In a 1987 textbook on the arms race, David Barash describes the action-reaction sequence this way:

> The U.S. has largely been running a race with itself, escalating the arms race first by achieving a quantitative lead and [later] by qualitative innovations, forcing the USSR

---

*from building a graphite-based nuclear reactor to make plutonium.
**which would have been used to make a heavy-water based reactor to make plutonium.

to keep struggling to keep up [which they always did], a few years later. Then the U.S. has reacted to the Soviet reactions, and the sequence has repeated itself.

One memorable case involves Soviet anti-aircraft developments of the 1950s in reaction to the large U.S. strategic bomber fleet; this in turn led to electronic countermeasures, low-level penetration capabilities, and air-launched cruise missiles. Another example is the American ICBM buildup of the early 1960s, beginning as a reaction to the alleged missile gap.

Barash also mentions the Soviet Tallinn air-defense system, which gave added U.S. impetus to penetration aids, the MIRV program, an ABM system, and the Poseidon and Minuteman-III. The initial U.S. response got started because intelligence agencies had incorrectly interpreted reconnaissance photos to infer that the Tallinn bomber defense was a ballistic-missile screen.

Although U-2 overflights of the Soviet Union in 1959 failed to detect newly operational ICBM launch sites, the initial inability of air-defense forces to bring the high-altitude craft down motivated the Soviet government to intensify its ballistic-missile and aircraft-defense programs.

Surveillance films from the first satellite (CORONA) in 1960 threw light on one worst-case-analysis incident, since clarified by declassified intelligence information. Before CORONA, American National Intelligence Estimates had predicted that the Soviets would have 140 to 200 ICBMs deployed before 1961; after CORONA, the estimate was reduced to 10 to 25 Soviet ICBMs. But this clarification came too late to stop the U.S. buildup.

Another action-reaction cycle commenced with the development of anti-satellite weapons by both sides. Satellite technology had significantly increased confidence in estimates of strategic weapons, and any threat to that stabilizing influence immediately encouraged defensive and offensive measures.

Outside the hardware action-reaction cycle, perhaps a more appalling testament to crisis "ratcheting" was the enablement of convulsive nuclear responses, including defense-alert and "dead-hand" automatic reaction sequences that were not fail-safe. Without candidly divulging the realities, the superpowers (and other nuclear-armed nations) sought to convince the world of a functional, assured, safe, prompt, and devastating response.

Even civil defense inspired reactive undertones: Misinterpretation of routine preparations by the respective rivals, including evacuation and sheltering, for any kind of disaster — natural or other — became the foundation upon which advocates based their support for response-in-kind.

Unilateral attempts at lifting the threat of nuclear destruction by creating a shield against ballistic missiles also led to predictable countermeasures, such as low-trajectory missiles, penetration aids, and an overwhelming array of new attack missiles. That was exemplified by the Soviet response to President Reagan's SDI program: Quite noticeably, SS-24 and SS-25 ballistic-missile production was stepped up.

Barash has tabulated sixteen developments in military technology that were initiated by one superpower and countered by the other: the first sustained chain reaction, the atomic bomb, the intercontinental bomber, a European military pact, tactical nuclear weapons in Europe, nuclear-powered submarines, ICBMs, orbiting satellites, supersonic bombers, SLBMs, solid-fueled missiles, MRVs, ABM, MIRVs, enhanced radiation warheads, and long-range cruise missiles. The United States started all but three of these, and the Soviet Union duly matched the initiative, the average lag time being a little more than five years.

Additive to these lists are the internal competitions that developed within the respective military-industrial establishments. On the Soviet side was the duplication of ministry resources to develop the SS-18 and SS-19 missiles, both of which had the same MIRV capability. On the U.S. side, one can cite parallel and mutually stimulating developments at the two nuclear weapons laboratories.

Herbert York has stated that "excessive technical conservatism" was probably responsible for too many crash programs, but that a valuable lesson can nevertheless be gleaned from the experience, "to be remembered the next time this sort of problem arises." He also is said to have remarked that arms-race promoters made up problems to fit solutions.

General Lee Butler, in 1998, described nuclear deterrence as a "dialogue of the blind and deaf. It was largely a bargain we in the West made with ourselves."

Ironically — possibly prolonging the agony of the Eastern European satellites — military pressures from the West bolstered the Soviet system of repression.

Reinforcing our perception of a positive-feedback loop is this quote from David Cortright's 1993 book about the citizen's role:

> The military establishments of the two sides were symbiotic, feeding off each other and sustaining a right-wind political climate that reinforced the war system on both sides. In the words of Russian poet Evgenii Yevtushenko, "Your hard-liners help our hard-liners, and our hard-liners help your hard-liners."

Why. One of many driving forces for superpower shadowboxing was national perception — about intelligence information, perceived disparities, deterrence strategy, nuclear "sufficiency," and mutual fear.

According to *Atomic Audit*, the capabilities of many existing weapons were downplayed and their vulnerabilities exaggerated in order to make new weapons appear both necessary and cost-effective.

A 1994 retrospective by the General Accounting Office found that there had been "systematic disparities between what the data showed and DOD's claims and estimates for (1) the Soviet threat, (2) the performance of mature systems, and (3) the expected performance and costs of proposed upgrades." The GAO considered the disparities "systemic" because

> they seem to follow a particular pattern, tending to overstate threats to our weapon systems, to understate the performance of mature systems, to overstate the expected performance of upgrades, and to understate the expected costs of those upgrades.... The capabilities of many existing weapons were downplayed and their vulnerabilities exaggerated in order to make the weapons ... appear both necessary and cost-effective.

Going back even further, *Atomic Audit* concluded that

> the most significant factor affecting the scale and pace of the U.S. nuclear buildup [was that the] unconstrained U.S.-Soviet nuclear arms race that actually occurred [until SALT in 1969] did not derive from any analysis of the size of an arsenal needed for

deterrence. Instead, a variety of other factors drove the process, including a lack of reliable information on the strength and disposition of an opponent's forces, interservice rivalry, and pork-barrel pressures for military spending.

In the Soviet Union, the military's emphasis on "correlation of forces," otherwise termed "equal security," obliged their government to insist on requirements that "depended strongly on the posture and capabilities of likely Soviet adversaries, especially the United States and its NATO allies." Thus, the Soviet Union was largely reactive to Western military developments and deployments, including nuclear forces.

The Reagan administration's policies, as viewed from Britain, led Cambridge University dons to the following observations: "The on-going arms race has resulted in less security and a greater likelihood of nuclear war," and "There can be no doubt that the greatest danger at present arises from the decision of the Reagan administration to step up the arms race by the introduction of weapon systems [that] must be perceived by the other side as preparation for a first strike."

The Soviets discovered through espionage that a U.S. biological weapons program from the late 1950s through the early 1970s had developed pathogens that could be "countered by antibiotics or vaccines to protect troops and civilians from potential accidents." This prompted the Soviets to set up their own program, which greatly exceeded the U.S. effort in duration and intensity.

The now-divulged enormous cache of biological weapons produced by the Soviet Union (contravening the Biological Weapons Convention which had no verification provisions) is explained by a former deputy director of their secret bioweapons project, Ken Alibek, as follows:

> In the early 1980s, relations between the Soviet Union and the West had plummeted to their lowest point in decades. The election of President Ronald Reagan had lead to the biggest American arms buildup our generation had seen. Our soldiers were dying in Afghanistan at the hands of U.S.-backed guerrillas, and Washington was about to deploy a new generation of cruise missiles in Western Europe, capable of reaching Soviet soil in minutes. Intelligence reports claimed that Americans envisioned the death of at least sixty million Soviet citizens in the case of a nuclear war.

According to Alibeck's revelations, Soviet newspapers

> chafed over Reagan's description of our country as an evil empire. It was easy to believe that the West would seize upon our moment of weakness to destroy us. It was even conceivable that our Army strategists would call for a pre-emptive strike, perhaps with biological weapons.

As a result, the Soviet program to develop biological weapons was accelerated, including mass production of weaponized anthrax, even though they had no actual evidence that the United States had defaulted on its own renunciation of biowarfare. By 1987, "the combined production of [Soviet] anthrax lines around the country [Kurgan, Penza, and Stepnogorsk] was nearly five thousand tons a year...." (Compare that with the few grams of mail-posted anthrax that produced death and terror soon after the 9/11 attacks on America.)

In any event, the Soviet response to the Reagan nuclear buildup was asymmetrical, in part due to Kremlin fear in the 1960s and 1970s of the earlier American biological-weapons program. Though the U.S. program had been sharply curtailed by President Nixon, it was done unilaterally without reciprocity or verification by means of a treaty. Thus, national mis-perceptions stimulated two rounds of biological-weapons shadowboxing — first, the initial Soviet bioprogram and, later, their grotesque mass production.

Aside from such specific situations and challenges, it has been recognized that "to a large extent the dynamics of arms races are internally generated, quite independent of the supposed external enemy."

> From the standpoint of nuclear deterrence, *Atomic Audit* found that there is no evidence that a huge nuclear arsenal is more effective than a small one in deterring a conventional attack. [Military and civilian strategists reasoned] that if a few nuclear weapons were good for deterrence, more would be better. There is clear historical evidence ... that even the possession of nuclear weapons-usable materials and actual or presumed nuclear expertise can, in some circumstances, serve as a deterrent. Similarly, there is also evidence that a state with a relatively small proven arsenal, such as China or France, can deter nuclear attacks by states with far larger arsenals.

In short, the excessive inventories of nuclear arms resulted more from complex internal bureaucratic processes and misunderstandings rather than simply a deliberate arms race.

When asked, whether he thought the driving forces for the arms race were "action-reaction" or "domestic," noted historian David Carlton responded "both, at various times."

The perceptions that drove the arms race were often in the minds of fallible human beings. As General Lee Butler once put it, "The capacity for human and mechanical failure, and for human misunderstandings, was limitless." He recalls particular instances of human error and mechanical flaws: bombers crashing during nuclear-war exercises, the conventional-explosive parts of nuclear-armed missiles exploding in their silos, and submarines going to the bottom of the ocean laden with nuclear missiles and warheads.

How. Various governments avoided limits or reductions in excessive stocks of armaments by dismissing the opportunities for negotiated arms control, by seeking weapons systems with unattainable performance, and by cloaking decisions and negotiations in secrecy.

In the early years, nuclear weapons development dominated U.S. policy, a conclusion with which *Atomic Audit* concurs: "Another important reason that the U.S. nuclear-development program proceeded as it did was the government's refusal, through the mid-1950s, to engage the Soviet Union in discussions on arms control or nuclear disarmament."

In later years, blatantly unacceptable bargaining conditions were used to stymie negotiations. Sergei Kortunov, who became an influential member of President Yeltsin's Defense Council, remembers that someone on the Russian negotiating team for START II told him

> those heavy [Soviet] missiles were unusable. They were really "political missiles." Their purpose was to get the United States to take us seriously at the bargaining table.

To obscure underlying strategic goals, as well as political ambitions, tight secrecy was maintained on the formulation of policy and negotiating tactics. When compounded with lack of information about the other side, little progress in treaty negotiations could be made. As the Cambridge faculty commented in 1987,

> there is hardly any direct information about corresponding developments in the Soviet Union. Indeed, this very secrecy — inherent in the regime — has considerably contributed to the arms race, as it enabled the military in the U.S. to demand funds for new weapon systems, allegedly to match advances in the USSR, which in almost every case turned out to be non-existent.
>
> Owing to secrecy and the indeterminate nature of deterrence, strict military requirements were not the only, or even the major, factor responsible for the types and numbers of weapons built and deployed. Diplomatic and domestic political and economic influences pushed and pulled the program in a number of directions, often contradictory.

Disinformation, propaganda, and ideology played a big role in prodding the public, and thus their representatives, to support the nuclear arms race. Similarly, these were driving forces in respective biological-weapons programs: A highly placed Soviet defector, commenting in 1996, reported that Soviet scientists were "constantly spurred on by being told the United States had a huge biological weapons program." The source of this incentive, according to arms control specialist Ray Garthoff, was a "U.S. disinformation and deception program designed to stimulate Soviet interest and investment in CBW." This "disastrous outcome" led to extensive chemical and biological munitions programs that "may still exist on a reduced scale a decade after the end of the Cold War."

Hardliners on both sides, usually those who had served in government and thus knew their way around, employed worst-case analysis to exaggerate apparent disparities, manipulate friendly bureaucracies, and — in the West — nourish gullible news media.

Those provocative practices served to exacerbate the arms race and hamper arms control negotiations, allowing time for quantitative or qualitative improvements in arsenals. Even when the Cold War was obviously coming to a close, U.S. arsenals continued to expand. *Atomic Audit* points out that "In fact, some argue that the U.S. buildup may have postponed the [Gorbachev reforms] including reductions in military spending ... thereby prolonging the Cold War."

****

Having been personally involved in many of these issues and some of the peace overtures, I can attest to historical assessment that progress in arms control and Soviet domestic reform took place not because of, but despite the Reagan buildup.

The Cold War ended when the Union of Soviet Socialist Republics collapsed from intramural failures.

Even though "victory" has been proclaimed for the West, it is doubtful that a single "winner" can be unambiguously assigned. Both superpowers paid a heavy cost and were perilously close to unthinkable mutual obliteration. The Communist hegemony collapsed mostly on its own. Internal rot doomed it from the beginning — it was just a matter of time.

Institutions in the United States have estimated the financial cost at close to $13 trillion, nearly half of which was spent on the nuclear arsenal.

Nor can "peace through strength" be credited with much more than getting President Reagan elected: The internal rupture of the Soviet Union was more likely delayed than hastened by militaristic policies of the West. What the slogan did induce was heavy spending by the Pentagon, which helped quadruple annual government budget deficits during Reagan's administration.

In fact, it was not U.S. President Ronald Reagan, but Communist Party leader Mikhail Gorbachev who won a Nobel Peace Prize for efforts to bring matters to a quiet end. The Reagan "peace through strength" policy had a negative effect, stiffening the resolve of Soviet hardliners.

The endless action-reaction cycles of the Cold War did not result from real dangers; the threats perceived by a security-conscious public were largely mythical. Embellished and exploited by vested interests, the dangers loomed large, with secrecy and nuclear mystique camouflaging weaknesses in data and analysis.

One result was that each side was "shadowboxing," with nuclear weapons as the figurative boxing gloves. The opponents could not discern each other's movements or motivations — only shadowy projections displayed between overt jabs and reflexive stances. Each side concentrated on fulfilling its own interim political objectives more than achieving long-term stability.

When the superpower dual came to an end, the public had a right to expect reduction in some tension-producers, such as threats of massive nuclear destruction, self-serving official secrecy, and tight government control over personal lives. Many myths about arms control and verification had already been punctured. Emerging was the realization of how close the superpowers had skirted to unthinkable doomsday. There were bills to be paid and belated lessons to learn.

While the United States was reacting to its own shadow, it might also be said the Soviet Union was fighting against that very same counterpart. Perhaps "phantom" is a better metaphor to characterize the imagined perils.

The initial threats from the East and the ensuing arms buildup were eagerly expanded by the U.S. military-industrial-laboratory-legislative atomic quadrangle. But now the question is, how do we keep these dangers from continuing to stalk the new millennium? One way is through informed and aggressive public participation: For example, the Pentagon, with a vested interest in expanding its sphere of influence and its built-in bias favoring hard military solutions, should not be prescribing national policy regarding ballistic-missile defense. Public involvement was key to tempering some excesses, and that type of pressure will be needed in the future.

****

Believers as they might be, the proponents of "peace through strength" will have a hard time convincing past or present citizen activists that anyone "won" the Cold War, or that the end was hastened by an uncompromising and militant standoff. **All** *of humanity won when nuclear Armageddon was avoided.*

It was primarily by worldwide public protests that the INF treaty in 1987, after a decade of fruitless negotiations, was accelerated to closure (with a surprisingly gratifying outcome, removing a mutual, immediate threat). The Soviets weren't forced to the bargaining table by President Reagan; they were ready for serious negotiations long before.

A couple of years later the Union of Soviet Socialist Republics dissolved— largely because of a combination of internal economic, sociological, and political rot, coupled with external governmental and non-governmental pressure. A nearly universal sigh of relief was uttered around the world.

The mortal ideological struggle, as well as stressful surrogate wars, had bled lives and quality of life. For half a century sidelined or stifled were progressive agendas on civil and human rights, medical and health care, workers' and farmers' legislation, education and science.

Although more than half a century has passed since World War II ended, the decision-making process that led to the bombings of Hiroshima and Nagasaki is just now being understood. Sharp extremes that separated understanding about the Cold War will no doubt take a long time to be reconciled. Meanwhile, many useful lessons could be mulled over in order to avoid unproductive confrontations in the future.

Activist David Cortright is disappointed that the peace movement was "ineffective in altering the larger structures of militarism." The Reagan administration was able to bolster the war system and reinforce the power of military institutions. Subsequent presidential administrations have had their way alternately with moderate decreases or increases of embedded militarism. Nonetheless, matters might have been much worse without public resistance.

Regarding the three strategic-policy algorithms of deterrence, compellence, and avoidance, it is our observation that citizen-stimulated avoidance helped to sidetrack escalated Cold War bellicosity.

Perhaps the most important conclusion for this Chapter (and the book) is that nuclear weapons won nothing.

Chapter IV:

PUBLIC INVOLVEMENT

(Irritant or *Force Majeure*?)

Chapter IV: Public Involvement

- A. PUBLIC-INTEREST ORGANIZATIONS 306
- **NGOs** 307
  - FAS 308
  - SANE 310
  - Pugwash 310
  - Council for a Livable World 311
  - NRDC 312
    - [Our First Meeting] 313
  - UCS 313
  - SESPA 314
  - PSR 314
  - CAS 315
  - ACA 316
  - CDI 316
  - NOMOR 317
  - Mobe 317
  - Nuclear Freeze Campaign 318
- **Programmatic Organizations** 319
  - Carnegie 320
  - Brookings 320
  - Hudson 321
  - Center for Security Policy 321
  - Henry L. Stimson Center 321
  - Nuclear Control Institute 322
  - Religious Associations 322
  - Extremists 323
- **Soviet Moderates** 325
  - Soviet Academy of Sciences 325
  - Committee of Soviet Scientists 326
  - Sakharov 327
- **American Funding Foundations** 329
  - W. Alton Jones Foundation 329
  - Ploughshares Fund 329
  - John D. and Catherine T. MacArthur Foundation 329
- **Educators** 330
  - Individuals 330
  - MIT 332
  - Stanford University 333
  - Princeton University 334
  - University of Illinois 334
  - Monterey Institute 335
  - IGCC 335
  - University of Maryland 335
  - Bochum University 335
  - Notre Dame University 336
  - The Advanced International Studies Institute 336
- **Professional Associations** 337
  - APS 337
  - JASON 338
  - AAAS 338
  - Academies of Science 339
- **International and Transnational Groups** 340
  - IPPNW 340
  - UN 340
  - UNIDIR 341
  - SIPRI 342
  - International Institute for Strategic Studies 342
  - Canadian Institute for International Peace and Security 342
  - VERTIC 342
  - I N S T E A D Center for Verification Technology 343
  - Peace Research Institute Frankfurt 343
  - B A S I C 343
  - INESAP 343
  - Labor Unions 343

European Peace Movements 344
   The Women of Greenham Common 345
   The Greens of Germany 346
Networks 346
   The Acronym Institute 346
   INES 346
   Abolition 2000 347
   MoveOn 347
   DontBlowIt 348
   Institute for Global Communications 348
   Safer World 348
   Nuclear Threat Initiative 348
Fourth Estate 349
   Scientific American 349
   The *Progressive* 350
   Dailies and Weeklies 350
   The Silver Screen 351
      [*On the Beach*] 351
      [*The Day After*] 352
      [*Dr. Strangelove*] 352
      [*Seven Days in May*] 353
      [*Fail-Safe*] 353

B. ARMS CONTROL CONTROVERSY 356
Fallout Shelters 356
Nuclear Testing 359
      [Impact of Public Opposition] 361
The ABM and Sentinel Debates 362
   Suburban Chicago Opposition 363
   Nationwide 364
   The Unraveling. 365
National Spending Priorities 366
Military-Industrial Complex 367
ACDA 368
SALT/MIRV 369
Proliferation 370
Euromissiles 371
Neutron Bombs 371
Freezing the Arms Race 373
SDI 376
ASAT 379
INF: Zero Solution/Option 380
Verification 382
MX 383
Missile Accuracy (and Reliability) 385
Nuclear Winter 387
B-1 and B-2 Bombers 388
Founding a Peace Institute 390
Conventional-Force Reductions 391
National Missile Defense 392

C. POLITICAL FORCES 394
U.S. Movers and Shakers 396
   Parties, Individuals, and Writers 396
      Joseph McCarthy 397
      Barry Goldwater 397

Chapter IV: Public Involvement

Zbigniew Brzezinski 397
Lyndon LaRouche 397
Robert Welch and the John Birch Society 397
William Buckley 397
The Alsop Brothers 398
Evans and Novak 398
Berrigan Brothers 398
Congress and Courts 398
Paid Lobbyists and Advisors 399
Private Institutions 400
    Cato Institute 400
    Heritage Foundation 401
    Hoover Institution 401
Western European Leaders 402
    Winston Churchill 402
    Charles DeGaulle 402
    Konrad Adenauer 402
    Willy Brandt 402
Soviet Policy Makers 402
    Party Leaders 402
        Josef Stalin 402
        Nikita Khrushchev 403
        Alexei Kosygin 403
        Leonid Brezhnev 403
        Yuri Andropov 403
        Konstantin Chernenko 404
        Mikhail Gorbachev 404
    Soviet Scholars and Organizations 404
        Petr Kapitsa 405
        Sergei Kapitsa 405
        Pugwash 405
        IMEMO 405
        The *Soviet Academy of Sciences* 405
        Andrei Kokoshin 405
Cultural Differences 406
    [Dead on Arrival] 406
Cultural Movements 409
    The Politics of Racism 410
    Gender Activism 411
    Student Radicalism 411
Religion vs. Atheism 412
Public Assemblies 413
    Conclaves 414
    Projects 414
    Conferences 414
    Demonstrations 416
    Summits 417
    Treaties 418

D. INTIMIDATION OF OPPOSITION 419
Prosecutorial Predominance 420
In the West 422
    Un-American Activities 423
    McCarthyism: Blacklisting 424
    [Korean Purges] 426
    Oppenheimer 426

The Atomic Spies 428
Pentagon Papers 429
Red Squads 430
The *Progressive* Case 431
In the East 432

In the first part of this book, the growth, tribulations, and culmination of the Cold War have been detailed. Now it is appropriate to describe the role of outsiders — public-interest groups and individuals who had a significant impact on the nuclear arms race. Because organized and individual dissent remains important in current events, that phase of Cold War history is positioned here, in the fourth of seven Chapters, the book's midpoint, with focus shifting to the new millennium. This Chapter starts the transition from the past to the present.

In Chapter III, Cold War excesses intensified by hardliners were described in detail. Countervailing — though underappreciated — resistance by the organized and unorganized public emerged, in a widespread effort to moderate the arms race. Public opposition, however, had to struggle with government suppression of the growing dissent to nuclear policy.

Because so many individuals and organizations — national and international — were involved, our list of participants in this Chapter is necessarily incomplete. Added for the historical record is a personal survey of the role of scientists and other dissidents. Historians can fill in the gaps, improve the perspective, or rectify misjudgements.

Overt judgments, when they occur, are limited to the question of whether public involvement helped shorten, prolong, diminish, or exacerbate the nuclear arms race. Although it seems clear that the evolved anti-war, anti-nuclear social movement played an appreciable role in taming militaristic governmental policies, it will be for future historians to stand in judgement as to whether public interveners were merely a passing irritant or remain as an abiding force majeure.[*]

For this book, the "public" consists of individuals involved in issues at a time when they were not government officials, because for some professionals there was a revolving door in and out of government service. Collaborative activism largely influenced by non-government organizations (NGOs), foundations, private enterprises, educational institutions, and newsmedia. Included are scientists who frequently banded together in professional groups, often exercising constitutional rights as private citizens in order to exhibit an *alter ego* in public affairs. Because the database for this book is biased by personal experience, suitable recognition might not be given to some individuals and organizations. This survey is more evaluative than comprehensive.

---

[*] an irresistible force

Disinformation, propaganda, and ideology played a big role in sustaining the nuclear arms race. Dangerous action-reaction cycles were implemented by headstrong government leaders and accepted by a passive public. Vested interests exploited and distorted the dangers, using secrecy and the nuclear mystique to advance the arms buildup.

Militarization of the political process has left key questions. Did the end of the superpower confrontation eliminate the risk of mutual destruction, the mindless official secrecy, or tight government control of personal lives? How do we keep the real threats and the perception of danger from continuing to haunt the new millennium? Who or what will motivate the drive to recognize and remedy the residual environmental consequences?

Public engagement surely was key to tempering some excesses of the nuclear arms race. That kind of vigilance and pressure, from outside the military-industrial-laboratory-legislative quadrangle, was (and still is) needed to provide institutional memory and prevent retrogression back to a state of increased mutual danger. Although it is difficult to accurately assess protest effectiveness, relevant activities are recounted.

Periodic episodes of public concern and activism punctuated the years following the atomic bombings of Japan. An initial debate in and outside the government was about institutional mechanisms for controlling atomic energy. But in the 1950s and 1960s, with nuclear fallout from atmospheric testing becoming the focus of public attention, possible health and environmental effects became a prominent issue; worldwide public protest culminated in the Limited Test Ban Treaty. In later years, nuclear war emerged as a major political issue, and anti-nuclear weapons movements burgeoned in the United States and Europe.

After the Korean War started, a surge of right-wing nationalism—reinforced by the fear of worldwide communism and influenced by the McCarthyite accusations—put the peace movement on the defensive. By boosting the pervasiveness of government secrecy, McCarthyism had a lasting and depressing impact on science and scientists. Nevertheless, public opposition to atomic weapons was catalyzed by laws in 1955 requiring citizen participation in air-raid drills. The first anti-nuclear demonstration in 1957 at the Nevada nuclear test site was initiated by the Committee for Non-Violent Action.

By 1960, new peace organizations, such as the Student Peace Union and Women Strike for Peace, became active. The civil rights movement (including Students for a Democratic Society [SDS], the Fellowship of Reconciliation [FOR], and the Congress of Racial Equality [C O R E]) broadened the peace movement to embrace fundamental social changes.

The Vietnam war brought out congressional critics like Senators Eugene McCarthy, Wayne Morse, and William J. Fulbright, and the war stimulated opposition from many individuals and groups, with the addition of especially vocal new-left radicals.

Disillusioned ex-military gathered together as Vietnam Veterans Against the War (VVAW)—including organizer John F. Kerry, who eventually was elected to the U.S. Senate and later became a candidate for president. Kerry's road to

Washington started with winter 1971 VVAW "hearings" in Detroit that elicited unsworn testimonials from Vietnam veterans about atrocities purportedly committed by American military (at a time when the Mi Lai prosecutions were underway). Long before Kerry became a Senator, his report at a U.S. Senate hearing in April 1971about alleged atrocities has been considered a defining moment that helped galvanize protests against the war. Government authorities never publicly challenged the allegations.

The Indochinese conflict also became a vehicle for criticism of society and what seemed to be a war-tolerant generation. A convergency of numerous peace and anti-Cold War organizations opposed the incursion into Vietnam by the West.

Some organizations spawned violence-prone factions, like the Weathermen that split from the SDS. The Black Panthers and the Black Liberation Army symbolized and acted on racial and societal issues intertwined with what they saw as war-making and civil repression actively or passively conducted by the "establishment." Mainstream peace organizations recognized injustices of the "system" and sympathized with grievances of the radicals, but less militant and more representative groups did not subscribe to the violent means advocated or exploited to bring about change.

As the war in Vietnam came to a close, activists — recognizing the broadly oppressive nature of the arms race — began to challenge the military budget and the concurrent neglect of human needs, such as education, health care, and housing for the poor. Other issues that began to attract renewed attention were grass-roots democracy, ecological awareness, social justice, equal opportunity, government decentralization, feminism, and human diversity.

Anti-nuclear movements, over time, emphasized six underlying themes about nuclear weapons and the arms race:
- immorality of targeting civilians,
- illegality under international law,
- dangers of nuclear instability,
- distortions of domestic and international policies,
- erosion of democracy, and
- self-destructiveness of becoming a nuclear target.

The movements also benefitted from the expertise and support of independent-minded scientists who were able to understand and interpret complex, often exaggerated or misunderstood, technical claims of the government.

In order to present the broad spectrum of public participation, this Chapter has been divided into four Sections. The array of *public-interest organizations* and entities — national and international — is covered in the first Section. Next, a number of hot-burner *arms control issues* has been picked because they induced prominent public dispute. Countering or sometimes abetting the public role were various organized *political forces*, both East and West, described in the third Section. An ominous aspect of public dissent was the degree of *governmental intimidation* that was invoked, and this is reviewed in the fourth Section of this Chapter.

A. PUBLIC-INTEREST ORGANIZATIONS

Individuals alone, with a few notable exceptions, usually have had little impact on political matters; but, banded together in public-interest organizations and networking with others, conscientious people did affect the political calculus, and thus had a major role in moderating Cold War excesses.

The number of lines of text allocated to any individual, group, or organization should not be interpreted as an allotment proportioned exactly with influence, importance, or impact. While striving for balance, sometimes the coverage has more to do with information in possession, our a personal vantage point during the events, or even an admittedly unpurged remnant of partiality.

Organized opposition to frightening excessiveness originated from two directions: knowledgeable nuclear scientists — many of whom had helped to develop the first nuclear weapons — and members of the traditional peace movement.

After World War II, nonviolent direct action — sometimes civil disobedience — became a prominent tactic for opponents of war, racism, and civil injustice.

The day after Hiroshima was destroyed, John Simpson, a physics professor at the University of Chicago and a group leader in the Manhattan Project, convened a meeting in which scientists discussed practical ways of informing the public and policymakers of the perils and opportunities that lay ahead. Among those who participated were Leo Szilard, James Franck, and Eugene Rabinowitch, all of whom worked on the Manhattan Project at the "Met Lab," as the Chicago operation was known.

A month later, the Chicago group jelled into the granddaddy of nuclear-scientists' public-interest organizations, the Atomic Scientists of Chicago. Simpson, was a key founder and the organization's first chairperson. Within a year they commenced publication of the *Bulletin of the Atomic Scientists*. During that formative period, the Chicago scientists united with like-minded groups at Los Alamos, Oak Ridge, and New York to form the Federation of Atomic Scientists (FAS — later renamed the Federation of American Scientists). The FAS and the Atomic Scientists of Chicago became the core of the "scientists' movement." Chicago's Chapter of the FAS eventually settled at Argonne National Laboratory, the Met Lab's successor in the southwestern suburbs of Chicago.

Szilard, who had proposed that the atomic bomb be harmlessly demonstrated to Japan before being dropped on a city, encouraged the scientists at Chicago to consider what to do with the new technology after the war. To get ahead of issue, he helped form in 1946 the Emergency Committee of Atomic Scientists, which — with the able enterprise of Bernard Feld, John Simpson, and Willie Higinbotham

— eventually led to the FAS. Acting on a freelance basis, Szilard managed to develop a personal and unprecedented dialog with Nikita Khrushchev. Szilard was practicing what he preached: that "intellectual preparation through informal private discussions" was needed before government negotiations would be anything but "pedestrian."

As a result of dissatisfaction with emerging government policies on peaceful uses of nuclear energy, many scientists stimulated their professional organizations to take an active public-interest role. Proposals to control nonmilitary applications and proliferation of nuclear technology originated with individual, private-citizen scientists. Niels Bohr, for example, proposed what became the Acheson-Lilienthal plan. Bertrand Russell and Albert Einstein (just a few days before he died) joined in the famous 1955 Russell-Einstein Manifesto.

The Manifesto, which gained the signatures of many internationally known scientists (including Academician Topchiev, vice chair of the Soviet Academy of Sciences), inspired others to gather and discuss such issues as the dangers of nuclear testing. Russell was a leader in launching the movement against nuclear weapons; in 1961 he also participated in direct action, getting arrested in a sit-down at the London Ministry of Defence.

The Pugwash movement can be traced to the Manifesto's call. Collaborating with the like-minded British Atomic Scientists' Association, the FAS in 1955 cosponsored with the Association of Parliamentarians for World Government an international conference that endorsed the Russell-Einstein Manifesto.

The Atomic Scientists Association had been set up in March 1946, but because its own officers were close to the British nuclear "establishment," the association was relatively ineffective in influencing policy, and it formally dissolved in 1959.

A partial list of public-interest and related organizations has been compiled by Gerald Holton, who has been setting up a database and studying the history and diversity of scientific activism.

This ledger of public-interest groups runs the gamut — domestic and international — from the vigorous activities of individuals to the unified pressures of organized groups, systematic programs, wealthy foundations, *ad-hoc* networks, and social institutions of modern nations. This report is necessarily influenced by personal experience and the views of the cited sociological and historical analyses.

NGOs

Through direct action and networking, some non-governmental organizations, empowered by a resolute constituency, stepped up to fill a void in public-interest advocacy and information exchange. NGOs were able to accomplish more than individuals could on their own, and they often developed expertise and standing that gave them some influence on the conduct of policy. A particular strength of public-interest organizations was their access to sympathetic experts — who were sources of competent opinions in such fields as technology, policy, and law, or who could be called upon to assist in developing independent assessments, innovative positions, or useful strategies.

Most of the NGOs were initially funded by individual contributions and subscriptions, but eventually they gained foundation support. Much of the leadership and staff effort was volunteered freely or at very low wages, still a hallmark of cause-committed workers.

Grassroots political action, a *modus operandi* of NGOs, intensified with improvements in communications and information technology. Mass mailings, leafleting, fax machines, pirate radio stations, international e-mail, and (eventually) commercial imaging satellites were exploited by the NGOs.

With post-Cold War expansion of the Internet, NGOs have found it inexpensive and productive to develop sites on the World Wide Web. Along with the ubiquity of cell phones, the strength and timeliness of networking and membership communication have increased — and since extended even into third-world nations.

Several already-existing, multipurpose organizations eased into early involvement with arms control problems; others were *ad-hoc* creations to concentrate on Cold-War-related issues; and a few were formally instituted for specific, long-term engagement in these newer 20-century problems.

Many public groups were impelled to vociferously oppose government policies. Among the earliest peace-action organizations were the United World Federalists, founded in 1947, dedicated to "world peace through world law." In the late 1940s, their president was Alan Cranston, who became a U.S. Senator and worked for arms control for the remainder of his life.

The National Committee for a SANE Nuclear Policy, founded in 1957, raised objections to the growth of nuclear weapons, emphasizing the possible health consequences of radiation fallout from bomb tests.

At academic institutions in the late 1960s and early 1970s, peace activism became noticeable. United Campuses to Prevent Nuclear War was established. The FAS and the newly founded (1969) Union of Concerned Scientists — both still-active national organizations — became instrumental along with other groups of scientists in expanding and educating the anti-nuclear-war movement. Pacifist gatherings such as American Friends Service Committee, Fellowship of Reconciliation, War Resisters League, and Women's International League for Peace and Freedom also became very much involved.

While most NGOs favored arms restraint, a few did not; the subsection following NGOs, entitled "Programmatic Organizations," identifies some opponents of arms control.

In the pages that follow are vignettes of relevant NGOs:

FAS. The Federation of American Scientists has been a leading advocate of nuclear arms control. Many original members had contributed to the development of the atomic bomb. They wanted to encourage the evolution of nuclear energy toward peaceful, rather than military purposes, while addressing the implications and dangers of the nuclear age. The FAS is the oldest organization dedicated to the goals of ending the worldwide arms race, avoiding the use of nuclear weapons, and achieving complete nuclear disarmament.

The FAS was and continues to be a strong advocate of open communications among international colleagues. In late 1945 it established a Committee on Foreign Correspondence, later sending packets of information to foreign scientists, including many in the Soviet Union.

The FAS headquarters is in Washington, D.C. Funded by membership dues and foundation grants, it is a nonprofit, policy-oriented organization whose Board of Sponsors includes over Nobel laureates in the sciences. With a dedicated staff, the FAS conducts analysis and advocacy on science, technology and public policy, including national security, nuclear weapons, arms sales, biological hazards, secrecy, and space policy. The organization has acted as a knowledgeable, independent voice in debates about the science and technology of global security.

Throughout FAS history, its publications and projects have addressed a wide range of science-and-society issues, including those of population, energy, agriculture, medical care, and ethnic conflict (such as within Cambodia, Peru and Yugoslavia). FAS combines the scholarly resources of its scientists with a knowledge of practical politics. Being organized for the public interest, the Federation has brought a scientific perspective to the legislative arena through direct lobbying, membership and grassroots work, as well as expert testimony at Congressional hearings.

The organization was administered for many years by Jeremy J. Stone, who received his early education in arms control at the Hudson Institute, working with Herman Kahn and Donald G. Brennan. Stone tried to use personal diplomacy in an effort to bring the United States and the Soviet Union together. Before becoming head of the FAS in 1973, he was independently active in elaborating the case against missile defenses, through publications and presentations — so much so that Stone made some enemies among Soviet officials; they tried to discourage his contact with the USSR by suggesting that he was a CIA agent.

The FAS Web site provides valuable and comprehensive information on arms control. Added has been their purchase and display of Ikonos satellite images of previously mysterious nuclear locations around the world. Because of rampant government secrecy, efforts to understand or challenge official policies have frequently been hampered, but recent public access to detailed satellite pictures has made it more difficult for governments to disguise clandestine or embarrassing activities. Official pronouncements about international military threats and potential arms control violations can now be confronted.

Several FAS staffers have become prominent experts in the arms control field. John Pike, who established his credentials with the FAS, was widely consulted for independent information and analysis on arms control agreements, intelligence activities, and space policy. Stephen Aftergood has been one of the few people in the nation with expertise on secrecy in governmental institutions. Other staffers, working at times with longtime members like Frank von Hippel, have dealt with nuclear arms control, nonproliferation, and disarmament. Members who have had leadership roles on these issues include Matthew Messelson, William Higinbotham, and Theodore B. Taylor, and Barbara Hatch Rosenberg.

The FAS has been influential in many public debates on arms control developments, such as ballistic-missile defenses, civil defense, and MIRVs. Daniel Patrick Moynihan credits the FAS with being one of the first to advise that there was no strategic need in the 1960s to imitate the Russian preference for large land-based missiles. Subsequently the United States chose to develop missiles that were smaller, less costly, and more accurate.

In early stages of the Cold War, the FAS was primarily an organization of local chapters, rather than the District-of-Columbia-based unit that it later became. Centralizing the organization at the nation's capital and hiring full-time staff specialists kept the FAS effective for more than half a century.

In cooperation with other NGOs, the FAS established a number of collaborations with Soviet scientists at a time when communications were at a low point.

SANE. The Committee for a Sane Nuclear Policy was created in 1957 to promote the end of nuclear testing — Lenore Marshall, Norman Cousins, and Clarence Pickett being among the founders. The National Student Council for SANE was organized the next year. In 1960 rallies calling for an end to the arms race were arranged in New York and New Jersey.

SANE soon shifted to encouraging arms control. Dr. Benjamin Spock became a prominent sponsor. As the Vietnam war persisted, SANE started to challenge militaristic policies. At one point SANE had about 150 offices nationwide, and it became politically active in electoral campaigns. SANE also linked peace and civil rights into its programs.

During its most vocal, activist span, SANE challenged many of the Pentagon's arms programs, including nuclear testing, ABM, B-1 bomber, neutron bomb, MX, SDI, and ASAT.

In 1986, the Nuclear Freeze movement merged with SANE, first to combine nuclear disarmament issues with a commitment to nonintervention and social justice. This morphed in 1992 to become Peace Action, which continued to lobby against nuclear armaments. Peace Action was the largest grass-roots peace and disarmament group in the United States, reporting more than 60,000 members belonging to 125 chapters.

One high moment in SANE history was in 1985 when its action coordinator David Cortright, with the Rev. Jesse Jackson, personally presented to the new Soviet leader, Mikhail Gorbachev, a petition of 1.2 million names calling for a ban on nuclear testing.

Pugwash. The Pugwash Movement has been a long-standing international discussion group among scientists. Its first meeting was in 1957, financed by Cyrus Eaton, a Canadian-American industrialist, who requested that it be held in his birthplace, the Nova Scotian village of Pugwash. The meeting was stimulated by the 1955 Russell-Einstein manifesto, which called upon scientists of all political persuasions to assemble to discuss the threat posed to civilization by the advent of thermonuclear weapons. In addition to the general mood of East-West mistrust and fear, radioactive fallout from atmospheric tests was a source of major anxiety at the time.

At a time when official talks were difficult, Pugwash was one of the first Western groups to gain participation from Soviet scientists. Their first meeting was an historic encounter between eminent scientists from East and West because nuclear and strategic issues were then extremely sensitive.

Leo Szilard, Bernard Feld, and Eugene Rabinowitch played pivotal roles in initiating that first Pugwash Conference on Science and World Affairs, and for years afterward they were important participants. Guests were invited in their personal capacity, representing no one but themselves (although their observations and opinions were often solicited by government officials).

Joseph Rotblat was a cofounder who received considerable recognition, including the 1995 Nobel Peace Prize. He and the Pugwash Conferences jointly shared the prize. Pugwash has continued to be an active forum, for participants, especially scientists, from a spectrum of nations. No doubt the various studies and reports emanating from Pugwash meetings profoundly affected thinking about what was possible to accomplish in arms control.

The Nobel award generated some harsh criticism in the United States by uncontrite hardliners, such as Frank J. Gaffney, Jr., a Reagan-administration official quoted in the *New York Times*: "[Pugwash cannot] be seen as anything other than, at the very least, an unwitting tool of the Kremlin, if not worse," but "[I will give Pugwash participants] the benefit of the doubt and think they were simply dupes."

Although Rotblat worked on the atomic bomb at Los Alamos, under assignment from the British government, he twice underwent changes of conscience. The first was when he withdrew from the Manhattan project, having found out that the Germans had not succeeded in developing an atomic explosive. The second change occurred more slowly: he now thinks he was wrong in participating in the development of nuclear weapons, because deterrence only works with rational people (and Hitler could not be considered rational).

Council for a Livable World. Founded in 1962 by Leo Szilard in the aftermath of the Cuban missile crisis, the Council for a Livable World became one of the first political action committees working on nuclear arms control and disarmament issues. Szilard himself had major influence on the creation of the first nuclear bomb and on subsequent efforts to control and outlaw nuclear weapons. He conceived the idea of a nuclear chain reaction in 1933, co-designed the first nuclear reactor with Fermi, and was instrumental in getting the Manhattan Project launched. But when Germany was defeated, Szilard tried to stall and prevent the use of nuclear weapons on Japanese cities; subsequently, he worked for international control of civilian and military uses of nuclear fission.

The Council for a Livable World, originally the "Council for Abolishing War," has operated as a traditional political-action committee by endorsing and contributing to political candidates. It also tracked congressional voting records on defense issues and lobbies members of Congress. During the arms control debates of the late 1960s and early 1970s, the Council supplied literature about various government weapon systems, an example being *abm — Point of no return?*, a pamphlet critiquing the Nike-X ABM system.

Leo Szilard died of a heart attack in 1964. Successor as executive director of the Council was John Isaacs, especially energetic and influential in criticizing ballistic-missile defense schemes and in promoting the elimination of nuclear weapons.

NRDC. The Natural Resources Defense Council (NRDC), especially through its Washington, D.C., office, has been a major NGO player in arms control issues.

Originally constituted to defend the environment from excesses of human activity, the NRDC decided that the nuclear arms race constituted the most urgent and menacing environmental threat of all.

For more than 25 years, NRDC has been active in helping to shape U.S. nuclear nonproliferation, arms control, energy, and environmental policies. They also have influenced the application of environmental laws to U.S. nuclear and national security programs. Their overarching goal has been "the reduction, and ultimate elimination, of unacceptable risks to people and the environment from the exploitation of nuclear energy for both military and peaceful purposes." The NRDC uses law, science, and the support of a large nationwide membership. Headquarters is in New York, and Washington has one of three regional offices. NRDC's *Nuclear Weapons Data book* series, assembled primarily by Thomas B. Cochran, William M. Arkin, and Robert S. Norris — beginning in 1984 — has been the standard public reference on nuclear armaments and facilities in the United States and the USSR/FSU.

Collaborating with the FAS, Cochran and Christopher Paine have were personally involved in joint verification projects with Soviet scientists, and they were active in bringing to Capitol Hill an independent perspective on arms control. When Mikhail Gorbachev rose to power, opportunities for cooperative projects with the West and with NGOs became far more promising. Frank von Hippel, Roald Sagdeev, and Evgeny Velikhov exploited this opening, working with Cochran. Some of the most remarkable private-sector ideas — seismic monitoring of underground nuclear tests, and measurement of gammas and neutrons emitted from a Soviet nuclear-armed cruise missile — originated from this collaboration.

Cochran's most celebrated initiative was to get University of California seismologists to set up seismometers in Kazakhstan, where they detected a U.S. nuclear test explosion that took place in Nevada. Later the NRDC participated in the joint FAS/CSS project on verification, and they organized a precedent-setting experiment in the Black Sea.

An NRDC/FAS team was among the first outsiders to visit the Soviet ballistic-missile-defense site at Sary Shagan and the nuclear-materials-production complex at Kyshtym. With the help of the Soviet Academy of Sciences, in July 1989 the team was allowed into these formerly closed facilities. They were given briefings on the activities therein, with unprecedented permission to take photographs. They were shown that the laser facility at Sary Shagan in Kazakhstan was used to evaluate aircraft and satellite tracking capabilities, and did not appear to have the higher power ASAT lasers that Westerners feared. The Kyshtym nuclear complex known as Chelyabinsk-40 had five graphite-moderated water-cooled production reactors, three of which were still operating, as well as a plutonium separation plant. Western intelligence agencies were eager beneficiaries of this first-hand information, never before available to them.

> **[Our First Meeting]** (*Simonenko and DeVolpi*): Two of the co-authors of *Nuclear Shadowboxing* first met in Moscow in 1989 in connection with the collaborative NRDC/FAS project. Vadim Simonenko was in charge of the theoretical division at one of the Soviet nuclear weapons labs (Chelyabinsk-70), and Alexander DeVolpi was professionally responsible for an arms control and nonproliferation program at an American non-nuclear weapons lab (Argonne); both were attending unofficially as members, respectively, of the Committee of Soviet Scientists and the Federation of American Scientists.
> The group also visited Kiev to discuss with Ukrainian participants and officials the removal of nuclear weapons still in the Ukraine.
> One personal memory clearly recalled — symbolic of gradually improving U.S.-USSR relations — is of weapons-expert Vadim extending an arm to help the less ambulatory verification-technologist Alex across the slippery ice-covered walkways in Kiev.

The NRDC team was active in challenging the usual mantra of "effective verification," so often invoked in the past to ward off arms control agreements.

Paine and Cochran were influential in encouraging timely Nunn-Lugar legislation to help the FSU dismantle its warheads. This activity was an outgrowth of meetings, cosponsored by the FAS, at which it became obvious that breakup of the FSU meant that the nuclear arsenal was no longer securely under the control of a central government.

They also raised timely questions and suggested remedies about conditions and incentives for unsafeguarded exports and the fear that highly trained nuclear engineers and technicians might emigrate and sell their skills. Paine and Cochran pointed to the unprecedented opportunity for cooperation to verifiably eliminate the vast bulk of the world's nuclear weapons.

The NRDC has continued to scrutinize the nuclear-stockpile stewardship program of DOE. At their web site has been a National Ignition Facility and Science-Based Stockpile Stewardship Resource page, "an archive of documents and related NRDC commentary intended to promote public awareness of both the U.S. Department of Energy's nuclear 'Stockpile Stewardship' strategy and the National Ignition Facility, a laser facility under construction that constitutes a major component of the department's strategy." Faulting a lack of external scientific peer review, the NRDC has found "serious problems with the National Ignition Facility [at] Lawrence Livermore National Laboratory in California, including enormous cost overruns as well as the likely inability of the project to achieve its primary mission — fusion ignition."

UCS. The Union of Concerned Scientists was founded in March 1969 by faculty members and students at the Massachusetts Institute of Technology who were disconcerted by the misuse of science and technology in society. The UCS was formed from a local FAS Chapter that broke off from the national federation, which had failed to take a proactive stand against the use of science and technology in the U.S. war against Vietnam.

One of the founders of the UCS was Henry Kendall, professor at MIT, who died in 1999 — nine years after he had been awarded the Nobel Prize in Physics. Kendall had been chairman of the UCS since 1973, and brought the organization into deep involvement with arms control and nuclear safety issues, with enlightened opposition to the SDI, and with increased awareness of global-warming risks. Kurt Gottfried, a graduate of MIT and Professor Emeritus at Cornell, was also a cofounder of UCS.

Organizing, coordinating, and pamphleteering members of the UCS were in the forefront of the March 4 (1969) audacious national work stoppage at campuses and laboratories in connection with the ongoing debate about ABM and other defense programs.

The UCS grew to become an independent nonprofit alliance of concerned citizens and scientists across the country committed to citizen advocacy for making a cleaner, healthier environment and a safer world.

SESPA. Created in 1969 at an annual meeting of the American Physical Society, Scientists and Engineers for Social and Political Action, also known as Science for the People, has ceased to exist. It was a voluntary association that was engaged in a wider ideological spectrum than most scientists' organizations. Martin L. Perl (later a Nobel Prize winner) and Charles Schwartz organized SESPA and urged the American Physical Society to conduct sessions on politically charged national-security issues. Following a debate on ABM in 1969, SESPA convened about 250 physicists to march to the White House and to call on members of Congress.

A popular speaker on behalf of SESPA was Dr. John Froines, a chemistry-physics professor from the University of Oregon, who had become one of the notorious Chicago Seven indicted under the anti-riot act for their raucous activities at the 1968 Democratic convention.

The organization, including scientists like Perl, Michael H. Goldhaber, and Schwartz, joined in the March 4 activities and subsequent Anti-Vietnam-War Days. Schwartz was the author of a pledge by engineers and scientists against participation in war research or weapons production.

PSR. The Physicians for Social Responsibility is an important example of an organization centering about a professional cadre, in this case medical practitioners who were disturbed in particular by the war in Vietnam and more generally by the Cold War.

PSR was founded in Boston in 1961, and became reinvigorated in the 1970s, especially under the leadership of Dr. Helen Caldicott. In 1980 PSR went "transnational" when American and Soviet physicians met in Geneva to inaugurate its international counterpart, IPPNW.

During the nuclear-freeze debate, by organizing symposia in nearly every major American city on the medical effects of nuclear war, PSR hammered on the visceral effects of policies going amuck.

Although founded out of growing concern about the health effects of atmospheric nuclear testing, the fledgling organization initially focused its attention on nuclear power. The doctors' organization, however, "quickly shifted its focus to the threat of nuclear war," because, explained one of its leaders,

> The more we looked into the nuclear dilemma, the more evident it became that the real problem was nuclear weapons.

(In the course of time, nuclear scientists helped medically trained PSR members and chapters to understand fundamental distinctions about comparative risks and benefits of radiation and nuclear energy.)

The number of local PSR chapters grew to more than 150 in 1985. Much of its credibility and success came about because of official sponsorship by prominent medical schools and state medical societies.

In sustaining its activities beyond the Cold War, PSR settled into a 15,000-member organization that has a public-interest nuclear and security program. Continuing issues include the environmental impact of nuclear weapons production sites and the futility of the proposed national missile defense. PSR favors abolition of nuclear weapons and has produced for public use a color brochure on the topic. Recognizing "that nuclear war continues to be the most acute threat to human life and the global biosphere," PSR has reaffirmed "its commitment of nearly forty years to the elimination of nuclear weapons and the reversal of the arms race and the national budgetary priorities which fuel that race, sacrificing our nation's health, social and economic needs."

CAS. The Concerned Argonne Scientists at Argonne National Laboratory was a local but very active Chicago-area organization of mostly scientists and engineers. It convened about the same time and for same reason as the MIT-based UCS.

The Argonne FAS group had been the medium for intercession in the ABM siting dispute of the late 1960s. Some members wanted to participate in the raging Vietnam-moratorium issue, but the FAS feared that its effectiveness would be diluted if it took an aggressive stand on social or political issues outside their area of expertise. As a result, the parallel CAS organization was organized in a collegial 1969 arrangement.

The CAS was one of the two-dozen institutional participants in the original March 4 "work stoppage" and protest against misuse of science for military purposes. Members of both the Argonne FAS and CAS chapters were volunteer speakers in mid-West academic institutions on that 1969 day.

During of the second year of nationwide March 4 activities and work stoppages, the Chicago Chapter of the FAS held a meeting in which it discussed efforts to bring new programs and initiatives to Argonne, the idea being to get the laboratory more involved in nonmilitary issues related to arms control. On the same day, to highlight a collegial, but contrasting view, the CAS had a parallel gathering as part of the national observance, with a discussion on the impact of science and technology on society.

Directly resulting from CAS activities and the interests and experience of the members through the 1970s, an official program in verification analysis and technology was created at Argonne in the early 1980s, funded by government agencies.

One CAS project became a definitive published analysis of technical means for verifying a possible agreement banning ASAT weapons. CAS members contributed their technical expertise to FAS and NRDC verification projects in order to compensate for the verification-averse U.S. nuclear weapons labs. The CAS engaged in joint undertakings with many Chicago-area activist organizations, such as NOMOR and Defense Priorities, and issued statements on such nuclear issues as MIRV, neutron bombs, B-1 bomber, MX, nuclear winter, and SDI.

All these weapon systems, especially MIRVed multi-warheaded missiles, come up again and again in this book when discussing controversial issues that ultimately involved the public.

ACA. The Arms Control Association, founded in 1971, is another of the many organizations that reacted to the threatening state of international affairs in the late 1960s and early 1970s. Original members included many individuals, in or out of government, who were active in national-security problems.

Herbert ("Pete") Scoville (1915–1985) was one of the founders of the ACA. Bolstered by his doctorate in physical chemistry, Scoville had a significant role in development of photo-reconnaissance at the CIA, and thus was familiar with Soviet military development and the beneficial potential of treaty verification.

ACA describes itself as a national nonpartisan membership organization dedicated to promoting public understanding and support of effective arms control policies. Through its public education and media programs and its magazine, *Arms Control Today*, ACA provides policymakers, the press, and the interested public with authoritative information, analysis and commentary on arms proposals, negotiations and agreements, and related national-security issues. In addition to holding regular press briefings on major developments, the Association's staff has provided commentary and analysis on a broad spectrum of issues for journalists and scholars both in the United States and abroad.

CDI. As a private, non-governmental research organization, the Center for Defense Information is an independent monitor of military spending. It holds that strong social, economic, political, and military components and a healthy environment contribute equally to the nation's security. According to their web site, "CDI seeks realistic and cost-effective military spending without excess expenditures for weapons and policies that increase the danger of war. CDI supports adequate defense by evaluating our defense needs and how to meet them without wasteful spending or compromising our national security."

CDI was launched in 1973 in Washington, D.C., by Rear Admiral Gene R. LaRocque. The Center is financed by voluntary, tax-deductible contributions from individuals and grants from foundations, accepting no funds from the military or from military contractors. Today, CDI is one of the foremost research organizations in the country that analyze military spending, policies, and weapons systems. Retired military officers and experienced civilians work together to examine military and security issues and inform the public, providing an independent and expert source of facts and alternatives.

CDI has earned a reputation as an authoritative and impartial monitor of the military, continuing to fulfill the public's increasing demand for information and independent ideas, free from the influence of the military-industrial-laboratory establishment. Congress, the media, and the public can access CDI's military analysis through publications, television and radio programs, as well as the Internet. CDI has published a comprehensive Military Almanac, which supplies members of Congress with independent facts and information regarding military matters. *The Defense Monitor,* their monthly bulletin, has been distributed for 30 years and is widely cited, including within this book.

In keeping with the prevailing anti-warfare mood, CDI distributed a "Nuclear War Prevention Kit" and offered a film, "War Without Winners" in 1980. The flyer stated,

> We are the most powerful nation on earth.... Yet we can be destroyed with less than 30 minutes warning — so can the Russians. Across the nation, citizens are asking "What can I do?"

CDI's responding three-part, 28-page kit included suggestions for citizen action, reminding that President Kennedy once said "The weapons of war must be abolished before they abolish us."

NOMOR. Using an acronym for <u>n</u>uclear <u>o</u>verkill <u>mor</u>atorium, NOMOR was organized in 1976 by Chicago-area citizens pressing for a unilateral U.S. pause in the production and testing of nuclear weapons.

During the height of the nuclear freeze debate, NOMOR prepared and distributed its "Primer" on "You, the Nuclear Arms Race, and Survival: Nine Questions and Answers." They described the "problem" as "nuclear roulette," a nuclear arms-race game being played by governments with human survival as the stakes. NOMOR challenged the basic assumptions of the Pentagon's program, which were (1) that producing ever more nuclear weapons enhances *national security*, and (2) that the nuclear race or a nuclear war can be "won."

For individuals to help in stopping the arms race, NOMOR suggested joining their campaign — calling on the U.S. government to enact a unilateral moratorium on all production and testing — and requested individual help in building a national and international campaign on a global framework for immediate cessation of the arms race by all nuclear powers.

NOMOR routinely arranged for Hiroshima-day observances at the University of Chicago (which are kept up annually by a solemn Hyde Park citizens' group at the Thomas More bronze mushroom-cloud memorial). NOMOR carried out a vigorous and successful campaign that called on the Illinois General Assembly to support the Mutual Freeze Resolution. NOMOR ceased operations in 1986.

Mobe. Mobilization for Survival was a mixed-issue "environmentalist" group. It held that "nuclear energy and nuclear weapons are merely different sides of the same coin." Sidney Lens, a Chicago-area radical, is credited with drafting the original call for the organization in 1976. Concurrent with demonstrations as part of "International Action Week" against the neutron bomb and "teach-ins" at universities and colleges in August 1978, Mobe arranged more than 100 marches, demonstrations and sit-ins, mostly at military bases and nuclear power stations in

36 states. In 1979 Mobe held a protest with other "affinity groups" and war-resisters at the headquarters of the Department of Energy in Washington.

Nuclear Freeze Campaign. Although halting and reversing the arms race was a primary goal of NGOs, the idea of simply freezing the competition at the then-current levels seemed in the late 1970s to be politically more palatable. A formal National Nuclear Weapons Freeze Campaign began in 1979 when Randall Forsberg, Director of the Institute for Defense and Disarmament Studies in Brookline, Massachusetts, drafted "The Call to Halt the Nuclear Arms Race," a four-page statement outlining a bilateral nuclear weapons freeze strategy.

Major religious, civic, and political organizations that became early endorsers of the Freeze included the YWCA, the National Conference of Black Mayors; the national-board and social-issues offices of the National Council of Churches, the United Presbyterian Church, the Unitarian Universalists Association; as well as the Bishops and diocesan conventions within the Episcopal and Roman Catholic Churches. These and other organizations provided educational activities on the Freeze and actively promoted it. Many influential pastoral letters condemning the arms race were written and given to parishioners.

Designating "Disarmament Week" for late October 1981, the Nuclear Weapons Freeze Campaign called on local organizations to create exhibits, show films, and hold lectures, press conferences, religious services, and teach-ins about the danger of nuclear war. The campaign also held a national "Call-In" on 26 October 1981, encouraging Americans to telephone the White House and urge President Reagan to propose a mutual freeze to Premier Brezhnev of the Soviet Union.

With the opening of a national office for the Nuclear Weapons Freeze Campaign in St. Louis on 2 December 1981, Randy Kehler became the first national coordinator. The national Freeze office in St. Louis acted as an information clearinghouse for thousands of similar Freeze groups around the country.

As the campaign took hold among activists, many groups around the nation began to band together. The Chicago Area Faculty for a Freeze on the Nuclear Arms Race issued its "Notice of Founding Meeting" on 19 January 1982, "To organize support ... for a freeze on all further nuclear weapons testing, development, production, and deployment." The North Shore Peace Initiative issued its "Call to Halt the Nuclear Arms Race," and other Chicago organizations made a collective request for air time on radio and TV. Robert Cleland, representing NOMOR, wrote in the *Chicago Tribune* that "It's time for a mutual arms freeze."

The Nuclear Freeze Political Action Committee, FREEZEPAC, was formed in April 1982. This bipartisan committee supported candidates for the U.S. Senate and House of Representatives who advocated a comprehensive and verifiable bilateral nuclear weapons freeze. During the 1982 elections, more than 50 percent of the FREEZEPAC-supported candidates won office.

In the spring of 1982, the Freeze Campaign reported that 20,000 volunteers were working in 140 offices throughout 47 states.

At their national conference in December 1984, the National Nuclear Weapons Freeze Campaign endorsed three nonviolent civil-disobedience actions: the Central America invasion contingency plan, an August "witness" at the Nevada Nuclear Test Site, and anti-apartheid demonstrations. Legislative priorities included ending the production of the MX missile and cutting off funds for weapons programs.

Kehler's files contain many documents about the movement to get the United States and the Soviet Union to adopt a mutual freeze on the testing, production, and deployment of nuclear weapons and missiles.

It is estimated that by 1985 the total number of local and national peace groups had increased to about 8000. In addition to groups connected with the Freeze campaign, others grew, such as Physicians for Social Responsibility, the Council for a Livable World, and SANE — the latter's membership growing from 4000 in 1977 to 150,000 in 1986. President Reagan's policies had instilled so much fear that it catalyzed widespread public protest.

Programmatic Organizations

For lack of a better term, "programmatic organizations" distinguishes entities that do not neatly fit into the category of *ad-hoc* public-interest NGOs. Some programmatic organizations had hawkish views toward national security, and they received generous financial backing from right-wing foundations. Strong-defense organizations often acted as incubators and waystations for potential and former government appointees who were biding their time during intervals between compatible presidential administrations.

Left-wing groups, though probably with far less in resources, counterbalanced the postures of right-wing groups. There were extremists on both sides of moderation.

While some think tanks were hardline supporters of what they called "strong" defense, others were long-standing liberal foundations that tended to support arms control. Nonpartisan public-policy organizations conceived by experienced professionals differentiated themselves from the think tanks. Single-issue extremists sometimes banded together in organizations, to advance their own agenda, often peripheral to the Cold War. Many religious groups became involved, and most were inclined toward moderation in defense issues.

Strong-defense organizations have usually been heavily funded by industrial interests. Allied with, or supportive of strong-defense groups has been the military-industrial-scientific faction. With overhead costs covered by the military services, many right-wing policy-analysis groups were able to subsidize an umbrella organization that would help rationalize the massive military build-up. In contrast, NGOs of all leanings tended to be rank and file groups with public subscriptions and limited foundation support.

Public-interest NGOs faced powerful opposition from well-financed hawkish organizations like the Center for Strategic and International Studies, American Enterprise Institute, Freedom House, American Security Council, Heritage

Foundation, Hoover Institute, Hudson Institute, and Cato Institute. The Committee on the Present Danger was an amply financed pro-defense organization with volunteer membership consisting of many former, then-current, or future government officials.

Carnegie. The Carnegie Corporation of New York is a grant-making foundation created in 1911. Via its Endowment for International Peace, Carnegie has funded many arms control endeavors, not the least of which was the FAS Cooperative Research Project on Arms Reduction. Since then, Carnegie's Nuclear Non-Proliferation Program has fostered numerous valuable studies. In the area of international peace and security it has worked largely on nonproliferation and post-Cold War developments in the FSU.

Carnegie's International Peace and Security program has had three formal research areas: Nonproliferation of Weapons of Mass Destruction; Russia and Other Post-Soviet States; and New Dimensions of Security. In recent years of grant-making, international events have served to underscore the importance of these program areas for global security. The Corporation's "Higher Education in the former Soviet Union" initiative is also administered under this program.

Publications from Carnegie's Nuclear Non-Proliferation Program are especially useful to policy makers and scholars. The journal *Foreign Policy* has a home page at the Carnegie web site, with many articles online. The Carnegie Moscow Center has online an extensive series of publications and excerpts from its new journal *"Pro et Contra."*

Brookings. Headquartered in Washington, D.C., Brookings Institution functions as an independent resource-incubator on a broad range of public policy. Founded in 1916, it is financed primarily through its endowment, supplemented by support from other organizations and individuals. Examples of some of its arms control studies and reports are *The Logic of Accidental Nuclear War* (1993), *Global Zero Alert for Nuclear Forces* (1995), *Atomic Audit* (1998), *The Nuclear Turning Point* (1999), and *De-Alerting Strategic Forces* (2003).

For the research it supports, Brookings functions as an independent analyst and critic, committed to publishing findings. Its conferences and activities serve as a bridge between scholarship and public policy, bringing new knowledge to the attention of decision-makers and affording scholars better insight into public-policy issues. Brookings was the first private organization devoted to public-policy issues at the national level. Its funds have largely been devoted to its own research and educational activities. Brookings also undertakes some government-contract studies, reserving the right to publish its findings.

The foundation has spawned independent specialists, such as Bruce Blair. Putting his background in applied mathematics to use in studying command and control, Blair found out that the entire nuclear retaliatory system was geared operationally to launch-on-warning. He asked himself rhetorically if the President could override this automatic response, and his search for answers to this important question has since led Blair (who took over at CDI) to become a leading expert on strategic command and control.

Also produced with Brookings support was *Atomic Audit,* a landmark scrutiny of Cold War nuclear weaponry, cited extensively herein.

Hudson. The Hudson Institute is a public-policy research organization that "forecasts trends and develops solutions." It was founded in 1961 by Herman Kahn and later based in Indianapolis, Indiana. Donald Brennan was its first president.

Funded by grants and donations, the Hudson Institute considers itself a "nonpartisan futurist organization" that "does not advocate an expressed ideology or political position." In 1969 the organization advocated deployment of ABM systems and in 1972 urged noncommital alternatives to S T A R T.

Many Cold War books produced by Hudson were written by Kahn, including *On Thermonuclear War.* The Institute garnered major government contracts to develop nuclear strategies, embracing scenarios on an outbreak of nuclear war and advocacy of civil defense.

Center for Security Policy. Modeled along the lines of a think tank, the Center for Security Policy was founded near the end of the Cold War by some prominent defense hawks of the Reagan administration. While not directly lobbying or funding candidates, the Center advertises that it "churns out a rapid stream of policy papers that are relied upon by conservatives in Congress." "Peace through Strength" is one of its highlighted mantras.

It is headed by the very vocal Frank Gaffney, an official in the Defense Department during the Reagan administration. "Advisory Council" members include Kathleen Baily, Robert Barker, Alan Keyes, Charles Kupperman, Henry Cooper, Douglas Feith, Richard Perle, and Fred Iklé.

Gaffney has been a vocal proponent of missile defense and, according to one columnist,

> has been characterized by his critics as the "messenger-in-chief" of a well-funded network of defense contractors, conservative foundations, and Reaganite true believers, for whom the necessity of a missile defense system is something akin to a religious conviction.

A protégée of hardliner Richard Perle, Gaffney has been unrestrained in his criticism of NGOs like the Council for a Livable World. He portrayed the Council head, John Isaacs, and his "like-minded friends" as

> ultraliberals who would foolishly disarm the United States in the face of danger. There are few belief systems to which people adhere more assiduously with less reason that than of arms control.

Gaffney also has said that such people are "[hopelessly naive to embrace] the utterly addled idea of ridding the planet of nuclear arms."

Henry L. Stimson Center. A relative latecomer, the Stimson Center was typical of independent, nonprofit, public-policy institutions involved in national-security issues. Formed by Barry Blechman and Michael Krepon, the Stimson Center (named after one of the foremost hawks of World War II) billed itself as trying to find and promote innovative solutions to security problems. The Center continued to address arms-related difficulties arising in the aftermath of the Soviet Union's breakdown.

Nuclear Control Institute. NCI has been an "independent research and advocacy center specializing in problems of nuclear proliferation." Its focus on activities of proliferant nations has provided a useful public source of information. However, because of an anti-nuclear-power emphasis, its blanket advocacy of eliminating what it indiscriminately considers to be "atom-bomb materials" has diminished its nonproliferation credibility.

Religious Associations. Religious groups — at clerical, denominational, and ecumenical level — brought a strong moral perspective into public protests against war and armaments. For instance, U.S. Catholic Bishops issued a pastoral letter on war and peace in 1983, addressing issues like "counterpopulation warfare," the initiation of nuclear war, limited nuclear war, and deterrence. To their congregations (the largest religious denominations in the United States), the bishops recommended support of arms control agreements and steps toward progressive disarmament.

The Reagan White House took the bishops' pastoral letter as a serious challenge to their nuclear buildup policies. Throughout the Bishops' drafting process, the administration attempted to influence or co-opt the thrust of the letter, enlisting the National Security Council, the military, the press, and partisan individuals into a fruitless attempt to diminish the bishops' message.

Some Protestant religious bodies that made pronouncements favoring restraint or reductions in nuclear arms included the United Presbyterian Church, the Episcopal Church, the United Methodist Church, the Unitarians, and the National Council of Churches.

Numerous community church, temple, and mosque congregations in many nations carried out their own moral evaluations of the Cold War and the struggles it spawned. One of the most active American multi-denominational groups was Clergy and Laity Concerned (CALC), which had its headquarters in New York City, and at one time had 37 local chapters in locales such as Buffalo, Chicago, Denver, Detroit, Louisville, South Carolina, and Tennessee. CALC was the object of extensive investigation by the FBI, now revealed after release of documents under the Freedom of Information Act. The FBI ascertained at the time that

> [CALC's] main purpose is the demilitarizing of the American way of life by way of education and direct action [and that CALC was an organization formed in the mid-sixties] to bring the Vietnam war to a sane and peaceful end.

FBI investigators confirmed that CALC:

> brings religious witness to bear on such policy and enables religious communities to become effective agents of such change.

Despite extensive surveillance and intrusive monitoring of the national office and local chapters, and despite casting innuendos about CALC's members and links, the FBI failed to find grounds for prosecution.

Churches were a prominent voice for ethnic minority frustration about domestic diversions caused by the arms race. African-Americans as a group strongly supported peace and nuclear disarmament, but their energies and commitments had to be focused "on the institutionalized violence of poverty and injustice within their communities." Through the Southern Christian Leadership Council, "black ministers played a prominent role in the peace movement." One of the most active African-American leaders was the Reverend Jesse Jackson, an early endorser of the nuclear freeze and a board member of SANE. Jackson's leadership role, on behalf of peace and nuclear temperance for the world community, ensured that African-American views were represented in the "predominantly white peace movement."

The First Presidency of the Mormon Church in 1981 opposed the MX basing decision (which included Utah), stating: "Our fathers came to the Western area to establish a base from which to carry the gospel of peace to the peoples of the earth. It is ironic, and a denial of that gospel, that in this same general area there should be constructed a mammoth weapons system potentially capable of destroying much of civilization."

Quaker affiliates, during the 20 century, continued their long-standing opposition to war. They often participated in local and national activities. In 1978 the Friends Committee on National Legislation reacted to President Carter's military budget for fiscal year 1979 by calling to public attention the negative domestic impact of the increased budget.

Activist/organizer/writer David Cortright recalls the American Friends Service Committee being in the forefront of NGOs presenting the nuclear-moratorium idea to Soviet officials in 1979, three years before the Kremlin adopted a modified version of the idea.

Cortright gives credit to religious organizations for casting a

> "mantle of respectability" over the peace movement and [giving] new legitimacy to discussions of disarmament.... The backing of the religious community made peace a mainstream issue and gave credibility and momentum to the disarmament movement.

Extremists. There were extremists on both sides of the Cold War divide. Often these were single-issue enthusiasts who were less concerned about nuclear armaments than their own peripheral crusades.

One of the most virulent right-wing organizations was the John Birch Society. Under the rubric of rejecting totalitarianism in any form, the organization had considerable influence on conservatives and hardliners. The Society has always been well funded. During his 1960 presidential campaign as a right-wing conservative, Barry Goldwater defended its goals and activities.

Many radicalized individuals threw their lives into opposing war (especially the one in Vietnam) and nuclear arms buildups. Outside of institutionalized churches, yet stimulating their conscience, were mavericks like the Berrigan brothers, Rev. William Sloan Coffin, and nuns that initiated Pax Christi.

Concurrent with opposition to nuclear arms and belligerency, militant extremists on the left carried out violent attacks on individuals, officials, and structures. Most left-wing radicals were directing their wrath at societal or government institutions and practices that were tangentially related to national military policies. The Symbionese Liberation Army (SLA) was targeting American police and law-enforcement individuals. The Red Army Faction and the Baader-Meinhof Gang, left-wing terrorists operating mainly in West Germany from 1968 to 1977, carried out political assassinations and other terrorist acts. They considered themselves Marxist revolutionaries who opposed capitalism and political injustice.

For the right-wing, hardliners were largely in seats of power during many of the more-acrimonious years, and zealots were entrenched in local law enforcement, police, FBI, and military-intelligence.

Just as some hardliners vigorously pushed for endless arms buildups and nuclear-use strategies, some extremists went so far as to oppose all things nuclear, whether for war or for economic growth. Anti-nuclear extremists, often with colorful antics, were usually able to get the attention of the newsmedia. Encouraged by prominent individuals — like Paul Ehrlich, Barry Commoner, Amory Lovins, and Ralph Nader — organizations such as Mobe, Friends of the Earth, Clamshell Alliance, and Greenpeace were vocal and visible in the front line of what was loosely described as the "Movement," also pejoratively labeled as part of the "New Left."

Ehrlich has been a well-intentioned environmental radical: one who recognized the real and potential ecological impact of the virtual conflict but has carried the sentiment to an extreme. As a doomsayer, in 1968 Ehrlich predicted that the world would undergo famines in the 1970s, with hundreds of millions of people (including Americans) starving to death. (Despite inequities in food distribution, production is still well ahead of population growth, and obesity kills many Americans yearly.) Ehrlich anticipated that smog disasters in 1973 might kill 200,000 people in New York and Los Angeles (but the air in those cities is getting cleaner, and the deaths have not occurred).

His predictions about scarcities of natural resources have not been any better: Reserves of key minerals and oil have been identified, while supplies, inflation-adjusted, became cheaper. Disdained by Ehrlich have been advancements in technology that are responsible for stretching out resources. Technology has made it easier to find new oil fields at lower costs, to extract more from those fields, and even to pump oil from fields once thought dry. As copper has become scarcer, industry has used new technology to switch to equivalent or even superior materials (e.g., glass fibers). Despite Ehrlich's admonitions about human-induced land degradation and crop shortfalls, corn production continues to increase all over the world, and progress in food genetics could dwarf even these increases. Perhaps he was too far ahead of the pack.

Some environmental extremists — like Amory Lovins — while vehemently opposing nuclear weaponry and the arms race, could not distinguish between the warlike aims of weapons and the peaceful applications of nuclear energy. Even more level-headed organizations, such as the Natural Resources Defense Council and the Union of Concerned Scientists, have chosen to cater to their anti-nuclear-power constituencies by taking a less-than-balanced approach to nuclear power.

Playing on public fears about radiation, anti-nuclear extremists have equated nuclear reactors with nuclear weapons. Although nuclear weapons were poised to destroy humanity, the "anti-nukes" were obsessed in their single-issue opposition to civilian nuclear power plants.

Despite myriad applications of radiation for medical diagnostics and other modern benefits, anti-nuclear extremists wanted to flush the baby (peaceful applications) down the drain with the dirty water (nuclear weapons). Even today, many such anti-nuclear extremists do not distinguish between peaceful and warlike nuclear roles. Nuclear reactors have been supplying copious pollution-free electricity for many nations — including France, Russia, Germany, the United States, and Japan. Some people with long-enough memories recall that insufficient natural resources were a major driving force for early-20-century Imperial Japan to embark on a war of aggrandizement in the East.

In democracies, irrationality and hyperbole often help candidates win votes. But winning an election campaign doesn't assure long-term peace and progress.

Soviet Moderates

Because of the Communist Party's tight and pervasive reign in the Soviet Union, there was little room for NGOs as in the tradition of the West. Instead, moderation in arms control had to be instituted either by government officials, working within the system, or by dissidents — outcasts from the *apparat*.

Among Soviet weapon scientists, there was common understanding of the need to reduce arsenals, while U.S. counterparts were eager to increase their size and sophistication. Many Soviet scientists privately opposed the increasing number of tests and armaments.

Some Soviet moderates were members of the scientific and defense establishment, with access to the highest level of decision-making authorities. They did not work through newsmedia or dissident public organizations. The moderates had to persuade authorities to accept their proposals, and after authorities made an official decision, the moderates usually accepted the decision without open reservation.

Essentially no overt or organized nuclear opposition surfaced in the Soviet Union. One reason was that intimidating methods of non-promotion or demotion were used: That was an effective means of pacification because all job opportunities were inside the government; no private employment was available.

Eventually some restrained public dissent arose, mostly on issues dealing with civil liberties and democratization.

Soviet Academy of Sciences. The Academy of Sciences was a prestigious organ of the Soviet government, and it included scientists who had independent knowledge of arms-race technicalities and risks.

Evgeny P. Velikhov, head of Kurchatov Institute, became vice-president of the Academy in 1977. He received the highest national award, Hero of Socialist Labor, and was accorded considerable respect within the Soviet military-industrial establishment. Some of these scientists, especially Velikhov, were sufficiently secure in their position to be able to express some independent, but guarded, opinions and were able to travel — occasionally outside the USSR.

Senior scientists such as Igor Tamm, Petr Kapitsa, Lev Artsimovich, Aleksandr Vinogradov, and Andrey Tupolov preceded Velikhov in developing a degree of independence for the Academy. They had a significant role in encouraging and tutoring other Soviet scientists in the public interest, and they helped bridge the superpower policy gap on arms control and disarmament, especially through interaction with a comparable U.S. National Academy of Sciences (NAS) Committee on International Security and Arms Control (CISAC). Talks between members of the two national Academies continued through the ebbing of the East-West conflict.

Velikhov was friendly to Roald Z. Sagdeev, director of the Space Research Institute, who became the youngest full member of the Academy. In 1982, informal talks began with arms controllers from both the Soviet and U.S. Science Academies, and Velikhov brought Sagdeev. Grouped with two influential social scientists, Evgeny Primakov and Georgi Arbatov, they were dubbed "the gang of four," and they later became closely associated with Mikhail Gorbachev in the early years of his ascendency to the Politburo.

The alarming and bellicose rhetoric and actions of Ronald Reagan, upon becoming President of the United States, can be most assuredly credited — if that is the correct word choice — with stimulating this informal protest of scientists, in the Soviet Union as it did in the West, all of whom fearing that Reagan's bellicosity had substantially increased the danger of nuclear war.

Committee of Soviet Scientists. The Committee of Soviet Scientists for Peace and Against the Nuclear Threat (shortened to CSS), a non-governmental group of senior Soviet scientists, was established in 1983 "for the purpose of studying the technical feasibility of disarmament agreements and discussing these questions with Western groups." The CSS became a key intermediary for the Soviet side in dialogs with Westerners.

Academicians Velikhov and Sagdeev were the primary organizers of the CSS. Formed during the early stages of Gorbachev's premiership, the Committee presented the boldest of unofficial Soviet efforts to bring in more moderate arms control policies. Hewing closely enough to the Party line, the activists were tolerated in carrying out approved activities with Western NGOs.

Velikhov and Sagdeev, through the CSS, accelerated the East-West side-channel dialog after Reagan's "Star Wars" speech in March 1983. Based on their own analysis, which indicated that the American SDI could be neutralized by much-lower-cost countermeasures, the Committee was successful in persuading Gorbachev not to launch a Soviet "Star Wars" program.

By having access to government leaders, Velikhov and Sagdeev were able to get Chairman Andropov to agree to a moratorium on testing Soviet ASAT. The Academicians also influenced Gorbachev's decision to begin a unilateral halt of Soviet nuclear tests, an initiative that was brushed off by the Reagan administration, which argued that the Soviet Union would test clandestinely. Velikhov then suggested that a non-governmental group set up seismometers in Kazakhstan; that lead to an extraordinary experiment carried out under the auspices of the CSS and the NRDC.

An even more remarkable accomplishment of the CSS occurred in 1989 when Velikhov was able to arrange the loan of a nuclear-armed cruise missile on a Soviet warship for use in a joint verification demonstration with an NRDC-organized U.S. team. American officials and hardliners had argued that it was impossible to verify whether SLCMs had a conventional or a nuclear warhead, but joint CSS-FAS technical analysis had concluded that radiation from the nuclear warhead could be detected externally. In July 1989, aboard the Soviet cruiser *Slava* in the Black Sea, the civilian teams indeed measured a characteristic nuclear-radiation spectrum of heavy-metal[*] gamma rays from the designated SLCM, thus confirming NGO expectations of verifiability.

Much credit for the unique opportunity aboard the Slava belongs to the Soviet Academy of Sciences and the Soviet Navy; no foreign scientists or officials had ever been that close to a Soviet nuclear weapon or been afforded an opportunity to make specific measurements. The SLCM verification demonstration can be credited with establishing a solid technical basis for future bilateral negotiations on warhead elimination.

Thanks again to arrangements by Velikhov, on the same trip Frank von Hippel and Tom Cochran (with three Congress-members and reporters from the *Washington Post* and *New York Times*) were able to visit Soviet sites previously out of bounds: one of these was Sary Shagan in Kazakhstan. Contrary to Western intelligence projections, the site only had low-power (100-watt) ruby lasers and a (20-kilowatt) CO_2 laser — not the megawatt-class lasers envisioned by DIA. The visit showed that on-site inspection could be an antidote to worst-case conclusions based merely on the external size of a building. The group also visited two Russian military plutonium-production reactors that had just been shut down, confirming their removal from service. Later Gorbachev embraced the idea of a ban on the production of fissile materials for weapons, a topic of long-continuing international negotiations.

The CSS was able to gain the attention of their government, spurring negotiation of nuclear-warhead elimination. The same techniques that were discussed in the joint project with the FAS and NRDC a decade earlier were being fitfully negotiated between the United States and Russia in 2001, but the G.W. Bush administration declined to enter into warhead-elimination negotiations.

Another joint activity was intended to demonstrate that ground-based high-power lasers, whether for military or civilian programs, need not be a threat to arms control and could be managed through provisions of a bilateral treaty.

Sakharov. Andrei D. Sakharov (1921-1989) was a Soviet physicist who was, in the words of the Nobel Peace Committee, a "spokesman for the conscience of mankind." Gradually Sakharov had become one of the most courageous critics of the regime and a defender of human rights and democracy. Because he became a dissident, he and his wife, Yelena G. Bonner, were exiled to Gorky in 1980 for five years.

---

[*]Heavy-metal is a term used for uranium and plutonium.

Sakharov was fascinated by fundamental physics and cosmology, but first spent two decades designing nuclear weapons. He came to be regarded as the father of the Soviet hydrogen bomb, contributing perhaps more than anyone else to the military might of the USSR.

In the central Volga region of the USSR there was a secret city known as the "Installation," where a special design bureau started creating nuclear weapons in 1950; some forty years later it was revealed to be Sarov, once the site of a famous Orthodox monastery. Tamm and Sakharov moved to the Installation in the spring of 1950. There they worked on Sakharov's *Sloika* design, which proved to be the first Soviet H-bomb, successfully tested on 12 August 1953.

In 1957, Kurchatov asked Sakharov to write an article denouncing American military development of a so-called "clean bomb" that would leave less radioactive debris. Some Western scientists had begun to warn about the dangers of radioactive fallout from any nuclear explosion, and passionate debate was under way in the United States and elsewhere. Ironically, Sakharov was responsible for carrying out the highest-yield above-ground tests. Using available biological data, Sakharov calculated that detonation of a one megaton "clean" H-bomb would produce enough radioactive carbon to have long-lasting global effects, predicting 6600 deaths worldwide over the next 8000 years (An overestimate; at the time, the simple "linear, no-threshold" interpolation of radiation damage had more credibility than it has today.

Sakharov's published papers, "Radioactive Carbon from Nuclear Explosions and Nonthreshold Biological Effects" and "The Radioactive Danger of Nuclear Tests," disputed the conclusions of many of his Soviet colleagues and of American weapon-specialist Edward Teller, who argued that tests of nuclear weapons were entirely safe. To Sakharov (given the then-current state of knowledge of the effects of low-level radiation) the death toll from nuclear testing in the atmosphere — however small compared to deaths from other causes — was simply a fact proved by science, with inescapable moral consequences.

Carl Sagan, sharing apprehensions about nuclear war, quotes Sakharov:

> A very large nuclear war would be a calamity of indescribable proportions and absolutely unpredictable consequences, with the uncertainties tending toward the worse.... All-out nuclear war would mean the destruction of contemporary civilization, throw man back centuries, cause the deaths of hundreds of millions or billions of people, and, with a certain degree of probability, would cause man to be destroyed as a biological species.

In June 1985 alarmed friends reported that the Sakharovs were "missing" from their home in Gorky. Following the election of Gorbachev and the setting in motion of perestroika, they were allowed to return to Moscow in 1986. In April 1989, Sakharov was elected a member of the Soviet Union's new parliament, the All-Union Congress of People's Deputies, becoming co-leader of the democratic opposition faction for a brief time before he died that year.

Although Sakharov's work was instrumental in creating the Soviet hydrogen bomb, he became "a courageous activist for peace and disarmament, as well as for human rights" because of his concern about the dangers of nuclear weapons and the threat of nuclear war.

Sakharov held as constants in his thinking: (1) nuclear deterrence is inescapable, (2) strategic parity is essential, (3) arms control negotiations are of primary importance, and (4) trust, developing from cooperation and openness, is a prerequisite for progress in peace.

He believed that

> nuclear war cannot be planned with the aim of winning it. Nuclear weapons cannot be viewed as a means of restraining aggression carried out by means of conventional weapons.

Sakharov felt that any use of nuclear weapons would amount to "collective suicide." These precepts contrast very strongly with RAND-led strategic thinking in the United States (and Great Britain). Moreover, Sakharov repeatedly said that "an effort to construct a protective shield against a massive nuclear attack would be both illusory and provocative."

American Funding Foundations

Some benevolent domestic foundations were more active in supplying funds than in having a staff involved with analysis of arms control or verification. One was the Ford Foundation which provided grants to specific public-interest organizations, such as the Soviet-American Disarmament Study Group. Here we mention three funds that specialized in arms control-related activities.

W. Alton Jones Foundation. Established in 1944 by W. Alton Jones, his objective was "to promote the well-being and general good of mankind throughout the world." Goals are to build a sustainable world (by developing new ways for humanity to interact sustainably with the planet's ecological systems) and to shape a secure world (by eliminating the possibility of nuclear war and by providing alternative methods of resolving conflicts and promoting security).

The Foundation, based in Charlottesville, Virginia, approximately two hours by road from Washington, D.C. was one of the supporters of the FAS Cooperative Research Project on Arms Reduction.

Ploughshares Fund. The Ploughshares Fund, founded in 1981, has been an organization "dedicated to preventing war in the nuclear age" and "stopping the spread of weapons of war, from nuclear weapons to land mines." Pooling donations from wealthy individuals, Ploughshares makes grants to projects designed to stop the spread of weapons of mass destruction (WMD), dismantle the worlds' existing nuclear arsenals, control international trade in conventional weapons, reverse environmental devastation caused by production of nuclear weapons, and prevent armed conflict.

John D. and Catherine T. MacArthur Foundation. The Chicago-based John D. and Catherine T. MacArthur Foundation has supplied grants to individuals and projects in the area of global security and sustainability. It has awarded funds to the FAS, the Lawyer's Committee on Nuclear Policy, the Monterey Institute of International Studies, CDI, the NAS CISAC, and to professional societies, university programs, Pugwash, UCS, etc. Among continuing topics that were supported, the MacArthur Foundation had an initiative in the FSU for peace and security. The Foundation has had research and writing grant competitions that can be accessed through the Internet.

The MacArthur Program on Peace and International Cooperation took the position that "to reduce the threat of nuclear war, we needed to break away from Cold War thinking and seek new ways, new directions of dealing with the Soviet Union and the nuclear weapons race." To that end, the Foundation made an enormous investment in education, public-interest organizations, and communications. Policies of the MacArthur Foundation stimulated other funding organizations to invest in peace activities.

Educators

In the academic community, individuals — students, faculty, researchers — as members of educational institutions and as participants within formalized university programs, took sides on the contentious issues. For the most part, the educational environment has fostered opposition to government policies based too much on nuclear weapons.

After World War II, scientists and engineers who had worked on the Manhattan Project choose to either return to universities or remain at weapons laboratories. Those that became academic scientists were freed from most wartime-secrecy restrictions. As a result of the expertise acquired during the atomic-bomb project, they were in a unique position to influence the dialectic of the arms race.

Suspending the nuclear arms race was once a popular cause, and many academics banded together in support. A typical local organization was the Chicago Area Faculty for a Freeze on the Nuclear Arms Race (CAFF), which was formed in 1982 by faculty at nearby colleges and universities. Many "teach-ins," especially on the environmental dangers of the nuclear age, were arranged by academic groups. One in January 1970 at Northwestern University had scientists Barry Commoner and Paul Ehrlich and political speakers Adlai Stevenson and Paul Simon.

Environmental teach-ins became a major tool for student activism. With growing institutional concern about pollution and environmental degradation, the gatherings caught on after students also became energized over civil rights and the war in Vietnam.

Perhaps one of the most valuable roles of arms control and nuclear-strategy programs at universities was to get governments to think of the longer-term consequences of their policies. These academic programs became a foundation for serious study of the impact of nuclear dependency in terms of both military force structure and humanitarian repercussions. The programs also became a source of appointees to influential government positions.

Individuals. Many early opponents and advocates of conflict nuclearization were drawn from academia.

Hans Bethe, an eminent Nobel-Prize-winning nuclear physicist from Cornell University, was widely respected for his socially responsible participation in complex technical issues dealing with the arms race. Born in 1906 in Germany, he helped the United States develop both the atomic and hydrogen bombs. Scientists held him in very high esteem.

After the Hiroshima bomb was detonated, Bethe joined Albert Einstein's Emergency Committee of Atomic Scientists, which wanted to establish international controls over atomic weapons. Their proposals became part of the Baruch Plan. When the Soviet Union rejected that proposal, Bethe became a leading advocate for "finite containment" of Russian expansionism. The containment policy sought to limit the nuclear arms race through international agreement. Just the opposite policy, "infinite containment," was espoused by Edward Teller, who argued that the only way to combat Soviet expansionism was to maintain an unlimited U.S. lead in the arms race.

In 1958, after studying a nuclear-test ban as member and chairman of the president's Science Advisory Committee, Bethe became a leader among those who argued that a detection system could be developed to verify the ban. He has also acted to counterbalance the heavy, hawkish hand of Edward Teller, opposed to ending the tests, who argued that a ban could mean "sacrificing millions of lives in a dirty nuclear war later."

Bethe, joined by IBM physicist Richard L. Garwin, published an article in the March 1968 issue of the *Scientific American* on the competing aspects of offensive and defensive missile systems, markedly raising the level of sophistication of subsequent debate on the ABM issue. As ballistic-missile defense challenges unfolded in the 1980s, Bethe and Teller continued to clash over the advantages and disadvantages of the SDI program.

Bethe had a leading role in forming the Union of Concerned Scientists. He continued to be an active mentor on arms control while teaching at Cornell University.

Among other proactive educators in the political arena, Frank von Hippel and Joel Primack had an early impact when they chronicled some of the commitments of individual scientists in national-security issues. They found that much valid technical advice had been ignored or misrepresented by government officials. Von Hippel, working out of Princeton University, subsequently became much more influential, becoming a leader in prying open Soviet-American personal relationships on arms control. Later he joined the Clinton administration for sixteen (self-described as frustrating) months in the position of Assistant Director for National Security in the White House Office of Science and Technology.

A particularly open-minded academic scientist that contributed to arms control, David Hafemeister, professor at California Polytechnic State University, has had numerous team-support positions, most recently facilitating study groups of the National Academy of Sciences. He has served on the FAS Council, participated in the FAS/NRDC/CSS verification workshops, and gained valuable experience in the executive and legislative branches of the federal government.

Allen Krass, from the University of New Hampshire, made a study of the enrichment of uranium, and he worked at ACDA. Others respected for major studies, in and out of government, are John Holdren, George and Matthew Bunn, Peter Zimmerman, Janne Nolan, and Ashton Carter.

Another individual that comes to mind is Linus Pauling, who criticized the government for downplaying the biomedical consequences of atmospheric nuclear testing. Many organizations later sprang up to educate the public about the dangers of fallout, especially contamination of milk by strontium-90 — which behaves chemically like calcium, lodging especially in children's growing bones.

MIT. Scientists and engineers at Massachusetts Institute of Technology have had close involvement with nuclear issues from the very beginning. For instance, Victor Weisskopf was one of the founding members of the Emergency Committee of Atomic Scientists. Donald Brennan, the first Hudson Institute president, was a member of Pugwash and the Cambridge-based American Academy of Arts and Sciences.

Political Science Professor Eugene B. Skolnikoff, taking note of growing public challenge to government actions (at that point focused on the ABM), defended academic protests because of the "growing demand that the universities in fact find ways to perform this task of public policy analysis and criticism." He pointed out that "for the first time since World War II, there was a major public challenge of a complex technological project, and a refusal to accept the usual assurances that secret data and intelligence would justify the project."

Another MIT professor, Jack P. Ruina, remarked at about the same time that "The national debate on Sentinel [antiballistic missile] is the first example I know of a military system being a matter of public debate not confined to a small group of experts or advocates of a special cause."

It was at MIT that the March 4 "Day Without Research" demonstration was formulated in 1969. A featured speaker for the initial convocation at the MIT auditorium was Harvard biologist and Nobel-Prize winner George Wald, who declared that

> this whole generation of students is beset with a profound uneasiness [particularly because] the Vietnam War is the most shameful episode in the whole of American history. The concept of War Crimes is an American invention. We've committed many War Crimes in Vietnam.

Wald added "so long as we keep that big an army, it will always find things to do."

Wald also pointed to a "military-industrial-labor union complex" that has helped institutionalize a big peacetime army. Lamenting that "our government has become preoccupied with death," the Nobelist remarked that "So-called Defense now absorbs 60 percent of the national budget, and about 12 percent of the Gross National Product." He observed that "all of us know there is no adequate defense against massive nuclear attack," yet as "we talk of deploying ABMs, we are also building the MIRV, the weapon to circumvent ABMs." Pointing out that "atomic bombs represent an unusable weapon," he reminded the March 4 audience that "a balance of terror is still terror. We have to get rid of those atomic weapons, here and everywhere. We cannot live with them.... Our business is with life, not death."

Other speakers for the research strike at MIT that day were Senator George McGovern (reconverting the economy from defense to domestic production), physicist Hans Bethe (the ABM), biologist Matthew Messelson (chemical and biological warfare), author Gar Alperovitz (scientists and the atomic bomb), and linguist Noam Chomsky (responsibility of the intellectual).

The post-World-War-II concentration at MIT of scientists who had participated in the Manhattan Project strongly influenced the campaign for peace and stability. The institute's staff became involved in many arms control-related activities, especially verification methods appropriate to the technical underpinnings of the university.

Leading activists early on at MIT were Philip Morrison, Bernard Feld (who edited *The Bulletin of the Atomic Scientists* from 1975 through 1984), Skolnikoff, and Kosta Tsipis. Morrison, when he was a Cornell Professor of Physics, expressed doubts about atomic warfare. (In 1952 he had to face the Senate Internal Security Subcommittee, which had a special, intimidating way of publicizing those "named" in anti-communist testimony; the process of accumulating names was based mostly on innuendo or unfounded allegations, rather than investigations that allowed due process.) Morrison was highly respected in the arms control community for his dedication to peace.

The MIT "Program in Science, Technology, and Society" maintained an staff, notably Theodore A. Postol (formerly at Argonne) and George Lewis, who had energetic arms control involvement. The Program's curriculum was developed by Tsipis in 1977 so that physics faculty and students could work together on controlling nuclear weapons and avoiding nuclear war.

George Rathjens, a political scientist at MIT, was a longtime leader in Pugwash. Because he feared creation of an American "Official-Secrets Act," Rathjens was the person who inadvertently triggered the controversial *Progressive* case by bringing an draft article to the attention of DOE.

Professor Marvin Miller of the Nuclear Engineering Department has been widely consulted on nonproliferation issues, and was especially diligent in trying to get whistleblower Mordecai Vanunu released from his Israeli prison cell (which took place in April 2004, but he was required to remain in Israel). Miller, as a member of the FAS, participated in various arms control verification projects and the Black Sea Experiment.

Stanford University. Stanford hosted two longstanding but contrasting institutions involved in arms control: The Hoover Foundation, a conservative think-tank and The Center for International Security and Arms Control.

Renamed the Center for International Security and Cooperation, Stanford's CISAC became part of the University's Institute for International Studies, a multi-disciplinary community dedicated to research and training in issues of international security.

Stanford's CISAC grew from the university's pioneering commitment to explore concerns about an escalating arms competition that marked the decades following World War II. When the Program was founded in 1970, Stanford University became one of the first academic institutions in the nation to commit faculty and resources to the study of the critical issues surrounding the Cold War and the ability of great powers to destroy each other's societies.

From Stanford University come two of the most eminent physicists that have continued significantly to sensible arms control. Wolfgang (Pief) Panofsky was a veteran of wartime Los Alamos where he monitored the *Trinity* test with devices he had developed; he was frequently been called upon to render his carefully considered opinions about nuclear issues. Panofsky chaired a number of important technical groups and committees for the government and the National Academy of Sciences.

Sidney Drell, a founding member of Stanford's CISAC, participated in and led many government-constituted panels and boards, not the least of which have been JASON and the President's Science Advisory Committee. Drell, who — like Panofsky — had received many awards, often wrote publicly about sensitive arms control issues.

Michael M. May, former director of Livermore Lab, applied his extensive weapons and advisory background at the Stanford CISAC. May had technical-support roles for the TTBT and SALT negotiations, was a member at one time or another of government defense and energy boards, and participated on the RAND Board of Trustees and the NAS CISAC.

The Center has been bringing together scholars, policymakers, area specialists, business people, and other experts to focus on a wide range of security questions of current importance. CISAC at Stanford has a resource base on weapons of mass destruction, scientific aspects of international security, great-power relations, and conflict resolution.

Princeton University. Princeton's Center for Energy and Environmental Studies has been the academic base for Frank von Hippel, an extremely active professor who initiated NGO arms control projects and shouldered a guiding role for the FAS. Von Hippel traveled to most centers and laboratories of arms control; he is best known for his proactive and organizational roles on nuclear issues, especially in reaching (as early as 1983) Soviet and later Chinese scientists on behalf of the FAS.

The Princeton Center has concentrated on a wide range of issues, but particularly nuclear nonproliferation and disarmament. Others on the staff who were heavily involved include Harold Feiveson and Oleg Bukharin, who specialized in nuclear matters dealing with his Russian homeland and with the Moscow Physical-Technical Institute's program in arms control. The Center also had Chinese students and scholars. Kenneth Luongo, formerly heading the arms control program at DOE, formed Princeton-based RANSAC, the Russian American Nuclear Security Advisory Council.

University of Illinois. A Program in Arms Control, Disarmament, and International Security (ACDIS) was established in 1978 at the University of Illinois. It continued at the Urbana-Champaign campus with its own quarterly publication, *Swords and Ploughshares*. The program has consisted faculty and students, emphasizing nuclear reactor and South Asia issues, the latter under Stephen P. Cohen. Two other faculty prominent in nuclear questions were Fred

Lamb and Larry Smaar. Jeremiah Sullivan was particularly active on defense matters, including JASON leadership.

Monterey Institute. Situated at Monterey, California, is the Center for Non-Proliferation Studies (CNS) at the Monterey Institute for International Studies. It professes to be the world's largest non-governmental organization devoted to combating the spread of weapons of mass destruction. The Center has published the *Nonproliferation Review*, a journal featuring the latest research on nonproliferation, and the Center maintains databases about the former Soviet Union and about nuclear weapons and missiles. In addition, under William C. Potter as the long-term Director, CNS worked jointly on several projects and activities with the Monterey Institute's Center for Russian and Eurasian Studies.

IGCC. The Institute on Global Conflict and Cooperation, of the University of California at San Diego, is where Herbert York settled (he is a former director of Livermore weapons lab). IGCC was founded in 1983, as a multi-campus research unit serving the entire University of California (UC) system, to study the causes of international conflict and help devise options for resolving it through international cooperation. IGCC's structure enabled research teams to be drawn from all nine UC campuses as well as the UC-managed Lawrence Livermore and Los Alamos National Laboratories, thus providing broad-based links to the U.S. government, foreign governments, and foreign-policy institutes from around the globe.

IGCC expressly involved the weapons labs in its research projects whenever policy challenges in promoting cooperation among nations require technical solutions.

University of Maryland. One of the most active academic physicists in the FAS/NRDC projects has been Steve Fetter, who became tenured at the University of Maryland. As already mentioned, he participated in the Black Sea Experiment.

The university had a Center for International Security Studies, of which John Steinbruner became director. The Center pursued policy-oriented scholarship on major issues facing the United States in the global arena. Established in 1987, the Center was based at the Maryland School of Public Affairs. Its research agenda has encompassed a range of issues, including (1) the use of traditional security instruments — such as arms control, peacekeeping, and the use of force — in the post-Cold-War context; (2) the nexus between international economics and security; cooperation and conflict between the United States and its allies; and (3) the process of foreign policymaking, including the role of particular institutions and the impact of public attitudes. Over the years, research projects have included the Nuclear History Program; Economics and National Security; and the Project on Rethinking Arms Control. More recent endeavors have addressed topics like the National Security Council and the making of U.S. foreign policy.

Bochum University. Bochum University in western Germany has the distinction of having had the most proactive NGO verification program anywhere. Under the enduring leadership of Juergen Altmann, the Bochum Verification Project carried out field experiments throughout Europe starting in 1988.

The Bochum Project over the years was funded by grants from (among others) the Volkswagen-Stiftung, the John D. and Catherine T. MacArthur Foundation, and the Ministry of Science and Research of the Federal State of Nordrhein-Westfalen.

The Project conducted applied-physics research for automatic sensor systems that could be used in peacekeeping and the verification of disarmament. Since 1989, several field experiments, most of them international, have been carried out with military vehicles in the Czech Republic, Germany, and the Netherlands. In addition, two one-year measurement programs were stationed at a military airport and at a test site for tanks and trucks in the Czech Republic.

International experiments were open to, and included, collaborating scientists from several countries within and across (former) Soviet-bloc boundaries: Canada, Czech Republic, France, Great Britain, Netherlands, Russia, Ukraine, and the United States.

The Ruhr-Universität Bochum, which nurtured the Verification Project, was notably engaged in interdisciplinary research and teaching pertaining to problems of peace-keeping, disarmament, arms control, and armed conflict. More-generalized aspects regarding science and disarmament were also studied. Analyses dealt mainly with prospective assessment and preventive limitation of new military technologies.

Notre Dame University. A latecomer in the academic scene has been the Joan B. Kroc Institute at Notre Dame University. The peace studies program was founded in 1986 at the South Bend, Indiana, campus. The Institute "conducts research, education, and outreach programs on the causes of violence and the conditions for sustainable peace." Its research agenda "focuses on the religious and ethnic dimensions of conflict and peacebuilding; the ethics of the use of force; and the peacemaking role of international norms, policies and institutions, including a focus on economic sanctions and enforcement of human rights."

"Inspired by the vision of Rev. Theodore M. Hesburgh, President Emeritus of the University of Notre Dame, the program attracts students from around the world to study peacemaking while building cross-cultural understanding among themselves." Leading scholars at the Kroc Institute included George A. Lopez and David Cortright.

The Advanced International Studies Institute. During the height of the Cold War, the Advanced International Studies Institute, "in association with the University of Miami," issued press reports on Soviet Affairs from its office in Washington, D.C. One such report claimed that the Soviet Union, not the United States, was the "instigator of the horrors of nuclear warfare competition." To arrive at this conclusion, they quoted a Soviet Academician, Anatoli Aleksandrov, who said that the "actual seeds of the Soviet A-bomb" began in the early 1930s.

This lesser-known institute is mentioned as an example of other mystifying and short-lived organizations funded by right-wing interests in an attempt to counterbalance liberal leanings of the more proactive educational institutions.

Professional Associations

Contemporary issues were intellectual grist for professional societies, especially mainstream physics associations, who often spoke out or got involved about Cold War issues. So did engineers and other professionals and academics through their affiliations. Active individuals and associations often became opinion leaders on the consequences of technical choices.

APS. The American Physical Society was the primary professional organization for most leaders of atomic-bomb development. In reviewing the history of the Society, a president of the APS summarized its involvement in national security issues:

> After [World War II] many physicists became heavily involved in arms control, and some continued working on military technologies. Government leaders and the public alike viewed physicists as a group that could provide expert advice on matters of national security and on other technical issues. Prominent members of the physics community had an entrée into Washington and could get the attention of government leaders. High-level scientific advisory committees were established, and after Sputnik, President Eisenhower established the post of science advisor to the President....
>
> However, many in the physics community believed that another route to change public policy was to work outside of government, especially when there was a perception that government policy was wrong. The turbulence and controversy that the Vietnam War created in American society spurred the development of social activism in the American physics community. This activism intensified the trend of working outside the government to affect political change, and ultimately changed APS. The activism that arose with the Vietnam War manifested itself in political radicalism on university campuses, fueled also by antipathy to defense-sponsored physics research. After the ruckus of the 1968 Democratic national convention, which included physicist John Froines as one of the "Chicago Seven," the APS began to schedule topics on sensitive defense issues. One of the first was a debate in 1969 on the Nixon Administration's proposed ABM system.

Actually, unprecedented intervention by a scientific society had occurred earlier, in 1948, when the APS strongly defended Edward U. Condon, then director of the National Bureau of Standards, against spurious charges brought by the House Un-American Activities Committee (HUAC). Condon — a renowned theoretical physicist who made some contributions in the wartime development of radar, the Los Alamos project, and uranium separation — became active in the atomic-scientists movement, which was subjected to HUAC investigations.

Resulting from the debate about social and political implications of physics was the controversial establishment of an APS Forum on Physics and Society, a membership unit akin to the scientific divisions. The forum, despite initial misgivings by APS Council members, contributed to the "advancement and diffusion of knowledge" by sponsoring defense-related sessions at APS meetings, by conducting studies, and by publishing a newsletter, *Physics and Society*. The Forum at one time had about 4500 members, more than 10 percent of the APS membership.

One of the largest APS technical sessions ever held — an overflow crowd of 3500, in Washington, D.C. — was the debate in 1969 about the Safeguard ABM system. Favoring deployment were Eugene P. Wigner, professor of physics at Princeton University, and Donald Brennan, senior staff scientist at the Hudson

Institute; opposing were Hans A. Bethe and George W. Rathjens, professor of political science at MIT. The exchange covered technical, military, and political facets of ABM.

The APS has supported comprehensive studies that make use of the broad expertise of physicists, and sometimes the panelists have been granted access by the government to relevant classified information. The Society also forms survey committees on matters dealing with controversial topics, including women in physics, minority representation, international affairs, and worldwide freedom of scientists. One of the more influential committees has been the Panel on Public Affairs, founded in 1975. Its main achievements have been to carry out studies and issue policy statements on the technical aspects of energy use, nuclear reactor safety, and directed-energy weapons. A report on the latter topic, "The Physics and Technology of Directed Energy Weapons," had considerable influence on the contentious SDI technical program of the 1980s, and also served as a model for other studies.

In mid-2000, the APS, FAS, and UCS jointly opposed a pending decision by the Clinton administration under pressure to deploy an immature missile-defense system, regardless of the outcome of ongoing tests. Eventually President Clinton, due to leave office early 2001, opted to leave that decision to the next administration.

JASON. Founded by academic physicists in 1959 and financed by the Defense Department, the government-advisory group JASON examines and analyzes prominent technical issues and long-term developmental goals. It does so at the risk of appearing to endorse military solutions to some problems that might be more socio-technical. In order to facilitate their reviews, members are vested with high-level security clearances.

In 1969, JASON concluded in a report to the Institute of Defense Analyses that the ABM system as then proposed was "ill-conceived."

AAAS. The American Association for the Advancement of Science, its broad-based membership transcending scientific disciplines, ordinarily steered clear of topics having political implications. However, in 1980 the AAAS Council set up a steering committee on nuclear arms control, citing the importance of science to human welfare and the dangers implicit in nuclear armaments. It urged support of U.S. efforts toward effective bilateral nuclear arms limitations and a Comprehensive Test Ban Treaty, opposed development by any country of new weapons systems that make verification difficult, and favored — under an arms control regime — conversion of nuclear weapons facilities to peaceful uses.

The AAAS maintained a Standing Committee on Scientific Freedom and Responsibility that delved into cases involving harassment or abridgement of the right of scientists to conduct research and publish their findings.

The AAAS publishes a weekly journal called *Science*, which has a tradition of printing news and articles that not only have a technical component but also bear significantly on public policy. During the ABM debate, the 21 March 1969 edition carried a detailed and balanced summary of the positions and statements of the government and the opposition. At the time, Senators Gore, Fulbright, Cooper, and

Hart were leading congressional resistance to ABM, and for expertise they called upon three scientists of unusual prestige and knowledge — James R. Killian, chairman of the board at MIT, George B. Kistiakowsky, professor of chemistry at Harvard, and Herbert F. York, professor of physics at the University of California, San Diego.

Academies of Science. The national academies of science in the United States and the Soviet Union, as well as in other nations, were sometimes asked by their governments, or otherwise motivated, to carry out studies of specific technical issues.

In 1961 the American Academy of Arts and Sciences started what became a tradition of promoting bilateral discussions: It established a Committee on International Studies of Arms Control — a vehicle for the Soviet-American Study Group, which included Paul Doty, Donald Brennan, Betty Lall, Frank Long, Academician Topchiev, Vasily Emelyanov, Petr Kapitsa, Lev Artsimovich, and other Soviet Pugwash members. In the mid-1960s one of the major technical topics for discussion was the question of on-site, challenge inspections to reinforce treaty verification.

Around 1980, the NAS created its Committee on International Security and Arms Control (CISAC — the same acronym as Stanford's Center). The NAS CISAC had been constituted after the dissolution in the Soviet-American Disarmament Study, an inter-academy initiative that had begun in the 1950s, morphing into the Study Group on Arms Control and Disarmament in 1960. Similar groups were formed in the United Kingdom, France, and Italy. Spurgeon M. Keeney, Jr., helped organize and guide the NAS CISAC.

By 1984, under the technical leadership of Marvin Goldberger from the California Institute of Technology (CalTech), the NAS CISAC had been meeting for a few years with a comparable committee of the Soviet Academy of Sciences, headed by Evgeny P. Velikhov. The meetings between Soviet scientists and CISAC received high-level attention in Moscow.

CISAC's discussions with Soviet counterparts covered issues such as survivable strategic forces, ballistic missile defense, force constraints, and fissile material production cutoff. At about the same time, the International Pugwash Group was sponsoring multilateral meetings on the technical basis for a nuclear weapons freeze, and the Federation of American Scientists was planning for a joint meeting with Soviet counterparts on strengthening the ABM treaty.

After the Cold War, the U.S. and RF Academies of Science continued jointly or separately to explore bilateral and multilateral arms control options.

Franklin A. Long of Cornell directed a long-term joint project of the American Academy of Arts and Sciences on weapons in space, which has been part of a broader program of public-policy studies on science, security and international cooperation.

International and Transnational Groups

Various individuals, associations, and organizations spanning international boundaries weighed in on the nuclear arms race. They were involved in "transnational relations," a term denoting regular interactions across national boundaries when at least one party is not operating on behalf of a national government or international organization.

Lord Solly Zuckerman, although not directly working for an arms control organization, was one of the most articulate British critics of the nuclear arms race. As a veteran advisor to the government since World War II on conventional and nuclear matters, Lord Zuckerman wrote extensively about what he called "Nuclear Illusion and Reality." Bertrand Goldschmidt was a comparable conscientious objector in France.

Influence on the tenor of the arms race was compounded when organized action took place. Despite post-Cold War neoconservative efforts to incarnate Reagan as a peacemaker, the number of individuals and organizations that became activated beginning with his campaign for the presidency testifies to a much more negative public reaction to his aggressive rhetoric.

IPPNW. Founded in 1980 as a transnational organization, the International Physicians for the Prevention of Nuclear War called for medical practitioners of all countries to work toward the prevention of nuclear war and for the elimination of all nuclear weapons.

Cofounders Drs. Bernard Lown of the United States and Evgeni Chazov of the Soviet Union organized a team to conduct meticulous scientific research on the medical effects of radiation trauma. This study was based on data collected by Japanese colleagues who had examined the effects of the atomic bombs dropped on Hiroshima and Nagasaki. Taking note of the experience in Japan with physical injuries and psychological shock, they regarded nuclear war to be the ultimate human and environmental disaster. The organization called for physicians to review information on medical implications of nuclear weapons and nuclear war and to take an activist role in educating the public, the medical profession, and policymakers about its dangers.

Chazov became Leonid Brezhnev's personal physician. Even though he headed a movement that implicitly criticized Soviet policies, as a cardiologist Chazov was able to communicate with, and speak out to, the top leadership because they depended on him for their health. It is reported that the international physicians' movement had a profound effect on top Soviet leaders partly because they personally agreed with the organization's ideas about the dangers of nuclear war.

IPPNW was honored with a Nobel Prize in 1985 for its efforts to increase public awareness about the danger of nuclear weapons. The award had a negative reaction in some right-wing quarters, who subjected the co-leaders of the organization, Lown and Chazov, to harsh criticism in newspapers.

UN. The United Nations was involved in superpower arms-race issues mostly through its special sessions or conferences on disarmament. One Such Special Session, attended by representatives of 149 member nations, ended in 1978 with a document setting forth goals and priorities for disarmament negotiations. The "Final Document" called for the United States and the Soviet Union to conclude a strategic-arms-limitation agreement at the earliest possible date; it also advocated a comprehensive nuclear weapons test ban and discussed the urgency of negotiations to control nuclear weapons. The Special Session urged increased UN involvement of non-government organizations, which took place in subsequent years.

During that 1978 session, the United States officially declared that it would not use nuclear weapons "against any state party to the Nonproliferation Treaty ... so long as that state was not engaged in an attack on the United States or its allies while allied or associated with a nuclear weapon state in carrying out such an attack."

Although domestic conservatives and neoconservatives, especially in the United States, have criticized and tried to shun the UN, the organization stands as the agreed forum for universal negotiation and implementation of international treaties.

UNIDIR. The United Nations Institute for Disarmament Research, headquartered with the UN in Geneva, Switzerland, compiles information on arms control and nonproliferation issues. The General Assembly took the view that sustained research and study activity by the UN in the field of disarmament would promote informed participation by all states party to disarmament efforts and considered it advisable to undertake more forward-looking research within the framework of the United Nations. It is against this background that the General Assembly established UNIDIR.

UNIDIR started work on 1 October 1980. In 1997, Dr. Patricia Lewis, founder of VERTIC, became UNIDIR's Director.

Recognizing that negotiations on disarmament and continuing efforts to ensure greater security must be based on objective, in-depth, technical studies, the General Assembly decided to sustain UNIDIR so that the international community would be provided with diversified and complete information on problems relating to disarmament. The Assembly has also pointed to the importance of ensuring that disarmament studies be conducted with scientific independence.

In 1990, on the occasion of UNIDIR's tenth anniversary, the Assembly reiterated the international community's need for access to independent and in-depth research on disarmament, in particular on emerging problems and the foreseeable consequences of disarmament. The Assembly encouraged the UNIDIR to continue its independent research on these issues.

The specific mandate of UNIDIR included:

(a) Providing the international community with more diversified and complete data on problems relating to international security, the armaments race and disarmament in all fields, particularly the nuclear field, so as to facilitate progress, through negotiations, toward greater security for all States and toward the economic and social development of all peoples;
(b) Promoting informed participation by all nations in disarmament efforts;
(c) Assisting ongoing negotiations on disarmament and continuing efforts to ensure greater international security at a progressively lower level of armaments, particularly nuclear armaments, by means of objective and factual studies and analyses;

(d) Carrying out more in-depth, forward-looking and long-term research on disarmament, so as to provide a general insight to the problems involved, and stimulating new initiatives for negotiations.

SIPRI. The Stockholm International Peace Research Institute, funded primarily by the Swedish government, has been the world's leading independent, continuing center for peace research. Staffed by individuals from many countries, it has produced highly objective and comprehensive reports. Its yearly compilations of world arsenals and treaties — nuclear and conventional — are widely cited major resources for researchers.

The SIPRI web site reflects a broad range of research on arms transfers and production, European security issues, as well as nuclear, conventional, chemical and biological weapons. The site includes an extensive list of publications back to 1969. SIPRI has continued its work well past the Cold War.

International Institute for Strategic Studies. This London-based NGO, launched in 1958, is a research institute dedicated to the study of international security, defense, and arms control. It has been supported by contributions from members in more than 50 countries and by sale of its widely read publications. Its yearbooks on armaments were and are eagerly awaited and referenced.

Canadian Institute for International Peace and Security. A quasi-government peace and security organization was established by the Parliament of Canada in August 1984. The Institute sponsored many studies on arms control and verification.

Governing legislation stated that

> The purpose of the Institute is to increase knowledge and understanding of the issues relating to international peace and security from a Canadian perspective, with particular emphasis on arms control, disarmament, defence and conflict resolution, and to foster, fund and conduct research on matters relating to international peace and security; promote scholarship in matters relating to international peace and security; study and propose ideas and policies for the enhancement of international peace and security; and collect and disseminate information on, and encourage public discussion of issues of international peace and security.

The Canadian Institute for International Peace and Security was dismantled in 1992 by the Brian Mulroney government, ostensibly for budgetary reasons.

VERTIC. Because of a perceived need to improve public and official knowledge about verification technology, Patricia Lewis in 1986 started a London-based NGO, originally called the Verification Technology Information Centre — now the Verification Research, Training & Information Centre. Its stated mission is "to promote effective and efficient verification as a means of ensuring confidence in the implementation of international agreements and intra-national agreements with international involvement."

VERTIC started to publish a monthly bulletin in June 1989 called *Trust and Verify*, intended to throw light on the sincerity of President Reagan's pronouncements about verification (which had become a political football).

By becoming a major resource with factual information on the technology, methods, and applications of verification, VERTIC gained access to key government officials, especially in Britain. In the early 1990s, VERTIC was coordinating six working groups on verification in connection with ongoing and future arms control and environmental agreements.

The organization participated and supported verification projects (such as those actualized by Bochum University) and helped organize relevant conferences. A series of yearbooks on arms control and environmental agreements were published under the auspices of VERTIC.

Funding for the independent organization came from foundation grants and trusts.

I N S T E A D Center for Verification Technology. Situated at the Freije University of Amsterdam, Netherlands, the I N S T E A D Center carried out analysis in arms control and defense, under the directorship of Brig. Genl. (Ret.) Henney van der Graaf. I N S T E A D stands for Interuniversity Network for Studies on Technology Assessment in Defence. The organization was one of the prime sponsors of a 1991 conference in Mosbach-Neckarelz, Germany, on controlling military research and development and on managing exports of dual-use technologies. I N S T E A D personnel, especially van der Graaf, participated in the Bochum-arranged experiments and in many verification studies.

Peace Research Institute Frankfurt. The main activity of PRIF at the University of Frankfurt, Germany, has been to research the causes of war and violence and to analyze current political conflicts. The Institute has published research studies on arms control and other topics and provides information to public interest groups, political parties and government agencies. PRIF had maintained in Bonn a Peace Research Information Unit, tasked to establish contacts with German and foreign organizations in the field of peace research.

B A S I C. The British American Security Information Council, which specialized in NATO nuclear issues, was intermittently active, with offices in London and Washington, D.C. B A S I C described itself as "an independent research organization that analyzes government policies and promotes public awareness of defense, disarmament, military strategy and nuclear policies."

INESAP. The International Network of Engineers and Scientists Against Proliferation was a nonprofit NGO with participants from all over the world. Most of the activities of INESAP were managed by IANUS, at Darmstadt University of Technology (Germany), a member organization in a wider network. Main objectives have been to promote nuclear disarmament, tighten existing arms control and nonproliferation regimes, implement unconventional approaches to curbing proliferation of weapons of mass destruction, and control transfer of related technology. INESAP has regularly published an Information Bulletin.

Labor Unions. Immediately after World War II, communist and non-communist unions around the world joined to form the World Federation of Trade Unions, but a few years later this organization split over implementation of the Marshall Plan in Western Europe. The 1955 merger of the U.S.-based AFL and CIO brought about a union confederation that took a strong anti-communist position. Other unions and union members were active against the war in Vietnam and participated in the subsequent anti-nuclear weapons movement.

During the Freeze movement, as well as before, many workers' organizations, including the AFL-CIO, formally endorsed peace campaigns.

European union confederations of Germany, Netherlands, and Britain provided support and endorsements for the large rallies of the 80s, especially in opposition to NATO deployment of nuclear missiles.

The single, most effective trade movement was Solidarity. Beginning around 1979, it gave impetus to open public resistance in Poland. The indigenous union had an impact on the end of European Communism. A significant rise of group action in Poland followed the open-air masses conducted by Pope John Paul II for millions of people. There were sporadic strikes over meat prices, as well as demands for free trade unions and release of political prisoners. For a full decade before Poland gained its independence, the Solidarity movement had struggled for religious and political freedom.

Financial support and other aid to the underground Polish union movement came largely from trade unions and leftist organizations, particularly those in France and Italy. Roman Catholic priests, in Poland and throughout Europe, also helped. Underground newspapers and clandestine mobile radio transmitters were important means of Solidarity messaging and public communication. Contrary to claims made by supporters of President Reagan, little or no aid came from the U.S. government.

European Peace Movements

Because they inhabited a potential nuclear battlefield, Europeans were especially provoked by the dangers and excesses of superpower confrontation. To contend with this viscerally felt threat, activists in Western Europe gathered together in what are best described as "movements" or alliances, rather than formalized NGOs as in the United States.

In 1958 the Campaign for Nuclear Disarmament (CND) was founded in England, in part to protest against nuclear-weapon tests in the atmosphere. It became one of the most influential anti-nuclear weapons pressure groups, particularly in the early 1960s and mid-1980s. In 1962 CND mobilized more than 150,000 people, including prominent British Labour Party members, to participate in an Easter nuclear weapons protest march.

The 1977 decision by the United States to deploy the neutron bomb aroused anti-nuclear sentiment in western Europe. Peace campaigns that had emerged in Europe were even more decisively catalyzed by NATO's 1979 determination to introduce a new generation of missiles and to abandon serious disarmament negotiations after the June signing of SALT II. When spontaneously coordinated mass demonstrations — with 2 to 3 million participants — convened in 12 European capitals and major cities in 1981, politicians took notice. Two years later, 12-million protestors in western Europe marched on a single day, October 23, demanding an end to INF deployments; London, Amsterdam, Bonn, Rome, and other major cities witnessed the largest political demonstrations in their histories.

CND was a major organizer of these protests and a supporter of the Greenham Common Women's Peace Camp.

Although at one time CND had tens of thousands of members, and branches in other countries, by the mid-1970s its membership had declined to a few hundred, even after the newly elected Labour prime minister decided to go ahead with the Polaris submarine program. Reawakening to the fact that nuclear weapons still had not been eliminated, CND grew back to nearly 30,000 members in the 1990s.

In 1980-81, the hawkish assertions of Margaret Thatcher and Ronald Reagan — which came across as brash and bellicose — confirmed existing European fears, as did disclosure of periodic nuclear alerts and accidents. Demonstrations became part of a transnational protest involving a broader political coalition that included Vietnam-war opposition, women's rights, environmental protection, and anti-nuclear-power movements. European religious groups and secular educators became active in grassroots peace education. The Dutch were the first of this wave to advance their peace campaigns, gaining momentum from the unifying issue of cruise-missile deployment in England and Italy. This single issue expanded to advocacy of unilateral initiatives for nuclear disarmament. Although the initiatives were widely mischaracterized by critics as "unilateral disarmament," much of the focus was on getting the nuclear powers to originate small, verifiable steps toward arms reductions.

The European Nuclear Disarmament (END) movement, which began in Britain, added a major dimension in 1980: transnational linkage and nonalignment. END campaigned in particular for nuclear-free zones in Europe. Massive nonviolent action to obstruct weapons deployment became a prominent tool. END also promoted human rights and the freeing of eastern Europe.

Although various treaties had already designated some nuclear weapons-free zones, grassroots anti-nuclear movements helped define additional exclusion areas of the globe, specifically in the Pacific.

Peace movements in France (CODENE) and Southern Europe (including Communist Parties, especially in Italy) were engaged in lively protest. In Great Britain, the Atomic Scientists' Association, and Scientists for Global Responsibility, were supportive in related peace causes.

In Soviet-dominated nations, overt protest was politically impractical. The Palme Commission, a European "Independent Commission on Disarmament and Security Issues," was able to engage in "élite-level" contacts with the Soviet Union, especially during the Brezhnev era. Among the Soviet participants or advisors were Academician Georgi Arbatov and retired General Mikhail Milshtein.

The Women of Greenham Common. Civil disobedience gained new respect from the symbolic witness of strongly feminist camps at Greenham Common, England, in opposition to deployment of nuclear-armed cruise missiles, which were stationed at the Greenham military base in Berkshire. This was part of a new wave of single-issue activism that changed the face of anti-nuclear groups. Eventually, regular or permanent camps were set up at almost 20 bases in the UK, as well as at cruise-missile bases in Belgium and Sicily. Not until the former base at Greenham was converted to civilian use in August 2000 did the 19-year protest end.

Although the Greenham women became a symbol of resistance to the "male-dominated world" of nuclear weapons, it was by no means the only form of female solidarity and expression during the Cold War. In November of 1981, more than 3000 women in America organized a "Pentagon Action," dramatically encircling the military headquarters with yarn and cloth; the participating women linked anti-war sentiment with a wide range of political issues, including the stalled Equal Rights Amendment to the U.S. constitution, imbalanced human services, and abiding racism.

These feminist peace actions reflected a rejection of the (white) male-dominated nuclear arms race — choosing instead solidarity with the broad-based peace movement. Women and feminism strengthened the humanistic touch, contributed leadership (such as Nuclear Freeze founder, Randal Forsberg), and equalized partnership in the activism of the 1980s.

The Greens of Germany. Opposition to nuclear power in West Germany fed nuclear weapons protests there, particularly under the Greens (*Die Grönen*). The opposition was strengthened by the realization that East and West Germany were likely to be the first victims and the atomic battlefield of any tactical nuclear war. The Greens obtained millions of signatures against NATO deployment of cruise and Pershing missiles. The Greens have had a broad approach to peace politics — synthesizing anti-militarism, ecology, feminism, worker democracy, and political decentralization.

Networks

Although many NGOs collaborated ("networked") with other organizations over the years, the more recent coalitions derive their strength almost entirely from electronic networking. The advent of the Internet has made it possible for networking organizations to pick up the slack from NGOs that have a constituency but no longer have adequate financial base or personnel for a brick-and-mortar operation. Many of these organizations were still active in 2009.

The Acronym Institute. Formed from a consolidation of other organizations, the Acronym Institute in the UK conducted research and published information with considerable depth on disarmament and arms control negotiations. Their journal, *Disarmament Diplomacy*, was funded by the W. Alton Jones Foundation, the Ford Foundation, and a charitable trust. The journal published articles to stimulate debate on topical disarmament and arms control issues, as well as opinion and analysis, documents and sources, and a review of current news. Rebecca Johnson was the long-term Executive Director.

INES. The International Network of Engineers and Scientists for Global Responsibility has supported work of INESAP, especially as part of the Abolition 2000 network. Leading nuclear arms control and nonproliferation organizations have worked together through the Coalition to Reduce Nuclear Dangers, aiming to build support for a practical, step-by-step program that would reduce the hazards of nuclear weapons and prevent new nuclear threats from emerging. The Coalition has maintained a virtual library, with key statements, analysis, technical information and practical strategies to reduce the dangers of weapons of mass destruction. Major focus areas are CTBT and nuclear-missile defense. As of late 2009 it was still active, launching a campaign for a nuclear-free world.

Some verification and disarmament research in Germany has continued, either through this or related networks, sponsored by various ministerial or parliamentary organizations.

Scientists for Global Responsibility was an affiliated UK-based organization promoting ethical science and technology.

Abolition 2000. Abolition 2000 was organized as an international citizen's movement to eliminate nuclear weapons, a global network calling for negotiations on a treaty to eliminate nuclear weapons within a time-bound framework. It was an outgrowth of a coalition called Project Abolition that former Senator Alan Cranston helped put together.

In April 1995, during the first weeks of the Non-Proliferation Treaty Review and Extension Conference, activists from around the world recognized that the objective of nuclear abolition was not on the NPT Conference agenda. Activist representatives from dozens of countries around the world came together to compose a statement that has become the founding document of the Abolition 2000 Network. More than 1000 NGOs on six continents signed it and were actively working in twelve working groups to accomplish eleven points that are listed in their statement.

> At a meeting held in The Hague in November 1995, Abolition 2000: A Global Network to Eliminate Nuclear Weapons was formally established as a "tight network" of regional and working groups, bridging a multitude of cultures, languages, and political systems.
>
> Abolition 2000 is not a membership body but is open to all organizations endorsing the Abolition 2000 Statement. The Network aims to provide groups concerned with nuclear issues a forum for the exchange of information and the development of joint initiatives.

In late 2009, Abolition 2000 continued its activities and web-site presence.

MoveOn. MoveOn was (and still in 2009) an organization that was started by a couple of Silicon Valley entrepreneurs when the impeachment of President Clinton was being considered. According to its Web site, MoveOn is working

> to bring ordinary people back into politics.... is a catalyst for a new kind of grassroots involvement, supporting busy but concerned citizens in finding their political voice. Our nationwide network of more than 2,000,000 online activists is one of the most effective and responsive outlets for democratic participation available today.
>
> When there is a disconnect between broad public opinion and legislative action, MoveOn builds electronic advocacy groups. Examples of such issues are campaign finance, environmental and energy issues, media consolidation, or the Iraq war. Once a group is assembled, MoveOn provides information and tools to help each individual have the greatest possible impact.

Regarding nuclear weapons, MoveOn's "Back from the Brink" campaign said it was time for the United States and Russia to "de-alert," that is, taking nuclear weapons off their hair trigger. Before President G.W. Bush launched his invasion of Iraq, the networking organization was focusing on trigger-happy nuclear postures. They reminded the public that, even with the virtual contest over for more than a decade, the United States and Russia continued to have thousands of nuclear missiles on high alert.

Many organizations, some already mentioned, were early participants in the "Back from the Brink" campaign: Alliance for Nuclear Accountability, Center for Defense Information, Federation of American Scientists, Global Resource Action Center for the Environment, Institute for Energy and Environmental Research, International Physicians for the Prevention of Nuclear War, Lawyer's Committee on Nuclear Policy, Peace Action & Peace Action Education Fund, Physicians for Social Responsibility, Snake River Alliance, Tri-Valley Communities Against a Radioactive Environment, and Women's International League for Peace and Freedom.

MoveOn concentrated for a while on the war brought against Iraq by President G.W. Bush. Because it has so many members, MoveOn has had an appreciable influence on U.S. politics.

DontBlowIt. Campaigning to convince the U.S. President that nuclear weapons should be a thing of the past was DontBlowIt, an Internet-based organization that sent free e-mail postcards to elected officials. Because there are more than 36,000 nuclear weapons in the world, enough to destroy the planet several times over, this network wanted the new president (G. W. Bush) to know that he has an opportunity to create a safer future. Funding was provided by The W. Alton Jones Foundation, the Ploughshares Fund, and the John Merck Fund.

DontBlowIt merged with an initiative of the Vietnam Veterans of America Foundation, the Nuclear Threat Reduction Campaign. The Campaign was aimed at educating and mobilizing key constituencies in order to reduce the threat posed by nuclear, biological and chemical weapons.

Institute for Global Communications. Another relative newcomer, IGC was formed in 1987 to manage PeaceNet and the newly acquired EcoNet. EcoNet was dedicated to environmental preservation and sustainability.

IGC regards international cooperation and partnership as essential in addressing problems of the 21 century. It broadened its reach in 1988 to include international membership, and began to collaborate with like-minded organizations outside the United States. The first international link was made with GreenNet in the UK. In 1989, IGC added Internet e-mail to its services.

The mission of IGC was to advance the work of progressive organizations and individuals for peace, justice, economic opportunity, human rights, democracy and environmental sustainability through strategic use of online technologies. As of 2009, it has continued its mission.

Safer World. The Act Now for a Safer World campaign, was a new-millennium U.S. cooperative entrant on the public-interest scene, "working to keep nuclear, biological and chemical weapons and materials out of terrorist hands." This nonpartisan effort had the backing of Sam Nunn, Pete Dominici, William Perry,

Susan Eisenhower, and Warren Rudman. It has since become a UK-centered organization with a special focus at preventing armed violence.

Nuclear Threat Initiative. Funded by wealthy Ted Turner and co-chaired with Sam Nunn, the new millennium startup NTI has been concentrating on reducing the risk of use and spread of non-conventional weapons by serving "as a catalyst for new thinking" and taking "direct action to reduce these threats." It is an "operational organization ... actively engaged in developing, shaping and implementing" projects that they fund. They have disbursed many, many millions of dollars in program awards and grants. The web site contains catalogs of in-depth reference information related to their focus on helping create a safer world.

As the "big gorilla" in the NGO world in terms of funding, NTI has been making direct inroads in reducing risks of worldwide highly enriched uranium and other nuclear materials, as well as in helping to prevent the spread of nuclear and biological weapons-related technology.

Fourth Estate

The newsmedia were split — pro, con, and indifferent — regarding the nuclear arms race. Many technical and nontechnical journals carried articles and opinion pieces. Scientific journals ran both technical and interpretive articles, particularly about arms technology. Many monthly magazines carried political analyses, and specialty journals published extensive political-science perspectives. Newspapers editorialized with a partisan slant, but little depth of coverage.

Bulletin of the Atomic Scientists. Topical magazines such as the *Bulletin of the Atomic Scientists*, headquartered at the University of Chicago campus, supplied in-depth studies and opinions about the arms race. It had its beginnings in December 1945, a few months after Hiroshima, when the Atomic Scientists of Chicago (with Eugene Rabinowitch as editor) began publishing *The Bulletin of the Atomic Scientists of Chicago*.

The *Bulletin* has served as a major forum for many of the early public exchanges between scientists about such issues as civilian control of nuclear energy and the controversy over ABM siting. One of the few publications that carried Soviet-authored articles, the *Bulletin* has been a widely sought voice of worldwide scientists. It has continued to be a valuable outlet for views on technical and nontechnical issues, broadly encompassing not just military but other topics that affect national and international security.

Magazines such as the *Bulletin* were very important to the communication-isolated Soviet scientists, informing them about Western debates regarding the nuclear arms race. Because libraries at the Soviet weapons labs had subscriptions, such periodicals were available to nuclear-weapon specialists, allowing them to become among the best-informed people in the USSR on international and military affairs. Moreover, those scientists had a certain freedom from internal censorship, allowing them to have broad discussions on international and domestic issues.

Scientific American. Typically publishing several articles each year related to nuclear or defense technology, the monthly *Scientific American* bridged the gap between the technical and popular literature with illustrated treatments of challenging issues. The March 1968 Bethe-Garwin ABM article has been the single most-cited treatise on the topic. Because *Scientific American* was available in many Soviet libraries, it was able to convey Western views on arms control issues to a wide audience behind the Iron Curtain.

The *Progressive*. Published in Madison, Wisconsin, the leftward-leaning *Progressive* enjoyed rubbing authorities the wrong way. A classic example is an article by Erwin Knoll and Theodore A. Postol, "The day the Bomb went off." It began, "It was a sunny morning in the Chicago loop...." [when a 20-megaton H-bomb hypothetically was dropped, with fallout reaching Detroit].

In the October 1978 issue of the *Progressive*, Editor Sam Day disclosed that "The neutron bomb lives after all," because production of that weapon was started despite President Carter's announced temporary ban.

Perhaps the most notorious moment for the *Progressive* was when its name was became synonymous with a federal case. The magazine was temporarily enjoined from publication of an article containing publicly available information on hydrogen-bomb design concepts.

Dailies and Weeklies. Daily newspapers, especially those at major cities, kept the public informed, although they were often dependent on public-relations releases from government agencies. As the protest movement grew, some dailies became more willing to question official doctrine regarding nuclear policies.

Newspapers split and vacillated over contemporary disputes. On the ABM issue, the Republican-leaning *Chicago Tribune* editorialized in 1969 that "The opposition to emplacement of a purely defensive anti-ballistic missile system around two of our launching sties has been largely silly, often emotional, and frequently patently political." The *Tribune* later moderated its tone, although it has tended to maintain its political character and support missile defense. The *New York Times* and *Washington Post* have attempted in their coverage to be more balanced and thorough than regional newspapers.

Weekly news magazines *Time* and *Newsweek* frequently carried distillations of ongoing controversies on the arms race, as well as contributions by opinionated journalists. One such was George F. Will, a prominent right-wing columnist, who invariably took a strong position against arms control. In a 1984 column in *Newsweek*, he argued that "the arms control process is injurious to U.S. interests." Some of his opinions are cited below at length because they are representative of the hardline view of the time.

Despite the huge arsenals, Will argued in 1984 that there was no "spiraling" arms race because the U.S. nuclear weapons inventory has been sharply reduced, he claimed, citing numbers of 8000 fewer warheads and 25 percent less megatonnage than in the 1960s. This, Will stressed, was the result not of arms agreements but of modernization programs that produced safer, more effective weapons — modernization, he charged, that arms control advocates were trying to block with treaties and agreements. At the same time, he argued the neocon position "the Soviet nuclear arsenal has grown quantitatively and qualitatively... in 41 categories of nuclear capabilities ... that the United States had been well ahead in every category in 1962 and was behind in all but two by the late 1970s." (We can now confirm that " arms control advocates" were much closer in estimating the actual balance of power.)

Belittling the results of the ABM treaty, Will said he believed it caused the United States to "spend many more millions on MX missiles, an unsatisfactory response to the fact that our undefended land-based ICBM's are vulnerable." Because of Soviet secrecy, limits on warhead numbers or megatonnage would be hard to verify, he claimed. He also blamed the SALT I treaty for accelerating the arms race, and attributed to the Soviets a big qualitative advantage. He went on to charge that "the Soviets see [arms control] as one arena in a comprehensive, unending competition [that exploits] the American thirst for agreements."

Will attached credence to five alleged Soviet arms control aims: (1) limit the wrong things, (2) make sure the limits on important things are ambiguous, (3) accept specific limits only if they are unverifiable, (4) evade even strict, verifiable limits, and (5) get the treaty to legitimize violations. He believed that "Because ours is an open society, our government cannot cheat on agreements," but that the Reagan administration was "apt to forgive Soviet cheating." The executive branch was run at the time by hardliners who agreed with George Will.

None of those fears materialized.

The Silver Screen. The motion-picture and television media produced shows that related to the nuclear arms race. Hollywood and pop-culture stars made a significant contribution to popular education about nuclear issues, adding recognition and a sense of legitimacy that encouraged citizens to speak out.

The controversy over telecasting *The Day After* is recited by an insider, David Cortright, in his book *Peace Works: The Citizen's Role in Ending the Cold War*. The televised movie resulted from "conscious decisions by writers, producers, directors and actors who wanted to make a statement on the prevention of nuclear war." The producers were "subjected to powerful political pressures from the ABC network, the U.S. military, and the [Reagan] White House staff." Right-wing writers like William Buckley editorialized against the film. The opposition, which did not want to scare European allies, succeeded in getting deleted a reference to Pershing-II missiles being moved forward in Europe as the scripted nuclear-war crisis unfolded.

[**On the Beach**]. A thoughtful 1959 presentation about Australians awaiting the effects of nuclear fallout from explosions that have destroyed the rest of the world. Because of an intermittent radio transmission, a U.S. submarine hopefully searches for survivors off the coast of California. Good performances by Fred Astaire, Gregory Peck, and Ava Gardner. Directed by Stanley Kramer.

Besides trying to edit *The Day After*, the Reagan administration "demanded the opportunity to respond on the air." After the movie was televised, a Ted Koppel roundtable discussion — heavy with former national security officials like Henry

Kissinger, Robert McNamara and Brent Scowcroft — also had Buckley, Elie Wiesel and Carl Sagan. Kissinger labeled the film "simpleminded."

> [*The Day After*]. Still considered one of the most controversial made-for-TV movies because of its unrelenting grimness. Chilling depiction of the aftereffects of the nuclear bombing of Lawrence, Kansas, contributes to a potent drama that humanizes the effects of nuclear war. Made in 1983 at the height of Cold War fears fueled by President Reagan's truculence, the presentation drew an estimated 100 million viewers, an extraordinary audience, thanks to the organized efforts of public-interest groups. Director was Nicholas Meyer, and leading actors were Jason Robards and JoBeth Williams.

Thomas C. Reed, the nuclear weapons designer who became a Reagan insider, says that "the movie was understated in its depiction of the horrors of nuclear war, but it had the unintended consequence of strengthening Reagan's determination to develop a shield against such attack." Reed adds,

> These activities, at home and overseas, were not entirely spontaneous. The archives of the Communist Party of the Soviet Union ... show that during the 1980s alone the CPSU distributed over $200 million in support of friendly political activities outside the Soviet Bloc.

Although polls within a few days following the broadcast of *The Day After* "suggested little had changed," the quick-response surveys "tended to miss the deeper impact of the film," which evoked and visualized the reality of nuclear war and "made the nuclear threat dramatically concrete."

> [*Dr. Strangelove*] *or: How I Learned to Stop Worrying and Love the Bomb.* A 1964 black comedy directed by Stanley Kubrick, starring Peter Sellers, George C. Scott, and Sterling Hayden. The U.S. President has to contend with both the Russians and his own political and military leaders, when a fanatical general launches an atomic-bomb attack on the USSR.

Governments at the time avoided balanced information and public education about the risks of nuclear-threat policies. Tendentious imageries of peace and security had been deliberately or routinely reinforced. Consequently, activist-artists associated with entertainment media, such as music concerts, television and cinema, set out to use their popularity and resources to offset the "business-as-usual" tendencies and dramatizations of daily life in a world of growing risk and fear.

> [*Seven Days in May*]. An absorbing, believable story of a military scheme to take over the U.S. government. Presented in 1964, two years after the Cuban missile crisis, the cast includes Burt Lancaster, Kirk Douglas, Frederick March, and Ava Gardner; directed by John Frankenheimer. Lancaster acts as the general (James Matoon Scott, JCS Chairman) planning the coup, and Douglas as the staff colonel who discovers the plot and reports it to Frederick March, the President.
>
> General Scott fears that an impending nuclear-disarmament treaty will be detrimental to U.S. national security; he feels than no treaty was ever worth anything, that Russia cannot be trusted, and that "one-worlders" don't give enough priority to "patriotism, loyalty, and sentiment." The anguished President, who suppresses the coup, says that "The [real] enemy is an age — the nuclear age."

****

As nuclear armaments spread and became more fearsome during the ebb and flow of Cold War tides, public resistance evolved with a multifaceted persona. The opposition to nuclear proliferation was initially comprised of individual scientists, academics, students and citizens, who became alarmed about the potential for reflexive and instantaneous obliteration of society. These individuals, finding common cause, learned to organize and network in order to increase their effectiveness. Chagrined by an apathetic public, the movement drew attention to the proliferation of nuclear weaponry, the growth of defense budgets, the horrors of atomic war, and the elusiveness of peace. Intensity of the opposition ebbed and flowed, usually reacting to the latest surge of armaments.

> [*Fail-Safe*]. First produced in 1964, *Fail-Safe* reflects a decision-making crisis that heads of American and Soviet governments are plunged into when a U.S. bomber is programmed to drop its nuclear weapons over the USSR. Suspense arises when the bombers are un-recallable due to "mechanical failure" in a "fail-safe" system of mechanical and human controls. A high-tension drama derived from a best seller by, the original version starred Sidney Lumet, Henry Fonda, and Walter Matthau.
>
> A "live" telecast remake instigated by George Clooney, was aired in April 2000, drawing upon fears that so many nations possess nuclear weapons and their technology. The suspenseful telecast featured actors Richard Dryfus and Brian Denehey agonizing over predicaments such as nuclear overkill and conflict resolution by war.

To get the message across, scientists and engineers organized themselves at local, national, and international levels. Most allied themselves with NGOs in order to weigh in on policy issues more effectively. Public-interest groups formed alliances with technical specialists, who used their prominence to gain attention and credibility, and assimilated their expertise so as to solidify arguments against military plans considered ill-advised. New NGOs popped up to deal more systematically with the emerging threat to humanity; institutes and foundations were called upon for direct and indirect support.

Driven at first by insufferable policies — unrestricted nuclear testing, mutual assured destruction, fallout shelters, and ballistic missile defense — vocal political activism became widespread. Government deafness to dissent intensified public resistance. Many sectors of society and the organized structure of science, education, religion, business, labor, and newsmedia combined to strengthen resistance to excessive militarism. The anti-nuclear movement became a transnational social organism.

Contrary to neoconservative efforts to paint a different picture, Ronald Reagan's combative approach to the Soviet Union was widely viewed with great alarm and foreboding from the moment he announced his candidacy. This stimulated the reluctant but inspired involvement of many individuals and organizations to counter his approach to world affairs.

Our historical review of individuals and public-interest organizations supports the "diaspora" analogy: As resistance to nuclear excess spread beyond the intellectual core of scientists, it was symbiotically reinforced. Formatively, the scientists had a leading role not only because of their analytical skills but because of their expertise in the new and exotic field of applied nuclear physics.

Geographically, the movement propagated over local, regional, and national boundaries, becoming transnational and eventually global. The protest and its substantive foundation reached all scientific and scholarly disciplines. Discontent encompassed national and international policies, with attention to specific issues associated with belligerency and war, overt militarization, environmental abuse, weapons proliferation, and unbridled nuclear expansion.

Some arguments put forth by hardliners and softliners of both sides were strong on hyperbole, driven by exasperating difficulties in responding to emotionalized claims and counterclaims,. This was especially true about projected harm from radiation, and especially from scientists who sometimes overstressed the physical consequences of radioactive fallout.

The Soviet participants in international meetings were approved by their government; they were duty-bound to report all contacts and relevant findings (which was the case for Americans too having contact with foreigners). Initially, Soviet scientists were expected to act as observers, to occasionally put pressure on Western governments, and often to encourage newsmedia and public-interest organizations to accept Communist Party views. Because the Soviet government tolerated few contacts with foreigners, increasing legitimacy was gained by loyal scientists who were allowed to travel to the West. Eventually high-level Soviet officials started to listen to the reports of Pugwash participants and to the more independent Committee of Soviet Scientists.

Regardless of constraints, Soviet scientists who participated with international groups had their own personal views of what should be accomplished, and they provided a rare private channel for Western scientists to feed information to Soviet policymakers and influence their decisions.

The effectiveness of the public movement in the West is difficult to gauge: it will have to be measured more in terms of what might otherwise have happened—rather than what can be proven. How far the nuclear arms race might have gone without their chorus of dissent is speculative, but terrifying. In any event, there are distinct instances where domestic and transnational networks had significant influence. This was specifically acknowledged by Mikhail Gorbachev, who in 1990 gave credit to the joint efforts of Soviet and American scientists, especially those who offered advice on basic principles of international security and strategic stability. Dr. Bernard Lown and the International Physicians for the Prevention of Nuclear War were specifically credited by Gorbachev with an "enormous contribution" to preventing nuclear catastrophe.

The glandular responses of government officials — East and West — to public resistance, anti-nuclear demonstrations, anti-war publicity, and dissenting statements made it clear that the opponents had a major—though gradual—impact by decelerating the tempo and constricting excessiveness of nuclear arms buildups and temptations.

Notwithstanding the Soviet Union's demise as political entity, ideologic force, and military power, huge stockpiles of potentially suicidal weapons remain. Hardliners have vigorously sought to maintain large stockpiles of nuclear weaponry; their opponents face a daunting task to defuse superfluous arsenals.

Cold-Warriors resented some coverage by journalistic media, particularly the cinema, which by making relevant thematic movies brought out a poignant reminder of contemporary inhumanity. To counter the so-called "liberal press," the right wing responded through well-funded journalists and publications within their ideological orbit.

Perhaps most disappointing was what Upton Sinclair and Studs Terkel identified as "brass-check journalism," a reference to the days when prostitutes received makeshift coinage from customers; at the end of the day, the brass checks were exchanged for cash. A shortage of qualified independent journalism was noticeable during the Cold War and persists to some extent.

B. ARMS CONTROL CONTROVERSY

Notable arms control issues that incited considerable public debate and shaped national policies are outlined below, roughly in chronological order. Intervention of the public, supported by knowledgeable scientists, particularly in the United States, was crucial to reining in government tendency (both executive and legislative branches) to see every dollar expended on military hardware as well-spent money. With government subject to little or no oversight and evincing a distressingly narrow view of what was in the national interest, it was principally public input that kept the more extreme and contentious nuclear-weapons proposals from being implemented.

Nuclear arms control was influenced and sometimes transcended by domestic and international problems. Perhaps foremost was the pervasive East-West conflict stemming from the disparate political, social, and economic systems. Some of this friction spilled over into surrogate wars, as in Vietnam and Afghanistan. United States involvement in Vietnam not only brought out strident partisan political polarization and protest: It also changed the style of domestic confrontation.

All of these controversies, along with the burgeoning civil-rights movement in the United States, increased the public's realization that many issues were linked — that there existed a shared concern among protestors about governmental arrogance and secrecy. This realization led to symbiotic relations among the expanding citizen networks, energizing and making more effective the opposition to nuclear arms buildup.

In 1969, in the months after coming into office, President Nixon secretly planned a massive military operation against North Vietnam, ordered a secret bombing campaign into Cambodia, and deliberately planted fake information about Viet Cong plans to blockade Haiphong harbor and invade North Vietnam. But, to Nixon's regret, "the doves and public are making ['Vietnamization' (turning the war over to the South Vietnamese)] impossible to happen." Also to his consternation, major public anti-war actions were scheduled by the Vietnam Moratorium movement and Mobilization Against the War, all of which clearly had an impact. Nixon was also fearful that the demonstrations would turn violent.

The long-running Vietnam war had a reinvigorating and synergistic effects on protest rallies and anti-nuclear dissenters: Heightened fear of nuclear war activated more citizenry to demonstrate against government disdain for arms control, and *vice versa*, anti-Vietnam demonstrations energized opposition to nuclear war.

Fallout Shelters

Although testing of nuclear weapons was one of the first and most enduring matters for world protest, civil defense became the most heated of controversies among scientists, and there was large-scale public involvement. Starting in the 1950s, civil defense — particularly proposals for fallout shelters in which to hide until the radiation subsided — brought forcibly to public attention the consequences of nuclear war. Public opposition won out eventually, frustrating the government's civil-defense program, while nuclear testing — wrapped in the mantle of national security — continued nearly unabated.

The public (and policymakers) were slow to comprehend the immediate devastation and overwhelming institutional stresses from nuclear detonations. There was much hope in less-informed quarters that widespread use of fallout shelters, coupled with emergency evacuation and recovery procedures, would ensure the survival of the infrastructure, leadership, and much of the population. Despite public dissent, considerable money was spent on civil defense and emergency preparations — resulting, however, in inconsequential protection against the effects of nuclear war.

Government efforts for civil defense are began in the 1950s and its revivals in the 1970s. Independent scientists and a skeptical public were crucial to rational and successful avoidance of a fig-leaf "shelter" against the havoc of nuclear war. Unaligned scientists were instrumental in deciphering the obtuse claims of the government and its hired specialists. While proposed defensive measures needed to be evaluated, and maybe some of them implemented, the government was not justified in stating or implying that they were a viable alternative to preventing war in the first place.

Civil-defense proponents, such as the respected physicist Eugene P. Wigner, favored increased planning, preparation, and stockpiling. But most experts, dating back to Vannevar Bush in 1951, believed that it was impossible to predict the extent of direct explosive or fallout damage from a nuclear attack and that it would be inordinately expensive and probably useless to try to use fallout shelters to protect the civilian population. Many antinuclear organizations actively campaigned against civil defense programs on the grounds that shelters fostered the belief that a nuclear war was survivable and winnable. Secretary of State John Foster Dulles opposed civil defense preparations for a different reason: he preferred that the nation rely on its "capacity for retaliation," not wanting the population's mood to shift from offensive to defensive.

Herman Kahn, because of fear of thermonuclear war that he himself helped generate, was able to exploit the controversy by obtaining contracts in the 1960s for the Hudson Institute to study civil defense systematically. Kahn conceived of a geographically small, $200 billion shelter system in caverns 1000-feet underground, theoretically capable of withstanding direct hits by nuclear bombs of up to five megatons. Such a shelter, he argued, would contain the seeds of an economic and industrial complex that would provide a base for rebuilding the country after the nuclear war he implied was nearly inevitable.

For a while, the program languished. Then, despite objections from within his own cabinet, President Carter responded to charges of a civil-defense gap by resurrecting the program. Conservative critics had evoked the long-discredited bomber and missile gaps of the 1950s; Rowland Evans and Robert Novak, two hardline columnists, decried what they called a "theology" that "[arms control] agreements are possible only if American citizens perceive that civil defense cannot prevent mutual destruction."

Long afterwards, an evaluation of cost and implications by the *Atomic Audit* concluded that

> While civil defense measures could mitigate the effects of nuclear weapons and save millions of lives, many tens of millions of lives at minimum were at continuous risk during the Cold War.

Scientific organizations were divided about the potential benefits of civil defense and emergency planning. Choosing to promote arms control rather than war planning, the Federation of American Scientists, and its chapters throughout the United States, began to speak out against the frantic and illusory appeal of civil defense in the late 1950s, early 1960s. The Chicago FAS group, under the leadership of physicist David R. Inglis (former chairman of the national organization), realized the limitations of civil defense in the advent of nuclear war. The Argonne physicists engaged in detailed analysis of advance preparations and emergency responses.

The government's vision of civil defense coupled to ballistic missile defense also brought criticism from the wider community of scientists, who saw that both notions would create false hope of survival following a nuclear attack. Cortright reminds us that "More than any other nuclear program, civil defense depends on the cooperation of the public, and to be effective, it requires massive public participation and volunteer effort."

Particularly energized by government efforts to plan for civil defense was the Physicians for Social Responsibility. PSR played an active role in community debates about civil defense and trying to cope with nuclear war.

That situation was coupled in the 1980s with the expressed belief by President Reagan that a sufficiently well-prepared nation "can survive and win a nuclear war," and with the views of his vice-president, G.H. Bush, describing how a nation could win a nuclear war:

> You have survivability of command and control, survivability of industrial potential, protection of a percentage of your citizens, and you have a capability that inflicts more damage on the opposition than it can inflict on you.

This callous government attitude provoked deep-seated humanitarian objections to the imposition of a Crisis Relocation Plan on local communities, who felt that they were not being consulted on the practicality or even the logistics of relocating populations. The relocation plan for New York City was particularly unworkable. As these realizations grew, prevention rather than protection against nuclear war became the choice preferred by citizens. Nearly 300 cities and towns rejected FEMA's relocation proposals, some preferring to sponsor public-education programs on preventing war.

The government had to retreat (no other word fits better) from its outlandish plans for civil defense. It was a major blow to the Reagan strategy of military confrontation. The extravagant plans were discredited by the derailing of administration and right-wing claims about surviving nuclear war. This was an important and precedent-setting victory for the peace movement — the combined efforts of the public, NGOs, activists, and scientists.

Nuclear Testing

The debate over nuclear testing can be traced to the mid-1950s, when the public and scientists reacted to wanton discharges of radioactivity into the atmosphere. Scientists had a strong hand in public test-ban discussions because it involved health and environmental effects and the reliability of technical means of verification. In response to the 1954 Bravo test on Bikini Atoll, a public campaign was organized by the FAS and SANE. Individual scientists like Ralph Lapp in the United States and Joseph Rotblat in Britain "provided the first public information on the health and environmental effects of radioactive fallout." In 1957 Nobel laureate Linus Pauling launched a campaign to gather signatures of prominent scientists opposed to further nuclear testing. Meanwhile, mass demonstrations and acts of civil disobedience were organized by SANE in the United States and the Campaign for Nuclear Disarmament in England. Later on, Women Strike for Peace was mobilized to give more of a family dimension in pushing the government toward a test ban.

The Soviets included a test ban in their comprehensive disarmament plan submitted to the UN in 1955. Although some alleged that it was intended only to score propaganda points, the plan itself was evidently genuine, incorporating many previous proposals from the West. Soviet scientists had advised their leaders that a test ban was a way of reducing the danger of nuclear war. At the time, the AEC perceived that the United States was far ahead of the Soviet Union in the development of thermonuclear weapons, but feared that the Soviets might catch up. In any event, Nikita Khrushchev took the initiative and issued a decree in 1958 announcing a unilateral suspension of Soviet nuclear tests. The United States agreed to a contingent test suspension after completing the program it already had under way, and the British — but not the French — joined the moratorium.

Fall 1961 was a time when political conflicts were leading up to the 1962 Cuban missile crisis, and little progress was being made in treaty negotiations. The Eisenhower administration, before leaving office (January 1961), had announced that the United States would not continue to abide by the test moratorium. On 30 October 1961 the Soviets broke out of the trilateral moratorium and tested a so-called "political weapon" — the largest thermonuclear bomb ever exploded, yielding more than 50 megatons and capable of being reconfigured to more than 100 megatons.

The original goal of public protest was a comprehensive ban on nuclear tests, but only piecemeal moratoria and limitations were forthcoming from governments intent on improving their weapons. While the Limited Test Ban Treaty in 1963 was the first major U.S.-Soviet arms control agreement, it helped demobilize the mass campaign that had been agitating for a full-fledged test ban. Keeping nuclear detonations underground — out of sight, out of mind — defused the bomb-banning movement. In the absence of popular pressure and transnational coordination, the prospects for meaningful East-West arms control declined.

Andrei Sakharov wrote that he and other Soviet scientists helped persuade Khrushchev to accept the Limited Test Ban Treaty.

Later in the 1960s, at a time of increasing public resistance to its ill-fated ABM planning, the U.S. government stirred up new controversy with its underground nuclear-testing program. In 1969, natural-gas-stimulation tests in Colorado brought out protests from geologists and citizens, who sued unsuccessfully in court to prevent the explosions. Later that year, a series of ABM-related tests was scheduled to take place at Amchitka Island, part of the Aleutian chain; these tests had been moved from Nevada because of the possibility of damaging buildings around Las Vegas. Expected to be the most powerful underground tests carried out by the United States, strenuous objections were raised on behalf of wildlife and the fear of stimulating earthquakes in Alaska.

Partly because of that protest which failed to stop the test, Greenpeace was started in 1971. An activist style in the cause of environmental protection and disarmament has become a hallmark of that organization. Since then, Greenpeace has tried to interrupt nuclear detonations elsewhere by the French, Soviets, and Americans elsewhere in the world. Greenpeace was founded and led by David McTaggart, who attracted some publicity in 1972 when he was harassed by the French upon sailing into their nuclear-test zone near Muroroa.

The Threshold Test Ban Treaty, consummated in 1974, was decried as a sham by those who favored a comprehensive halt to all nuclear weapons testing. The generous limit of 150 kilotons for permitted explosions (underground) had little or no adverse effect on the work of weapon designers, who were able to continue the development of small, efficient nuclear weapons. Appeals of public-protest groups were ineffective, lacking receptive government leaders.

In 1978, President Carter's Secretary of Energy, James Schlesinger, brought the directors of two weapons laboratories, Harold Agnew of Los Alamos and Roger Batzel of Livermore, to the White House to tell of their objections to a comprehensive test ban. The head of DOE's Office of Defense Programs, testified before a congressional committee that "confidence in the U.S. stockpile, in our best judgment, would degrade." Contradicting that statement, however, were Norris Bradbury (former Los Alamos director), J. Carson Mark (previously head of its theoretical division), and Hans Bethe and Richard Garwin (Manhattan Project scientists).

Vigorous public objection, especially by NGOs, against nuclear testing — whether above or below ground — has persisted ever since radioactive fallout from the Bikini detonations began to visibly affect humans and the habitat. In 1985, Mikhail Gorbachev announced a unilateral Soviet moratorium that was not reciprocated by the Reagan administration. In fact, when public-interest groups demonstrated that it was possible to verify a test ban with seismic monitoring, the administration shifted its arguments for continued testing: the ostensible purpose became maintaining the safety and reliability of the existing U.S. arsenal.

> [Impact of Public Opposition]. The mindset of former insiders is becoming more and more apparent as they write their memoirs. In his 2004 book, *At the Abyss: An Insider's History of the Cold War,* former Secretary of the Air Force Thomas C. Reed complains about the "political constraints" of "testing moratoria and bans" that "always have serious consequences." He takes issue with the October 1958 moratorium on nuclear tests, objecting that "This 'agreement' was not a treaty." Rather,
>
> > it was referred to by some as a "gentleman's agreement," illustrating by that name alone the folly of such an undertaking. A "gentleman's agreement" with a nuclear-armed dictatorship totally lacking legitimacy, led by murderers, few of who could be called gentleman? Absurd. Yet the moratorium had the potential of stopping the [U.S.] Minuteman warhead development in its tracks....
> >
> > The development and deployment of Minuteman saved U.S. taxpayers hundreds of billion dollars. Would those savings have been possible if the Soviets had not unilaterally resumed nuclear testing in 1961? Such treaties are important, and they can be beneficial to mankind, but their terms and conditions need to reflect fiscal and technical reality.
>
> Reed, in his statement above, is unwittingly supplying a backhanded compliment to public opposition. Warhawks were seething because their momentum was being slowed down. The moratorium was a verifiable three-year timeout for negotiations about the rationality of two grown-up nations spending endlessly and recklessly in order to amass a larger capability of destroying each other. Reed, the technocrat, was more concerned about fiscal and technical "reality." (He doesn't explain how Minuteman deployment saved "hundreds of billion dollars" within his context of fiscal "reality.")
>
> Opponents of testing were not allowed at the time to realize how effective public opposition was and how much it frustrated Cold Warriors.

In promoting the moratorium, International Physicians for the Prevention of Nuclear War was an especially influential organization, according to Academician Evgeny Velikhov. The transnational movement against nuclear testing persisted with its struggle in the face of U.S. refusal to join Gorbachev's moratorium. In April 1985 the State Department claimed, not surprisingly, that it was "deeply concerned about the desirability of an uninspected testing moratorium and the verifiability of restraints on nuclear tests." To debunk this political position, the USSR allowed on-site monitoring of the moratorium by the team composed of Western and Soviet scientists.

Not until after Mikhail Gorbachev encouraged openness (*glasnost*) in Soviet society did nuclear testing become an issue that mobilized the largest popular antinuclear activity in Soviet history: the Nevada-Semipalatinsk movement, which brought tens of thousands of protesters into the streets, contributing to shutting down the main Soviet nuclear test range in Kazakhstan. A popular Kazakh poet was leader of the NGO movement that linked American and Soviet citizens.

Post-Cold War debate about the CTBT is illustrative of the conflict between hardliners (who persist in supporting nuclear tests) and antinuclear partisans (who see the counterproductive aspects of continued testing). Public and congressional support for a halt in nuclear explosions led Congress in September 1992 — over the administration's objections — to impose a nine-month moratorium on testing, a moratorium that still continues.

The ABM and Sentinel Debates

The first major Cold War U.S. military program to suffer reversal because of public protest was a proposed antiballistic missile system. Missile defense continues to this day to be controversial. Because public-interest scientists were successful, we provide here more detail about the disputes. Having personally been involved with ABM opposition, especially in the Chicago area, an in-depth account can be furnished on how the turnabout was pressured. The debates are now well documented and stand as a model of successful resistance to a governmental steamroller.

Ballistic missile defense became a prominent military objective after *Sputnik* (1957), but a presidential scientific-advisory committee shortly afterwards cautioned that the (Nike-Zeus) technology for a U.S. ABM system was not ready for deployment. Meanwhile, the Soviets were working on what would eventually become their *Galosh* ABM system for defense of Moscow. Senator Barry Goldwater, the Republican candidate for President in 1964, politicized the lack of U.S. ABM progress. Because a missile-defense arms race was beginning to break out, discussions were held among Pugwash participants that same year about a possible arms control accord that included missile defenses.

A year later the U.S. Joint Chiefs — recognizing the futility of defending against a Soviet attack — changed objectives and recommended deployment of an anti-Chinese-missile system. The Johnson administration's version of the ABM — known as Sentinel — was aimed primarily at defending the nation's cities against a possible light ICBM attack that *might* come a decade later from Communist China.

Pressure for missile-defense deployment increased after information was made public in 1966 that the Soviet Union was deploying their *Galosh* system. The Republican presidential aspirant in 1967, Richard M. Nixon, used his campaign to pressure President Johnson on hastening the deployment of a U.S. ABM.

In any event, for the proposed Sentinel ABM system, Defense Department officials Paul Nitze and John S. Foster, Jr., favored a large metropolitan-based deployment, as did Congressional hawks. Consequently, the Army announced in 1967 that, of ten sites altogether, eight locations near major cities were to be surveyed for installation of the nuclear-armed Sentinel missiles: Boston, Chicago, Dallas, Detroit, Honolulu, New York, Salt Lake City, and Seattle.

Soon after, opposition developed in diverse locales, such as Seattle, Chicago, Detroit, Pittsburgh, and Boston. In a study aimed at evaluating the role of scientists during the ABM debates of the late 1960s, singled out were graduate physiology student Newell Mack of the University of Washington (near Seattle), physicist John Erskine of Argonne National Laboratory (Chicago area), and physicist Alvin Saperstein of Wayne State University (Detroit) for inspiring opposition at major metropolitan areas.

Starting early in 1967, under the leadership of Mack, graduate students and scientists at the University of Washington became entangled in the controversy, as did the Seattle branch of the FAS. By early 1968, national organizations like the Council for a Livable World and the FAS adopted positions against the ABM system, especially its urban siting. Articles in magazines such as *Scientific American* publicized the analysis of independent scientists.

Suburban Chicago Opposition. Discovery that the Army was surveying Spartan ABM sites in the Chicago area triggered increased opposition by the local FAS affiliate at Argonne National Laboratory, a center for nonmilitary nuclear research just west of Chicago. In late 1968, Erskine visited the proposed ABM location near his home. He brought his concerns up at one of the regular FAS lunch-group gatherings, where "The implications of ABM deployment had been discussed and dissected by the group for some time." David Inglis, who was the group's intellectual leader, had previously written articles opposing deployment of an ABM system, coining the phrase "H-bombs in the back yard." The Argonne scientists thus became "the nucleating agents for protests against the ABM." They routinely held weekly lunch-table discussions about the arms race and arms control.

Based on studies carried out on their own time, the Argonne activists formulated a plan to publicize ABM-deployment issues. In order to avoid unauthorized linkage with the Laboratory, they formed an *ad-hoc* organization, West Suburban Concerned Scientists. Nuclear physicists Erskine, Stanley Ruby, John Schiffer, Roy Ringo, and George Stanford began to address various local groups.

Nationally, protests propagated from the suburbs to the city centers. For example, in Chicago, a May 1969 rally of the ad-hoc organization Chicagoans Against the ABM drew 1200 people to Orchestra Hall. Speakers included Tennessee Senator Albert Gore, Nobelist George Wald, and Chicago Congressman Sidney Yates. There were 34 cosponsoring organizations — religious, labor, student, political, scientists (FAS Chicago Chapter; Scientists and Engineers for Social Responsibility), and peace (Chicago Committee for a Sane Nuclear Policy; North Shore Women for Peace; Women's International League for Peace and Freedom). Specific action items recommended were letters, telegrams, and telephone calls to members of Congress; local and statewide resolutions; advertisements in papers; public meetings and debates; and delegations to Congress-members. Thus, the proposed ABM system soon became the object of nationwide grassroots opposition.

Other contemporary national-security issues emerged. As one scientist pointed out, first we had the bomber gap, later a missile gap, and then a national-security credibility gap: now, there was a credulity gap. The credulity gap grew out of a government effort to push forward with the improbable in the form of ABM technology, capability, threat, and cost.

President Nixon contended that the Safeguard ABM system would not provoke an escalating arms-race response from the Russians. Moreover, he argued that if the United States did not start immediately with ABM, American diplomacy would not be credible in the 1970s. He was wrong on both counts: Soviet military planners let it be known that they would consider ABM an escalation in threat and a move toward a first-strike capability; and the already-existing huge nuclear deterrent had kept the United States in a formidable power-policy position.

Nationwide. Resistance to the ABM spread rapidly. The FAS Chapter in Los Angeles supported opposition to nearby county sites. FAS members in Detroit were drawing attention to the dispute. Some government scientific advisors, like Bethe, Wiesner, Drell, and York took the issue to the public and to Congress. Scientists and Engineers for Social and Political Action, using the slogan "Science for the People," entered the fray. The Universities Committee Against ABM collected donations and paid for advertisements in newspapers. A summer program on national priorities was organized by SANE. ABM opposition then became part of a growing general protest against what was perceived to be an inverted order of national priorities.

As protests against defense policy spread at campuses and laboratories, students and faculty at MIT and around the nation, as at Chicago-area universities, began to express opposition. A "research strike" was organized for 4 March 1969. An article in *Electronic News* the day before the strike was headlined "Colleges Brace for Stoppage Waves." Speaking engagements, many organized or coordinated by the UCS, and other activities took place at more than two-dozen universities and other locales. Statements about the misuse of scientific and technical knowledge were signed by participants. The UCS prepared and widely distributed an information pamphlet "abm abc," supplying detailed answers about ABM, MIRV, and the arms race.

"Counterprotests" supportive of the government took place on March 4 at some campuses and laboratories. A few scientists and engineers around the nation carried out a "work in." Pro-ABM scientists, prominently Edward Teller and Eugene Wigner, entered the dispute. One supportive organization was the Committee to Maintain a Prudent Defense Policy, co-founded in May 1969 by Dean Acheson (former Secretary of State), Paul Nitze (prior Deputy Secretary of Defense), and Professor Albert Wohlstetter (formerly of RAND, then at the University of Chicago).

John A. Wheeler of Princeton, an outstanding physicist who helped in the atomic and hydrogen bomb projects, testified before Congress that a limited ABM was a timely first step in constructing a "shield." He based this conclusion partly on his observations of "the subways of Leningrad and Moscow, two hundred feet below the surface," and what he heard of the Soviet ABM, all of which meant to him that the Soviets took defense seriously. He considered the Soviet ABM to be effective because "Without firing a single shot it has forced us into making a 20-percent cut in the effectiveness [i.e., the warhead yield] of our [MIRVed Minuteman] missiles." (In truth, the United States military chose to reduce the warhead-explosive yield in order to permit more missiles per launch vehicle, thereby increasing their net effectiveness. And the subways would have been tombs rather than shelters in the event of a nuclear attack.)

After accumulating facts and analyzing arguments, scientists normally prefer to publish reports that can be examined by peers. One of the more influential studies was prepared by A. Chayes and J.B. Wiesner, *ABM: An Evaluation of the Decision to Deploy an Antiballistic Missile System*. In the book, problems identified with the ABM were listed: insufficient reliability, inadequate fraction of incoming warheads intercepted, deficiency in mission-performance, and excessive acceleration of the arms-race. In the Introduction, Senator Edward Kennedy judged that "The material in this [book] indicates that the wise choice would be a deferral of a decision to deploy Safeguard at this particular time."

The independent evaluation made explicit note of several ABM-adverse findings. For instance, "five demonstration firings of ICBMs for Congressmen have failed." And, "the Ballistic Missile Early Warning System once misidentified the moon as a bunch of incoming warheads."

The evaluation had been requested by Senator Kennedy. The group of eminent scientists who prepared it was under direction of Jerome B. Wiesner, science advisor to the late President John Kennedy. It stung the government, causing an immediate "sharp rebuttal" from the Pentagon, Foster protesting that "One does not obtain a meaningful technical judgement by taking a vote of the scientific community or even of Nobel laureates."

On the issue of vulnerability to countermeasures, the ABM evaluation found that

> The ... Spartan missile defense "can be easily penetrated" by fooling the system with decoys, or by blacking it out with nuclear explosions or electronic jamming techniques.

In rebuttal, Foster claimed (fallaciously) that "the system could always be exercised against the latest Soviet weaponry." (If that really happened, it would be too late and too awful to designate it an "exercise.")

The Unraveling. In February 1969, Secretary of Defense Melvin Laird, recognizing the groundswell of opposition, ordered a halt to further investigations of ABM sites pending a project review. Except for the brief and expensive (about $5.5 billion) deployment in 1974 of a single Safeguard installation at a North Dakota ICBM site, the ABM system never materialized. Research and development continued, however, until the next phase in the history of ballistic missile defense: incarnation as SDI during President Reagan's administration.

The retrospective analysis of Primack and von Hippel concluded that:

> The fifteen Sentinel ABM bases initially envisioned might have come into being if it were not for the impolitic enthusiasm of [Pentagon officials] who decided to place several of these ABM bases in major American metropolitan areas.... The most important reason for the success of the Sentinel opposition lies in the fact that the arguments against "bombs in the backyard" struck such a responsive chord with the public.

Opposition scientists were successful primarily through a "multiplier effect." By bringing to public attention drawbacks of the proposed nuclear-armed ABM system, especially its technical limitations, they were able to catalyze and nourish opposition. Also, they created

> excellent relations with the press [reaching key local media, and] by doing their homework ... they could not be caught in careless errors.... [The Argonne scientists] prepared clear and well-written statements of their views for public distribution.

Other useful lessons for future dissidents to keep in mind include the following:

1. Executive branch spokesmen in an important national debate [cited selected] experts — while suppressing their [own experts' official] reports.

2. The ABM debate shows that even the general capabilities of advanced strategic systems can be publicly debated without the disclosure of classified details of hardware or tactics.

3. [In the debate], government officials publicly misrepresented confidential advice [about the effectiveness of the Sentinel and the Safeguard ABM systems].

4. The battle over the ABM sites in the suburbs served effectively to raise the entire issue of missile defense to a level of visibility where Congress was able to act for once as an equal branch of government in setting national defense policy.

National Spending Priorities

While specific objections to U.S. military and nuclear adventures occupied public attention, spending on defense became a major undercurrent issue, starting in the 1960s. Public concern over spending priorities became more pronounced as the "domino theory" — fear of susceptible nations falling under worldwide Communist expansion — rationalized America's involvement in the Vietnamese civil war. In early 1969 the liberal coalition that gelled around the peace movement began to direct its opposition both to the war in Vietnam and to excessive defense spending.

In 1967, military costs consumed 7 percent of the world's production of goods and services, supporting 50-million persons under arms. The United States, with the largest gross national product (GNP), easily outspent other nations in military expenditures. The USSR allocated a smaller amount than the United States to military purposes, but still a large fraction of its GNP. The military outlays of developed nations were growing twice as fast as their GNP. Public doubts grew about the argument that the United States could afford to allocate a large proportion of its resources to increasing military strength, particularly since there was little evidence that nonmilitary approaches to national security and international peace were being pursued.

The U.S. federal government was

> devoting more resources to the war machine than is spent by all federal, state and local governments on health and hospitals, education, old-age and retirement benefits, public assistance and relief, unemployment and social security, housing and community development, and the support of agriculture.

Of total 1969 federal outlays, $148 billion was available after social security trust funds were subtracted. War-related outlays, including the Defense Department, foreign military assistance, nuclear weapons, Veterans Administration, and interest on the public debt, amounted to $104 billion, taking 70 percent of that year's discretionary funds. At the time $29 billion per year was being spent to support Vietnam operations.

On 15 October 1969, and a month later, nationwide work stoppages took place among scientists and other professionals in and out of government, to demonstrate the growing opposition to the war in Vietnam and to distorted federal-spending priorities. Town meetings and rallies in cities were held throughout the United

States. A broadly based Coalition on National Priorities and Military Policy was formed.

The Chicago Committee on National Priorities, a regional group, organized public hearings in the late 1960s. They emphasized that two thirds of the federal budget was then being spent on the war, the defense establishment, and accrued interest from past wars. After taking into account the cost of operating the government, only 15 percent of the budget was left for health, education, and welfare. This was at a time when the Defense Department was allocating billions of dollars for the multiple-warhead MIRV weapon system. It was pointed out that if the United States deployed MIRV, the Soviets would also have to do so — as indeed they later did.

In the Chicago area during this interval, members of the Concerned Argonne Scientists and NOMOR jointly challenged the military view that "continued high-level spending on nuclear weapons and delivery systems would contribute to national security." They pointed out that it was unlikely that Soviet leaders would be "so myopic as to think they could win a meaningful victory in a nuclear war."

Newspapers like the *Chicago Tribune* disapproved "a complete and unconditional United States withdrawal from Viet Nam." They editorialized against "well-meaning and sincere opponents of war" who might be "sucked in on such revolutionary designs" and are "contributing to the breakdown of all order in this country." Moratorium-demonstration plans for November 1969 were drawn up, according to the *Tribune*, at a university meeting where "virtually every seditious group in the nation was represented." The newspaper derided the "dogma of the far left," warning that "[it might not be] far from the time when there will be barricades in the streets."

National priorities and defense spending triggered even more significant public struggles because they opened up broader issues: a panorama of festering social ills — civil rights, human rights, police repression, women's indignities, etc. Misplaced government spending stood in stark contrast to displaced human priorities.

Military-Industrial Complexes

A new generation of critics, many young and opposed to the draft, questioned whether universities like MIT should continue to do research in support of the military. Critics in Congress became emboldened and began to sharpen their ax for cuts in the Pentagon budget. Senator William Proxmire conducted hearings on "priorities," raising questions about the amount spent on defense compared to education and other nonmilitary activities. Perhaps a more abiding consequence was the move by organizations to get their own arms experts and to scrutinize the Pentagon budget more closely.

A favorite target of the national-priorities realignment movement was the U.S. military-industrial complex, which gained special notoriety during the yearly March 4 observances. The tacit partnership of the complex with the weapons laboratories and Congressional supporters did not escape criticism. Even

Republican Senator Barry Goldwater took note of the "scientific-military-industrial complex," somewhat reminiscent of Eisenhower's farewell caution.

In 1969, as a sign of the times, *Newsweek* wrote that Sunnyvale, in California's Santa Clara Valley, had forsaken its prior title as "the prune capital of the world," instead being saddled with the reputation that "Death is our favorite industry." Ironically, this is the same Sunnyvale that after the Cold War prospered in nonmilitary endeavors, especially computer technology. But at the end of the 1960s, Sunnyvale was home to Lockheed Missiles & Space and had payrolls for United Technology and other defense contractors.

Jobs were always a major consideration, accompanied by fear that an end to the war in Vietnam or the signing of international arms control agreements would lead to unemployment.

Of course, the Soviets had their own military-industrial complex; military industrialists were powerful because of their position in the ministries. Soviet citizens did complain of difficulties in obtaining food and consumer goods, and Soviet allies in Eastern Europe were dissatisfied with economic conditions, but that had little influence on the Politburo or the Central Committee. The Soviet military-industrial complex readily favored big expenditures on a ballistic missile defense to match the U.S. SDI. The complex and its leaders, like "Big Oleg" Baklanov, resisted conversion of military production to peaceful uses, even going so far during the final throes of the USSR as to participate in an abortive August 1991 coup against Gorbachev and Yeltsin.

ACDA

When John F. Kennedy became president, he moved quickly to fill a "gaping hole" in American foreign policy — the "lack of a concrete plan for disarmament" — by establishing in 1961 the U.S. Arms Control and Disarmament Agency. It was an attempt to institutionalize the disarmament process within the U.S. government. ACDA was given responsibility for coordinating government policy on nuclear testing and disarmament. Created after considerable opposition from conservatives in Congress, the agency — headed by political appointees — usually mirrored the views of the current administration. ACDA became the principal advisory agency on arms control and disarmament, and was directly involved in negotiations.

The institutionalization of ACDA resulted from considerable grass-roots pressure. Key individuals helping draft the implementing statute were Arthur Dean and John McCloy. William C. Foster was director during ACDA's formative period, from September 1961 to January 1969. Paul Warnke, who later became chief negotiator at the SALT II, served as President Carter's director of the agency.

ACDA's director was the principal advisor to the President and the Secretary of State on all issues related to arms control and disarmament, and usually was personally involved in the arms control negotiations with the Soviet Union. Many appointees to ACDA were seconded (assigned) by other agencies, including the Departments of Defense and Energy.

In 1999 ACDA ceased to exist as an independent agency when its functions were integrated into the U.S. State Department. Although disestablished, the functions ACDA once performed continued for a while, albeit with considerably less visibility.

SALT/MIRV

The ABM Treaty, part of the SALT I package, put an official end to deployment of ballistic missile defenses at the time. However, many strategic-offense projects were left unchecked, stirring considerable public dissent regarding the inadequacies of the SALT I and SALT II agreements. The treaties were perceived as licenses to increase warhead limits and as a clear tipoff that the underlying purpose of the "bargaining-chip" negotiating strategy (build offensive systems to use as chips in "bargaining from strength") was to buy time for additional weapons buildup. The SALT I agreement permitted American deliverable warheads to increase from 4600 to 9000 and the Soviets to go from 2100 to 4000. The SALT II Treaty would have legitimized continued expansion. Moreover, lack of curbs on MIRV deployment made it permissible for each side to develop and deploy missile systems with improved lethality and accuracy.

MIRV was
- unnecessary (the United States already had a secure retaliation force),
- dangerous (the Russians would also install them),
- avoidable (by negotiation),
- and imminent (because the U.S. testing program was already well advanced).

When MIRV was originally touted and developed by the United States, the ABM Treaty was still several years away. As a result, the looming perception of improved offense and effective defense was taking on ominous strategic overtones. The combined development of MIRV and anti-ballistic missiles, despite Secretary Laird's hints in 1969 that we would negotiate one of them away, was identified as "another step in the action-reaction cycle—just one more drink."

One of the earliest physicists to write often on nuclear-defense issues was Ralph Lapp. In April of 1969 he addressed the military-technical issues of the Soviet SS-9 warhead and its MIRVed ability to hit Minuteman silos. Concluding that Secretary of Defense Melvin Laird had "apparently misjudged the technical factors," Lapp's independent appraisal did not substantiate the case for an alleged Soviet first-strike capability.

In 1970, as the United States began MIRV deployment, it was described as the "Gorgon Medusa of the nuclear age." Suggested were two "ugly consequences of MIRV: the revival of first-strike concerns, and increased hazards to the civilian population from accidental launch." To avert these dangers, an immediate moratorium was suggested for MIRV testing, development and deployment.

Herbert F. York, who had headed Livermore weapons lab and preceded John Foster as DOD Director of Defense Research and Engineering, testified before a Senate Committee in 1970 that "ABM and MIRV require and inspire each other; together they will lessen our national security." He identified a "cycle of action and reaction" in the funding of these programs, observing that Soviet MIRV developments were "probably stimulated by our program." York's concern, therefore, for the SALT I negotiations was that both ABM and MIRV would have "multiplying and ratcheting" effects on the arms race and that they would introduce destabilizing uncertainties into the strategic equation.

An influential and timely article within the context of strategic-arms limitations appeared in the *Scientific American*. Written by G.W. Rathjens and G.B. Kistiakowsky, the article addressed not only the weapons systems but also the "worst-case analysis" paradigm. They reviewed strategic-weapons delivery systems, comparative ability to destroy property and slaughter people, and beneficial impact of possible arms restraints and reductions.

Right-wing columnists, on the other hand, railed against SALT negotiations. For instance, Walter Trohan, in a report to the *Chicago Tribune*, wrote that such arms-limitation talks could hurt the United States. He feared, for one thing, that the SALT I Treaty might permanently leave the United State in second place, militarily, behind Soviet Russia. Trohan chastised those who "would transfer American defenses to the United Nations." That "might put them under Russian command" and "might mean a surrender of U.S. sovereignty and eventual U.S. subjugation to world rule dominated by Russia." (Trohan had concocted a red herring, because arms controllers did not advocate that extreme path or extent of internationalization.)

Other skeptics perceived SALT as a negotiation and ratification process that "[provided] the political conditions for an increase in military expenditures and overall destructive capability ... more of a charade than an arms reversal program." The "quality of the charade" was revealed by the appointment of George Seignious, a career military officer and on-record opponent of arms control, to head the Arms Control and Disarmament Agency. The appointment was seen as an attempt to get the support of Pentagon officials for SALT II.

It was apparent that SALT II would not prevent the ongoing increase in overall destructive capability. One who doubted its arms-limitation value noted that unabridged qualitative improvements "make weapons faster, more deadly, more accurate, and more difficult to detect or defend against."

Although MIRV slipped through with minimal public outcry, its realization eventually turned out to be costly in terms of increased mutual risk; public resistance was ultimately mobilized. Despite the organized opposition, it was too late to stop MIRV — a lost opportunity for mutual self-restraint that has haunted the world ever since. And, ironically, hardliner antagonism to arms control prevented SALT II from being ratified, even though it would have done little to temper the arms race. Score another one for the hawks, who blocked further arms control progress and enabled the continued arms buildup that still jeopardizes nuclear stability and safety.

Proliferation

Anxiety that nuclear weapons could proliferate has had, for decades, a way of working up public fear. While nuclear weapons states strenuously argued against and inhibited the spread of weapons technology to other nations (horizontal proliferation), they were increasing the number and quality of their own nuclear weapons and delivery systems (vertical proliferation) — proliferation just the same. Although measures to restrain horizontal proliferation were easily justifiable, the illogical, self-serving, and inconsistent practices of the weapon states did not exactly reinforce the nonproliferation objective.

Considerable funding went into analyses by defense apologists, who decried the risks and dangers of nuclear proliferation while justifying a nuclear arms buildup by their own nations to protect against such contingency. Half a century later, time has shown that the spread of nuclear weapons to other nations has proceeded at a pace that was much slower than the alarmists predicted.

Some NGOs, seeing hypocrisy in the actions of the weapons states, almost universally opposed both horizontal and vertical proliferation thrusts. Other NGOs broadened or focused their agenda to challenge the spread of peaceful nuclear energy, erroneously attributing weapons propagation to technology availability rather than to paramount national aspirations.

This book began by scoping out the birth and circumstances of the A-bomb; yet, even today, the skewed diffusion and role of nuclear weapons remains an abiding public issue. The topic has continued as the primary agenda for NGOs.

Euromissiles

NATO's plans to base American nuclear missiles in Europe roused considerable dispute in the late 1970s and early 1980s. The additional nuclear weapons to be assigned were 108 Pershing-II ballistic missiles and 454 GLCMs. Public protests against the expansion flared up in Europe and the United States. Many individuals and organizations, such as the Union of Concerned Scientists, registered their disagreement with the deployment. George Kennan, made the profound statement in the about Euromissiles when he accepted the Einstein Peace Prize in 1981.

To protest the deployment of intermediate-range missiles in Europe, one of the largest American protests by anti-nuclear-weapon activists took place when more than 700,000 marched in New York City in June 1982.

Although the NATO deployment was carried out despite the protests, the perceived arrogance of the militarists irrevocably angered an increasingly larger sector of the passive public, resulting in expanding fundamental dissent against larger issues of worldwide nuclear arsenals and uninhibited military spending.

Neutron Bombs

The Carter administration's eventual decision in mid-1977 to produce neutron weapons (enhanced-radiation warheads — ERWs) ignited a controversy that flared up when the Reagan administration chose to deploy them in Europe.

Compared with regular battlefield nuclear weapons, ERWs derived more of their energy from fusion and less from fission, so that there would be wider spread of lethal neutron radiation and comparatively less destruction from blast. They were touted as "more humane" because their reduced fission-product yield limited radioactive fallout, supposedly sparing civilians by restricting their lethal effects to the battlefield.

But still they were nuclear weapons, with explosive yields dialed down. The trigger for detonation was a fission explosive — attended by unavoidable fission products, x-rays, gamma rays, and neutrons. Heat and blast effects were comparatively diminished.

An all-important battle predicament was that ERWs could not be distinguished in the field from any other type of nuclear weapon — thus making it unlikely that "limited" use of ERWs would avert nuclear warfare. Almost certainly battlefield use would invite escalation in kind, possibly even beyond Europe.

Newspapers and columnists were often impressed with the government's rationalizations about the military value of ERWs as gravity bombs, artillery shells, or missile warheads — three options for battlefield delivery. The *New York Times* straddled the fence, agreeing that relatively reduced blast and heat were qualities "that make neutron warheads attractive to our allies and to the Pentagon." Even so, the *Times* granted that the long-term solution to security in Western Europe was in arms control negotiations that would "do away with the need for tactical nuclear weapons altogether."

Right-wing columnist, Patrick Buchanan, a presidential candidate in 1996 and 2000, called the neutron bomb "humane" because it was targeted against military personnel rather than civilians. Supportive of Carter's decision to go ahead with production, Buchanan quoted a senator as saying that ERW opponents "have been almost irrational." Buchanan bought the argument that "neutron bombs in the NATO arsenals [drastically limit] damage to friendly territories and population." He felt the United States needed this added punch because

> We are virtually defenseless against bombers, and cruise missiles launched from Soviet planes, warships, and submarines. [The] threat of a U.S. retaliatory strike upon the Soviet heartland ... is less and less credible.... The presence of neutron bombs ... is a credible, invaluable deterrent against any Soviet attack ever being launched.

Equally hawkish were major U.S. newspapers, one editorializingd in March 1978 that "the neutron bomb very well might be the only weapon capable of saving Western Europe." A month later, objecting to President Carter's initial decision to delay development, the newspaper added that "This is one more serious administration mistake which could prove devastating to a future effort to contain a Russian military tide."

Both NOMOR and the Concerned Argonne Scientists challenged these views in letters to the editor, pointing out that conventional antitank weapons could counter the perceived threat of a powerful Russian armored attack into Western Europe. In contrast to a defense relying too much on nuclear weapons (including neutron warheads), the organizations gave four specific reasons why conventional weapons were preferable and would suffice. They were:

(1) selective in destroying targets, without extraordinary damage to the environment,
(2) a believable form of deterrence,
(3) dispersible, to avoid vulnerable concentration, and
(4) directly controllable and usable by nations that needed to defend themselves.

The rebuttal concluded that conventional defense was a better investment (of a billion dollars) than nuclear devices that indiscriminately irradiate large battle areas.

In Europe, considerable popular resistance to ERWs was aroused. The World Peace Council in 1977 distributed a pamphlet calling for the neutron bomb to be banned; they pointed out that "The entire population of Paris could be exterminated by 10 or 12 small neutron bombs."

Despite expanding public protest, President Reagan gave his approval for European deployment of neutron weapons in 1981. The ERWs were to be installed in Lance missiles and 8-inch artillery shells. Their yield could be preadjusted from two kilotons on down, and their primary tactical purpose was to defeat a massive Soviet tank and artillery attack.

Foreign government officials and press had mixed views about the neutron bomb. Typically NATO members praised it, with some reservations, while nations outside of NATO almost universally considered it inhumane.

Freezing the Arms Race

Around 1978, in response to frustration over the uninhibited nuclear arms race, a sweeping public movement began in an effort to slow down or "freeze" the military competition. Randall Forsberg, founder of the Nuclear Freeze Campaign, explained it in her *Call to Halt the Nuclear Arms Race*:

> To improve national and international security, the United States and the Soviet Union should stop the nuclear arms race. Specifically, they should adopt a mutual freeze on the testing, production and deployment of nuclear weapons and of missiles and new aircraft designed primarily to deliver nuclear weapons. This is an essential, verifiable first step toward lessening the risk of nuclear war and reducing the nuclear arsenals.

Earlier roots of the movement are traceable to public concern about nuclear power, especially in connection with the nuclear arms race. The accident at Three Mile Island, coupled with a visceral fear of radiation, gave a boost to anti-nuclear causes — which, for the most part, began to focus on the weaponry. Forsberg's *Call* combining emerging proposals for a nuclear weapons moratorium was based in part on published analyses showing the danger of nuclear arsenals. The formalized Freeze Campaign encompassed

- the suspension of underground nuclear tests, pending final agreement on a comprehensive test ban treaty;
- a halt to testing, production and deployment of strategic missiles and new strategic aircraft;
- a lid on the number of land- and submarine-based launch tubes;
- and cessation of further MIRVing or other changes to existing missiles or bomber loads.

Verification of a freeze would primarily be by national technical means, but measures were suggested to confirm a halt in production of weapon-grade fissile materials and the manufacture of nuclear bombs.

The case for a broad nuclear moratorium was based on the existence of nuclear parity. Freeze proponents saw the following benefits:
- avoidance of nuclear warfighting developments,
- reduction in pressure to build counterforce weapons,
- avoidance of weapons systems that would be difficult to verify (such as cruise missiles),
- preservation of European security,
- stopping the spread of nuclear arms, and
- the economic benefits that would follow.

The timing seemed right, and the idea was popular.

Contrary to the contentions of hardliners, the proposed nuclear freeze would do nothing to reduce armaments or weaken deterrence. It was a comparatively modest proposal that did not go far enough in the minds of those who called for significant reductions or abolition of nuclear weapons, and went too far in the minds of strong-defense advocates who favored an arms buildup. Nevertheless, at the time the Nuclear Freeze Campaign fostered policies that become a lightening rod for much of the peace movement.

Upon taking office in early 1981, President Reagan had "launched the largest peacetime military buildup in U.S. history," increasing the military budget of $130 billion in 1979 to nearly $300 billion in 1985. Reagan's aides had threatened nuclear strikes against the Soviet heartland, talked of a "nuclear warning shot" in Europe, and prepared for "prevailing" in a "protracted" nuclear war, which might include "limited nuclear exchanges."

A *pro* and *con* matchup of freeze arguments was presented in the January 1983 issue of *Physics Today*. Favoring the moratorium, physicists Hal Feiveson and Frank von Hippel argued that

> [The] reckless superpower competition to develop "counterforce" weapons — that is, weapons designed to destroy the nuclear weapons of the adversary — has finally provoked, in the United States, a popular demand to "freeze" the nuclear arms race.

Taking the contrary view, University of California physicist Harold W. Lewis, argued that the freeze rested on "the assumption that it is somehow the availability of weapons that leads to war, rather than the international conflict over national interests." In Professor Lewis's mind, "The inventory of nuclear weapons has nothing to do with [progress toward peaceful means of resolving international questions]."

Reflecting a hardline position, Lewis asserted that "the verification problem has turned out to be technically very difficult." A government spokesman echoed this concern, claiming that "A freeze on all testing, production and deployment of nuclear weapons would include important elements that cannot be verified," based on the vague and undefined standard that arms control must include "effective" means of verification. Feiveson and von Hippel countered: "Although it is not surprising that opponents of a freeze focus on the weakest points of its verifiability, this should not obscure the fact that methods of verifying the most important elements of a freeze have already been worked out in considerable detail." Incidentally, it was controversies of this type that inspired many scientists to volunteer their technical knowledge for improving verification technology and meeting other arms control challenges.

Lewis added his opinion that nuclear war was "far more likely to occur through the inexorable proliferation of nuclear weapons to parties less responsible than either we or the Soviets."

Central Park in New York City became the site of the largest political demonstration in U.S. history, as nearly one million people gathered on 12 June 1982 to conspicuously protest the nuclear arms race. Organized by SANE/Nuclear Freeze, participants "trekked from every part of America and from around the world to demand an end to the arms race and commemorate the opening of the United Nations Second Special Session on Disarmament." People came by charted buses, special trains, or marched in from the boroughs. The rally enjoyed overwhelming popular, celebrity, and scholarly support. Speakers included Dr. Benjamin Spock, Freeze founder Forsberg, Congress-members Ron Dellums and Ed Markey, and veteran peace activists David Dellinger, Cora Weiss, and Seymour Melman.

A California "Peace Sunday" brought 90,000 to a stadium to hear political speeches and performances by musical artists. The speakers included Muhammad Ali, Dolores Huerta, Reverend Jesse Jackson, Petra Kelley, and Patti Davis (youngest daughter of President Reagan).

Fall 1982, in the closest equivalent to a national referendum in the history of American democracy, 30 percent of the American electorate voted on a bilateral freeze proposal put on local ballots through the efforts of the Freeze campaign.

On a local level, in October of 1982, the *Chicago Tribune* came under fire from readers for repeatedly asserting that

> the supporters of a "mutual and balanced freeze" on Soviet and American nuclear weapons don't understand the problem very well; if they did, they would abandon the freeze campaign.

The *Tribune* equivocated its editorial position that month by opining, "So the President [Reagan] is right, and the nuclear freeze advocates are wrong. It is too bad that neither seems to understand why."

Taking issue with the newspaper's intransigence, one reader retorted,

> Despite a barrage of reader protest to its editorial position, the *Chicago Tribune* continues to oppose a freeze on the nuclear arms race.... By associating the term "unilateral disarmament" with the freeze, the Tribune is misrepresenting the movement. The [actual] proposal is for a mutual freeze, to be followed up by negotiations and verification.

After elections a month later, the *Chicago Tribune* got "The freeze voters' message" and modified its editorial position: The newspaper wrote that President Reagan, who had called the nuclear freeze movement "wrongheaded" could not "for one minute ignore the freeze movement" because "American voters made a powerful statement with their overwhelming approval of freeze referendums all across the country."

As a matter of fact, "nuclear freeze referenda were on the ballot in nines states and dozens of major cities, with 11 million people, 60 percent of those voting, supporting the proposal for a mutual halt to the arms race."

At its fourth national convention held in St. Louis in December 1983, the National Nuclear Weapons Freeze Campaign established "Freeze Voter '84," a political action committee, to campaign for candidates supporting the Freeze and work toward the defeat of candidates opposing it. Conference participants called on Congress to pass a "quick freeze" to halt funding for testing and development of nuclear weapons. They also expanded their platform to include:

- getting the U. S. and the Soviet Union to adopt non-intervention policies in Third World countries;
- adopting a "no first use" policy on nuclear weapons; and
- banning the use of satellite and space weapons.

While no moratorium directly resulted from the freeze movement, the momentum of the arms race began to be systematically slowed as a result of that public campaign and other ongoing challenges to unbounded nuclear armaments.

A backhanded salute to the nuclear-freeze movement has been furnished by Thomas C. Reed, Reagan's national-security counselor. He elevates the significance of the Freeze campaign, when he complains in his book that "the Reagan administration faced a propaganda offensive calling for a "nuclear freeze." Evidently the public movement caused considerable consternation at the highest levels of the Reagan administration. (Incidentally, the word "propaganda" is usually associated with governments who use taxpayers funds for organized information campaigns, not with citizens who voluntarily supply their own money to support public protests.)

These and other attention-grabbing public challenges in the 1980s had a significant impact on government policy, profoundly challenging the Western political establishment. As interpreted a decade later, "Although the nuclear freeze as a specific proposal to halt the arms race was never adopted, the larger social movement it spawned had an enormous impact in paving the way for an end to the Cold War."

Years later, and after many other arms initiatives that were challenged — some discussed below — the 1987 signing of the Intermediate-Range Nuclear Forces (INF) Treaty was a forerunner for the cooling down of the arms race. Dramatically improved East-West relations were followed by demilitarization and political toleration. Even so, hardliners would not give credit to grass-roots opinion and activism:

> When the INF Treaty was signed, Randall Forsberg, founder of the Nuclear Weapons Freeze Campaign, declared, "This is a victory for the peace movement," [but] Secretary of State George Shultz countered that "if we listened to the freeze movement there never would have been an INF treaty."

SDI

Highly exaggerated claims made by proponents of the Strategic Defense Initiative made it easier for the public to identify its absurdities and derail the program. The SDI concept took shape in the summer of 1981 at a right-wing Heritage Foundation meeting, organized by Edward Teller, Lowell Wood, and George Keyworth. It included "an influential group of scientists, industrialists, military men, defense executives, and close friends of President Reagan.... Members of this group first met with Reagan in the White House on 8 January 1982." The proponents included members of High Frontier and other right-wing organizations.

The extraordinary influence of the two proponents, Teller and Wood, both from Livermore, eventually became a liability to the SDI cause once their role became obvious to opposition scientists.

The earlier debates in the 1960s and 1970s about ABM proposals had led to widespread awareness and skepticism, and critics were well prepared to analyze SDI and mobilize against its flaws.

Doubts that independent scientists had about SDI were amplified by a long list of contemporary technological disasters in the late 1980s: the Challenger space-shuttle explosion; production problems with the B-1 bomber and the Titan and Delta rockets; delays in the Stealth bomber and Trident-II missile programs; a flawed mirror in the Hubble space telescope; temporary grounding of the space shuttle; and nuclear accidents at Three Mile Island and Chernobyl. Reports by independent panels of the Office of Technology Assessment and Defense Science Board in 1988 strongly criticized the Reagan vision (derisively called "Star Wars" by critics, a label that infuriated the administration).

In the Soviet Union, reformists such as Evgeny Velikhov, Roald Sagdeev, and Andrei Kokoshin — who kept in contact with Westerners like Bernard Lown, Jeremy Stone, and Frank von Hippel — successfully urged that Gorbachev unlink SDI from the INF negotiations.

Scientific opposition to SDI was widespread and organized; the UCS and the FAS produced many reports criticizing technical underpinnings of the program. The FAS's John Pike was a particularly well-informed source for the media. Analyses, statements, sentiments, or voices from technical groups or organizations — the OTA, the APS, the NAS, university faculties — bolstered by a widely circulated "Pledge of Non-Participation" in the SDI program — turned into a boycott of SDI work by many scientists and engineers.

The credibility of SDI was dealt another serious blow in July 1989 when a group of ten American physicists, congress-members, and journalists were allowed inside the heretofore highly secret Soviet proving ground located at Sary Shagan in Kazakhstan. The Pentagon had portrayed a huge laser there in the most threatening terms. But the team, which included von Hippel and Tom Cochran, reported that the lasers were "very ordinary" and "1000 times less powerful than those of SDI," and the flap was "much ado about nothing." An historian who studied the Pentagon claims said that the "laser gap" was a "lie."

SDI proponents tried to bootstrap the program by claiming that the Soviets were ahead in antiballistic missile research, but their arguments were unconvincing.

Another hallmark of the Reagan administration was the "bargaining chip" approach. Had SDI succeeded as a program to trade for Kremlin concessions in ICBMs, it would have been, as Reagan's national security advisor later admitted, "the greatest sting operation in history."

Trading away a viable offense for a defense dream was never realistic. Instead, successful political and moral pressure was exerted by the anti-nuclear weapons movement, aroused and outraged by the Reagan administration's bellicose rhetoric and its relentless trajectory toward nuclear brinkmanship. The Star Wars fantasy of a leakproof population defense against nuclear attack was derided by expert consensus, and an historical study confirms that

> some administration officials, right-wing activists and military and scientific figures [shamelessly misled] the public and Congress in an effort to prop up budgets and circumvent commitments made by Washington in the ABM Treaty.

Ironically, President Reagan himself never visualized SDI as a bargaining chip; it was his national-security advisors who thought of it in those terms. Reagan insisted that he "had to tell the Soviet leaders a hundred times that the SDI is not a bargaining chip." Ronald Reagan was a true believer in missile defense from as far back as the early 1970s. In the minds of his administration, though, SDI was a way of countering the Nuclear Freeze movement while offering a positive vision that would be based on "the traditional American belief in technological superiority."

The (George) H.W. Bush administration later embraced the concept of a space-based "brilliant pebbles" defense system — another brainchild of Teller and Wood, who had personally marketed to the Reagan administration the concept of shooting down missiles with x-ray lasers. Teller and Wood were eventually accused by their Livermore colleagues of "overselling" and "potentially misleading" Reagan administration officials about the prospects for a nuclear-driven x-ray laser. Further, Roy D. Woodruff, a dissident at Livermore, successfully challenged — at great personal risk — the "overly optimistic, technically incorrect" interpretation of secret x-ray laser results from an underground nuclear test. The magic wand never materialized.

Another dead end for the Reaganites was their effort to distort the existing arms control framework (without the consent of Congress). First, the White House tried to reinterpret the ABM Treaty so as to allow some SDI testing. Next, it sought to breach the limits of weapons development contained in the unratified SALT II agreement in order to start phased deployment of SDI components. Both of these arms control derailments were averted by public and Congressional resistance.

Shortly after coming into office, President Clinton reversed the Reagan and Bush policy on ballistic missile defense. This was in good part due to the urging of U.S. public-interest groups and individuals, such as Chris Paine and Tom Cochran of the NRDC and Frank von Hippel of the FAS. Not only did they and others influence members of Congress and the administration, but in some cases they and like-minded colleagues joined the Clinton administration in positions where they could have a direct effect on policy. Even so, they found that it was difficult to counter the military and corporate interests that promote BMD systems.

Still disputed is the impact SDI had on Soviet policy. "Peace through strength" supporters credit Star Wars with acceleration of Soviet political and economic reforms. But, according to Sovietologists, if SDI had any effect at all, it delayed the onset of reforms and heightened East-West tensions. Soviet initiatives for nuclear disarmament were not a response to Star Wars; the initiatives were pursued by reformers within the system despite opposition by Soviet hardliners, who exploited the SDI program to further their own agenda. Moreover, *perestroika* (restructuring of the Soviet economy) resulted from internal pressures, not pressure from Ronald Reagan or SDI.

ASAT

Antisatellite projects have gone through several incarnations. These programs are examples of technologies, driven by greed and narrow interests, that are inimical to the broader concept of national security. The ASAT idea was to develop a means to destroy satellites, despite the very important strategic stabilizing role of satellite reconnaissance, and despite the ability of the other side to reciprocate by destroying objects in orbit. To a significant extent, nations, especially the United States, relied on satellites for military, economic, environmental, communication, and other purposes. They still do, more so than ever.

Considering the risk of U-2 flyovers, and numerous other incidents that infringed on air space in order to gain information about warlike intentions, supports the value of satellite reconnaissance in time of peace or conflict.

Some ASAT concepts involved exploding a nuclear weapon near the targeted orbiting satellite; in fact, a common thread in the ABM, SDI, ASAT, and even some of the new-millennium Ballistic missile Defense (BMD) proposals has been a defense that depends on nuclear detonations to destroy moving targets that cannot be reliably intercepted with non-nuclear weapons. ASAT opponents also fear that nuclear weapons would be stationed in orbit and later directed to bombard land-based targets.

Beginning in October 1968, the Soviet Union tested co-orbital satellites — objects in close orbit to each other. The West characterized these as "interceptors." Through 1981 there were 19 identified tests in the Cosmos series that were taken to be ASAT interceptions. It is claimed that, in the 18th test, the (non-nuclear) interceptor was exploded in the proximity of the Soviet target satellite, which was thereby disabled. Thus, development of a Soviet ASAT was reported in the Western press, with allegations of successful standby deployment.

In the early 1980s, published information about Soviet advances in directed-energy devices raised alarm in the U.S. defense community. The devices could have been adaptations of scientific advances in lasers and nuclear-particle beams conceivably could be stepped up to destroy objects at a distance. However, there was no way to know whether the Soviet directed-energy technologies were intended for ASAT or any land-based military purposes.

The alleged Soviet satellite interceptions took place at a time when lasers and particle beams were being developed for military and peaceful applications. To deter the Soviet Union from attacking Western satellites, countermeasures were claimed to be necessary for countering perceived ASAT capabilities. Even so, any threat would have been limited only to satellites in low orbits, which did not include "the vital early warning and communication satellites in geosynchronous and median orbits." Enormous boosters would be needed to launch ASATs that could threaten satellites in higher orbit.

Clearly if ASAT weapons did exist, they would have serious impact on crisis stability and uncertainty. Because accidents and ambiguous activities often occur, it was feared that ASAT warfare — especially in a time of heightened tension — could escalate to exchanges of nuclear weapons. Antisatellite capability endangers satellites that keep each side informed about the other's strategic preparations and operations, and its development could also lead to potential offensive roles. ASAT programs thus drew sustained opposition from scientists and NGOs — such as the UCS — who proposed verifiable treaties banning deployment of ground-based and orbiting antisatellite systems.

Aside from trying to match the perceived ASAT threat, the simpler alternative of making satellites more survivable and redundant was considered (and eventually carried out). In the meantime, bilateral and international arms control talks were proposed to deal with the issues. In 1983, Chairman Yuri Andropov informally promised a unilateral Soviet moratorium on testing ASAT weapons. The following year, Andropov's successor, Konstantin Chernyenko, suggested banning antisatellite weapons in outer space.

The laser ASAT program was challenged by NGOs — in particular by a collaboration that included the FAS and the CSS, with support from volunteer scientists. The NGO coalition concluded that it was possible to set up instruments to verify that high-powered ground-based lasers being developed for peaceful applications were not operated in ASAT mode.

An original impetus for ASAT, nuclear weapons in orbit, had evidently not materialized. Nuclear-armed ASATs would have had major operational limitations, the main one being that orbiting debris and electromagnetic interference from their detonation would damage not only the intended target satellite but also other satellites as well.

ASAT programs were linked to SDI, as both a defense component and an arms-race escalation. In 1986, during the height of political and budgetary pressure on President Reagan's SDI, a temporary ASAT moratorium — indefinitely extended — was passed by Congress under the lead of Representative George Brown.

INF: Zero Solution/Option

The landmark Intermediate Nuclear Forces (INF) Treaty did not suddenly happen, nor was it solely crafted at the behest of government officials. Transcending more than a decade of arms control issues — including Euromissiles, neutron bombs, a global freeze, SDI, and ASAT — the zero solution (getting rid of all INF missiles in Europe) is traceable mostly to public reaction and action.

Contrary to neocon assertions, the INF Treaty was mostly a triumph of the peace movement in western Europe, played out against the backdrop of prior and concurrent actions of American and Soviet-bloc academics, physicians, scientists, and other activists.

"Zero solution" was a logical phrase for a concept that gained political currency in Europe: no intermediate-range nuclear missiles for either NATO or WTO members. The term "zero option" was used by U.S. officials — reluctant to acknowledge non-deployment as a "solution" and quick to remind others that they, the officials, had the power to choose a viable alternative.

Peace mobilization and public objection in 1977 to nuclear militarization in Europe surged in reaction to the U.S. proposal to assign neutron bombs to NATO. The Dutch Interchurch Peace Council led one of the first influential protests. The Campaign for Nuclear Disarmament got into gear, and thousands of new peace organizations were being formed all over the continent. Although much of this dissent was focused on nuclear missiles deployed by NATO and WTO, the peace movement brought unbearable pressure against the entire inbred East-West arms competition.

NATO had justified its intent to deploy nuclear-armed Pershing-IIs and GLCMs on the grounds that the missiles would counterbalance Soviet SS-20s. The street people wanted neither; they campaigned for the zero solution. Perceiving that the NATO deployment was accurate enough to knock out command bunkers, the Soviets became alarmed too. All of this led Europeans to fear that their homelands would become a nuclear battlefield. Backyard dread like this has stymied many contemplated official programs and misadventures.

While military leaders spoke reassuringly of "flexible response" and "limited war," nuclear phobia was spreading to Europe. Few people had confidence that a war could be limited to a small battlefield. Even President Reagan indirectly acknowledged that a limited nuclear war could expand. In 1981, mass rallies directed equally at the Soviet Union and the United States started to take place in Brussels, Paris, and London. When it was realized that NATO cruise missiles would be based in Sicily, Italian citizens joined the wave of protests too.

> For many people, the culprit was not one side or the other but the entire Cold War system of military competition and nuclear escalation.

These pressures dragged the public-relations-wary Reagan White House into at least appearing to favor a zero "option." They thought that the Soviet Union would not accept the proposal, and they were right. So, deployment proceeded despite millions of strenuous protesters who poured into the streets of Europe in 1983. Yet, as the missiles were being deployed in 1984, protests continued, becoming more embedded, and broadened in the public disarmament agenda. The apparent victory for American hardliners was to be short-lived.

The main breakthrough came in 1986, when Mikhail Gorbachev — who listened to peaceniks of the West and East — proposed broad-ranging disarmament to take effect by the end of the millennium. In the following year, the Kremlin agreed to decouple the INF talks from a growing problem with SDI; this forced the White House to take *yes* for an answer to their dormant proposal for the zero option.

Hardliners were aghast that the INF Treaty might actually be consummated, and they inveighed against the zero option because it might "decouple" the United States from European defense. Henry Kissinger and Richard Nixon wrote a warning that the proposed agreement might lead to "the breakdown of the NATO alliance." Brent Scowcroft, John Deutsch, and R. James Woolsey were concerned that the zero solution would be a "step toward denuclearizing Europe" and would weaken the "linkage between the United States and Europe."

Meanwhile, President Reagan's administration was under public siege, starting in 1986, for its role in what became the Iran-Contra scandal. Possibly to deflect attention, the White House got serious about arms control. In fact, the administration's turnaround regarding the INF negotiations resulted in a considerable boost to Reagan's public-approval ratings.

The INF Treaty became a reality in 1987, a delayed victory for peace activists, showing that pressure on policy makers can be exerted by public opinion, peace movements, and sympathetic insiders. "[The] old policy of nuclear buildup and East-West military competition was no longer possible." Blanket claims about national security were not enough to muffle domestic political dissent.

Verification

Verifiability became a litmus test for acceptability of a proposed arms control treaty. In the disputes over verifiability, technical-policy issues surfaced, such as intrusiveness of on-site inspection, difficulty of counting nuclear-armed RVs on a MIRVed missile, proof of warhead dismantlement, and ambiguity of dual-capable SLCM warheads.

Scientists at U.S. universities and non-weapons laboratories in the 1980s found the government's unrealistic standards of compliance and its arguments about nonverifiability to be incompatible with experience and principles of physics. Individually and together, the scientists brought the techno-political inconsistencies to the attention of the executive branch, Congress, and the public. Laboratory scientists, by affiliating with NGOs, were able to bring practical experience into the debate.

One technical controversy embedded in verification polemics was the question of nuclearized SLCMs. Technological breakthroughs had made it possible to replace conventional high-explosive (HE) with nuclear warheads without changing the appearance of the missile's exterior or its launch tube. SLCM verifiability thus became a serious obstacle in 1987 to the S T A R T treaty, then under discussion. Complicating matters further, "insertable" nuclear warheads had been developed by the Livermore, allowing military forces in the field or aboard ship to convert conventional SLCMs to nuclear.

However, occasional inspections of SLCMs officially declared to have conventional HE could assure that nuclear warheads had not been installed. While shipboard inspections would not eliminate the treaty-breakout risk, it would demonstrate ongoing compliance with the proposed treaty. Opponents of a treaty alleged, *inter alia*, that it was not possible to verify SLCMs aboard a ship unintrusively, nor to discriminate between types of warheads. Nevertheless, independent scientists understood from long experience that, by looking for the characteristic radiation from the heavy metals (fissionable materials) in nuclear warheads, those warheads could be distinguished from conventional ones by using instruments already developed.

Frustrated by government intransigence in the debate over verification, public-interest foundations funded and NGOs carried out a series of joint experiments.

One of them was a specific research project shared by FAS, NRDC, and CSS. (Both FAS and NRDC maintained full-time professional staffs, while the CSS was an *ad-hoc* non-governmental group comprised mostly of senior Soviet scientists.

NGOs made a major contribution by assembling, editing, and publishing reports on verification conferences and workshops. VERTIC, under the leadership of Patricia Lewis, issued a series of yearbooks on arms control, environmental agreements, and verification.

As a result of these interventions by independent scientists and NGOs, it became increasingly difficult for governments and hardliners to undermine arms control agreements by alleging — without foundation — that the treaties were not verifiable. The Reagan administration's mantra, "trust but verify," based on an old Russian proverb, inadvertently gave rise to renewed international interest in verification technology and procedures. As a consequence, a more subtle fallback strategy was then adopted by the hardliners, who either tried to raise the bar for verification standards or insisted that precious U.S. secrets would be compromised.

Government research, development, and information exchange continued; however, official activities were initially restricted by security constraints, and participation by outsiders was limited. In addition, the U.S. weapon labs attempted to coopt the entire agenda for verification R&D, especially on topics dealing with nuclear munitions. Gradually, the Department of Energy, through its contractors, supported conferences and workshops that were opened to outsiders; and the Defense Nuclear Agency-sponsored verification conferences, deliberately arranging for foreign arms control-policy specialists and verification experts to attend.

MX

President Nixon decided early in his administration to equip American land-based missiles so that each had maneuverable-warhead delivery systems, known as MIRVs. This had a profound impact on the arms race, laying the foundation for a decision to proceed with the largest multiple-warhead ICBM that was allowed under SALT II. The Carter administration had authorized full-scale development of the MX-missile (loaded with ten MIRVed thermonuclear warheads), an ICBM that was to be 2½ times as heavy as the Minuteman III it was replacing. In the course of time, the MX was "designed to be accurate enough to destroy Soviet missiles in their silos." It was also to be mobile so it could be moved from one base to another around the country.

Pressure to replace the Minuteman had come from several directions. One was the notion that "bigger is better," stressed by those who believed the United States should have the capacity to threaten destruction of the Soviet land-based ICBM force. Such "silo-busting" capability became an important factor during Carter's presidency. His national-security advisor, Zbigniew Brzezinski, argued that the strategic power of the MX would give the United States an ability to respond in kind, and that it should improve Soviet incentives to negotiate a hoped-for SALT III treaty.

Critics recognized that SALT II legitimized building up to higher ceilings, as well as permitting new weapons being constructed outside the ceilings. The high ceiling of 1320 MIRVs in SALT II encouraged a 50-percent increase in deployment, allowing the 10-warhead MX to be deployed as a mobile missile.

Former CIA official Herbert "Pete" Scoville, Jr., determined that the United States

> never sought [in SALT II] to negotiate seriously the differences in position of the two countries, because the government had already decided not to seek a MIRV ban in SALT I. This was a critical mistake for which we are paying dearly today [1981]; now it is the Soviet MIRVed missiles that are threatening our ICBMs and are being used to justify the MX deployment in multiple protective shelters.

How to distribute the MX among bases, a controversial issue, was not announced until September 1979. The original plan was for "racetrack style" deployment, with 23 shelters available per missile, together with an ability for the missile to dash from one shelter to another on warning of an attack. Mostly, the missiles were to be shuttled around wide-open spaces west of the Mississippi River. This didn't go over well with the American public, particularly out West.

It was Livermore weapon-lab scientists who in 1978 had proposed moving the MX between hardened shelters, fearing "continued modernization of Soviet missile systems" that "will soon lead to a serious strategic imbalance."

Livermore also suggested that "adding ballistic missile defense was a cost-effective method of enhancing MX survivability in the 1990s." The "imbalance" they alleged was based on projections of an unchecked and uncluttered "Future Soviet ICBM threat (1990s)" consisting of 908 heavy missiles with 17,540 RVs (an estimate that turned out to be twice the actual number attained by the USSR).

MX deployment decisions did nothing to help the SALT II Treaty ratification process, which had started to collapse. The complex interaction between MX and SALT is a striking illustration of the consequences of President Carter's vacillation. Even after the election of President Reagan, the MX program remained uncertain; there was no question, however, that priority development of the missile would be continued.

When he came into office, Reagan's way of persuading the Soviets to enter into arms limitation agreements was to negotiate *after building up U.S. nuclear capabilities*. He invoked the possibility of an "arms race" as a "card" (or bargaining chip) to be played in nuclear arms negotiations. Reagan's top Pentagon scientist enumerated the President's wish list for the arms race: MX, D-5, BMD, GPS, ALCM, GLCM, and SLCMS, B-1, and the "Stealth" plane. Except for the ballistic missile defense, these systems all came to fruition. At the time, Congressional critics estimated the cost to be between $200 billion and $500 billion. When asked if the buildup, particularly the MX-SDI double-whammy, was actually being designed and built to persuade the Soviets to enter into arms-limitation agreements, a Pentagon scientist responded, "There has got to be a pony in this pile [of horse manure] somewhere." The "pony" he had in mind was an arms-reduction treaty.

The Reagan arms buildup was billed as a "coherent package" to modernize aging hardware in the atomic defense triad of air, sea, and land-based arms. Instead of spreading out 4600 MX silos in the western states, the plan called for 1600 silos to be "defended" by an ABM system. Critics immediately recognized that there was a problem if a nuclear missile was detonated over one's own territory to stop somebody else's missile. There was also concern that the radiation bursts could detonate warheads still in silos. Moreover, as estimated by Herbert Scoville, an attack against MX sites in the West would cause extensive fallout: 100 rem (total) radiation on East-coast cities; 10,000 rem on Denver (mean acute lethal dose being 600 rem); and the Midwest farm belt would be contaminated for years.

One counterproposal, attributed to Rep. Albert Gore, Jr., was to load only a single warhead on each missile to make the MX a much-less attractive target for a preemptive strike.

A massive campaign for and against the MX unfolded in the press and among concerned citizens. NOMOR, as part of its reach-out program, issued "A Layman's Guide to the Mobile Missile (MX)," describing its rationale, history, dimensions, armament, accuracy, basing mode, cost, location, and context.

Even Harold M. Agnew, a former director of Los Alamos weapons laboratory, advised against MX deployment, suggesting instead that the Minuteman missiles' silos should be hardened and many more warning satellites could be placed in orbit.

Among the senators leading the fight against MX was Adlai Stevenson of Illinois, stating for the Congressional Record that President Carter's decision to develop the MX missile "signals a shift in U.S. policy away from an assured second-strike capability to counterforce and the first strike."

Am MIT physicist challenged the Minuteman-vulnerability assumptions that had provided a basis for development of MX. He calculated his own probabilities of silo kill, and concluded that "to claim a Soviet attack in 1982 would destroy 90 percent of the Minuteman silos is nonsense without any theoretical, and even less pragmatic, basis."

The FAS declared the MX to be not just an inflationary multi-billion-dollar strategic mistake, but an arms control disaster.

While "Peacekeeper" MX missiles did eventually get built and deployed, they were treated like "new wine in old bottles," being assigned to existing silos throughout the nation, rather than to any of the mobile-basing modes that heightened the controversy. The public protest had succeeded in moderating the buildup and deployment of this untimely weapons system.

Missile Accuracy (and Reliability)

In the 1970s, fear that the opposing side had a knockout punch was a strong driving force for new, more accurate ballistic missiles. In the early 1980s, the Reagan administration announced the existence of a "window of vulnerability," during which the U.S. retaliatory forces were assumed to be subject to destruction by massive Soviet preemption. This "window" could only be closed, hardliners argued, by deploying the highly precise MX missile.

Because assumptions about the adversary's capabilities could not be validated, the estimates of missile and warhead capabilities had to be based on indirect and incomplete information. Even hardliners such as Senators Barry Goldwater and Orrin Hatch were skeptical that guided missiles would hit targets with pinpoint accuracy. Goldwater noted that intercontinental rockets had never been fired along realistic routes against specific land targets (nor have they been to this day).

> The problem of missile accuracy over untried routes, termed "bias" [errors due to gravitational fluctuations, electromagnetic forces, and atmospheric turbulence] by experts, threatens a major assumption behind the U.S. drive to build the MX missile — that Soviet missiles are now so accurate that the enemy could destroy all the American Minuteman missiles [6000 miles away] in a pre-emptive strike [landing within 500 feet].

It was to counter this perceived threat that the Reagan administration planned to unpredictably move MX missiles from silo to silo in Utah and Nevada. This would make Nevada and Utah a "nuclear sponge," a phrase used by the Air Force Chief of Staff. As with the abandoned mobile-basing mode, this again did not sit too well with Westerners in general, to say nothing of environmental groups, pacifists, and others.

According to former CIA official, Herbert Scoville, Jr., although "accuracy is much more important than yield in determining the effectiveness of a warhead in destroying hard targets," other factors were important too. The question of accuracy really could not be separated from such decisive ICBM parameters as system reliability, warhead yield, and — above all — "nuclear sufficiency" (a term coined by the Nixon administration to refer to an assured ability to retaliate, i.e., to the preservation of MAD — Mutual Assured Destruction). Unless reliability and coherency are built into the entire missile-launch system, as well as command and control, better accuracy of a few missiles would have added little strategic value. Nor would going to large warhead yield compensate adequately for inaccurate or unreliable missiles.

Keeping in mind that the trajectory of a ballistic missile consists of three phases: the powered-flight portion (boost-phase), the ballistic-flight portion (mid-course), and reentry (terminal-phase) of the vehicle into the atmosphere as the missile approaches its target,

> One must first be able to distinguish the target from its surroundings; one must then be able to determine the exact position of the target with respect to the launching point of the missile; finally, one must be able to launch and guide the projectile in such a way as to make the [unavoidable] "error radius" ... smaller than the [desired] "kill radius."

Parameters subject to errors or uncertainties that contribute to target-miss include: initial position, velocity, vertical alignment, azimuthal alignment, accelerometer bias, accelerometer calibration, misalignment of accelerometer, vibration of accelerometer, gyroscope drift, thrust termination, gravitational anomalies, target location, and reentry effects. These result in accuracies (defined as CEP — circular error probable) limited to 150 meters for land-based and 400 meters for submarine-based intercontinental missiles. Effectiveness also depends on the "hardness" of a target and on the explosive yield of the weapon.

In the 1970s, arrays of microelectronics and microsensors were being developed. "The new developments make possible what essentially is a new set of methods for accurate delivery of strategic weapons ... including stellar-guidance systems, terrain-matching systems and satellite-based global-positioning systems." These factors gave birth to the nuclear-armed air-breathing cruise missile as a strategic weapon.

Technologies to improve accuracy "will almost certainly be interpreted as reflecting the adoption by the U.S. of a strategic doctrine that anticipates the efficacious destruction of Russian land-based missiles as one of its options." Consequently, Soviet response would be to either shift to "launch on warning" (which they did) or eventually to "achieve comparable performance for their strategic weapons" (which they also did).

Thus "two flaws" result from trying to develop higher-accuracy weapons:

the action taken will undoubtedly spur an adversary ... to continue the competition in strategic weapons [and] the action taken will ultimately lessen the security of one's own land-based missiles.

In other words, success simply in the appearance of improved accuracy would negate the military advantage to be gained by greater accuracy even before it could be achieved.

(To the preceding glitches, we add the ultimate uncertainty about nuclear sufficiency: It would never be known for sure whether technical improvements, such as improved accuracy, helped or hindered the assurance of deterrence. Any development that causes the opponent to go to a heightened state of launch-on-warning is a detriment to nuclear stability.)

Nuclear Winter

As nuclear arsenals and rhetoric were being escalated during the Reagan administration, scientists began to study the secondary effects of all-out nuclear warfare. A 1983 series of articles in *Science* magazine presented technical evaluations of the consequences of multiple nuclear explosions. Calculations suggested that a major nuclear war would cause a long-lasting "nuclear winter" that would spread around the globe, affecting everyone. In other words, there might be a short-term "winner" of a nuclear war, but there was concern that in the long-term the entire world would suffer devastating effects from a major impact on the climate.

Needless to say, these calculations stirred up controversy. The Pentagon did not take into account that atomic war could lead to "nuclear winter" until 1985, long after scientists like Carl Sagan postulated that "exploding nuclear bombs would set fires that would burn for months, sending up so much soot that sunlight would be blocked for months, lakes would freeze and temperatures could drop by as much as 75 degrees." An influential study predicted weeks or months of sub-zero temperatures resulting from a 5000-megaton exchange, one-fifth targeted on cities.

The potential atmospheric and climatic consequences of nuclear war were analyzed using models previously developed for volcanic eruptions. Significant hemispherical solar-radiation absorption, ozone attenuation, and sub-freezing land temperatures could be induced by fine dust and smoke generated in widespread nuclear explosions. An acceleration of particulate and radioactivity transport from the Northern to the Southern Hemisphere was predicted. Similar predictions of "a strong temperature drop" came from a computational study performed by the Soviet Academy of Science.

Some of the uncertainties relate to technical deficiencies in modeling, but a major source of doubt was connected with assumptions regarding the nature of nuclear exchanges.

Paul. R. Ehrlich and colleagues took a look at the potential long-term biological consequences, finding the postulated nuclear winter could severely reduce productivity of natural and agricultural ecosystems and could destroy the biological support systems of civilization.

Carl Sagan, in an article published in *Foreign Affairs*, described potential climatic implications of a large-scale nuclear war. "The central point," he stated, "is that the long-term consequences of a nuclear war could constitute a global climatic catastrophe." A popularized version of his article, published in the mass-circulation magazine *Parade*, reached Soviet scientists, who earlier became involved because of Victor Weisskopf's Pugwash contacts with Evgeny Velikhov.

Sagan reached his conclusion by looking at the sizes of nuclear arsenals if exploded and their immediate and long-term effects. At the time, the total number of nuclear weapons (strategic plus tactical) in the arsenals of the two nations was close to 50,000, with an aggregate yield near 15,000 megatons. Estimates of immediate deaths from blast, prompt radiation, and fires in a major exchange in which cities were targeted "ranged from several hundred million to 1.1 billion people." The four major climatic effects — smoke in the troposphere, dust in the stratosphere, fallout of radioactive debris, and partial destruction of the ozone layer "constitute the four known principal adverse environmental consequences that occur after a nuclear war is 'over'."

Perhaps the most striking and unexpected conclusion was that "even a comparatively small nuclear war can have devastating climatic consequences, provided cities are targeted." Surface land temperatures would drop by as much as 60 degrees Fahrenheit (35 degrees centigrade) if the centers of 100 major NATO and Warsaw Pact cities were burning. Reasonable apprehension evolved that severe and prolonged low temperatures could follow a nuclear war.

While the climatic studies of the time were based on admittedly pessimistic assumptions and unvalidated science, the forebodings they stirred up had to be taken seriously because it was not possible to debunk the claims — governments had not taken a systematic look at the potential dangers of all-out nuclear warfare. Irrespective of scientific validity, the controversy over nuclear winter aroused considerable debate on nuclear arms control issues and on the specific concern of wartime climatic effects; the debate also resulted in improvements to global-climate modeling and data accumulation.

B-1 and B-2 Bombers

Military-industrial-scientific lobbying for strategic bombers was vigorous throughout the bitter conflict. Nuclear-armed bombers have an enduring role in the strategic triad, and pressure for new, larger, more expensive bombers in enormous quantities has never subsided. The defense complex's crusade for B-1 and B-2 strategic bombers highlights this quest.

In 1981, writing in support of the B-1, a former assistant director of ACDA from 1977 to 1979 favored President Reagan's going ahead with the B-1 bomber, partly because it would not have the "political liabilities" of the MX, partly because of the perceived vulnerabilities of land-based ICBMs, and partly because he considered the bombers to be cost-effective.

The United States ended up spending hundreds of billions to build 100 of the controversial B-1 bombers, designed for intercontinental nuclear strikes deep inside the Soviet Union. But in the post-Cold War environment, they became so expensive to maintain that 27 of them had been mothballed by 1998.

In mid-1970s, a drive to "Stop the B-1 Bomber," under the National Peace Conversion Campaign, was led by two organizations with regional offices: the American Friends Service Committee and the Clergy & Laity Concerned. Framed with emphasis on national priorities, the Campaign issued flyers calling for public action against development of the bomber. The flyers pointed out that U.S. leadership in military power was at the expense of such factors as health, literacy, infant mortality, and life expectancy. In tackling the military's arguments, the Campaign retorted that the B-52 did not need near-term replacement, that the nuclear-triad system would provide robust deterrence without the B-1, that manned bombers for strategic war were not needed, and that the USSR was not trying to catch up with the United States in the number of strategic bombers.

(All of the arguments against the B-1 about societal expense and weak justification proved to be valid.)

Typical of public assemblies throughout the nation, a Peace Conversion Conference was held in 1974 at Northwestern University, with speakers drawn from the Chicago area and from the nation's capital. Workshops were held on such topics as military conversion, new developments in the arms race and arms control, multinational corporations, stockholder action, peace studies, and war resistance.

However, the B-1 program went ahead, albeit at a reduced scale. Actually, bombers are a more stabilizing element of the strategic triad than land-based missiles. But governments seem to want more of each, which ratchets up the arms race. The B-1B production version has been in service since 1986, with 100 of them manufactured, and some being assigned to close support operations in Irq.

Next came the B-2 stealth-bomber program. According to the Council for a Livable World,

> The saga of the B-2 began in the late 1970s when the United States commenced a top secret program to construct a fleet of radar-evading bombers dubbed "Stealth." These aircraft were originally intended to penetrate undetected deep into Soviet airspace to deliver nuclear payloads. Development of the B-2 program began in 1981. Initially, the Pentagon sought to acquire 132 planes at an estimated cost of $21.9 billion. Because of

the highly clandestine nature of this weapon, each phase of initial research and development was restricted to top secret clearance.

The Reagan administration funded the B-2 through the Pentagon's "special access" budget. Even details unrelated to technology such as the annual budget, program management, production, and testing timetables were considered classified information. This extreme secrecy served to prevent knowledge of the B-2 from reaching Congress and the general public, as well as the Soviet Union.

In 1988, Congress was shocked to learn that over $22 billion had already been spent on the B-2. Pentagon figures revealed that the estimate for 132 planes had skyrocketed from $21.9 billion in 1981, to $70.2 billion in 1990. Most members of Congress knew nothing about the B-2 program, and this revelation intensified the debate as the unit cost of each B-2 kept climbing.

By 1993, the cost for 20 B-2s had reached $44 billion, and the following year supporters argued for retrofitting the bombers for conventional-war missions. In 1999, operating against Serb positions in Kosovo, the B-2s flew 30 hours nonstop from Whiteman Air Force base in Missouri, refueling twice in midair each way. They were capably deployed, especially during bad weather, against well-defended targets such as air defenses, command posts and communications facilities.

The advantage that the B-2 has over other bombers in the Air Force's arsenal is its stealth ability, which works best at night. Both the B-1 and the B-52 fly faster and carry more weapons, but the Pentagon was unwilling to send the B-2 into action without protection. At a unit cost up to $2.2 billion per aircraft, it is unlikely that anyone in the Pentagon wants to be responsible for losing a B-2. Twenty have been in active service, including in Operation Iraqi Freedom.

The B-2 *Spirit* has a high maintenance cost, and needs to be kept in special stateside climate-controlled hangars to prevent deterioration of its radar-absorbing skin. Other aircraft, including the B-52, are deployable from forward bases and can be refueled while in flight.

In contrast to the Senate, which had steadfastly opposed the B-2 since 1993, parochial members of the House of Representatives have attempted to extend its production. However, the B-52 has at least 40 years of service life, according to the Pentagon.

Founding a Peace Institute

Just as the creation of ACDA in 1961 was controversial, so was the formation of the United States Peace Institute in 1985. There was considerable hardliner resistance to establishment of a government organization dedicated to peace and conflict resolution; yet the idea of developing effective options to control international violence and conflict finally did take the form of a national institute that serves the public and government in education, research, and information services. Passage of an enabling act was the result of a long-term bipartisan effort led by members of Congress and backed by the National Peace Academy Campaign.

The United States Institute of Peace was chartered to strengthen the nation's capabilities to promote peaceful resolution of international conflicts. In the post-Cold War political environment, the Institute has sponsored a number of initiatives that address issues of special concern: e.g., Bosnia in the Balkans; Religion, Ethics, and Human Rights; The Rule of Law; and Virtual Diplomacy. The Research and Studies Program designs research projects on a broad range of current issues affecting international peace. The Institute also attempts to bridge the gap between academia and government by convening meetings to bring academics and former officials together with current policymakers.

Conventional-Force Reductions

Counterpoised conventional forces in Europe were the focus of a major East-West Cold War dispute that dragged on and on, partly because the problem was intertwined with tactical nuclear forces. Fifteen years (1973–1989) of MBFR (Mutual Balanced Force Reduction) negotiations, aimed at lessening the quantity and offensive capability, were going nowhere. Before and throughout the negotiating period, East- and West-bloc troops and armor were braced to attack or defend, as required, at frequent high levels of alert.

After Mikhail Gorbachev came into office, he revised Soviet military doctrine to emphasize defensive operations, implemented a unilateral reduction of troops and equipment, and renounced military intervention. Partial credit for Gorbachev's initiative belongs to a transnational coalition of Soviet and Western supporters of alternative security measures and "non-offensive defense." By 1988, living conditions in the USSR had deteriorated so much that their own skeptics were convinced that conventional-force reductions would benefit their economy. This situation made it realistic for Western peace researchers and Soviet civilian reformers to advocate changes in Soviet and Western military postures. Such views about alternative defense and common security made their way into the debate within the Soviet government.

Among Soviet reformists was Andrei Kokoshin, who in 1985 had written favorably of Soviet unilateral initiatives to reduce the ongoing military threat in Europe. Kokoshin and Evgeny Velikhov, working with Westerners like Frank von Hippel and the FAS, actively promoted proposals on non-offensive defense — a force posture to provide security without appearing to be threatening. To bolster this case, Randall Forsberg's Institute for Defense and Disarmament Studies, with von Hippel and retired U.S. Ambassador Jonathan Dean, held a joint seminar on conventional-force reductions with IMEMO experts in Moscow.

The civilian advocates prevailed when Gorbachev announced his unilateral reductions in December 1988. Military officials went along reluctantly and belatedly with the official Soviet policy. This broke the MBFR stalemate on conventional troop reductions, soon leading to European agreements on Confidence- and Security-Building Measures (CSBMs) and the Conventional Forces in Europe (CFE) Treaty.

Despite efforts by the military on both sides to scuttle MBFR, CFE, and CSBM negotiations, public-interest groups kept pressure on the governments to engage in serious talks. To defang nonverifiability arguments, multilateral NGOs like the Bochum Project engaged in serious cooperative conventional-armament verification experiments on both sides of the receding Cold War. With the concurrence, at least tacit, of officials in Eastern and Western Europe, the Bochum Project carried out realistic verification tests starting in 1988 using NATO and Warsaw Pact military vehicles and aircraft, demonstrating that technical measures were available to provide confidence in stabilization measures and force reductions.

> **The Walls Are Tumbling** (*DeVolpi*). It was early December 1989, and our multinational team, cold and shivering, was carrying out conventional-force verification-demonstration experiments in an open field at Baumholder, West Germany. During breaks our attention was riveted on the radio because, just a month earlier, the Berlin Wall had been breached, and it was still being torn down.
>
> Our Czech colleagues had another reason to huddle around and listen to their mobile radio: A popular uprising in Czechoslovakia was demanding the resignation of their Communist Party leaders (which happened on the last day of our experiments). A new government, a new era, awaited our jubilant colleagues when they returned to their homeland.

National Missile Defense

Although the U.S. arms control movement remained alive after the Cold War, it endured a serious setback in 1999 when the U.S. Senate refused to ratify the CTBT. However, the movement scored a (temporary) "win" in the year 2000 when the Clinton administration postponed a decision to start deploying a national missile defense (NMD), which would have required abandoning the ABM treaty and possibly other arms control agreements as well.

The postponement of NMD deployment probably would not have come about without its flaws being spotlighted by individuals, organizations, foundations, and coalitions — particularly, the Coalition to Reduce Nuclear Dangers, which drew attention in June 1999 to the insufficient number of tests and their unrealistic nature. Major analytical support came from the UCS, the MIT Security Studies Program, and the FAS. Under the leadership of John Isaacs of the Council for a Livable World, the Coalition encouraged the appointment of highly respected "validators" who advised the Clinton administration to put off a deployment decision until the next administration.

This pressure, along with public disclosure that tainted test results were being covered up by the Pentagon, delayed for a time the decision to initiate deployment. However the G.W. Bush administration decided to ignore BMD's fundamental technical and strategic deficiencies, and the ABM Treaty was voided in 2002. The ABM Treaty had been a major bulwark in controlling of strategic arms. With the reincarnation of "true believers" in the G.W. Bush administration, NMD remains a divisive problem.

Section B: Arms-Control Controversy

****

Of the contentious East-West arms control issues that we have chosen to discuss, the majority involved new technology for nuclear armaments or weapons-delivery systems — most of them proposed when mutual assured destruction had already reached the saturation point. Other debates raged over inadequate government efforts to counterbalance nuclear escalation with overtures for peace and arms control. Verifiability of proposed treaties and agreements was a major sticking point, a matter of personal and professional concern to many scientists.

Inducing intense public aversion were other faulted government actions and programs: misplaced national-spending priorities, increased danger of proliferation, and the environmental consequences of an unchecked arms race.

As one looks at controversial events that defined the Cold War, a feeling of deja vu *creeps in: So many publicly contested unproductive arms developments have yet to be squelched.*

A fundamental struggle, still continuing, has simmered about passive and active defenses against nuclear attack. At first, defense advocates thought of nuclear weapons in the same terms as conventional weapons, but the advent of thermonuclear explosives made the unavoidable horrors visible to all but the hardliners. The short flight time of ballistic missiles makes them extremely difficult to intercept; scientists realized quickly that engineered hardware for unilateral ballistic missile defense could not be a dependable basis for national security.

After China in 1964 successfully tested a nuclear weapon (but had not yet tested a missile to deliver it), the fear of a growing Chinese nuclear force was used by the United States government to rationalize its initial attempt at ABM deployment, which later had two reincarnations: SDI and NMD. Forty years later, the threat from China has yet to materialize — their nuclear arsenal is still far smaller than that of the United States. Hindsight has validated the foresight and logic of ABM critics.

Independent scientists, some of whom volunteered to work with NGOs, challenged the wisdom of the government on the issue of treaty verifiability. The professionals realized from their technical experience that means and procedures were already available to carry out non-intrusive confirmation of critical agreements. The NGOs soon became savvy to the political distortions of the hardliners about treaty verifiability.

The beguiling dream of security offered by active or passive defense against nuclear attack dampened endeavors to reduce armaments; this recurrent fundamental conflict is still to be resolved.

Transnational coalitions of public-interest groups played a major role in stimulating arms control agreements between the adversarial governments — including a one-sided Soviet testing moratorium and the demonstration in Kazakhstan of on-site seismic monitoring. Ironically these interim successes thwarted the longer-range goal of a comprehensive test ban: When the governments agreed in 1987 to enter into negotiations on additional measures for verifying the Threshold Test Ban Treaty, the transnational movement lost much of its steam. It was nearly another decade before the CTBT was signed by the superpowers and most other nations. (Although approved by Russia's Duma, the CTBT was still not ratified a by the U.S. Senate two decades after the negotiations started.)

Because of growing public opposition, supported by knowledgeable scientists and reinforced by coalition-building activists, government policy increasingly was no longer taken as sacrosanct, to be meekly accepted without protest. Were it not for public intervention, many of the unfettered arms buildups and pseudodefenses would have run an even more expensive course, bringing humanity closer to suicide: national security turned on its head.

Few military programs were killed as a result of public protest, but some projects — like missile defense, nuclear testing, ASAT, MX, and Euromissiles — were trimmed down to less-threatening proportions. Public intervention was successful at least in slowing the unwarranted quest for unlimited killpower and in moderating the deployment of untimely weapons systems.

President Reagan's benevolent flip-flop about the realities of the Cold War could be ascribed largely to pressure from the peace movement. In October 1983, he and his senior advisors adopted the concept of offsetting asymmetries: the idea that one category of weapons could compensate for another: Not everything had to be matched item by item. Moreover, the Reagan administration eventually took a more conciliatory posture. Saber-rattling slowly subsided and negotiated arms reduction followed along the pathway promoted by peace activists and other moderates.

As for the highly touted Reagan "bargaining chip" approach to negotiations, there is no evidence that it had any substantive role, especially since playing a chip often resulted in asymmetric responses to the stakes. For example, Star Wars was of little value to the United States as a bargaining chip because the Soviets could easily and much less expensively thwart it by installing countermeasures and by increasing offensive forces.

Independent scientists and engineers were able to render a major public service in being able to understand, analyze, and explain the pros and cons of complex weapons proposals. SDI is an example of a program that failed the technological credulity test.

Successful protest against early versions of civil defense had an enduring aspect: It emboldened dissenters to maintain their vigil and continue to speak up against other travesties. The peace movement eventually became schooled in the means of carrying on resistance — of rallying support, and of networking with scientists, academics, other professionals, and the public. Invigorated by a whiff of success, activists saw that on a selected issue the federal government could be held to a standstill.

By challenging and diluting government-managed dissuasion, peace-activist coalition defiance in the 1980s eventually had a substantial role in derailing warlike policies of the Western-bloc establishment.

C. POLITICAL FORCES

Shaping the confrontational politics of the time were the bi-polarity of antagonists and the multi-polarity of nations that were bystanders, co-conspirators, or pawns in the Cold War. This state of affairs became a self-fulfilling dynamic, goaded even more by the pervasive action-reaction linkage and by the psychology of worst-case planning. There were also those who viewed the trying situation as an active battle between good and evil; so we must include such ideology in the spectrum of relevant political forces.

Up to now, the focus of this Chapter has been on groups, organizations, and issues. Yet, specific individuals were extremely influential during the twists and turns of the arms race. Also influencing government decisions were autonomous opinion leaders and dissidents, as well as some diffuse cultural factors.

To partisans on the American political right (usually labeled hardliners or hawks), the major driving force in the arms race was the Soviet (communist) threat. They saw the Soviet Union as a dangerous, aggressive, expansionist power committed to worldwide revolution and, ultimately, to world domination and the destruction of Western values. Manifestly aggressive Soviet conduct included repression in eastern Europe; provocative behavior in Berlin; support of communism in North Korea, Cuba, North Vietnam, and Nicaragua; the invasion of Afghanistan; and the continuing nuclear military buildup. Not only Western hardliners, but also middle-of-the-roaders saw these actions as mandating countermeasures, especially an increase in nuclear weapons, possibly striving for preemptive capability.

The USSR rationalized its actions on the basis of two centuries of French and German invasions; therefore, an Eastern barrier zone became unwillingly subjugated for over 40 years by barbed wire, armed guards, and puppet governments.

In the USSR, because of the monolithic Communist Party and government structure, the "political" motivators and leaders centered about the Party, its Central Committee, the Politburo, and the General Secretary. Legislation was usually enacted by decree, typically of the Council of Ministers and the Party's Central Committee, with essentially no role or objection from the public. There was a huge military-industrial clique within the ministries. The centralized, secretive, and authoritarian nature of the Communist government made it difficult for Western outsiders to understand the reasoning behind Soviet national-security policies. That opacity spawned in the West a veritable industry of "Sovietologists" and "Kremlinologists."

Whether communism constituted a mortal threat to the Western economy or value system is not simple to resolve. The spread of communist ideology, distinct from classical power politics, was contemporaneous with aggressive actions of the Soviet Union. The presence throughout the world of Communist Parties, sympathizers, and Marxists complicated political analysis. Many European refugees with first-hand experience, having suffered under Soviet political and military repression, became rabid anti-communists.

With these pressures and counter-pressures, nuclear and conventional arms kept on growing to massive and highly destructive proportions on both sides of the Iron Curtain.

The American political left (liberals or doves) placed some blame for the arms race on right-wing militancy. While not condoning repressive Soviet behavior, some excessive military and international actions were responsive in part to American attempts that went beyond "containment." Liberals, at least, realized that Soviet foreign policy was partly a defensive reaction to the pace set by the U.S. weapons buildup. Having rejected communism as a legitimate form of government, the Americans and British appeared to be out to destroy it. Also, to the Soviets, the United States seemed to inherit a style of European imperialism, using an immense nuclear arsenal for intimidation, and trying to exercise economic and political control over third-world nations.

Although U.S. constitutional structure tolerates public dissent, the actual implementation of foreign policy is firmly vested in the executive branch, with limited input from Congress. But right-wing zealots, egged on by hardline émigrés from Eastern Europe, were well financed and tightly connected to the military-industrial-laboratory-congressional quadrangle. Because of the role of funding, U.S. foreign policy, regardless of the party in office, has tended to reflect propensities of the militant complex.

U.S. Movers and Shakers

As mentioned, the dynamics of arms races are well rooted in domestic institutions. Here the political forces that explicitly promoted or moderated the arms race are outlined.

In a constitutional democracy, decisions are influenced by an amorphous, frequently shifting mélange of many individuals, groups, and other entities. But it is somewhat easier to understand the decision-making process in an authoritarian, structured state like the USSR, even though motivating factors are often obscure.

Parties, Individuals, and Writers. In the United States, political options are in principle chosen by the mainstream parties, Democrats and Republicans. They themselves are bombarded by input from individuals, groups, and the writers. On nuclear issues, party platforms seemed to be more opportunistic than ideological.

Political decisions and actions of the elected shift more often with realities of office than consistency of ideology.

Political alignments on Cold War national-security issues were polarized, but not necessarily along party lines; often members of each party tried to outdo the other in strong-defense advocacy. Congressional votes were characteristically along ideological lines: For example, in an August 1969 vote on the Smith Amendment to prohibit development and deployment of the Safeguard ABM, 21 southern Democrats joined 29 Republicans in favor of deployment, while 36 Democrats and 14 Republicans were against; the tie vote resulted in rejection of the amendment.

Some specific individuals in the United States had a disproportionate influence on political and military decisions. Mentioned here are a few noteworthy political players.

- *Joseph McCarthy.* In the early 1950s, attitudes toward the Soviet Union — and thus the possibilities of nuclear conciliation — were strongly influenced by Senator Joseph McCarthy's anti-communist witch hunt, which in 1953 helped the Republicans win in the Congress and Senate and put Dwight Eisenhower in the White House.
- *Barry Goldwater.* One of the most ultraconservative Republicans was Barry Goldwater, a U.S. Senator who ran for president in 1964. He strongly supported Senator McCarthy's crusade and defended the ultra-right-wing John Birch Society. On foreign policy issues, Goldwater opposed the nuclear test ban and proposed the use of nuclear weapons in the Vietnam war. In his final years in the Senate, through 1986, he softened some of his hardline views.
- *Zbigniew Brzezinski.* A national-security advisor to President Jimmy Carter, Zbigniew Brezinski was regarded as that administration's leading hardliner, although he did support ratification of the SALT II treaty. Because defense hawks in government position are not politically susceptible to charges of weakness, Brzezinski typifies someone who could deliver more treaties, agreements, and unilateral arms reductions than could moderates in office.
- *Lyndon LaRouche.* During the mid- to late-1980s, members of the right-wing political committee of Lyndon LaRouche strongly supported SDI. They provoked public confrontations. LaRouche was a political activist who ran for President of the United States, but never gained significant electoral support. He was probably the best-known exponent of conspiracy theories, even those involving right-wing conservatives.
- *Robert Welch and the John Birch Society.* Since 1958, members of the John Birch Society have underwritten a movement that, in their words, is intended "to restore and preserve freedom under the U.S. Constitution." Using educational and publicity campaign tools, the Society attempted to shape public policy through "concerted education and activism."

  Businessman Robert Welch founded the organization and led it for more than 25 years. The Society was named after John Birch, a Christian missionary, who "was murdered by the Chinese Communists just after the end of WWII." The Society considers Birch to have been "an early casualty in the struggle between freedom and world government." Virulently anti-communist, membership of the John Birch Society consisted of thousands of individuals who rejected totalitarianism "in any form." This led them to agitate against negotiated arms reductions with the Soviet Union.

  Fearing a "new world order," the John Birch Society plastered America's roadside billboards with "Get US Out of the United Nations" (still to be seen along highways). The Society has fought against the UN because it "does not recognize the supremacy of God" and views itself as the source of "rights." Birchers believed that "subversive forces and agendas" gave us the tragedy in Vietnam, and that America betrayed its soldiers and people. The Society considered the Panama Canal to have been one of the

United States' most important military assets, and that the transfer to a foreign power (Panama) was illegal.

Major news organizations often swallowed the militant and plausible press releases of the John Birch Society without checking the facts.

Unlike many other groups, the John Birch Society had heavy backing from industrialists and corporations, particularly oil companies and the defense industry. This money made the Society the best funded of all radical-right organizations, allowing them to set up chapters across the country and, by 1963, having at least 80,000 members.

- *William Buckley.* Exemplifying arch-conservative journalists, William Buckley founded *The National Review*, one of the most prominent publications of the American right wing. As a vigorous opponent of the communism, he repeatedly stated, "Better the chance of being dead than the certainty of being Red," often shortened during debates to "better dead than Red." Buckley founded Young Americans for Freedom, an organization of conservative college students that supported nuclear testing.
- *The Alsop Brothers.* Joseph Alsop was a prominent conservative journalist who became a superhawk, although he adamantly opposed McCarthyism. His brother, Stewart, was a widely read spokesperson for right-wing organizations. While acknowledging that the missile gap of the 1950s was "an inflated intelligence estimate of Soviet missile production," Stewart Alsop, rather conveniently, found intelligence derived from spy-satellite pictures of the late 1960s to be reliable indicators that the "sneaky and sophisticated" Soviets were "making really impressive strides in strategic weaponry and nobody seems to care." The "main reason," he wrote, was the "national disease, Vietnam," which produced "such a revulsion of all things military that absolutely hard intelligence is dismissed as mere propaganda from the military-industrial complex."
- *Evans and Novak.* Roland Evans and Robert Novak were widely syndicated conservative political columnists, who strongly boosted Ronald Reagan.

It is no accident or bias that all of the names listed so far have been on the political right; that's because prominent, influential people on the left were few and far between. Senator *William Fulbright* was perhaps the most effective individual opposing the war in Vietnam. One of the more liberal columnists was Chicagoan Mike Royko. *I. F. Stone*, an eloquent opponent of the arms race and the Vietnam war, was a muckraking reporter in Washington; his *I. F. Stone's Weekly* newsletter had a small but loyal following.

Some reporters that cited in this book because of their balanced views are James Coates of the *Chicago Tribune*, William Broad of the *New York Times*, and Jeff Smith of the *Washington Post*.

Bereft of individuals with staying power (in contrast to the many that worked through NGOs), one particularly noteworthy pair must be remembered:

- *Berrigan Brothers.* On the political left, among the individuals who were not institutionally based, were the Berrigan brothers, Catholic priests who were outspoken in their activism against the Cold War and the Vietnam involvement. Daniel Berrigan devoted himself in the 1960s to civil rights and anti-poverty work, eventually becoming a leading activist against the Vietnam War. Philip Berrigan, who served in World War II as an artillery officer, turned in the 1960s to peace activism. In 1968 the Berrigans were arrested for destroying Selective Service files in Catonsville, Maryland. Both Berrigans served prison terms and after being paroled in 1972, and they continued their involvement in such actions as "Plowshares" protests at weapons plants. They were repeatedly arrested and imprisoned, and continued to write prolifically.

In 2002, Phillip Berrigan left a final statement:

I die with the conviction ... that nuclear weapons are the scourge of the earth; to mine for them, manufacture them, deploy them, use them, is a curse against God, the human family, and the earth itself.

Congress and Courts. The U.S. Congress often attempted to influence the administration's foreign policy, and the federal courts were sometimes needed to adjudicate controversial issues involving actions of the legislative or executive branches.

Testimony before congressional committees or subcommittees offered the public a small window into matters of policy; hearings on strategic issues were often held by the Senate Committees for Armed Services and Foreign Relations. For example, during debates about President Nixon's ABM system, the sympathetic Senate Armed Services Committee heard Defense Secretary Melvin Laird challenge the prior administration's assumptions regarding Soviet first-strike capability. Laird offered detailed testimony that the growing missile force of the Soviets, especially the SS-9 with a 25-megaton warhead, made possible a "preemptive Soviet blitz" that could destroy "substantially all of our land-based Minuteman capability in hardened silos."

In front of the Senate's skeptical Foreign Relations Committee, Laird's blunt accusations that the Soviets were going for a first-strike capability encountered sharp rebuttals and open suspicion. Senator Albert Gore (Democrat, Tennessee) retorted that the ABM was a "weapons system searching for a mission" and that it "surely has not yet found it."

As chair of the Senate Foreign Relations Committee from 1959 to 1975, Arkansas Democrat William J. Fulbright was a leading internationalist who saw foreign affairs more in terms of national political aspirations than ideological conflict. He was considered the leading congressional "dove," both in matters dealing with the war in Vietnam and in the bitter engagement with the Soviet Union. Fulbright was a strong critic of the proposed ABM system.

The "legislative embrace" that the peace coalition was eventually able to engineer at one time "conveyed a sense of mounting political opposition to the Reagan military buildup." However, the House turned around later and voted in favor of the MX missile. Congress passed a non-binding watered-down nuclear-freeze resolution, but failed to use its "power of the purse to cut off appropriations for nuclear weapons...."

This experience should serve as a lesson for future arms control activists: "Politicians seeking office were eager to wrap themselves in the righteous mantle of concern about nuclear war," but the Cold War "grass-roots movement for disarmament was turned into an instrument of conventional politics." Legislation drafted by well-intentioned leaders like Edward Kennedy and Edward Markey fell far short of the goals set by the Freeze Campaign.

Paid Lobbyists and Advisors. In order to influence policy decisions, various interest groups, especially the well-heeled ones, hired lobbyists and advisors — often individuals who had served with a prior administration. The military-industrial-scientific complex is populated with individuals who shuttle between their enterprises and their government assignments. This opportunity arises especially in democracies, because elected governments are transient and have to depend on individuals who are willing to work for periods of a few years and who are interested in furthering a cause consistent with the administration's political leanings.

While paid lobbyists concentrate on influencing Congress, they also know their way around the executive branch. The lobbyists are more conspicuous than salaried advisors and appointees who enter the executive branch and Congress through timesharing arrangements. Non-career individuals can get into government from either of two pathways: (1) via contract with beltway bandits who hire them after the individual's government position was terminated for personal or political reasons, or (2) via industrial, academic, or laboratory positions, voluntarily choosing to accept temporary assignments at federal agencies.

While laws in the United States prohibit government employees from taking new jobs involving a direct conflict of interest, former civil servants still can be valuable assets for their new employers, especially beltway bandits seeking insight into forthcoming procurements. Also, personnel assigned from industry and the weapons labs to government positions are valuable to their parent organizations — so much so that assignees are frequently subsidized and often assured of a safe haven on termination of their assignment.

Academics, having less of a policy axe to grind, tend to be preferred in the public-service sector. While their experience might be theoretical rather than practical, they undertake many roles with less bias than other temporary appointees. Teaching institutions facilitate sabbaticals that result in the government being able to fill positions at little or no budgetary expense. If lacking industrial experience, laboratory credentials or political backing, educators going into government are less influential except at high policy levels.

Some presidential administrations tend on principle to favor appointees who have had more practical experience in industry or management. However, working in the government is quite different from working in private industry, and new appointees will have things to learn no matter where they are from. Industry has profit incentives to set aside resources to support individuals in government positions. Various private organizations that favored arms control attempted, with limited success, to counterbalance the incursions of personnel from the military-industrial-scientific complex.

An early liberal anti-communist lobby was the Americans for Democratic Action (ADA), originally formed to help the administration of Harry Truman. During the Nixon administration, the ADA threw the weight of its organization behind the peace movement.

With the Cold War over, arms control organizations have continued to oppose large or regenerated nuclear stockpiles. The Council for a Livable World and PeacePAC were "committed to ridding the world of weapons of mass destruction and eliminating wasteful military spending." They registered as political lobbies that endorse Congressional candidates. The Council formed a Center for Arms Control and Nonproliferation, located on Capitol Hill. The Coalition to Reduce Nuclear Dangers, consisting of leading nuclear arms control and non-proliferation organizations, worked together "to reduce the dangers of nuclear weapons and prevent new nuclear threats from emerging."

Private Institutions. Some *ad-hoc* organizations were formed to deal with specific issues, and some philanthropic institutions funded studies on arms control problems. In addition, there were a number of politically oriented right-wing funding institutions that were extremely proactive.

- *Cato Institute*. The Cato Institute has for more than two decades been promoting public policy based on individual liberty, limited government, free markets, and peace. For national security, its vision "includes a national defense based on strategic independence, which resists military intervention unless vital interests are at stake." Their defense-related research has been in four areas: (1) restructuring U.S. military forces "to fit the U.S. geostrategic situation and to better defend American vital interests," (2) reducing the "bloated" defense budget, (3) cutting "unneeded or redundant" weapons systems, (4) increasing funding for weapon systems that are "inadequately" supported. Now that the Cold War is over, Cato Institute analysts favor cutting the military, emphasizing that they are "[not stampeded by] Washington's nonproliferation hysteria." Cato's director of defense policy, who has recommended "a policy of military restraint," believes that historical data "show a strong [cause and effect] correlation between U.S. involvement in international situations and [the occurrence of] terrorist attacks."

  On key hot-button issues — BMD, CTBT and stockpile stewardship — CATO has sponsored its own studies. They advocated "a limited land-based national missile defense system...." Regarding the CTBT, an analysis by hardliner Kathleen C. Bailey, argued that "a treaty that bans the testing of nuclear weapons will neither constrain the modernization of nuclear weapons by countries that already possess them nor prevent the spread of such weapons to additional states." Bailey maintained that "the U.S. Stockpile Stewardship Program, which substitutes computer simulations for the testing, is expensive and technologically risky." She preferred continued testing of nuclear weapons to make them safer and more reliable.

- *Heritage Foundation*. Formed in 1973 as a nonprofit, nonpartisan, tax-exempt charitable organization, the Heritage Foundation categorizes itself as a "think tank" whose mission is "to formulate and promote conservative public policies based on the principles of free enterprise, limited government, individual freedom, traditional American values, and a strong national defense." Because it had more than 200,000 contributors, it has claimed to be "the most broadly supported think tank in America." Its primary audiences are members of Congress, key congressional staff, executive-branch policymakers, news media, and the academic community.

  Edwin Meese III, former Attorney General for President Reagan, has been listed as a "Heritage Expert on National Security." Another was Jack Spencer, a young policy analyst on the staff who identified himself as an expert in missile defense and other national security matters. Pointing to missile tests or acquisitions by Iran, Syria, Libya, India, Pakistan and Russia, Spencer wrote that the ballistic missile threat was growing rapidly as we started the new millennium. (Regarding Russia, though, he ignored the fact that its single-warhead missile is a replacement for a more threatening multiple-warhead system being phased out, and he disregards the actuality that many of the other nations are developing missiles that have short or intermediate range.) Heritage's national-security analyst simply assumed that an "effective missile-defense system" could be developed "[if only the President (Clinton) supported] deployment...."

  Before President G.W. Bush withdrew from the ABM Treaty, the Heritage Foundation made the legalistic argument that the Treaty was no longer in force because Russia was not a legitimate successor to the USSR.

  One of Heritage's goals has been "restoring U.S. military strength."

- *Hoover Institution*. Standing in contrast to CISAC at Stanford University is the Hoover Institution on War, Revolution and Peace, a right-wing think tank. One of its best-known militants was Edward Teller. The Institution has been involved not just in funding but also in active public-policy research. It has devoted itself to study of politics, economics, and political economy, both domestic and foreign, as well as international affairs. Founded in 1919 by Herbert Hoover, who later became the thirty-first president of the United States, the Institution originated as a specialized repository of documents on the causes and consequences of World War I. Today, with its group of scholars and ongoing programs of policy-oriented research, the Hoover Institution considers itself a contributor to ideas that define a free society. It claims to have extensive scholarly resources with particular strength in Soviet and Russian areas.

Western European Leaders

Many Western European leaders had significant influence; here we single out a few individuals.

- *Winston Churchill,* one of the greatest political leaders of the 20th century led opposition to Soviet expansionism, originating the term "iron curtain." While his most enduring legacy was a staunch struggle against Nazi aggression, Churchill also tried to stand fast against Soviet moves into Europe and elsewhere. He authorized accelerated development of Britain's self-sufficient nuclear weapons and delivery systems. Churchill did not see the East-West conflict in purely ideological terms, but more as a form of traditional 19th-century power politics.
- *Charles DeGaulle* was France's leader during the early stages of the Cold War. Despite his good World War II relationship with the Soviet Union, he later turned against the French Communists, who unswervingly supported the Stalin regime. DeGaulle shepherded his nation through its successful and largely independent development of an atomic bomb. Withdrawing from NATO, France devised its own nuclear deterrent, the *force de frappe*, largely because DeGaulle did not believe that the United States would use nuclear weapons to defend European soil.
- *Konrad Adenauer* was the first Chancellor of West Germany; he led the political rehabilitation of West Germany and brought it into the NATO Alliance. Adenauer achieved *rapprochement* with both France and the USSR.
- *Willy Brandt* was Chancellor of West German from 1969 through 1974. His *Ostpolitik* (Eastern Policy) established a greater degree of normalcy in relationships with Eastern Europe; this won him a Nobel Peace prize. Brandt, after leaving office, opposed deployment of American missiles in Western Europe.

Soviet Policy Makers

Policymaking in the Soviet Union centered around the Communist Party, its Central Committee, the Politburo, and the General Secretary. The domestic structure was highly centralized, secretive, and authoritarian, all factors that made it difficult for change to be introduced either from within or from outside.

But Soviet national-security policy was eventually altered as a result of input from respected scholars in the academies of science and institutes of study. Partly through these quasi-governmental organizations, foreign transnational individuals and groups gradually gained access to Soviet decision makers. Meanwhile, citizen dissidents had little real effect on the Soviet hierarchy.

Party Leaders. Usually Soviet party leaders had multiple positions that invested government authority.

- *Josef Stalin*, perhaps the most powerful political personality of his time, was a brutal absolute ruler of the Soviet state from 1929 until he died in 1953. Stalin made promises at the Yalta and Potsdam Conferences that he did not keep. Because of the trauma of World War II, a major goal of Stalin was to continue the economic and political destruction of Germany; this policy invariably became a source of post-war dispute with Britain and the United States, both of which favored reconstruction.

In 1947 Stalin launched Cominform, an information bureau intended to direct the activities of Communist parties elsewhere in the world.

Despite successful testing of a Soviet atomic bomb in 1949, Stalin became paranoid about a potential American attack on Russia's homeland.

Stalin's legacy of mistrust, secret police, stalag camps, and rubber-stamp legislative bodies continued long after his death.

Following Stalin's demise, there was a series of short-lived premierships and party leaders. *Georgi Malenkov* briefly headed the Soviet Union starting in 1953. Then *Nicolai Bulganin,* one of the key Party officials, became Premier in 1955. He lost a power struggle with *Nikita Khrushchev* in 1958.

- *Nikita Khrushchev*. Khrushchev remained Soviet leader until 1964. He reversed the trend of Stalinist communism, built up the nuclear arsenal, and presided over the Soviet Union during the building of the Berlin Wall, the Cuban missile crisis, the Sino-Soviet split, and the launching of sputnik. Khrushchev also agreed to establish the US-USSR "hot line" and signed the Nuclear Test Ban Treaty; he was the first Soviet leader to visit the United States. Overall, Khrushchev can be credited with being a reformist, initiating a thaw that permitted greater contact with the outside world.

  Up through 1961, when the Cuban missile crisis mushroomed, Khrushchev was behind a significant unilateral reduction in conventional forces, although the United States failed to reciprocate. Concessions by Khrushchev in troop reductions and in willingness to participate in nuclear disarmament agreements ran counter to the "negotiation-from-strength" position that was touted in the United States, which had imagined "bargaining chips" in the form of new weapons. (The truth, revealed in archival documents from the Eisenhower period, is that the U.S. government was already dominated by opponents of arms control and was not interested in a negotiated outcome). Ultimately, Khrushchev's de-emphasis of conventional forces, coupled with his brandishing of nuclear projectiles — "missile diplomacy" in Cuba — provoked subsequent U.S. missile expansion. (Another cycle of nuclear shadowboxing was stirred up.)

- *Alexei Kosygin* succeeded Khrushchev in 1964 as prime minister, being part of the governing triumvirate that included *Leonid Brezhnev* and *Nikolai Podgorny*. As a result of an opportunity that arose because he was attending a 1967 UN meeting in New York City, Kosygin participated in a summit meeting with U.S. President Johnson at Glassboro, New Jersey.

- *Leonid Brezhnev* was the longest serving post-war Party leader of the Soviet Union, his stewardship lasting from 1964 until his death in 1982. By 1970 Chairman Brezhnev had effectively taken over control of the government, and his initial policy was to seek parity with the United States, despite the sacrifice it caused in Soviet production of domestic goods. Although Brezhnev supported détente with America, Soviet relationships deteriorated with England, China, and the Middle East. Brezhnev supported the doctrine of peaceful coexistence, which led him to agree to the 1972 SALT I Treaty, and to the 1975 Helsinki Accords at the Conference on Security and Cooperation in Europe. With SALT I came official bilateral acceptance of mutual limitations on ABM defenses.

  Toward the end of Brezhnev's rule, Ronald Reagan was elected to the office of the Presidency of the United States. By then, over-centralization and heavy industrial subsidies were already leading to Soviet economic stagnation, and Brezhnev did not want to continue exhausting the economy by increasing military expenditures. He attempted unsuccessfully to face down the Soviet military-industrial complex, which insisted (considering Reagan's truculent tone) on further buildup, although they would not give him firm guarantees that the spending would make the "correlation of forces" more favorable to the Soviets.

- *Yuri Andropov.* When Brezhnev died in 1982, Yuri Andropov, who had spent 15 years as head of the KGB, became leader of the Soviet Union for 15 months before his own death. Andropov was tapped to succeed Brezhnev because earlier he had experience in economic and social reforms. He proposed decentralization and production incentives to stimulate the stagnated economy. Realizing that the Soviet Union could not afford to continue sacrificing its domestic economy to the East-West arms race, Andropov pressed for mutual reductions in nuclear arms. After his death in 1984, his policies suffered a temporary setback under Konstantin Chernenko, but a year later Mikhail Gorbachev set Andropov's policies back on course.
- *Konstantin Chernenko.* The leadership of the Soviet Union was in the hands of Chernenko for only a year before he died, in March 1985. During that time he attempted to restore the policies of Brezhnev, and he moved to silence dissidents, such as *Yuri Orlov*, who was sentenced to internal exile. Chernenko took a hard line in nuclear weapons negotiations.
- *Mikhail Gorbachev* became president of the USSR and General Secretary of the Communist Party in 1985. During his premiership, Gorbachev launched a series of initiatives to improve East-West relations dramatically and to reform the Soviet Union politically and economically — initiatives that inadvertently accelerated the dissolution. One of his first foreign-policy initiatives was to announce a unilateral moratorium on Soviet nuclear testing, to take effect on 6 August 1985, the tenth anniversary of U.S. atomic bombing of Hiroshima.

Being a reformer, Gorbachev emphasized the need for rapid internal economic development, a termination of the arms race, and "a freeze of nuclear arsenals and end to further deployment of missiles." His foreign and domestic policies were summed up by the terms *perestroika* (economic restructuring) and *glasnost* (political and social openness). Because defense spending was the biggest nonproductive sector of the economy, Gorbachev moved to reduce expenditures significantly, beginning with the withdrawal of large contingents of troops from outside Soviet borders. He agreed to the INF Treaty and opposed SDI. Gorbachev was awarded the Nobel Peace Prize in 1990 for his contributions to international cooperation and peace.

Because President Reagan had witnessed a succession of hard-line Soviet leaders, it was initially difficult for him to understand and accept the moderate policies of Gorbachev. However, in October 1986 at Reykjavik, Iceland, Gorbachev had gotten President Reagan to agree to eliminate all ballistic missiles and nuclear bombs in 10 years, an agreement promptly recanted by his White House staff.

Gorbachev met Reagan's successor, President H.W. Bush, in November 1990 at the Conference on Security and Cooperation in Europe. Their talks resulted in limits on deployment of convention arms that were tantamount to a formal end of the Cold War.

After the Warsaw Pact was dissolved, in July 1991, Gorbachev and Bush signed the START I Treaty. At the same time, state-owned enterprises in the Soviet Union were denationalized; this and other events led to a bungled coup by Communist Party conservatives in August 1991. For a while, members of the coup apparently had control of the Soviet "football," which was the briefcase used to put into effect commands for nuclear war. The coup fizzled, partly because of defiance in Moscow led by Boris Yeltsin, then President of the Russian Republic.

Beginning with the election of a functioning parliament in 1989, the influence of the Communist Party diminished, but it has never disappeared. Nevertheless, Russia became a federation with weakened central control, "quasi-corporatist" policy networks, and an active civil society. This helped in formation of independence-minded republics that led to the breakup of the Union at the end of 1991.

In October 1991, as a result of the growing political instability, Bush and Gorbachev agreed to take strategic bombers off alert, destroy ground-based tactical nuclear weapons, and remove short-range nuclear missiles from ships.

Soon afterwards, most of the republics and the Baltic states seceded from the USSR, leaving the three Slavic republics of Russia, Ukraine, and Belarus to form the Commonwealth of Independent States. Gorbachev resigned on 25 December 1991, and the USSR passed officially into oblivion.

Soviet Scholars and Organizations. Because the Party and the Soviet government exercised strong control, very few outsiders could affect national policy. The insiders who had influence were usually scholars and scientists, especially independent security analysts based in Moscow. While some dissidents were well known outside of the USSR, they probably had little influence on the course of internal affairs.

- *Petr Kapitsa*, who believed that scientists should be involved in social and political aspects of averting war, was a prominent Soviet physicist who, as early as 1955, was able to participate in the international scientists' movement.

  Most Soviet leaders held scientists in high regard and were usually receptive to their ideas. Some of the leaders were particularly sympathetic to the alarms raised by the international movement of physicians, who had special access because a couple of them were treating the health problems of the Kremlin leaders.

- *Sergei Kapitsa,* physicist son of Petr Kapitsa, undertook to popularize science and related issues for Soviet TV, just as did Carl Sagan in the United States. Control of the press and other news media in the Soviet Union made it rare that accounts of the horror of nuclear war reached the Soviet public — the whole area of military planning was barred from discussion. This meant that it was up to scholars like Kapitsa who were at least nominally Communist Party members to take up the slack.

- *Pugwash.* The international Pugwash organization prided itself on having the most direct line of communication with the Soviet government and scientists, a contention that was borne out by many examples. However, Soviet Pugwash participants had stronger connections with their bureaucracy than Western participants had with their own governments. By 1968 many Soviet scientists, partly as a result of their contacts with Pugwash and their access to Western literature, came to advocate a ban on ballistic missile defenses. They tried to sway their government by publishing articles, contacting key political leaders, and explaining the principal limitations and instabilities. By describing in a sympathetic manner the American debate about ABM and MIRVs to Soviet citizens in the mass-circulation newspaper *Izvestia*, Pugwash participant *Georgi Arbatov* reached a broad audience. Indeed, Academician Millionshchikov credits the Pugwash movement with helping energize the first U.S.-USSR accords (SALT I and ABM) in bilateral arms control.

- *IMEMO* (the Russian acronym for the Institute of World Economy and International Relations), along with the Institute of History, carried out scholarly work on disarmament problems. With contributions from *Aleksei Arbatov* and *Andrei Kokoshin*, IMEMO published the yearbook *Disarmament and Security 1986*, which showed the influence that unofficial transnational discussions had on European security. The IMEMO analysis became the framework for new Soviet proposals and concessions, particularly in conventional arms, in order to encourage nuclear weapons reductions. The IMEMO yearbook Chapter on conventional forces contained the seeds of subsequent Soviet proposals and unilateral initiatives.

- The *Soviet Academy of Sciences* was a quasi-independent organization. In the early 1960s Academician Millionshchikov facilitated high-level contacts with U.S. colleagues, especially those associated with Pugwash. *Georgi Arbatov*, Director of the Academy's U.S. and Canada Institute beginning in 1967, was a Soviet spokesman in East-West relations. Arbatov became one of the chief advisors to Mikhail Gorbachev, and his son *Alexei* interacted with the FAS/CSS NGO project on arms control and verification. Academy members who took part in Pugwash included Lev Artsimovich, Nicola Bogolubov, Petr Kapitsa, Nikolai Semanov, Dmitri Skobeltzen, Igor Tamm, Alexander Tupolov, and Alexander Vinogradov.

- *Andrei Kokoshin*, who also participated in CSS activities, was a major proponent of military reform during the Gorbachev era, and later became deputy defense minister in Boris Yeltsin's government. Kokoshin had been deputy director of the Institute of the USA and Canada, headed by Georgi Arbatov. Alexei Arbatov worked as a political scientist at IMEMO, which allowed him to travel to the West and become one of the most articulate and knowledgeable civilian analysts who proposed alternatives to official military policies.

Soviet arms control scientists — like the Arbatovs, Kokoshin, *Evgeny P. Velikhov*, and *Roald Z. Sagdeev*, all of whom had strong personal and professional contacts with Western counterparts — had a number of formal and informal ways of reaching officials in charge of Soviet national-security policy.

Cultural Differences

Two cultures, two nations, two traditions, two histories.... Many geographical and societal differences between the United States and the Soviet Union were a source of misunderstandings. Overriding, fortunately, was commonality of interest that kept the arms race from escalating into a nuclear conflagration.

Clinging to a socialized political and economic structure, the Soviets endeavored to make their cultural and governmental institutions compatible and interdependent. Under communism, a stable and secure socialist structure was viewed as much more important than individual freedom, and the economic consequences were secondary to Marxism.

Americans, however, are by tradition more oriented toward a looser societal structure, with less reliance on government institutions: Individual freedom and the economic results of personal activity tend to be valued more highly than stability and uniformity of communal organization. Thus, while in the United States individualism has been cherished more than values of the entire nation; in the former Soviet Union the rights of an individual were subordinated.

> **[Dead on Arrival].** In the late 1980s, the INF Treaty had been implemented, and Argonne staff had to translate some of the American-drafted verification procedures into the Russian language. Somebody on the American side had decided to distinguish the Russian collaborators by putting their transliterated names into boxes with black borders, as shown here. This was perfectly OK in English linguistic tradition, but in Russia black boxes are only used around the name of a person who had passed away. It was a comical incident that could have become embarrassing.
>
> **Dr. Ivan M. Ivanov**
> **Senior Consultant**
> *Ministry of Foreign Affairs*

Cultural differences of the past manifested themselves in many ways. In the context of this book, embedded traditions influenced nuclear arms development and control. Traditional attitudes of the Soviet Union had profound manifestations, some of which are mentioned below:

▸ **Personality cults.** The personality cult dominant throughout Tsarist and Soviet history led not only to acceptance of authoritarianism, but an active desire for it in preference to codified law.

▸ **State ownership.** The centrally managed economy, with no private ownership, led to demand that was constantly unsatisfied, with supply lagging behind in most areas of the national economy.

▸ **Central planning.** Under a nationalized economy, central planning was unable to stimulate private initiative; instead, it further reinforced conformity and engendered hostility to deviation.

▸ **Lack of initiative.** The preceding attitudes fostered widespread lack of initiative in the workplace, disincentives for improvement in quality, and a tendency to shift responsibility to others.

For most Soviet citizens all those traditions became so natural that contemporary republics of the former Union are trying reflexively to preserve them because they fear change. Nevertheless, political and social perceptions in the FSU are changing, if ever so slowly. It is not clear yet whether these changes are fundamental or just superficial. That's why, for this book, we are largely highlighting Soviet tradition as it affected the Cold War.

▸ **Governmental authority.** Soviet society chose to be governed by committee, expecting the government to take care of literally everything, including such items as food, housing, clothing, recreation, military service, and foreign affairs — without meaningful constitutional checks and balances. Many citizens of the Russian Republic still believe that government is created to tell people what to do.

In contrast, U.S. citizens elect representatives to a government that is supposed to carry out constitutionally mandated functions, including managing taxpayers' money effectively. In principle, a democratic government listens to people and does what the electorate wants (In practice, it often doesn't work exactly that way).

This raises the question of what a group of concerned citizens can do to move things in the direction of their convictions — in particular in the area of arms control and nonproliferation. In the USSR, the only way was to work quietly through official channels; in the United States — if that didn't succeed — it was possible to pressure the government through mass media and other demonstrative means.

David Cortright, who carefully evaluated changes in public attitude, came to the following conclusions:

> The peace movement did indeed have a major impact on public opinion. Strong political opposition to the nuclear buildup created significant news coverage, which led to greater public awareness and a shift in public opinion. Public opinion, in turn, created pressure on the Reagan administration to moderate its policies.... For much of the 1980s the initiatives of the peace movement set the pattern for and significantly influenced news media coverage, public opinion, and ultimately government policy.

Within the confines of the USSR, the situation was diametrically opposite for citizens and officials, largely as a result of nationalistic, cultural, and ideological differences:

▸ **Permissible behavior.** The Soviet citizens were accustomed to thinking that what was not explicitly permitted was forbidden. That was a so-called "closed door" attitude: If a Russian sees a closed door with no sign permitting entrance, it means access is forbidden. For an American, a door without a label does not necessarily prohibit entrance.

Concerning arms control/non-proliferation issues, Soviet negotiators preferred an agreement that described both what was allowed and what was prohibited, while U.S. negotiators could be satisfied with an arms agreement stating only what was prohibited — what was not mentioned being permitted.

▸ **Laws.** Soviet citizens assumed that laws were created primarily to control and punish people. Soviet management, convinced that little good could result if all the regulations were followed, considered it an important task to evade business laws and regulations. In addition, Soviet authorities blatantly enriched themselves.

Americans have traditionally believed that laws are created to protect people and property. To start a business, Americans tend to learn the relevant regulations and laws, knowing it will be necessary and beneficial to follow or take advantage of them for the future of their enterprise. Of course, many less-ethical entrepreneurs learned how to circumvent or transgress the law, with or without the help of compliant politicians.

Concerning arms control/non-proliferation issues, Soviet negotiators considered in advance the potential consequences for their side or the opposite side if agreement were to be violated. The U.S. negotiators tended to concentrate on potential legalistic loopholes in the agreement.

▸ **Fairness.** The traditional idea for the Soviet Union was that equal distribution of economic wealth was valued. However, the farmer in a collective tended to resent a neighbor's success, even wishing harm or — as a Russian philosopher said — that both of them become equally poor (or wealthy).

In traditional American thinking, equal distribution of opportunity is more valued, although this may bring economic inequities. The American farmer might try to discover the reason for a neighbor's success in order to be at least equally profitable.

In arms control and non-proliferation issues, Soviet/Russian negotiators would prefer an agreement with equal economic consequences for both sides (e.g., benefits resulting from the conversion of Russian weapons-grade uranium into nuclear power-plant fuel and selling it to the United States), whereas U.S. negotiators might prefer a profitable economic opportunity for their country (reduction in military spending, new opportunities for private businesses, etc.).

▸ **Justice.** The traditional Soviet thinking was that it is better to punish innocent people than to let a guilty person go unpunished. Even today, many Russians believe that the millions of people arrested and killed by the Stalin regime were at least partially guilty.

Traditional American jurisprudence is that it is better to let a guilty person go unpunished than to punish an innocent person — although many individuals believe otherwise. (However, no indication of such an American attitude was evident in arms control and non-proliferation negotiations.)

▸ **National pride.** Soviet-era thinking was that it is humiliating to expose foreigners to your own weakness or misfortune — better to leave it undiscussed even among yourselves.

The attempt to cover up the Chernobyl accident is a good example of this. Immediately after the accident, because of the Soviet tradition of secrecy, little progress was made by foreign governmental or domestic non-governmental entities in obtaining adequate information related to the condition of the reactor and the environmental and health consequences of the accident. Later on, under pressure from the international community, the information was released.

In contrast, traditional thinking in the West is that it is beneficial to disclose individual or national failings that have public consequences. Despite this intention, some governments still use excessive controls to manage information that has no real national security implications.

In regard to arms control, even an unintentional Soviet violation of a signed agreement was more difficult to characterize, one case being the partial construction of the radar station at Krasnoyarsk, which — if completed — would have been in violation of the ABM Treaty.

▸ **The value of human life.** The traditional Soviet approach was that it is honorable to die defending social, moral, and material values, even a piece of government property: Societal values and government property are worth more than human life.

The American traditional attitude has been that it is much better to save a human life than to prevent or mitigate damage to property.

▸ **Non-failure vs. achievement.** In the USSR, it was better not to try new, even attractive, ideas because the personal cost of failure was much greater than the possible benefits of success (science being an exception). The American traditional approach has been that it is worthwhile to try a new social or technical idea, because that is the way to advance.

This difference made Soviet negotiators more inclined to be satisfied with less accurate, less advanced, but more proven equipment for treaty verification.

▸ **The military.** The traditional Soviet approach was that military and intelligence personnel were the most trusted part of society, protecting people, country, and government, and they could do on their own whatever they believed was right for the country. In regard to arms control and non-proliferation issues, this perception made the Soviet people less likely to question or challenge treaties or lack of treaties by their government.

The traditional American approach has been that the military/intelligence arm of government should be subordinate to civilian authority and under strict control.

Once approved by the Supreme Soviet, arms control agreements were routinely ratified. In the United States, sometimes it was difficult, occasionally not possible, to secure Senate ratification of a presidentially signed treaty, even if the military and intelligence staffs endorsed it.

Cultural Movements

Intertwined with the issue-oriented campaigns and external political forces were (and are) various cultural movements that transcended immediate issues. Repugnant events stirred up cultural changes which themselves aroused public and official reaction.

A useful slice of time is recaptured in a book — *1968!* — about the tumultuous events of those 366 days. Perhaps every bit as significant was 1981, the first year of the Reagan presidency, which initiated a large number of weapons-buildup decisions and reactively stimulated an enormous and enduring public backlash.

The year 1971 was not without lasting consequences either. The United States backed an invasion of Laos by the South Vietnamese, the voting age was lowered to 18, Lt. William Calley was found guilty (and later pardoned) for 22 murders in the Mi Lai massacre, John Kerry became extremely active in the veterans' anti-war cause, 13,000 anti-war protesters were arrested in Washington, D.C., *The New York Times* started to publish the Pentagon Papers, and President Nixon's advisor Charles Colson suggested that the administration should "destroy" Kerry before he became too popular and effective.

Of profound embarrassment and revulsion to liberals were left-wing extremists who engaged in acts of violence (robberies, bombings, arson, manslaughter) in their fervent opposition to the "establishment" and to war.

Crosscutting cultural movements grappled with racism and sexism, issues that peaked near the end of the 1960s, but persisted during the entire Cold War.

Hippies were a phenomenon concurrent with, maybe a consequence of, the conflict-ridden 1960s and 1970s. They indulged in "mind-expanding experiences" with free love, flower children, mood-altering drugs, antiwar protests, communes, and civil-rights marches. The hippie movement took root in Europe, and words like "peace," "love," and "dope" gained new meanings everywhere. The movement abandoned conventional values and offended the societal and political mainstream.

The Politics of Racism. In the United States, racism was a unique, combustible, and transitional issue — not because of the Cold War, but despite it. Although racism was primarily a domestic cultural issue, it also had an impact on foreign policy and overseas aid. An American deprecatory attitude about orientals has been often cited as a factor in President Truman's decision to use the atomic bomb on Japan; yet it was Truman who appointed a Committee on Civil Rights and made early headway in desegregating the armed forces.

While Soviet Union and its satellite states had some internal ethnic conflicts, the strong arm of the government suppressed them.

Civil-rights activists in the United States linked domestic struggles against racism to anti-colonial movements abroad, while third-world nationalists hoisted the defender of the free world on its own petard when it came to the U.S. domestic treatment of racial minorities.

Some might argue that this book is going astray here in by addressing these phenomena, however briefly. Yet, societal conditions at the time indeed had an ultimate influence on the superpower conflict and, arguably, even on setting the nuclear temperament. Passive and active racism certainly were important latent factors, but they did not directly influence the significant nuclear events of the time.

Conversely, as some have noted, black activism was "beaten back by postwar repression and domestic politics." Too often, overriding "national security" issues (e.g., Korea, Vietnam, the red "menace," and the "war against terrorism") eclipsed and diverted attention from fundamental domestic shortcomings.

At least one perceptive writer insists that the effort of the U.S. government "to contain and manage the story of race in America" was a component of the policy of containing communism. Certainly America's civil rights struggle developed within a global environment shaped in part by ongoing decolonization and liberation of the third world. A few have argued that the positive government response was designed to hide previous failings in dealing with racism.

In any event, racial lenses helped shape U.S. relations with the outside world, possibly even to the point of spurring civil-rights reform in order to improve foreign relations. Domestically, presidential efforts were directed at placating civil rights agitators without alienating white Southerners who had prominent government positions. Overseas colonial legacies, such as Angola — drawn into a surrogate East-West war — helped highlight U.S. racial inequality. The apartheid state in South Africa — a resolute bulwark against communism — was antithetic to the U.S. civil-rights movement.

In the first years of anti-communist hysteria, black leftists had become early victims of McCarthyism. Segregationists like Senator Herman Talmadge (Democrat, Georgia) insisted that American civil-rights agitation was inspired by communists. The southern Dixiecrats were ardent Cold Warriors, but hostile to decolonized nations. Retention of racial segregation was pivotal for Dixiecrats.

Later on, third-world liberation captured the imagination of liberals. Yet, it was Lyndon Johnson, a white Southerner, who eventually brought the force of the presidency behind civil-rights reform.

Gender Activism. Macho tradition in disputed issues did not escape the attention and remedial efforts by gender activists. Some NGOs and issue-oriented organizations mentioned in Section A were founded or constituted by women in order to promote feminist values and concerns not otherwise expressed.

Gender-conscious repulsion and reaction to rape, abuse, abduction, honor killing, sexist religious fundamentalism, and disenfranchisement were gradually introduced into the protest milieu. Feminist activism also disturbed complacency about traditional male-oriented mores favoring militarism, savoring glory, exercising power, reveling in dominance, reinforcing hierarchy, and prolonging economic dependency.

Gender activists have striven to encourage settlement of issues through dialog, cooperation, nurturing, equality, and empathy. Keeping gender activism fired up in the new millennium have been ominous presidential threats to use nuclear weapons, especially against non-nuclear states, and governmental neglect of social needs in favor of military funding. Gender conformity had yet to become a prominent issue.

Student Radicalism. Student rebelliousness in a number of countries peaked in 1968. U.S. students occupied college campuses and took to the streets, especially to protest a war for which they might be drafted to fight. In West Germany, thousands braved water-canons; outside Tokyo, snake-dancing demonstrators opposed a new airport; in France, it was a strike against President Charles de Gaulle; and in Prague, the debate over socialism was later drowned out by invading Soviet tanks.

Noteworthy too in 1968 were the first radical feminist demonstration (protesting the Miss America pageant) and of the massacre of South Vietnamese peasants at the village of Mi Lai.

Cold War realities made for clashing political agendas: Radical students in many countries alternately favored or opposed the Communists in Vietnam, supported the Palestinians or the Israelis, or rallied for the Soviets or the oppressed. The students were also involved in their own cultural revolution about drugs, love, and intellectual liberty.

Student radicals scored many successes by strengthening opposition to renewed fascism, Stalinism, Jim Crow, and hypocrisy. While they gave time, energy, and vision to mainstream movements against nuclear war and weapons, they also incited police suppression and public backlash on both sides of the Iron Curtain. Although a few former student rebels eventually rose to prominence in the East, it was the Ronald Reagans, Margaret Thatchers, and Helmut Kohls who reaped political harvests sowed by discontented students.

Religion vs. Atheism*

Religious beliefs were on the front lines during the Cold War. Although the superpower struggle overtly followed secular lines, it was often cast in terms of religious survival.

Some adherents to religion viewed the Cold War as a crusade against atheistic communism. Conversely, Marxism rejected traditional religion. Atheism was the official stance of most communist countries, including the People's Republic of China and the Soviet Union. Karl Marx, an atheist, wrote that religion is the "opium of the people," often interpreted to mean that religion exists in order to blind people to the true state of affairs in a society, and thus make them more amenable to social control and exploitation. Others have suggested that communism itself could be seen as a form of religion.

Communist doctrine aside, many dictatorships have regulated or forbidden religious groups which were viewed as possible centers of opposition against their totalitarian rule. On the other hand, western intelligence agencies often cooperated with local religious groups in order to build up opposition in hostile countries. (An extreme example was CIA training and support of radical fundamentalist

---

*lacking belief in a god

Mujahideen — including Osama bin Laden — in Afghanistan in the 1980s). In the Soviet Union and in the People's Republic of China, some churches that submitted to strict state control were tolerated.

Because communists wanted to eradicate traditional religion, which they perceived to be an irrational belief system, powerful religious groups such as the Catholic Church were among the strongest enemies of communism.

Doctrinaire fanatics, especially in the United States were not only fearful of communism and atheism, they were convinced of the necessity of fighting a righteous war against the god-less communists, whatever the cost. Some still think that religious resistance won the Cold War for the West. Others believe that communism as an ideology never got "nuked": Although Karl Marx's economic ideas fell, his theological ideas remained intact. In other words, the unconvinced admit that the Cold War is over, but the battle of minds was lost.

In February 1950, Senator Joseph McCarthy made a speech in which he warned of what he considered to be a communist threat from within the U.S. government. His rallying call included an appeal to religious militancy.

> Ladies and gentlemen, can there be anyone here tonight who is so blind as to say that the war is not on? Can there be anyone who fails to realize that the communist world has said, "The time is now" — that this is the time for the showdown between the democratic Christian world and the communist atheistic world? Unless we face this fact, we shall pay the price that must be paid by those who wait too long.

That religion played a significant role in the Cold War might seem self-evident, given the atheistic nature of communism and the powerful influence of Christianity and on the lives of millions of people on both sides of the Iron Curtain. It must surely have seemed obvious that religion mattered to those who actually experienced religious persecution in Eastern-bloc countries, or whose education or employment was in jeopardy by virtue of their faith. There is no doubt that the governments of communist countries saw Christianity and Muhammadanism as a threat and responded accordingly. Members of Hebrew congregations in Eastern Europe suffered a devastating *pogrom*, though much more because of their ethnic label as Jews than their religious affiliation.

Some scholars have viewed the Cold War as one of history's great religious wars, a global conflict between the god-fearing and the god-less. It is said that the Vatican's alliance with the Western bloc contributed toward containment of the Soviet Union, as well as to the post-war triumph of Christian Democratic parties in Italy and Germany.

The Polish-born Pope's election is sometimes credited with instigating the public process leading to the end of the Cold War and the breakup of the Soviet Union. The Pope's warning that it was not possible to be both a Catholic and a communist is said to have struck home.

Public Assemblies

Inasmuch as organized activities tend to center about groups of individuals, officials, or organizations, major gatherings that had a role in public engagement over Cold War issues. Here's a case where the means were often as significant as the ends.

While small Cold-War-opposition entities formulated and carried out their own activities, they were better able to expand their scope by interacting with other individuals and organizations at larger gatherings. This was necessary because governments are empowered to devise and carry out policy through their internal structure and seemingly endless resources. Government agencies — with comparatively deep pockets — fostered research on weapons, publication of favorable results, and meetings of scientists and decision makers. On the other hand, resources for public activities usually came directly from individuals or organizations themselves, occasionally supplemented by private foundations or other donors.

Sometimes public projects, workshops and conferences ironically became the primary means through which agencies (and surrogates) of the superpower governments could indirectly monitor ongoing activities without official acknowledgment, responsibility, or influence. Direct unofficial contact between East and West was normally banned; so private channels often ended up being "back-channel" conduits for curious but stymied government bureaucrats and officials.

Extraordinary group actions like radical civil disobedience, or more conventional activities like voter drives, also took place and helped reduce public apathy to Cold War dangers.

Conclaves. Private, usually organizational, meetings were crucial to the development and evolution of organized dissent. Public protest was often expressed in group meetings, arranged using "snail mail" and telephones (although since the advent of the Internet, much organizational work is being done by e-mail and through Web sites).

Some typical formative conclaves focused on national-defense-spending priorities, stopping the ABM, and freezing the arms race.

In March 1981, American peace groups and arms control experts held a national conference to approve a nuclear-freeze strategy. More than 350 representatives from at least 30 states met at Georgetown University in Washington, D.C., to call for broad and visible public pressure on Congress to work toward a comprehensive nuclear arms freeze between the United States and Soviet Union.

Noteworthy examples of organized people-to-people contacts between American and Soviet citizens are a Mississippi Peace Cruise in 1986, similar voyages on the Volga and Dnieper rivers of the Soviet Union, and various citizen exchanges, sports competitions, youth camps, scholarly conferences, and professional exchanges.

Projects. Projects with specific goals were sometimes arranged under the auspices of private sponsors. Four such undertakings were: (1) the Cooperative Test-Ban Seismic Monitoring Project, (2) the FAS/NRDC/CSS Collaboration (which included the Joint Disarmament Project and the Black Sea Experiment), (3) Bochum University's CFE Verification Experiment Group, and (4) the Laser

ASAT Project. Each of these had either a specific or general arms control goal, often dealing with the technology of treaty verification.

Usually these projects were conducted in multinational workshops that had defined goals. Multilateralism was the norm; so scientists or military personnel from both the East and West commonly participated.

In a merging of interests, some DOE-laboratory participants were often able to attend meetings or participate in projects as "individuals," not in an official capacity. They had to issue a nominal disclaimer of official endorsement, especially the project or meeting included a "communist" (e.g., a Soviet scientist, bureaucrat, or citizen). But there was always an eager audience in the government for debriefings on these back-channel contacts.

Usually such gatherings were funded by private institutions, although travel expenses for laboratory personnel occasionally were paid from government resources. More often, participants from the DOE laboratories had to use their personal vacation time and depend on the NGO sponsors to compensate for travel expenses.

Conferences. The Pugwash series attracted invaluable participation by Soviet scientists, in turn giving rise to meetings between the respective Academies of Sciences on matters dealing with mutual security. These direct contacts were among the very few approved by the respective governments.

To avoid governmental displeasure and ensure continuation of the series, the published results had to be reduced to common denominators, acceptable to participants and their sponsors.

Pugwash Conferences have carried on annually. An International Student Pugwash was formed in 1978, initiating youth organizations from more than 18 countries.

Growing out of Pugwash in the 1960s has been ISODARCO, which stands for the International School on Disarmament and Research on Conflicts. The first ISODARCO "course" was held in 1966 in Italy, under the leadership of Edoardo Amaldi, Francesco Calogero, and Carlo Schaerf. This was one of the earliest fora to benefit from Soviet participation, helped by its location in a relatively neutral venue.

Directed by Carlo Schaerf of the University of Rome Tor Vergata, the tradition — as much conference as course — has been continued, held winter and summer in Italy. A series of ISODARCO sessions began in Beijing, China, around 1978.

Another forum for similar topics was been held biennially at Castiglioncello, also in Italy, under the auspices of the Union of Scientists for Disarmament.

At Erice, Sicily, starting in 1963, a series of U.S./USSR academic meetings took place on possible global consequences of nuclear confrontations and new types of defenses against nuclear destruction — a favorite topic of frequent participant Edward Teller. As the ideological conflict waned, a broader dialog developed around the cross-cultural scientific program at Erice.

Through many of the icy years, annual meetings of the American science association AAAS were a major forum for balanced presentations on technical and quasi-technical issues associated with conventional and nuclear armaments. Competent session speakers were drawn from government, think tanks, and other sectors. Because of its resources and high standing in the scientific community, the AAAS was often successful in bringing in Soviet participants (who bootlegged on the trip to visit other organizations in the United States).

In addition to direct experimental activities arranged by NGOs. several international meetings about verification were organized; for example, in order to accelerate a peaceful end to the ongoing virtual hostilities, a workshop on nuclear, conventional, and chemical weapons convened in London in 1989. NGOs and leaders from VERTIC, I N S T E A D, and Ruhr University, Bochum, were major organizers of these optimistic and constructive meetings.

A workshop the next year in Vienna focused on monitoring reductions in conventional forces that had started in earnest following the CFE Treaty. Events such as the breakup of the Warsaw Pact were reducing tension, but non-conventional weapons were not being eliminated; so the meeting was held in the context of increasing fears for proliferation of nuclear, chemical, and biological weapons. By the time this workshop took place, many governments had endorsed or acquiesced to the idea that NGOs were important to the arms control process; as a result, governments sent participants directly, to take part in the workshop or attend as observers, or indirectly, by underwriting their attendance as unaffiliated individuals. Many conference speakers were chief delegates to the CFE and CSBM negotiations, and some attendees were military and ministry representatives from European nations and alliances.

The third and final conference in the series, Geneva 1993, was organized to deal with verification after the Cold War. Plenary sessions reflected the past, present, and future of verification — addressing nuclear disarmament, chemical weapons, the CTBT, confidence-building measures, military R&D constraints, and environmental agreements. A nuclear-weapon-free world became an overt topic and object of hope. Reflecting the times, an unresolved question addressed was, "How much verification is sufficient?" The response of the conference was that verification is necessary, but no amount is sufficient by itself to ensure security and enduring peace.

Demonstrations. Public demonstrations, especially at the Nevada and Semipalatinsk test sites symbolized a popular outpouring to put a lid on nuclear testing. The Campaign for Nuclear Disarmament in 1962 mobilized more than 150,000 people in an historic march protesting nuclear weapons.

Warranting special attention were the mass demonstrations of 1968, because of the expanding anti-war movement around the world.

When President Johnson withdrew from the presidential race, he took specific note of the public division in America.

Many local and national rallies against the ABM took place from 1968 through 1969. The war in Vietnam had jarred many Americans out of complacency. Originating at the Massachusetts Institute of Technology, a "Day Without Research" demonstration occurred on March 4 of 1969. Anti-war demonstrations and resistance activities proliferated; the largest peace march in American history up to that time took place when 500,000 came to Washington, D.C., on 15 November 1969. A massive nonviolent attempt in Washington to shut down the machinery of unresponsive government followed in May 1971.

Coordinated mass demonstrations — with two to three million participants — convened in twelve European capitals and major cities in 1981. Two years later, 12 million protestors across western Europe marched on a single day, October 23, demanding an end to INF deployments; London, Amsterdam, Bonn, Rome, and other major cities witnessed the largest political demonstrations in their histories.

In June 1982 nearly a million people gathered in New York City to protest the nuclear arms race in what became the biggest political demonstration to take place in America.

A landmark political rally in European history — involving three million people — took place in October 1983, to demonstrate against deployment of nuclear missiles in Germany.

Particularly troublesome to the Reagan administration was a campaign of civil disobedience that began in 1986 at the Nevada testing grounds; it drew attention to public demands for the United States to join a year-old Soviet unilateral moratorium on nuclear testing.

Peace-activist Mary Kaldor summed up her (insightful, but overstated) views:

> The Cold War was ended by a wave of popular movements in both East and West that ... discredited the Cold War idea. It was the Eastern European democracy movement, not Western governments, that brought about the final collapse of Communism. And it was the Western European peace movements that first challenged the *status quo* in Europe.

Even more credit is granted to the world disarmament movement by historian Lawrence S. Wittner, who says that "the nuclear arms control and disarmament measures of the modern era have resulted primarily from the efforts of a worldwide citizens' campaign...." He, too, somewhat overstates the case.

Summits. Superpower summit meetings that drew the attention, efforts, or intervention of NGOs included Geneva in 1985, Reykjavik in 1986, and Washington in 1987.

The Reagan administration hoped that it could distract attention from "mounting political difficulties at home," but delegations from SANE, the Freeze Campaign, and Women for a Meaningful Summit were present to demand progress in arms control and peace.

While the president was still pushing SDI, arms control activists were gaining momentum in Congress against funds for testing nuclear weapons, against full-scale testing of ASAT weapons, against production of new binary chemical weapons — as well as votes favoring SALT II weapons ceilings, cutting the overall military budget, and balking at Reagan's budget for SDI.

At the 1985 Geneva summit, progress was made in framing an INF agreement and also — in principle — toward a 50-percent reduction in nuclear arms. SDI, though, intervened as a stumbling block that foreclosed further progress.

By the 1986 meeting at Reykjavik, the Reagan administration had to plead for a free hand in negotiating with the Soviets, because the ground swell of public and congressional opposition to his arms control policies had become a major factor. Reagan actually got carried away for the moment and agreed to Gorbachev's proposal to eliminate all ballistic missiles within a decade. They were even talking seriously of eliminating all strategic nuclear weapons. The deal collapsed, largely because of Reagan's SDI obsession and because his core of hardliners were shocked and achieved a retraction. Nevertheless, significant progress had been made toward an INF agreement.

The Soviets under Gorbachev also made concessions, willing for the first time to decouple INF from the SDI impasse. This resulted in an INF agreement that was signed by Reagan and Gorbachev at the subsequent December 1987 Washington summit, and entered into force after June 1988 ratification in Moscow.

Treaties. Negotiations for bilateral agreements and treaties were normally limited to government delegates and advisors, but multilateral discussions often gave opportunities for NGO input. The treaties on nuclear-test limitations and non-proliferation were heavily favored and pushed by the public and by NGOs acting for their public constituencies. NGOs were allowed as observers at UN Conferences on Disarmament.

In support of non-proliferation, the five-year Non-Proliferation Treaty review conferences drew considerable public scrutiny; in particular, the April 1995 NPT Review and Extension Conference gained outside attention and attendance by NGOs.

Enduring credit must be given to President Reagan and his administration for proposing the INF zero option and closing the Treaty before the end of his two terms in office. External political factors — especially public demonstrations in Europe — certainly had a major influence on that outcome by stimulating the Reagan proposal, encouraging Soviet decoupling of the treaty with SDI, and adding pressure to advance the difficult negotiations. In any event, the compromise became a landmark arms control measure, yet to be emulated for other nuclear forces.

****

Political pressure and power were essential ingredients in pushing governments to butt heads with each other. In the United States, right-wing individuals and their organizations were extremely influential and messianic, exhorting the government to respond to every hint of Soviet capability. Standing up to the hardliners were some American presidents, many European leaders, and indeed, in the USSR, a few of their technocrats. The latter were eventually buttressed by scholars and quasi-government organizations that constituted the only influential independent thinking in the Soviet Union.

When all is said and done, public-interest groups — in the West and East — typically found it difficult to compete with well-established and better-funded military interests and government bureaucracies. Because of this frustration some transnational peace activists focused on trying to modify Soviet rather than Western policies, while domestic activists kept up the pressure at home. The influence and insistence of outsiders and insiders were sufficient to cause both superpowers to gradually restrain themselves from going even more overboard in nuclear armamentaria and confrontationalism.

The exceedingly closed nature of the Soviet system allowed its leadership to promote belligerent policies in the absence of popular support; but that very system left a small opening for outsiders to bring alternative viewpoints and counter-arguments almost directly to Communist Party leaders.

D. INTIMIDATION OF OPPOSITION

A Chapter on public involvement could hardly end without discussing practices of intimidation exerted at all levels of government, in the West and especially in the East. Fear, risk, uncertainty, ostracism, or outright banishment was very personal and evocative. Sometimes individuals were simply shunned by neighbors, co-workers, or employers. Other times, matters became more serious in the workplace: being called on the carpet by management, loss-of-job opportunity, revocation of security clearance, slighted in promotion, or outright demotion. Graver yet were public, neighborhood, or school harassment; worse were judicial indictment, police arrest, interrogation, even jail time or gulags.

Government leaders, aided and abetted by hardliners, attached the "patriotism" label to almost every action or thought connected with the Cold War. Having just emerged from a nation-threatening world war, citizens in both East and West were already trained to salute and follow their leaders. But much of the war-weary populace wanted reconstruction and peace. So misgivings soon arose, especially in the West, about maintaining belligerency against a wartime ally. In the early 1950s, to many it seemed senseless and remote from reality that the United States had to build up and brandish its nuclear weapons in preparation for hostilities of unprecedented magnitude.

On the other hand, citizens of many nations surrounding the Soviet Union, as well as those trapped within, were being governed by either a foreign power or a foreign ideology, a fact underscored by the large number of refugees. Also, because of the geopolitical changes that resulted from World War II, the seeds of independence were being planted for the eventual revival of ancient ethnic divisions within territorial boundaries of prewar nations.

Because so much more has been publicized or divulged about intimidation in the West, especially in the United States, this Section starts with that venue. The Section closes with a brief overview of coercion of dissidents in the East, leaving more disclosures and details largely to emerging post-Cold War narratives.

There were many shades of opinion on how to deal with frictions that led to overt clashes between East and West — conflicts over military occupation, political ideology, economic systems, cultural differences, national boundaries, and oppression of human rights. Specific nuclear policies — civil defense, testing, arms buildups — were contested by the public. These disputes did not remain polite and unheated: they took on a virulent polarization that resulted in much governmental intimidation of the opposition. Dissenters were labeled as unpatriotic, even treasonous.

Governments in both the East and West, feeling it necessary to protect their very existence, more-often-than-not attempted to turn their countries into police states. Some succeeded.

Intimidation of the opposition took many forms. In the East, dissidents were easily cowed by blatant police-state tactics. In the West, the nonconformists ostensibly had constitutional protections, but these were often overlooked in the name of national security. Skeptics and critics of official nuclear policy faced public derision, and worse, in their employment settings, institutional associations, personal affiliations, and community activities.

Constitutional and unconstitutional means were used to suppress dissent and to deny access to information. Secrecy was imposed over the entire decision-making process, not just sensitive military information. Laws were rigidly applied on matters dealing not only with nuclear data, but also decisions affecting domestic and foreign policy. Bureaucracies resented intrusion into their deliberative process; they rationalized that outsiders — à la *Catch-22* — were not well enough informed to deal with the Cold War complexities. The public was expected to trust professional bureaucrats to know what was best.

Besides denying information and imposing secrecy, direct coercion and less-direct fear-inducing threats were utilized. The more subtle forms of oppression were applied in the law-enforcement process by depriving constitutional rights, imposing economic burdens, inflicting indignities, and attaching stigma. Protections of the U.S. Bill of Rights eroded as modern-day witch hunts harassed those who were not considered suitably adherent to the symbolism of the nation's flag.

While some communist spies in the West were unmasked, mostly participants in the Manhattan Project, nary a one was identified via the 1950s McCarthyism extravaganza.

Here are a few of the more blatant, extralegal governmental excesses in the United States, spurred by self-appointed patriots.

Prosecutorial Predominance

Taking on the "establishment" has always been risky, even in constitutional democracies, since the odds tend to be stacked against the interloper. Passive resistance or active objection to official policies often triggers administrative or judicial reaction.

At places of employment, managers rarely were sympathetic or supportive when an employee came into disfavor with the civil authorities: most supervisors were more concerned with corporate or bureaucratic image or with how a dissident employee's action might reflect on their business. Administrative repercussions could be expected in situations where personal activities were associated with the employer's name — regardless of disclaimers that thoughtful objectors were acting on their own — or where there were suspicions that national-security information had been compromised, even if no foundation for the suspicion had been stated.

Because of the government's hierarchical system, where authority devolves from bureaucratic departments to field offices, contractors, and eventually to subordinate staffs, the pressure from rules and regulations can be overwhelming. Intimidation is achieved simply by an implied threat of reducing funding or of recording detrimental information in employee-performance evaluations.

While "innocent until proven guilty" is a mantra under the American judicial system, the personal and financial cost of proving innocence can be ruinous. Although public-defender groups exist, notably the American Civil Liberties Union, the prosecutorial side has a preponderance of institutional weapons once an accusation enters into the legal process.

Prosecution in the United States starts officially with drawing up and serving subpoenas or warrants for wiretapping, injunction, court appearance, or arrest. A full-time staff, public funding, legal teams, and other resources strongly favor the state or federal authorities. The judiciary, largely composed of former law-enforcement personnel, is inclined to favor the prosecution. Government officials have the benefit of a presumptive case long before the defense made its rebuttal.

Those who bucked the system, especially in the early Cold War days, were at a severe disadvantage: Politicians used their bully pulpits to whip up public opinion against dissenters. Immunity to repercussions often protected the accusers, while the accused faced a monumental task of clearing their name, defending their actions, or protecting their jobs.

The more rebellious individuals faced police restraint, expensive bonds, and immediate incarceration. Public groups that were challenged had to contend with systematic legal and extralegal demands from (ironically) taxpayer-supported law-enforcement institutions. Government resources — at the local, state, or national level — were almost always overwhelming. Moreover, "national security" was frequently invoked to justify spying on dissenting organizations, infiltrating them, and disrupting lawful protest.

Because educational institutions were relatively independent of government funding and control, and because faculty often achieved tenure, academic dissenters had more latitude. Some national laboratories were operated under contract by universities, so that a measure of tolerant and protective spirit was available to scientists at Argonne, Brookhaven, and — in principle — at Los Alamos and Livermore.

In any event, individuals who actively protested governmental excesses faced extraordinary institutional pressures.

In the West

Dissent was often suppressed, inhibited, or penalized throughout the West. In the United States, which saw itself as a bulwark against communism, domestic intimidation against protest was particularly virulent — though inexplicable in terms of the protections supposedly offered by the Constitution and the Bill or Rights.

It was difficult enough for the American public to deal with revelations of successful wartime espionage by the allied Soviet Union, but allegations of Cold War spying on behalf of international communism cascaded into an American anti-Soviet/anti-communist backlash that had a chilling effect on dissent over nuclear arms policies.

It was also difficult for the U.S. government to sort alleged treasonous activities, such as espionage, from constitutionally protected expression of sympathy toward communism as a social and economic movement. The greatest contradiction was that several scientists who led the successful U.S. development of the atomic bomb were suspected of not being patriotic.

Some of this internal-security paranoia was reflected in the first war plan, designated NSC-20 ("Dropshot"), drawn up by the National Security Council in 1948. Besides having an objective of avoiding impairment of the economy, it included the goal of protecting "the fundamental values and institutions in our way of life." This plan thus went well beyond provisions for strategic security: It rationalized an *a-priori* presumption of wrongdoing for many who objected to the expanding nuclear arms race. Targeted more than usual were "intelligent people" or "various youth and women's organizations." No doubt immigrants were a matter of special concern. Dropshot authorized secret plans for mass arrest of Americans suspected of disloyalty.

Secrecy began to be used as a weapon even within the government. It was, for instance, a means to prevent dissident CIA opinions about Soviet capabilities from being expressed. Employees in nuclear-weapon plants who criticized working conditions were subjected to systematic harassment and intimidation, including threats to have their security clearances revoked.

Section D: Intimidation

The mantle of protective security kept individuals from meeting with others or visiting relevant facilities. Government officials were always particularly touchy about programs that might incur public inquiry or challenge. Even as the Cold War was winding down, the Pentagon put a tight lid of secrecy on its intelligence and missile-defense programs. This constraint was not limited to its own programs: it was applied as well to public meetings. For instance, in May 1985, the Pentagon pressured the organizers of an international conference on SDI-related laser and electro-optic systems to keep certain unclassified papers away from Soviet and East European participants. Moreover,

> The Pentagon [traded] heavily on the fact that many SDI skeptics are scientists and engineers with security clearances who are legally barred from revealing classified information about limitations of SDI, [while] Pentagon advocates have access to all classified information about SDI and can say whatever they like without the risk of being charged with security violations. The public [got] a very one-sided view of SDI.

Un-American Activities. Even before the end of World War II, the government's fear of communist spying spread into many sectors of public life. American communists were distrusted as possible "subversives." In June 1945, the FBI arrested six people associated with *Amerasia* (a journal about Asian affairs), accusing them of espionage on behalf of the Chinese communists. Two of the six were convicted of unauthorized possession of documents.

In 1947, when pressed by the House Committee on Un-American Activities (HUAC), President Truman imposed a Loyalty Order on all federal employees and ordered FBI security checks, extending to applicants for AEC fellowships. In 1950 an FBI investigation of Albert Einstein was opened, to find out whether he might be a communist or Soviet agent.

As early as 1947 the Justice Department considered prosecuting Leo Szilard (who had been attempting to write to Stalin about arms control) for violating the antiquated 1799 Logan Act, which "prohibited private citizens' correspondence with a foreign government on a subject of dispute between it and the United States." Ironically, Szilard was —perhaps more than anyone — responsible for getting America to develop the atomic bomb that led to Japan's expedited capitulation and gave the United States an immense military superiority over the Soviet Union after World War II. Upon being threatened by the Justice Department, Szilard made an appeal to scientific societies, invoking "the principle of the lesser evil," reminding them how German scientists gradually caved in to Hitler's purge of Jews.

The following information, found in an FBI file released under FOIA to a member of the organization now called the Federation of American Scientists, shows that, besides Congress, some state governments had their own investigative committees, which acted in coordination with Congress:

> In connection with his testimony before the Joint Fact-Finding Committee on Un-American Activities, State of Washington, on July 20, 1948, Dr. J.B. Mathews, former Research Director for the Special Committee on Un-American Activities, U.S. House of Representatives, submitted a list of "COMMUNIST Front Organizations," which included the FEDERATION OF ATOMIC SCIENTISTS.

So it was that a legislative committee of the westernmost state, which had the Seattle Chapter of the FAS, was looking for communists.

The most intensive focus was on Hollywood, perceived as a shaper of public thought, but other targets were government workers, college professors, artists, musicians, gays, and Jews. During 1947, HUAC witch hunts, victimized the "Hollywood Ten, a group of Hollywood screenwriters and one director. Targets were made to "take the pledge": "Are you now, or have you ever been, a member of the Communist Party?" Many who refused on principle to take the pledge were blacklisted by movie producers.

The McCarran Act (Internal Security Act of 1950) became one of the most controversial laws of the time. It was drawn up in part from earlier draft legislation submitted by Congressman Richard Nixon, who argued for fingerprinting and registration of all "subversives." The new version overcame Truman's veto and inspired the Senate Internal Security Subcommittee to work closely with Herbert Hoover's FBI. The Act also authorized concentration camps for "emergency situations." As one of the least understood, and most abused laws in the history of the republic, the Internal Security Act had a significantly detrimental impact on individual liberties.

The McCarran Act was only the tip of the inquisitorial iceberg. HUAC was always in operation, although it was relatively quiet for a while after the original Hollywood Ten hearings. In 1951, to prove its continuing relevance, HUAC began a second wave of show-business hearings, far outstripping its 1947 predecessor in scope, fanfare, and shamelessness. People who refused to name names of communists or sympathizers when called before HUAC were added to a blacklist. More than 320 people were placed on this list, blackballing them from working in the entertainment industry.

An aftermath of the 9/11-terrorism attacks against the United States, the Homeland Security ("Patriot") Act embodied peremptory features of the McCarran Act, in the form of administrative detentions without *habeas corpus* and other unbridled restrictions on citizens and foreigners. In addition, judicial restraints were loosened for electronic surveillance by the government. Visitors and non-citizens to the United States were newly subjected to McCarthy-era sanctions.

McCarthyism: Blacklisting. Senator Joseph R. McCarthy became the leading anti-communist crusader of the late 1940s and early 1950s, when he made the U.S. Senate a forum for charges similar to those levied in the House. Senator McCarthy's campaign against communist "subversion" ruined many careers and contributed substantially to the anti-communist hysteria of the time. His tactics gave rise to the abiding derogative term "McCarthyism."

In 1950 McCarthy denounced the FAS specifically as "heavily infiltrated with communist fellow-travelers." A year later, he accused the revered General George C. Marshall, then secretary of defense, of near-treason for his role in formulating America's China policy.

In February 1950, McCarthy made his biggest move, declaring, "I have here in my hand a list of 205 known to the secretary of state as being members of the Communist Party and who nevertheless are still working and shaping the policy of the State Department." The list had shrunk to ten names by the time McCarthy testified before a special subcommittee of the Foreign Relations Committee. The subcommittee concluded in mid-July1951 that McCarthy's accusations lacked substance and constituted a "fraud and a hoax" upon the United States Senate.

Senator McCarthy was an early member of the Senate Permanent Subcommittee on Investigations (under the Senate Committee on Government Operations). He started to chair the Subcommittee in 1953. No one was safe from his probing; people's careers were destroyed by just knowing the wrong person.

Many writers and performers moved to Mexico or Europe to avoid being put in prison. There was great pressure to abstain from controversial subject matter in films or on TV, with the result that productions that might have criticized national-security policy were not sponsored.

McCarthy's list of 205 people was drawn from previously screened State Department federal employees; it was not a secret. Although some had been communists, others had been fascists, alcoholics and sexual deviants. McCarthy's own drink problems and sexual preferences would have resulted in him being put on such a list, if he were investigated.

McCarthy began receiving information directly from J. Edgar Hoover, head of the Federal Bureau of Investigation (FBI). William Sullivan, one of Hoover's agents, later admitted that: "We were the ones who made the McCarthy hearings possible. We fed McCarthy all the material he was using."

Here is one explanation for this controversial period:

> With the war going badly in Korea and communist advances in Eastern Europe and in China, the American public were genuinely frightened about the possibilities of internal subversion. McCarthy, was made chairman of [a Senate Subcommittee], and this gave him the opportunity to investigate the possibility of communist subversion.
>
> [In the House of Representatives], fearing communist infiltration, HUAC sought the return of nuclear research to military control, but in a classic turf battle, the Atomic Energy Commission attempted to protect the civil rights of its scientific staff. During this period, anonymous panels remained arbitrary and capricious, in one case denying clearance because of membership in the American Association for the Advancement of Science. It also became difficult to hold international scientific meetings, because many foreign invitees were denied visas, and passports for overseas travel by Americans were withheld. The FAS Los Angeles Chapter became involved in these controversies.
>
> A common tactic used by investigators was to cut a deal by pressuring a suspect to inform on others: if the suspect did not give names, he or she would be thrown in jail or branded as seditious, and could not find work at all.

The subcommittee's 1954 probe of the U.S. Army, during which respected government officials and Army personnel were accused of being communist sympathizers, finally caused Senator McCarthy's downfall. It was toward the end of the hearings in which he took on the U.S. Army, that the Army's well-respected attorney, Joseph Welch, made his famous rhetorical query, "Have you no shame, sir? No sense of decency?" He added, with TV cameras watching, that McCarthy was a "lout" deserving no further attention. The tide of public opinion turned against him, and he died shortly thereafter of alcoholism.

No charge by McCarthy against a government official was ever proven. No spy was ever uncovered through his efforts. Nevertheless, often branded as communists or fellow travelers in the intimidating political climate generated by McCarthyism were scientists in the West who were willing to sit down with counterparts from the East to talk about peace and arms control — such as at Pugwash meetings.

A half-century after McCarthy embarked on his vilification campaign, a journalist summed up impressions of his own profession: "The nation's elite newspapers and broadcasters conspired to prolong McCarthy's witch-hunt by routinely reporting his accusations, but seldom investigating the truth of those charges." In bringing the problem up to date, the journalist's review goes on to note,

> With the end of the Cold War ... McCarthy's rehabilitation has begun.... But these efforts at refurbishing McCarthy's image rest upon selective historical memories.... He was a very late arrival in the pack of Red hunters who combined did more damage to the American Bill of Rights than Joe Stalin and all his Comintern agents.

According to one historian, "the search for potential spies motivated by ideology was by its very nature a low-probability enterprise, one that would identify hundreds of false positives for every true positive," while missing altogether those motivated by simple greed.

[Korean Purges]. The long reach of McCarthyism stretched overseas too. An example, recently unclassified, occurred as North Korea invaded South Korea in late June 1950. Kim Soo Im, a translator working for U.S. occupation forces, was executed by the South Korea military as a "very malicious international spy." The Seoul regime may have summarily put to death over 100,000 leftists and communist sympathizers in 1950.

Classified reports released a half-century later indicate that Kim was apparently forced to give a false confession after waterboarding torture. A teleplay in the 1950s introduced by host Ronald Reagan, depicted her as "Asia's Mata Hari."

Oppenheimer. Despite patriotically leading the successful and momentous scientific success of the Manhattan Project, J. Robert Oppenheimer's security clearance was revoked by the Atomic Energy Commission in 1953. Although Oppenheimer's reasoned opposition to development of the hydrogen bomb was given as the principal reason, the political climate of the time is more likely the reason for yet another casualty to red-baiting.*

The investigation that led to revocation of Oppenheimer's clearance was perhaps the most traumatic moral event involving physicists during this bitter conflict. Oppenheimer was a celebrated and revered leader in the development of the atomic bomb.

After the AEC ruled Oppenheimer to be loyal but a security risk, Hans Bethe — then president of the American Physical Society — issued a statement deploring the decision. The APS Council found it particularly disturbing that the charges arose from Oppenheimer's opposition to the H-bomb that was vigorously advocated by Teller and Luis Alvarez. The council noted that it was a very difficult

---

*Accusing someone of being communist, socialist or, in a broader sense, leftist, mainly with the intention of discrediting his/her political views.

Section D: Intimidation

technical and policy matter, in which opinions varied widely among many scientists of assured loyalty and competence.

Edward Teller had acknowledged in 1947,

> It is not even impossible to imagine that the effects of an atomic war fought with greatly perfected weapons and pushed by the utmost determination will endanger the survival of man.

The AEC's General Advisory Committee (GAC), in its 1949 report under the leadership of Oppenheimer, saw a higher degree of risk to humanity:

> The extreme danger to mankind inherent in the proposal by [Edward Teller and others to develop thermonuclear weapons] wholly outweighs any military advantage.

In the same report, Enrico Fermi and I.I. Rabi added their views:

> The fact that no limits exist to the destructiveness of this [Teller's proposed thermonuclear] weapon makes its very existence and the knowledge of its construction a danger to humanity.... It is ... an evil thing.

As those quotations show, Teller and Oppenheimer had somewhat different assessments of the societal risks in developing ever more potent nuclear weapons — hardly grounds for challenging loyalty. In fact, Teller's testimony alienated him from a large number of his colleagues.

Since then, Teller has written his own version of some of the damming information that had not been brought out. However, historian Richard Rhodes points out that Teller "did not recall these 'facts' when their subjects were alive to rebut them." Teller has denied having made statements that appear in FBI "documents, legal depositions and even a letter in his own hand." Teller has also been challenged for distortion of facts about an alleged proposal of Niels Bohr to share "all of our [nuclear-weapon] knowledge with the Soviets."

A new-millennium right-wing view of the GAC report and its members is described by Thomas C. Reed, political advisor to President Reagan:

> [The 1949 GAC report] was an amazing display of political and technical naiveté. It was not the first time during the Cold War that a distinguished group of scholars would wander far beyond their technical competence to advocate political courses of action. Nor would it be the last. But the miss distance of the 1949 GAC was historic.
>
> ... during much of the Cold War the musings of Nobel prize-winning scientists seem to have been exempted from [a] reality check....
>
> Thirty years later [despite the advice of the 1949 GAC] the Western alliance was still intact. The principal American deterrent protecting that alliance was its fleet of nuclear-powered submarines, each carrying sixteen Poseidon missiles. Each of those missiles, in turn, carried ten warheads with a yield in the kiloton range. The yield of those Poseidon warheads was not much larger than that of the Nagasaki Fat Man, but by using thermonuclear technology, the warhead weighed far less than that early device. In addition, robust safety and security devices were built into the Poseidon warhead. Miniaturization and hardening put the American nuclear deterrent at sea, away from the American mainland and immune to surprise attack.

Hard-liners such as Reed evidently remain convinced that scientists should not "wander far beyond their technical competence" because the consequences are better left to the politicians (who in 1949 decided that it was better to protect the United States by exposing it to the potential of a quantum jump in massive nuclear retaliation).

During Oppenheimer's 1950 security hearings, Teller testified using carefully chosen nuances that many in the scientific community felt constituted an "unforgivable betrayal."

> I would prefer to see the vital interests of this country in hands that I understand better and therefore trust more.

HUAC, with the support of Senator McCarthy and the executive branch of the U.S. government, "hounded many people suspected of having attachments to the Communist Party or even of associating with others who possibly did." Oppenheimer admitted, even when he headed the Manhattan Project, that he had ties to left-wing organizations and individuals. As a result, during the McCarthy era he was placed under 24-hour FBI surveillance. Although Oppenheimer divulged the names of friends who once had Soviet sympathies, he was "up against a vast and widely supported social force that had absolutely no tolerance for communism in any form and that persisted until the end of the Cold War."

Many scientists, sometimes at risk to themselves and their careers, spoke up for Oppenheimer. Because the removal of Oppenheimer's clearance in 1954 was widely regarded as excessive, even irrational, it proved to be a turning point and led to revisions that included genuine due process in the security code.

Bethe and Robert Christy, veterans of Los Alamos, posthumously commended Oppenheimer as a "brilliant" scientific leader of the Manhattan Project: Not only, they said, was "Oppie" fully informed at all times of the technical developments, but he kept other scientists and engineers current on the project's progress, insisting on freedom of communication within the laboratory — very much against the wishes of General Groves, who wanted information strictly compartmentalized. Although that might have contributed to his downfall with security officials, his flexible technical-communication policies contributed to the very success of the A-bomb project.

The Atomic Spies. Spying on enemies and friends, to the betterment or detriment of the parties involved, is a time-honored tradition. During and after World War II nearly all the major nations were spying on each other. Sometimes this was done by trained and recruited professionals, but often by amateurs who contributed their efforts for financial, political or ideological reasons.

Perhaps no spy did more to inform the Soviets on technical details of nuclear developments than Klaus Fuchs. Born in Germany, Fuchs became a naturalized British subject after escaping from Hitler's Germany. While in Germany, he had joined the Communist Party. In 1942 he was given a job in the British atomic-research project, and in 1944 he was seconded as part of the British scientific team to work on the Manhattan Project in the United States. There he had access to top-secret information, which he passed to the Soviets.

Fuchs was a top-flight, innovative nuclear physicist, who made important contributions to the development of fission and fusion weapons. Because of his close technical involvement in the atom bomb project, his information was presumably of inestimable value to the Soviet Union. In 1945 he returned to Britain and was appointed head of the theoretical physics division of the Atomic Energy

Establishment at Harwell. Fuch's spying was found out in 1949, and eventually he was sentenced to 14 years' imprisonment.

Another key source of nuclear information for the Soviet Union seems to have been Theodore A. ("Ted") Hall. While working at Los Alamos, Hall divulged technical information on how to detonate a nuclear device by implosion. Igor Kurchatov, head of the Soviet bomb project, after reviewing the report, is said to have responded, "The materials are of great interest." In 1962 Hall moved to England and was never charged with espionage or imprisoned. Judging from the Venona* code transcripts, he might have been the mysterious spy suspect with the code name *Mlad,* or possibly even the shadowy *Perseus.* On the other hand, a former American intelligence official believes that *Perseus* "is a U.S. national, very much alive [in 2004], and now living in California."

Hall's brother, Edward, was a Colonel in the U.S. Air Force who was instrumental in developing the solid-propellant rocket for the Minuteman program.

Two Americans, Julius and Ethel Rosenberg, were convicted in 1951 of selling and transmitting atomic-weapon secrets to the Soviet Union. Ethel's brother, David Greenglass, who worked as a machinist on the atomic-bomb project at Los Alamos, testified that he passed them secret sketches and drawings of the detonating mechanism. It now appears, however, that the information they passed was too general to have been of any particular value to the Soviet Union. The Rosenbergs were executed in 1953, despite widespread pleas for clemency.

With espionage files of the 1940s and after — including those of the KGB, the Stasi,** and the CIA — being gradually opened to public scrutiny, considerable care and circumspection must be exercised in evaluating the information disclosed. Sometimes agents or supervisors exaggerated their successes, or repeated speculation to their case managers. Before reaching conclusions or making accusations about suspected spies, substantive independent confirmation should be obtained, a practice not always exercised by overly eager and sometimes libelous accusers.

The impact of Soviet spying on the Manhattan Project and subsequent nuclear developments was mixed. While clearly violating national-security laws and U.S. interests, its effect was to expedite Soviet development (rather than to make it possible). With or without spying, independent development of nuclear weapons was inevitable.

Pentagon Papers. Residing in the Department of Defense was a top-secret history of U.S. decision-making in Vietnam, stored in safes at the Pentagon and its contractors. A copy of this history, dubbed the Pentagon Papers, was given to the press in 1971 by Daniel Ellsberg while he was at RAND. The report was pessimistic about the prospects of winning in Viet Nam.

Ellsberg's unauthorized release resulted in considerable official efforts at intimidation and retribution. Although charges were brought against him for his

---

* VENONA: The code name used for the U.S. Signals Intelligence effort to collect and decrypt the text of Soviet KGB and GRU messages from the 1940s.

** *Stasi*: the East German secret police.

role in the release of the Pentagon papers, the case was dismissed in 1973 by a judge who cited government misconduct.

Another government quarry was Morton Halperin, a member of the National Security Council staff at the time; it was later revealed that, under Nixon's order, Halperin's home was wiretapped from 1969 to 1971 as part of the effort to find out who had leaked the report.

One of the Defense Department's coauthors of the Pentagon Papers, Leslie Gelb, had moved over to the Brookings Institute; he too became a suspect. Nixon's staffer, Charles Colson, suggested that Brookings be firebombed — a proposal that, along with some others, was not carried out.

Henry Kissinger is quoted as saying, "Nixon often said exalted things that people didn't think would have to be done." Apparently all members of the National Security Council — Kissinger, Melvin Laird, John Mitchell, and H.R. Haldeman — knew of the wishes that Nixon expressed. The President devised and encouraged extralegal or illegal activities, and the Council failed to take responsibility for holding such "exalted things" in check.

Red Squads. In cities like New York, Los Angeles, Chicago, Denver, Seattle and Detroit, so-called "Red Squads" were set up by police departments to carry out investigations. Sometimes, as in Chicago, the Red Squad operation was led by the city government, while federal authorities, such as the U.S. Army, maintained their own surveillance at the same time. Although the stated objective of these investigations was to uncover communist spies and conspirators, the underlying purpose was to find and intimidate those who were opposed to such governmental policies as the war in Vietnam and the nuclear arms buildup. Organizations such as the Chicago Peace Council and other activist anti-war organizations were special targets.

The Security Section of the Chicago Police Department's Intelligence Division was found to have set up an alphabetical index system of over 200,000 cards. The cards contained references to information supplied by informers, police investigations and surveillance, newspaper items, and transmittals from other agencies. Simply having one's name printed on a leaflet as a sponsor was sufficient to get the name indexed.

In 1974, two organizations, the Alliance to End Repression and the American Civil Liberties Union, sued the City of Chicago for carrying out illegal surveillance. The organizations eventually (in 1985) won the litigation and gained an injunction against further illegal spying. The lawsuit resulted in a court decision that the Chicago Police Red Squad had violated First Amendment rights of the plaintiffs (in particular, the Alliance and the Chicago Peace Council) by publically disseminating derogatory information and by infiltrating them at a decision-making level.

The court also found that the police had violated the rights of an individual plaintiff by maintaining a detailed dossier containing extensive private information about her. According to the plaintiff's attorney, this was the first time a court had decided that surveillance by any local or state police department had been found to violate the First Amendment. It was the first time there was a ruling that domestic security investigations had to be founded on reasonable suspicion of criminal activity. In the wake of the 9/11 attacks, the legal restrictions imposed on Chicago police were eased after city officials argued that their "ability to investigate gangs and terrorism had been compromised."

Highlighting success of the Chicago litigation, the *New York Times* pointed to the court's finding that the Red Squad "put plaintiffs in the case under surveillance without any evidence that they had broken the law." The *Times* mentioned "the use of informers to infiltrate the Alliance and Peace Council" as a "striking example" of impermissible surveillance. (Two police informers had become members of the Alliance board.)

Even small organizations like the Concerned Argonne Scientists were monitored at their workplaces and deliberately infiltrated by federal officials, according to FOIA records released long after. The CAS was found to have been secretly cited for

> promoting the cause of a group of Indians who on July 30, 1971, illegally took possession of an unused Nike [formerly nuclear-armed ABM] site ... [and participating] in anti-war activities... during the period 1970-1971 in collecting signatures on various petitions and in writing letters to editors of local newspapers.

In 1971 a "full field investigation" of one of the CAS leaders was requested by the Department of Energy because of "anti-war activities."

The extent of surveillance on citizens in Denver, Colorado, has only more recently come to light, partly because the secret investigations continued at least into the 1980s. From FOIA files it has been found that more than 3200 people and 208 organizations in Denver were being closely watched. "Many of these people did nothing more than attend peaceful protests at the state Capitol or go to meetings of groups that police decided might represent a threat to public order." Two of the groups involved the Nobel Peace Prize winners Amnesty International and the American Friends Service Committee.

Not disclosed was the admission that the FBI never warned a Native American activist that they had uncovered a plot to kill him.

A nun who helped establish a coalition to protest Mexican government treatment of townspeople in Chiapas (Mexico) found that her group was labeled a "criminal extremist" organization. An activist discovered from Denver's spy files that the same label was attached to a citizens' police-proctoring group to which he belonged.

Government surveillance extended from the local to the federal level. Simply joining an organization was enough to trigger a seemingly ominous information entry in government dossiers. For example, using FOIA, the author of this book found that, because he became a member of the FAS, the FBI had added the information to his formerly secret file.

Some federal organizations known to have gathered files were the FBI, Department of Justice, CIA, Department of Energy, and the Army. There were more investigative federal officials, policemen, and subcommittee members than there were actual Communists in the United States. No "Red" espionage agents at all were found in the course of these investigations.

These secret homeland-security operations (grudgingly disclosed through FOIA applications and court suits) shortchanged constitutional rights of the public — evocative of the maxim that justice delayed is justice denied.

Government surveillance of public groups has been justified as being in the overarching national-security interest by proponents who have often lost sight of the Bill of Rights and the Constitution.

The *Progressive* Case. In March 1979 the Energy and Justice Departments of the Carter administration attempted to permanently enjoin the *Progressive Magazine* from publishing an article about government secrecy. The article by freelance reporter, Howard Morland, was intended to prove his point about the shortcomings of relying on government secrecy. Although he reported publicly available information that he collected about the H-bomb, it was considered to be in violation of the concept that nuclear information was "born secret," even if it could be found in open sources.

With the support of dissident expert scientists, the American Civil Liberties Union helped sustain the First-Amendment rights of the magazine. Despite the government's copious resources — full-court press, adamant cabinet members, and a stable of prominent scientists — the temporary injunction was lifted upon appeal to federal courts.

The ill-fated government course of action, in attempting to stifle the article, effectively confirmed information that otherwise would have gone without official validation.

Some individuals who worked for government contractors but sided with the *Progressive Magazine* were threatened during the case with security-clearance revocation. That would have severely hampered their ability to function within the highly classified government environment. The revocations, however, did not occur.

Later on, the government considered prosecuting those who sided with the defendants. Three years after the case was settled (1982), William Grayson, on behalf of DOE, compiled a list of names of every person mentioned in any court document, plus names mentioned in a book, *The Secret That Exploded,* that was published afterwards. This list of more than 300 hundred names (including this book's author) was printed under the title, "Possible Violations of DOE Regulations, Court Orders, or the Atomic Energy Act in Connection with *The Progressive* Case."

In the East

The paternalistic structure of the Soviet government and Communist Party made it difficult and perilous for individuals in Soviet-dominated nations to speak or act in a manner not consistent with official policy. The usual means available to Westerners for communication of ideas were not available. Dissent was often met with loss of privileges, banishment to gulags, or worse — with few or none of the constitutional protections found in democracies. Those caught spying for the West were subject to torture and summarily executed.

Individuals, for the most part, could express their candid political opinions only with their most trusted friends or had to disguise their views in the form of gallows humor.

Some independent peace-oriented groups and individuals did surface in Communist-dominated Europe. An independent East German pacifist movement emerged in the early 1980s; one of its members became the last defense minister of the German Democratic Republic, and was instrumental in having Soviet troops withdrawn from East Germany in 1990. Czechoslovak human-rights activists worked with END in 1980, but great difficulties had been imposed after the Soviet Army suppressed the "Prague Spring" reform movement in 1968.

Andrei Sakharov, considered the "father of the Soviet hydrogen bomb," later became his nation's most famous dissident. He and Yuri Orlov were among many who were persecuted for their nonconformist views. After many years of criticizing the Soviet educational, scientific, and political processes, Sakharov refused in 1965 to do further research in nuclear weapons development. Although not directly involved in the Pugwash exchanges, he was acquainted with Western views and had come to recognize how unilateral ABM defenses undermined the stability of mutual assured destruction.

In November 1970 Sakharov helped found the Committee for Human Rights. Despite receiving the Nobel Peace Prize in 1975, his continued dissent from Soviet policies led Chairman Leonid Brezhnev in 1980 to exile him to the closed city of Gorky. Gorbachev politically rehabilitated him, which was taken in the West as a sign of genuine commitment to policies of democratization. Another scientist exiled to Gorky was Lev Altshuler, a close friend of Sakharov.

At great personal risk, Soviet dissidents exchanged *samizdat* — a form of illicit newsletter publication — among themselves. One such illegal collection was edited by Roi Medvedev, a prominent *refusenik*.[*]

The most prominent Soviet dissident author was Alexander Solzhenitsyn, who won the 1970 Nobel prize in literature. After writing a letter critical of Stalin, Solzhenitsyn was sentenced to eight-years imprisonment. Although he was trained as a physicist and mathematician, he became well known as a novelist with a social conscience, and was widely published in the West. He joined other prominent dissidents, such as Sakharov, in forming the Human Rights Committee. In 1974 the Brezhnev regime expelled Solzhenitsyn from the Soviet Union.

****

The 1945 Amerasia incident and subsequent security-obsessive events reminds us of how little it took to foment hysteria in the 1940s and '50s. Even though the West "won" the Cold War, the tremendous societal costs should not be forgotten. Reputations and careers of innocents were destroyed in the

---

[*] The *refuseniks* were individuals — most of Jewish heritage — who were denied permission to leave the Soviet Union. The excuse often was that they had been given access at some point in their careers to information vital to Soviet national security and could not be allowed to leave.

name of anti-communism, and the U.S. government sometimes made a mockery of civil liberties under the guise of national security.

A shibboleth of patriotism was wrapped around official nuclear policy, and any implied or explicit dissent was tabbed as a threat to national security. The full force of law was often invoked to bolster the official notion of security. Sometimes the U.S. government used enforcement procedures that were outside the Constitution. When hardliners were in power, the morality of their cause was frequently enough to rationalize extralegal means of enforcement.

As in other nations, the U.S. government was not inclined to tolerate public questioning of assumptions that drove foreign policy — assumptions that led to seemingly unlimited military expenditures on nuclear weapons, chemical and bacteriological stockpiles, and conventional weaponry.

And, as for the Soviet authoritarian state, it lacked constitutional protections for unfettered public dissent.

In the United States, McCarthyism and the resulting witch hunts provoked frightening assaults on civil liberties. Unfounded charges and the inevitable stigma irreversibly damaged reputations and careers. Even Robert Oppenheimer was impugned — a most prominent scientist, who led the successful technical program to invent atomic bombs for the United States.

To this day, peace-oriented Nobel-prize-winning scientists are considered by hardliners to be meddlers when it comes to "political courses of action." The hardliners say scientists should stick to their areas of "technical competence," allowing more realistic decisionmakers to decide, for example, whether survival of a nation or of civilization should be put at risk.

Although the Soviet Union, a wartime ally, was denied official access to U.S./UK nuclear information, that denial was neutralized during World War II by an eminently successful espionage network. Soviet success, however, ultimately elevated U.S. post-war paranoia over spying. A huge secrecy apparatus was put in place, which even tried to restrict access to information already in the public domain.

When the war in Viet Nam was raging, suspicions over the possible influence of communists in the American peace movement caused the government to engage in unprecedented spying on citizens and attempts at suppression of constitutional activities. Federal and local government agencies reacted impulsively and forcefully against burgeoning societal changes that spawned the free-speech movement; law enforcement officials misunderstood the slogan "off the pig" ("kill the policeman") as a literal threat, and reacted with immoderate ferociousness. Constitutionally protected speech and association suffered frequent infringement at the hands of cavalier officials who treated vocalized foreign-policy opposition to be a violation of civil law.

The arrogance of office and the trappings of power often seduced officials into exceeding their public-service mandate. The judiciary, composed primarily of former prosecutors, frequently upheld civil-rights trespasses by law-enforcement officials. Applying the full force of law, and then some, against legitimate dissent created a perception of injustice and helplessness, resulting in amplified resistance through civil disobedience. The commonality of anti-war, anti-nuclear weapons, and civil-rights causes brought diverse interests together in a united front, exclaiming "We shall overcome" — which they eventually did.

In Eastern Europe, peaceniks and dissidents fared even worse, often being jailed or exiled.

It was not enough to be a law-abiding citizen in the United States, loftily protected by the Constitution. An aura of fear often attended the exercise of First Amendment rights. To protest excesses of the nuclear arms race risked being stripped of personal and institutional safeguards, meanwhile overt furtherance of national security was cloaked in the flag of patriotism.

Governmental intimidation of opposition — in the East and West — became the rule, especially when the opposition was aimed at fiscal policy, foreign intervention, or excessive nuclear armaments. Deplorably, investigative powers that were energized by good intentions often became politicized. Trawler-net surveillance data has been accumulated in secret dossiers, without the consent or knowledge of the subjects — a process that has been compounded in new-millennium anti-terrorist investigations.

The author of this book had his own scrapes with the excesses and temperament of national security, imparting added incentive for compiling into this book our personal retrospective on participative citizen action.

****

Public involvement was instrumental in constraining excesses. Without the scientists, NGOs, and foundations that coalesced — bringing independent information, analysis, and pressure to bear on the complex nuclear issues — civilization might have moved closer and more frequently to the brink of nuclear disaster. A large number of ad-hoc organizations emerged to deal with frightening dangers, and some still continue to push for reductions or eliminations of the residual, but still grotesque, nuclear arsenals.

The issues that were addressed by public-interest groups included civil defense, nuclear testing, missile defense, treaty verification, strategic bombers, and weapons proliferation. Some of these issues remain unsettled. That those matters were at the heart of national-security perceptions shown by the resolve demonstrated and means used by the political establishment to suppress dissent, and by the degree of intimidation suffered by those who opposed brinkmanship.

Many individuals were hounded by the government, causing considerable emotional stress and financial burden. Only in the years following, especially through law suits and enlightened legislation, was it possible to unearth abuses by government officials, who relied heavily on legalistic, dubious, and illegal application of information-classification and investigative functions.

Lest it be thought that these misapplications of the law have subsided, it is worthwhile to take note of excessive zeal in surveillance and detainment following the 9/11 terrorist assaults. For an example at the local level, the city of Denver — a year after 9/11 — admitted that it had an ongoing program of police surveillance, information collection, and record keeping about religious, peace, and protest groups. After being sued, the police department released 3200 "spy files" it had collected and computerized. The dossiers were about citizens of the United States uneasy about post-9/11 government overreaction.

More circumvention of due process has lately occurred to immigrants from Middle-Eastern (and look-a-likes from Far-Eastern) countries (whether they were undocumented individuals, green-card bearers, or American citizens).

****

Surely the scientific community and the public, during a half century of public-policy intervention, had some (possibly significant) influence on the gentle subsidence of the Cold War. Specifics include pressuring the Reagan administration to negotiate with the Soviet Union, shaping the zero-option proposal for the INF Treaty, stalling MX missile expansion, rejecting a civil-defense mentality, ending anti-satellite-weapons testing, limiting SDI funding, and preventing overt military intervention in Central America.

Perhaps more noteworthy than anything else is our reminder that **nuclear weapons won nothing***. They certainly didn't accelerate freedom for subjugated Eastern Europe which was under the yoke for about 45 years. More credit is due to citizen action than to nuclear weaponry or aggressive nuclear policy. What little restraint was shown by the West fared better than "bargaining chips" and "peace through strength" policies. The registry of abusive (East and West) policies that failed to dislodge Soviet and communist adventurism is lengthy, and it includes many repressive actions to limit public dissent.*

Notwithstanding the growth of doomsday technology — especially nuclear — the dire post-Hiroshima predictions of catastrophe did not come to pass. In fact, modern communications technology, which enabled the dissemination of protest, helped undermine the totalitarian Soviet state. Public intervention made the dangers of nuclear war clear enough on both sides of the Iron Curtain. Outsiders probably had a major role in averting the worst and in maintaining pressure to lessen the risk.

As the new millennium has unfolded, missile control and weapons security have improved; arsenals are now being sporadically reduced. This too must be credited in part to public engagement because nuclear governments are still moving at a clam's pace to reduce arsenals.

A good bit of this progress can be attributed to public opposition against the unbridled growth of nuclear arms. Aside from sharing information, public-interest groups who studied the operations and risks of nuclear armaments, appealed for adherence to humanitarian norms. They also stimulated benign policy initiatives. In their struggle against the establishment-embedded desire for "strong defense," it was not easy for peace activists to succeed. They did, however, have an advantage, in holding "principled ideas" — avoiding nuclear war, sharing common security, and promoting non-aggressive defense. Citizen activism and demonstrations for peace and human rights must be at least partly credited for ending the protracted state of belligerency without a nuclear showdown. Militant government policies were often frustrated. Things could have ended up much worse!

On the other hand, despite the peace movement, military force (accompanied by nuclear threats) still prevails as a major instrument of foreign policy.

As for improvements inside the iron curtain, some scholars who studied transnational movements attribute considerable constructive influence to Western activists. They provided aid to unorthodox Soviet scientists in the form of information, arguments, ideas, and legitimacy of association. Their counterparts within the USSR sought to influence government policy, and may have had more of an impact than realized by analysts who single out American pressure or the Soviet economic decline.

Some credit for the non-radioactive outcome of events probably is owed to the existence and growth of international institutions, such as the United Nations. Neutral organizations provided the means for input from nation-states not directly in the line of fire, as well as a forum for discussion by the adversaries that were not talking directly to each other. The UN arranged a platform for NGOs to speak out about the peace process, or lack thereof. International organizations like the IAEA became the means to legitimize nonproliferation and to make nations reasonably comfortable with adherence to the NPT. Treaties of multinational or international scope required an international body — a neutral intermediary — for administration and oversight. A few analysts argue that the existence of the UN helped prevent a third world war; in any event, the UN furnished a formalized assembly that could apply societal and diplomatic pressure against nationalistic excesses.

Explanations for the causes — and certainly the timing — of the end of Cold War correlate with Mikhail Gorbachev's openness to outside cultural and political contacts and his antipathy to nuclear weapons. Because of the high degree of Soviet centralization, General Secretary Gorbachev was able to use his enormous power to foster innovative policies. Many ideas to overcome internal opposition were adapted directly or indirectly from Western and Soviet scientists and public-spirited citizens.

Even though the undeclared state of hostility ended gracefully, it is necessary to remember that scientists created — in the face of growing public opposition — the possibility of killing millions or billions of people, and that enormous arsenals still remain ready to destroy cities and other targets. While these nuclear risks are subsiding, new dangers are being unleashed by unchecked advances in technology and the means to use them for military purposes.

Paramount among these new dangers is a surge in non-state terrorism—organized into nearly impervious cells and coupled through disparate nodes spread around the world. The militants are well-funded and adept at the use of modern technology, and they desire destructive weapons to convey their messianic message.

Coercive restraint against terrorism has limited utility. Perhaps the lessons, values, and alliances of NGO networking could be enlisted to address present-day disillusionment and to mediate economic and social relief. In addition, NGOs and individuals have proven that they are sometimes more credible than governments in communicating with dissidents.

The invasion of Iraq in 2003 caused some expected consequences to nuclear nonproliferation and the security of nations. In regard to the public-involvement, the invasion re-aroused an otherwise moribund peace and justice movement. Moreover, information-control tools that previously had given governments an advantage in the propaganda war are being negated by the growth of web-, satellite-, and cell-based networking among dissenters.

POSTERIOR PAGES

COLD WAR REDUX 440
    The Clash of Scorpions 440
    Hypothermia and Hyperthermia 441
    Learning Experiences 442
    The Unthinkable 445
    Cost of the Cold War 446
    Loyal Opposition 447
    Contentious Arms Control Issues 449

WHAT'S IN VOLUMES 2 AND 3? 451

GLOSSARY 453

ACRONYMS AND ABBREVIATIONS 457

MORE ABOUT THE AUTHOR 463
    [Independent Streak] 465
    [Paper Trail] 466
    [Cold War Professional Access] 467
    [Public-Interest Activities –1] 468
    [Public-Interest Activities –2] 469
    [Author's Summary of Professional Qualifications] 470
    [Credential Overkill?] 470

TABLE OF CONTENTS FOR VOLUME 1 471

INDEX FOR VOLUME 1 483

COLD WAR REDUX

This first volume has been largely about Cold War history and an analysis of events. Here is a summary of highlights that warrant special attention. (Volume 2 addresses the implications, the Cold War legacies and challenges for present and future generations; Volume 3, how to bring about reductions.)

The Clash of Scorpions

An apt metaphor for the Cold War is two scorpions in a bottle (without implying moral equivalence). From 1945 through 1991, the United States struggled with the Soviet Union until the latter's collapse. That conflictual period is referred to as the "Cold War" because the two countries never actually fought each other directly with weapons as they would have in a "hot war." Instead, the hostility was expressed by political and psychological posturing, carrying out surrogate wars, and stockpiling conventional and nuclear weapons — all in an effort to be able to outgun each other in the event that war did break out between them.

Those symbolic spectacles can be given little credit for what turned out to be peaceful resolution of the bilateral confrontation. Instead, the non-violent anticlimax probably resulted more from gradual economic exhaustion, increasingly pervasive communications technology, and grudging restraint by a leadership under unrelenting public pressure — along with some good luck.

****

The atomic bombing of Japan, although rationalized as a military operation to end the war, was probably determined more by domestic considerations and international politics. The newly vested U.S. military-industrial-scientific complex had a major stake in proving the value of its weapon and its investment. An important lesson: having committed so much, it was too difficult for the government to avoid using the new weapon.

At best, the first use of nuclear weapons shortened a war whose outcome was already decided. Allied forces had trapped most of the Japanese army overseas, and it would have taken at least two or three months to mount an invasion of mainland Japan. As it was, American occupation troops did not begin landing in Japan until three weeks after the devastation of Hiroshima.

Mindful of suffering by occupied civilians and military prisoners, other options to shorten the war existed prior to the atomic bombings. For example, a nonlethal demonstration of the bomb's awful power was suggested by scientists who created the bomb. Also, the Japanese government several weeks earlier had tendered the very same face-saving surrender offer that later was accepted by the United States.

The atomic devastation was tantamount to the first jab thrown in what turned out to be the Cold War. The bombing signaled a willingness to use nuclear weapons, and thus implied credibility for future conflicts.

Immediately after World War II, Soviet imposition of the virtual "iron curtain" in Europe started an intractable phase of the ensuing confrontation. Accelerated development of nuclear arsenals followed, first by the Americans and before long by the Soviets, who remained technologically close behind the Americans in the resulting arms race. A strategy of containment was devised in Washington and pursued for decades, urged on by influential interest-groups with access to official corridors.

Gradually a period of East-West *détente* emerged, spanning the 1970s and 80s. After some blustering by the two sides during the Reagan mid-80s, the Union of Soviet Socialist Republics began to fade away, a process completed in 1991.

Hypothermia and Hyperthermia

The long-running confrontation marked a chill (***hypo***thermia) in international and bilateral relations. But for the nations directly involved, this contest induced domestic ***hyper***thermia, that is, overheated political and ideological expressions of insecurity — lots of internal friction, but frosted external relations.

In the name of national security, various mechanisms were implemented at great expense, including a buildup of massive strategic forces. While the accumulation of nuclear (along with conventional, biological, and chemical) weapons went almost unchecked, the arsenals did no more than maintain a precarious stalemate.

Civil defense against nuclear attack was a dangerous illusion, as was missile defense. No defense — active or passive — could have done more than save a few distressed lives, while millions would have been lost.

In the 1950s and 1960s, political and psychological confrontation was more the norm than accommodation or negotiation. Only after massive public protest in the West generated pressure from the electorate did the governments try to engage in serious dialog across a negotiating table. Hardliners on both sides of the East-West divide frequently stalled the negotiations, often contending that proposed arms control treaties were unverifiable.

Mistrust made it difficult to negotiate pivotal verification protocols, but it also kept each side wary of deceit. Frequent disputes over shortcomings in verification were deliberate red herrings placed by those who considered treaties a sign of weakness. Although treaty verification could have been quite effective and reliable, knowledgeable outsiders were unable to convince policymakers. During the Reagan administration, possible technical initiatives to improve arms control verification technology and analysis were ignored, and years went by before polemic tirades about Soviet trustworthiness subsided. For what it's worth, we now have a better understanding of what motivated the obdurate warhawks.

Reviewing salient facets of the Cold War reveals that the action-reaction cycle which characterized the arms race was driven more by domestic political and institutional issues on both sides than by real threats, either "communist" or "imperialist." The precarious competition took the form of *Nuclear Shadowboxing* because each side was goaded more by its internal politics, economics, military strategy, culture, and ideology than by the actions of its perceived adversary. While both antagonists strategically sparred with ever more threatening nuclear feints, it turned out that they were symbolically boxing with their own shadow, not directly with their opponent.

That is not to say there were no real threats to Western nations from the Soviet monolith. Having "been there, seen that," it is not difficult to understand the reality and temper of the times. Among the coauthors of *Nuclear Shadowboxing*, the forerunner to *Nuclear Insights*, one author served with the American military during the Korean War, one with the Soviet Navy, one helped make Soviet nuclear weapons, and the fourth strove for international arms control. We all experienced the fears and hopes of the extended struggle. We do not minimize the suffering of many who were bridled by the Soviet yoke, nor do we agree with the bellicosity and brinkmanship of the governments.

Even the regional clashes at distant battle lines represented proxies assigned by the United States and the Soviet Union as a stand-in for direct bipolar competition.

Learning Experiences

From that traumatic period, when concepts of humanity nearly vanished, one must ask: Did our respective communal institutions gain any fungible and lasting insight?

Some Cold Warriors rejoice in "victory"—having "won" without the weapon of last resort being used. *Nuclear Insights* gives ample reason to be less sanguine, finding that the Soviet Union broke apart not so much because of "containment," "resolve," or "military power," but *despite* the compulsive gamesmanship and brinkmanship that characterized the excesses of the time. Having edged close to Armageddon is hardly something warhawks should brag about.

Through four U.S. presidential administrations, the costly U.S. hard-line approach in Vietnam failed to achieve success. Refusal to recognize expiration of the colonial era was one reason; another was the domino-theory fallacy that nations would sequentially fall to communist ideology unless, it was argued, military intervention took place.

The Cuban missile crisis, a prime example of brinkmanship, ended as a "success" story only in the sense that it didn't flare up into a nuclear exchange that was as close as human error.

At least one "true believer," who was in government office when the superpowers were moving toward another nuclear abyss, has taken autobiographical comfort in his estimation that President Reagan's policies put an end to the Cold War. But that's highly disputable: After all, the Soviet Union did not disassemble until 10 years after Reagan became President.

Historians concur that, long before Reagan was in office (in the 1980s), the USSR was already declining—economically, politically, and culturally. Perhaps Reagan accelerated the demise, but less-confrontational policies might very well have made it happen at smaller cost and risk to the West — and, who knows, maybe even sooner.

Here's another Cold War lesson of contemporary importance: National security is closely related to economic health. Nations now depend less on armies, territory, or natural resources, and more on the ability to adapt and integrate into the global economy — a transition that ideological communism and the arthritic Politburo were unable to assimilate (while China seems to have learned). *Successful governance requires engaging in a mutually reinforcing, but often ideologically unpalatable dynamic in economic and security relations with the world community.*

Both sides engaged in confrontationalism by feeding half-truth and speculation to their constituencies, and worse, they believed their own palaver. Both sides were impelled by fear created by hardliners, and both sides sanctified bellicosity with minimal attention to restraint and arms control. The public purse was a convenient and deep reservoir.

"Worst-case" methodology almost led to our mutual undoing. With each military or technological advance inducing the other side to follow suit, excessive reaction was deeply embodied in the development cycle for nuclear systems — fission weapons, thermonuclear weapons, ICBM, MRV, MIRV, ABM, SLCM, MX, etc. *This action-reaction cycle ratcheted the participants to dangerous encounters and outrageous budgetary excesses.*

In any event, American and Soviet political leaders during the contentious decades deserve at least some kudos for avoiding a nuclear clash.

A preponderance of nuclear weapons did not help the nuclear states prevail in regional conflicts like Berlin, Korea, Vietnam, or Afghanistan. Nor did the atomic bombing of Japan and the subsequent four-year U.S. monopoly keep the Soviet Union from taking over Eastern Europe and Mongolia.

Pervasive myths propagated about arms control, treaty compliance, and military strength. At the time, the myths were detached from accessible realities. The fallacious nature of their underpinnings was confirmed once the lengthy struggle terminated — but some of the fantasies are still promoted.

In the 1950s through the 1980s, three specific and dramatic claims of "gaps" in military strength and "windows" of vulnerability gave rise to a series of U.S. political decisions that upped the arms-race stakes by promoting more or better nuclear weapons. We know now that these claims were overstated rationalizations to further the goals of the hawks. If anyone was "ahead," it was the United States. The forces of deterrence were at such high levels that additional weapons added nothing to national security, but rather increased the risk of mutual nuclear annihilation.

The Soviet Union engaged in wholesale infringement of both freedom and boundaries inside and outside its borders on the Eurasian land mass. In the West, too, there were atrocious examples of overseas human-rights abridgement, such as American violations of foreign sovereignty in South and Central America, callous abductions in South America in the name of anti-communism, and Western-supported mercenary forces in Africa.

The Mi Lai massacre in Vietnam by U.S. troops helped turn public opinion against the American endeavor to prop up an anticommunist regime. Recording tapes finally released disclose the Nixon administration's initial attempt to suppress photos of the massacre. World opinion was as indignant then, as it has been after the wars in Afghanistan and Iraq, when photos and reports surfaced depicting American abuses of detainees in custody. The Cold War corpse has only recently been buried, but the ghost of lessons-never-learned is being resurrected.

Other neglected lessons should be drawn from the efforts of post-Cold War governments to sacrifice human life and rights in their pursuit of security. Outrageous new-millennium national policies still divert attention from long-existing violations in distant parts of the world, and encourage others to violate the rights of prisoners and civilians.

That important lessons have not been learned can also be discerned in the aftermath of President G.W. Bush's 2003 invasion of Iraq. A year after the unsanctioned invasion, the *New York Times* admitted that it had been hoodwinked into accepting unconfirmed reports of WMD and Al Qaeda in Iraq. The newspaper's explanation was that their reporters and editors were suckered by a "hall of mirrors," where one creative image was reflected many times to make it appear like a cascade of truth.* The hall of mirrors brings to mind the analogy of shadowboxing.

Neoconservatives at the Pentagon, not surprisingly, were among those who helped amplify the image of the Iraqi threat. Many are former Reagan-administration conservatives or hardliners, who resurfaced in the G.W. Bush administration trying to finish their unconsummated agenda.

Cold Warriors tend to be dismissive of public dissent. Worst than that, warhawks enabled or condoned suppression of constitutional rights in order to sustain nuclear pugnacity. Dissident individuals, groups, and organizations had to remain vigilant and maintain pressure against the excesses of the time. While public dissent failed to halt the massive nuclear buildups and confrontations, think what might have happened without such outside prodding.

Some legacies can be considered good, some bad, some tolerable. A few might have appeared anyway, perhaps at a different time. *For future peace and stability, it is important to recognize the lemming-like Cold War parade toward an unfathomable chasm.*

Even World War I was unexpected, unpredicted, and lacked a substantial *casus bellum*. In 1914, alliance leaders and strategists assumed that it would be no more than a localized, low-intensity conflict. But the forecasts and scenarios were proven wrong when the tragic war propagated throughout Europe and elsewhere.

It is dismaying that these admonitions might not be widely enough understood to preclude recurrence of near-fatal nuclear conflict, or worse — by design,

---

*Physicists have an apt analogy with lasers, which rely on mirrored surfaces to reflect and amplify stimulated light emission.

carelessness, error, or accident. Some traces of the virtual hostilities continue, especially dependence on nuclear umbrellas and "tripwire" deployments. NATO entered the new millennium with nuclear weapons assigned to its bases in Europe and forward-based American troops positioned in locations such as South Korea, where the nuclear temptation might come up if they were overrun in an invasion. Even today, prominent individuals hold onto the dogma that nuclear weapons remain useful components of national policy.

Hopefully, this book will contribute to clarification of the frightening and self-fulfilling consequences of perilous nuclear buildups, and a realization that their recurrence is not unimaginable.

The Unthinkable

American weaponry was qualitatively far ahead of the Soviet Union, but that advantage did not prevent a stalemate. The inability to prevail was frustrating to the warhawks, though it did not stop them from putting into place procedures and weapons that might have led to "the unthinkable."

Thanks to bragging rights, we can now read memoirs by overly proud American and Soviet Cold Warriors who rationalize the huge weaponry buildups. Their writings expose the ascendency of fear. Even during the 30 years of dominance in warheads and delivery systems, U.S. insiders continued to beat the drums of fear. Little of this would have passed open scrutiny had the real inventories been public knowledge. Eventually the Soviet Union caught up and passed the United States in some of the statistics, but never did that lead to the feared Kremlin domination. A nuclear standoff, even today, continues, but with more participants.

In 1962, after both superpowers had deployed nuclear-armed ballistic missiles outside their own national boundaries, the Cuban crisis was one of the first episodes to escalate to the brink. More crises followed, some artificial and transient, others scary and lingering.

Paranoia over national security pervaded public and intra-governmental debate (a problem exacerbated by the fact that the public was not kept well informed). The paranoia was incorporated into the "worst-case analysis" paradigm that was used to justify expansion of nuclear arsenals. Whether through ignorance or error, the fearsome assumptions about the adversary were usually imaginary.

Much of this can be attributed to the "nuclear priesthood," a cabal of hardliners who knew how to gain leverage through the use of an embedded, self-perpetuated, and expanded military-industrial-laboratory-congressional quadrangle. The faction that maintained militant momentum consisted of career officers, business entrepreneurs, influential scientists, optimistic technologists, operations analysts, and ambitious policymakers.

Individuals in the United States who promoted brinkmanship were often self-styled defense intellectuals, who banded together in "think tanks" that reinforced the primal fear that the nation was vulnerable to the threat of communism. The Soviets had their own fearmongers, whose monster was Western imperialism.

Those bloated fears resulted from a process of assessment and planning that was war-oriented, with little historical basis and few checks or bounds. War-gamers simulated nuclear warfare as though it were a programmed contest between reasoned and informed adversaries, both of whom understood the rules and agreed to play by them. For computerized war simulations, escalation control was a presumed parameter, despite possible failure of communication links and despite nearly instantaneous strategic-weapon delivery and unprecedented firepower. Illusions about the utility of nuclear war, and expectations of a favorable outcome, were widespread.

The intractability of President Reagan and his staff can be traced to advisors who were mostly very conservative (Republicans and some converted Democrats): Richard Perle, Edward Teller, Paul Nitze, Richard Pipes, Jeane Kirkpatrick, Fred Iklé, and Frank Gaffney. The Committee for the Present Danger ingratiated itself into the government, almost swallowing the executive branch whole.

Devastating weaponry in the hands of unpredictable leaders and nations — which had preemptive or escalatory policies or delegated control of nuclear weapons — risked all of civilization, holding it hostage for nearly a half century. The durability of societal institutions and survival of humanity as we know it was wagered in a form of high-stakes, "no-limit" poker — although the final round of the contest did end with more tractable "pot-limit" stakes.* The long period of marginal conflict reached closure, landing softly rather than convulsively.

Cost of the Cold War

The economic cost of that half century of quasi-military engagements was severe for both superpowers.

The CIA found that the collapse of the Soviet Union had more to do with domestic economic and social problems than with inability to keep up with U.S. military advances. The CIA had been aware since the 1970s — before Reagan's ascendency — that the Soviet Union was barely staving off economic and political collapse. American government leaders had been briefed on these findings.

In the United States, long periods of economic recession or inflation correlated with periods of heightened conflict (like the war in Vietnam and the Reagan buildup).

The United States alone charged about $19 trillion to the public purse, with almost $6 trillion of it going to nuclear arms. The Soviet Union went essentially bankrupt. Allied nations spent huge parts of their GNP; subjugated or war-burdened nations came out impoverished.

Balanced against the cost, both financial and societal, are reputed "benefits" of the Cold War. The West blocked further incursion of communism and Soviet tyranny (although the Soviets were able to prolong their hegemony over people and lands for many decades).

---

*For those not familiar with poker, it was all or nothing for some tense occasions during the Cold War.

Logically and properly, however, an important question comes up. Given that the West "won," and that "freedom" was protected for those outside the iron curtain, what was the role of nuclear weapons in bringing this about?

Actually, *nuclear weapons won nothing during the Cold War.* They have never really "won" anything — not a war, not the peace.

Even Japan, for all practical purposes, was defeated before atomic bombing hastened its surrender.

The Cold War got off to a bad start for the West, as mentioned, when Stalin took over Manchuria and retained Eastern Europe — despite the brandishing of nuclear weapons by the United States. And that was when the Americans had an authentic monopoly.

True enough, further Soviet expansion was stymied by the deterrent effect of nuclear weapons. In particular, some would argue that nuclear weapons kept Stalin at bay immediately after World War II.

Eastern Europe, though, was not freed by nuclear weapons. They remained and suffered under the Communist yoke for more than 40 years.

None of the hallmark proxy wars — in Berlin, Korea, Cuba, Vietnam, Afghanistan — was decided by nuclear weapons.

The Berlin airlift occurred in 1948 through 1949, when the United States still had a monopoly in nuclear weapons. The Korean war ended in a stalemate (the country still divided by the military demarcation line and demilitarized zone after half a century). Vietnam was taken over by the Viet Cong.

The Carribean/Cuban confrontation, an edgy situation where nuclear weapons were flaunted, settled down to a draw. It ended when the Soviet Union recalled its nuclear weapons from Cuba, and the United States withdrawing them soon afterwards from Turkey.

The Soviets lost in Afghanistan; the West later ousted the Taliban, but with no help from nuclear weapons.

Although the definition of a "hot" war has sometimes been stretched beyond the classical meaning, wars in the past have involved the systematic and organized shedding of human blood. In this sense, Japan notwithstanding, nuclear weapons have won no war. While they had a significant influence during the "Cold" War, they were brandished but not used. *Much smaller nuclear arsenals could well have had the same mutually deterrent outcome.*

The ineffectiveness of nuclear weaponry in the past as an instrument of policy, rather than for simple military deterrence, casts doubt on the political and military value of nuclear arsenals in the future.

Loyal Opposition

How effective was public intervention in moderating the confrontation? Ongoing efforts of hardliners and apologists to ignore, downplay, or contest claims of public effectiveness suggest that the loyal opposition was more than merely unimpressive and ineffectual.

Solitary private individuals rarely impacted the truculent Cold War pathway, but — when networked with others and with public-interest organizations — conscientious people did have a significant role in mitigating excessive militarism. Public-interest organizations were highly effectual in consolidating and focusing the opposition.

While not implying that public pressure brought the superpowers to their knees, it did help curb some of the more extravagant buildups and threats. Declassified tapes and records are proving that public opposition was indeed very worrisome to government officials.

While public influence was highly visible in the West, it was relatively inconspicuous in the Soviet Union, although a few "refuseniks" and high-level "apparatchiks" tried to bring about rapprochement.

Public resistance caused few U.S. military programs to go unfunded, but some — missile defense, nuclear testing, ASAT, MX, and Euromissiles — were trimmed to less-threatening proportions. In the West, public intervention was successful at least in slowing the quest for unlimited killpower and moderating the deployment of untimely weapon systems. It was different, though, in the East; once Soviet officials had decided on a course of action, they sometimes hedged by undertaking parallel paths of development.

All of this occurred despite the impact of Soviet Stalinism and American McCarthyism, when every possible official tool was invoked to instill fear in individuals and groups who spoke out in opposition to militaristic national-security policies. While dissidents in the United States enjoyed far more latitude than their counterparts in the Soviet Union, they often were squelched by the authorities. Anyone who protested the nuclear arms race risked being stripped of personal and institutional safeguards. Conversely, overt promotion of national security through nuclear weapons was clothed in the trappings of patriotism.

Contrary to the abiding convictions of "true believers," President Reagan's flip-flop about the realities of the Cold War was largely due to pressure from peace activists. He and his senior advisors belatedly promoted the concept of offsetting military asymmetries — an idea that one category of weapons could strategically nullify another. Eventually the Reagan administration went to the arms control negotiating table with a more conciliatory stance. Saber-rattling slowly subsided and agreed arms reduction ultimately followed the direction promoted by peace activists and other moderates.

No convincing evidence exists that Reagan's highly touted "bargaining chip" approach to negotiations had any substantive role in a world of asymmetric response. SDI was of little value to the United States as a poker chip, because the Soviets easily and much less expensively thwarted it by installing countermeasures and by increasing their offensive forces.

If hardliners had their way, those whom they derisively labeled as "specialists" (e.g., Nobel-prize-winning scientists) were supposed to avoid meddling in "political courses of action" — scientists were expected to defer to politicians and political appointees the decisions that jeopardized the survival of nations or civilization. But civilization might have moved closer and more frequently to the brink of nuclear disaster were it not for the scientists, NGOs, and foundations that coalesced to bring independent information, analysis, and pressure to bear on complex nuclear issues.

Rapidly advancing communications technology enabled protests to propagate, helping to undermine the totalitarian Soviet state. Dire post-Hiroshima predictions of nuclear catastrophe did not come to pass. Although the dangers of nuclear war were obvious to governments, both East and West, public involvement still was necessary, to maintain pressure in the direction of lessening the risk.

Contentious Arms Control Issues

Hardly any buildup — in contrast to build-down — of nuclear arms met broad public approval.

Both governments often strove to avoid, to delay, to drag out, or to water down arms agreements. American administrations used their bully pulpit, unfavorably portraying arms accords, to convince trusting constituencies of negativity in treaties and international agreements.

Even so, the cessation of atmospheric testing is attributable in part to the existence and influence of international institutions, who overcame the efforts of rabid opponents to stymie them. Neutral structures provided a forum for uncommitted nations and for adversaries who sometimes could not or would not negotiate directly.

****

To the oppressed of Eastern Europe, the Cold War meant more than 40 years of hard time. To the West, the standoff involved high risk, a long-lasting martial environment, and a costly waste of resources. In the United States, constitutional protections were abridged in the name of anti-communism. Surrogate wars were fought with blood and flesh to contain communism's expansion or suppress indigenous pressure against old colonial bonds.

Little, if any, of the hawkish approach can be called a success, except by defining it as such. Undeniably, for more than four decades nuclear force was unable to free eastern Europe or to stop the spread of communist ideology.

Realizing that it is Monday-morning quarterbacking to argue that a more nuanced and conciliatory approach would have been better, some interesting parallels have evolved.

The West now coexists with residual, homegrown communist nations — Cuba, China, and North Korea.

For Cuba, the United States is still trying containment, but Fidel Castro has lasted through the terms of nine U.S. presidents, and is working on his tenth. The Cuban economy is struggling without subsidies that the Soviet Union once supplied, but it keeps going on and on, irritating Mariel refugees and American politicians.

North Korea has been a thorn in the side of Western democracies and its neighbors in South-East Asia. This struggling communist country is seeking to guarantee its national security by developing nuclear weapons.

China — now that's a different story. Still a People's Republic, it has been subjected to both the sword and the feather approach by the West. Maybe patience will win out — gentle diplomacy over nuclear bluster. In an outcome that is at odds with the Western hardline doctrine, rapprochement with China seems to be working, without the brandishing of nuclear blades. China's atomic arsenal remains petite, but potent.

There are parallels and orthogonals in comparing China with the former Soviet Union. Compromise seems to be in favor and working: China is now a major trading partner with the United States and Russia, even though neighbors or satellites, such as Mongolia, Tibet, India, Hong Kong bring to mind some of the problems with Soviet expansionism.

Cold Warriors might not accept what is deduced from these experiences, but it seems that the adaptive, conciliatory approach is working. In any event, to contain the FSU and China (and Cuba, and North Korea), the United States certainly did not, and does not need huge nuclear arsenals.

Moreover, one can't help but repeat a remarkable observation: nuclear weapons won nothing outright or explicit during the entire Cold War. At best they served to sustain an over-caliber, pro-active nuclear Maginot line.

There are worthwhile nuclear insights and lessons to be heeded by future generations.

WHAT'S IN VOLUMES 2 AND 3?

While Volume 1 contains a retrospective view of the Cold War, Volume 2 is a forward-looking assessment; it is a logical continuation, bringing the reader well into the new millennium. Volume 3 is somewhat of a handbook for nuclear reductions.

In Volume 2 legacies and challenges left over from the Cold War are discussed. Logical ways to avoid national and international trauma from these legacies are suggested. Since emphasis is on nuclear weapons, constructively ways are put forth to avoid recurrence of the Cold War tragedy, or anything similar to it. This leads us to indicate what can be done to encourage worldwide reductions in new or old nuclear arsenals.

Here are some specific legacies knowledgeably addressed in Volume 2: radiation, institutional consequences, nuclear security in Russia, nuclear-armed nations, and the impact of nuclear technology (for better or for worse).

National challenges include political and military conflicts that might bring nuclear weapons into play, and the relationship of nuclear power for peaceful purposes with the potential for proliferation of nuclear weapons. Facing the world at large include are broader societal issues that are further complicated by terrorism, including the possibility that radiation and nuclear devices might become a means for furthering terrorist goals.

Are you curious about security and insecurity trends in this new millennium? Volume 2 examines the current nuclear balance among the original superpowers, the status of the other nuclear weapons states, and the advent of new ones. Addressed are issues like the role of conventional forces in sustaining national security, how to benefit from mutual security arrangements, what's going on in military development and spending, new nuclear hazards and proliferation risks, and how to go about improving security.

Heightened attention to transnational terrorism and international dealings in so-called weapons of mass destruction has upped the stakes, especially in protecting sensitive nuclear materials.

Do you have questions about whether nuclear reductions or disarmament are on the horizon? That's what Volume 3 is about. It is a technically-informed explanation of what's behind the ongoing policy debate, the global structure of treaties and negotiations, how to stuff the nuclear weapons genie back into the bottle, the feasibility of making deep cuts in nuclear weapons, how to dispose of nuclear materials, the reliability of verification, and what can be done to negotiate weapon-state reductions. These are topics for which even the wisest politicians need technically experienced advice.

The conclusions of the entire *Nuclear Insights* series are summed up in Volume 3's final section, the Dénouement.

Also, the following is consolidated in Volume 3: a list of all figures; a glossary; detailed biographies of the authors; some retrospective comments; and a Table of Contents that encompasses all three volumes. This is effectively a "roadmap" — a topical outline of the book — that guides us as authors, and can give you as readers, an overview of topics covered.

A final note: Please keep in mind that *Nuclear Insights* has been winnowed (by a factor of about two) from *Nuclear Shadowboxing*, which contains substantial supportive details. The latter contains thousands of citations and references, too much for most readers, but of value to scholars and researchers.

GLOSSARY

atomic bomb	A bomb that uses a fission chain reaction to create a devastating explosion that results in blast, heat, and radiation effects
atomic weight	The number of protons plus neutrons in the atomic nucleus
beltway bandits	Contractors in the Washington, DC, area who seek government contracts
boosting	A process of increasing the explosive yield of a nuclear weapon by inducing fusion reactions
brinksmanship	The pursuit of an imperious and dangerous policy to the boundary between safety and recklessness; a policy that brings a nation to the brink of war.
broken arrows	Accidents involving nuclear weapons that do not cause a nuclear explosion
calutron	An early means of enriching uranium by bending a beam of uranium ions in a magnetic field
chain reaction	A series of nuclear fissions linked by the neutrons released
counterforce	A policy of attacking military rather than civilian targets — opposite of *countervalue*
countervalue	Application of military force against targets of economic value, rather than military targets
critical mass	The minimum amount of nuclear material that supports a self-sustaining chain reaction
criticality	A condition reached when a mass of fissile material sustains a nuclear reaction
curie	An amount of radioactive material that decays at the rate of 3×10^{10} disintegrations per second
deterrence	A policy of dissuasion by threat of military retaliation, threatening as much as unlimited damage
doomsday machine	A hypothetical apparatus that endangers all of civilization
fast neutrons	Neutrons that have comparatively high velocity, such as those released from a fissioning nucleus

fertile isotope	An isotope whose atoms become fissile after absorbing one neutron; e.g., Th-232, U-238, Pu-240
first strike	A (nuclear) attack made without warning
first-strike capability	The ability to launch such a devastating attack that the opponent's retaliatory means is destroyed; the opponent therefore is essentially denied a *second-strike capability*.
fissile isotope	An isotope that can be fissioned by thermal neutrons; e.g., U-233, U-235, Pu-239, Pu-241
fissionable isotope	An isotope (either fissile or fertile) that can be fissioned by fast neutrons; all fissile atoms are fissionable, but not all fissionable atoms are fissile
fission (nuclear)	An energetic breakup of a nucleus, resulting in the release of neutrons, other radiation, and energy
fission products	Lighter elements formed when nuclear fission occurs — the "ashes" of the fission process
fission weapon	A weapon based on an uncontrolled nuclear chain reaction
fusion (nuclear)	Nuclear reactions in which nuclei are forced (fused) rapidly together to form heavier elements, plus release of energy and radiation
fusion weapon	A weapon that derives much of its energy from nuclear-fusion reactions
gun-barrel weapon	A nuclear weapon design in which two subcritical fissile masses become supercritical when slammed together (rapidly compressed) inside a tube
glasnost	Openness (Russian); the opening of political, commercial, and communication channels with the West beginning about 1986
Gulf War	The war in the Persian Gulf that began with Iraq's invasion of Kuwait in 1990 and ended after a UN-coordinated invasion of Iraq in 1991 (Operations Desert Storm and Desert Shield)
horizontal proliferation	An increase in the number of nations that develop or obtain nuclear weapons
implosion	A process in which fissile materials are rapidly compressed inward in order to quickly reach explosive criticality
initiator	A means of generating neutrons within a nuclear assembly in order to begin the chain reaction

Glossary 455

Iraq Invasion	The invasion of Iraq in 2003 by a U.S.-led coalition (operation Iraqi Freedom)
isotope	A nuclear species consisting of a specific number of protons (the atomic number of the element), and a specific atomic weight; e.g., two of the isotopes of uranium (atomic number 92) are U-235 and U-238
natural uranium	Uranium as found in nature, usually having 99.3% fissionable U-238 and 0.7% fissile U-235
neutron weapon	A nuclear weapon that sacrifices explosive yield in order to emit more neutrons and other radiation
nuclear	Pertaining to the atomic nucleus
nuclear weapon	A fission or fusion weapon that is designed and qualified for use by military forces
pit	The central part of a nuclear weapon — the fissile core
perestroika	Restructuring (Russian); policies of USSR transformation initiated by Gorbachev in 1986
production reactor	A nuclear reactor operated so as to produce weapons-grade plutonium
proliferation	An increase in the number of nations that have nuclear weapons — horizontal proliferation
radiation	Energy emission from an atom or its nucleus in the form of rays or particles
reprocessing	Treating spent fuel from a nuclear reactor so as to be able to reuse some of the fissionable components after separating them from fission products and other materials
second-strike capability	The ability to make an assured and devastating response after absorbing a first-strike attack
secondary	The second stage of a thermonuclear weapon in which most of the fusion reactions take place
strategic (nuclear) weapons	Nuclear weapons deliverable over long distances
tactical (nuclear) weapons	Nuclear weapons intended for battlefield use — variously called theater, battlefield, non-strategic, or sub-strategic weapons
thermal neutrons	Neutrons that have slowed down until they are in approximate thermal equilibrium with their surroundings

thermonuclear weapon	Another name for a fusion weapon, usually an explosive device that consists of a fission trigger and a separate fusion stage to amplify the explosive yield
throw-weight	The total load (usually RVs with nuclear warheads) that can be delivered to a target by a ballistic missile
vertical proliferation	An increase in the number of nuclear weapons in a nation's arsenal
weapons grade	Nuclear material that meets specifications for manufacture of a military-quality nuclear weapon
weapons-grade plutonium	Plutonium that contains at least 94% fissile isotopes (primarily Pu-239) — the balance being mostly fertile isotopes (primarily Pu-240)
weapons-grade uranium	Uranium that contains at least 93% fissile isotopes (usually U-235) — the balance being the fertile isotope U-238

ACRONYMS AND ABBREVIATIONS

AAAS	American Association for the Advancement of Science
AAM	air-to-air missile
ABACC	Brazilian-Argentine Agency for Accounting and Control of Nuclear Materials
ABM	antiballistic missile
ACA	Arms Control Association
ACDA	Arms Control and Disarmament Agency (U.S.)
ACHRE	Advisory Committee on Human Radiation Experiments
ADM	atomic demolition munition
AEDS	Atomic Energy Detection System (network for detecting nuclear explosions)
AFB	Air Force Base
AFTAC	Air Force Technical Applications Center (performs treaty verification functions)
ALCM	air-launched cruise missile
ANL	Argonne National Laboratory
ANP	aircraft nuclear propulsion
ANWFZ	African Nuclear Weapons Free Zone Treaty (Treaty of Pelindaba)
APS	American Physical Society
ASAT	antisatellite
ASM	air-to-ship missile
ASROC	antisubmarine rocket (launched from a surface ship). See SUBROC
ASW	anti-submarine warfare
BAMBI	Ballistic Missile Boost Intercept
BMD	ballistic missile defense
BMDO	Ballistic Missile Defense Organization (predecessor to the MDA)
BMEWS	Ballistic missile Early-warning System (U.S. radars in Alaska, Greenland, and England)
BNCT	Boron neutron capture therapy
BWC	Biological Weapons Convention (entered into force 1975)
C&C	command and control
C^3	command, control, and communications
C^3I	command, control, communications, and intelligence
CAS	Concerned Argonne Scientists
CBO	Congressional Budget Office
CDI	Center for Defense Information

CEP	circle of error, probable—the radius of the circle, centered on the target, within which a warhead has a 50% chance of landing
CFE	Conventional Forces in Europe (treaty signed in 1990)
CIA	Central Intelligence Agency (U.S.)
CIS	Commonwealth of Independent States (former republics of the Soviet Union)
CISAC	Center for International Security and Cooperation (at Stanford University) (formerly Center for International Security and Arms Control)
CISSM	Center for International Security Studies (at the University of Maryland)
CND	Campaign for Nuclear Disarmament (England)
COCOM	Coordinating Committee on Multilateral Export Controls
COMECON	Council for Mutual Economic Assistance — an economic organization of communist countries
CPD	Committee on the Present Danger
CPSU	Communist Party of the Soviet Union
CSBM	confidence- and security-building measures
CSCE	Conference on Security and Cooperation in Europe
CSS	Committee of Soviet Scientists for Peace and Against the Nuclear Threat
CTR	Cooperative Threat Reduction
CWC	Chemical Weapons Convention (entered into force 1997)
CTBT	Comprehensive Test Ban Treaty (rejected by the U.S. Senate, 13 October 1999)
D-5	Trident II D-5 (a U.S. MIRVed SLBM)
DARPA	Defense Advanced Research Projects Agency (U.S.)
DIA	Defense Intelligence Agency (U.S.)
DNA	Defense Nuclear Agency (now DSWA) (U.S.)
DOD, DoD	Department of Defense (U.S.)
DOE	Department of Energy (U.S.)
DSWA	Defense Special Weapons Agency (formerly DNA) (U.S.)
ELF	extremely low frequency
EMP	electromagnetic pulse (caused by a nuclear explosion)
END	European Nuclear Disarmament (NGO peace movement, 1980–1989)
ENDS	enhanced nuclear detonation system (reduces the chance of a warhead's detonators being fired electrically in an accident)
EPNW	earth-penetrating nuclear weapon
ERIS	Exoatmospheric Reentry Interceptor Subsystem (ballistic missile interceptor)
ERW	enhanced radiation weapon (the "neutron bomb" is an ERW)
FAS	Federation of American Scientists
FDA	Food and Drug Administration (U.S.)
FEMA	Federal Emergency Management Agency (U.S.)
FMCT	Fissile Material Cut-off Treaty (also called "fissban")
FOBS	fractional-orbit bombardment system
FOIA	Freedom of Information Act (U.S.)
FRG	Federal Republic of Germany (West Germany)

FRP	fire-resistant pit (one that can withstand prolonged exposure to a jet-fuel fire—a nuclear warhead safety feature)
FSU	former Soviet Union
FY	fiscal year
GAC	General Advisory Committee (1949)
GAO	General Accounting Office—the audit, evaluation, and investigative arm of the U.S. Congress
GEO	Geosynchronous orbit (satellite)
GLCM	ground-launched cruise missile
GPS	Global Positioning System (U.S.) (Navstar), consisting of dozens of satellites
GRU	The Soviet Main Intelligence Administration, operated by the Army
HE	conventional high explosive, used in early designs of U.S. nuclear weapons
HEO	Highly elliptical orbit (satellite)
HEU	highly enriched uranium
HOE	Homing Overlay Experiment — a missile-interception test series (made a hit 10 June 1984).
HTDS	Hanford Thyroid Disease Study
HUAC	House UnAmerican Activities Committee (U.S., 1934-1977)
HUMINT	human means of information gathering
IAEA	International Atomic Energy Agency
ICBM	intercontinental ballistic missile
ICF	inertial-confinement fusion
IGCC	Institute on Global Conflict and Cooperation (at the University of California at San Diego)
IHE	insensitive high explosive (for a nuclear warhead: a safety feature used in some of the more recently designed warheads)
IMEMO	Institute for World Economy and International Relations (Moscow)
IMS	International Monitoring System (for the CTBT, if ever ratified)
INESAP	International Network of Engineers and Scientists Against Proliferation
INF	intermediate-range nuclear forces
INSTEAD	Interuniversity Network for Studies on Technology Assessment in Defence (at the Freije University of Amsterdam)
IPPNW	International Physicians for the Prevention of Nuclear War
IRBM	intermediate-range ballistic missile
ISODARCO	International School on Disarmament and Research on Conflicts
JCAE	Joint Committee on Atomic Energy (U.S. Congress, 1947–1977)
JCS	Joint Chiefs of Staff (U.S.)
JDEC	Joint Data Exchange Center, in Moscow, "to ensure the uninterrupted exchange of information on the launches of ballistic missiles and space launch vehicles." Established June 2000; suspended by the G. W. Bush administration.
KGB	Committee for State Security (Soviet)—tasked with ferreting out potential threats to the state and preventing the development of unorthodox political and social attitudes among the population. [*FAS Web site*]
LANL	Los Alamos National Laboratory

LEO	Low earth orbit (satellite)
LEU	Low-enriched uranium
LNT	Linear No-Threshold—the theory that biological damage by radiation extends linearly all the way down to zero dose.
LoAD	low-altitude defense
MAD	mutual assured destruction
MBFR	Mutual Balanced Force Reduction (in Europe)
MDA	Missile Defense Agency (formerly the BMDO)
MED	Manhattan Engineering District (the Manhattan Project)
MEO	mid earth orbit (satellite)
MeV	million electron volts (energy unit for sub-atomic particles)
MILC	military-industrial-laboratory complex
MILNET	MILitary NETwork. Part of the Internet that deals with U.S military matters that are not classified.
MINATOM	The Ministry for Atomic Energy of the Russian Federation
MIRACL	Mid-Infrared Advanced Chemical Laser (a deuterium-fluoride chemical laser)
MIRV	multiple, independently targetable re-entry vehicle
MIT	Massachusetts Institute of Technology
MOD	Ministry of Defense (Russia)
MOX	mixed-oxide fuel for power reactors (blend of oxides of uranium and plutonium)
MPC&A	Materials Protection, Control, and Accountability
MRBM	medium-range ballistic missile
MRV	multiple re-entry vehicle (not independently targetable)
MTCR	Missile Technology Control Regime (an agreement among missile suppliers)
MWe	Megawatt-electric
MX	Missile-Experimental (a U.S. ten-warhead ICBM, dubbed "Peacekeeper" by Ronald Reagan)
NAOC	National Airborne Operations Center (U.S.)
NASA	National Aeronautics and Space Administration (U.S.)
NATO	North Atlantic Treaty Organization
NCRP	National Council on Radiation Protection and Measurement
NGO	non-governmental organization
NIE	National Intelligence Estimate
NIS	Newly Independent States (of the former Soviet Union)
NMD	national missile defense
NNWS	non-nuclear weapons state(s)
NORAD	North American Aerospace Defense Command
NOMOR	**N**uclear **O**verkill **MOR**atorium (a Chicago-based anti-arms-race organization)
NPR	Nuclear Posture Review (quadrennial U.S. military review)
NRC	Nuclear Regulatory Commission (U.S.)
NRDC	Natural Resources Defense Council
NPR	Nuclear Posture Review (initiated by the Clinton administration, 1994)

Acronyms and Abbreviations

NPT	Non-Proliferation Treaty (1968)
NSA	National Security Agency (U.S.)
NSC	National Security Council (U.S.—a Presidential advisory body)
NSDM	National Security Decision Memorandum
NTM	national technical means (for detecting the activities of an adversary)
NTS	Nevada Test Site
NWFZ	nuclear-weapon-free zone
NWS	nuclear weapons state(s)
OPS	one-point safe (a nuclear warhead safety-design feature)
OSI	on-site inspection
OSS	Office of Strategic Services (U.S.)
PAL	permissive action link (locking device to prevent unauthorized firing of a nuclear weapon)
PD	Presidential Directive
PNE	peaceful nuclear explosion; Peaceful Nuclear Explosions treaty (signed 1976; ratified 1990)
PSAC	Presidential Scientific Advisory Committee
PTBT	Partial Test Ban Treaty
PSR	Physicians for Social Responsibility
R&D	research and development
RBE	relative biological effectiveness (of ionizing radiation)
RERTR	Reduced Enrichment for Research and Test Reactors program at Argonne National Laboratory
RF	Russian Federation
RV	re-entry vehicle (the ballistic missile stage containing the warhead as it re-enters the atmosphere)
RV	remotely piloted vehicle
SAC	Strategic Air Command (now SC) (U.S.)
SALT	Strategic Arms Limitation Treaty, for which negotiations between the U.S. and the SU started in November 1969, leading to the treaties SALT I (1972) and SALT II (1979).
SAM	surface-to-air missile
SANE	Committee for a SANE Nuclear Policy
SC	Strategic Command (formerly SAC) (U.S.)
SCC	Standing Consultative Commission (set up to resolve SALT I disputes)
SDI	Strategic Defense Initiative (the "Star Wars" BMD program started by President Reagan in 1983)
SESPA	Scientists and Engineers for Social and Political Action (also known as "Science for the People")
SIGINT	signals intelligence
SIOP	Single Integrated Operational Plan (U.S. list of potential nuclear targets)
SIOP-ESI	SIOP–Extremely Sensitive Information (a U.S. secrecy classification for SIOP information)
SIPRI	Stockholm International Peace Research Institute
SLBM	submarine-launched ballistic missile
SLCM	submarine-launched cruise missile

SNDV	strategic nuclear delivery vehicle
SNTRA	Soviet Nuclear Threat Reduction Act (U.S.), passed in 1991
SPOT	Satellite Pour l'Observation de la Terre (a French high-resolution imaging satellite system that has provided images available to the public)
SRAM	short-range attack missile
SRBM	short-range ballistic missile
SSBN	ballistic missile submarine, nuclear-powered
SSM	surface-to-ship missile *or* ship-to-ship missile
START	Strategic Arms Reductions Talks (successor to SALT)
SUBROC	submarine rocket (anti-submarine weapon launched underwater from a submarine's torpedo tube). See ASROC
TEL	transporter-equipment-launcher (for land-mobile missiles)
THAAD	Terminal High-Altitude Area Defense (also called Theater High-Altitude Area Defense)
TTAPS	study predicting nuclear winter (authors: Richard P. Turco, Owen B. Toon, Thomas P. Ackerman, James B. Pollack, and Carl Sagan)
TTBT	Threshold Test Ban Treaty (signed 1974; ratified 1990)
TLI	treaty-limited item
UCS	Union of Concerned Scientists
UK	United Kingdom
UNIDIR	United Nations Institute for Disarmament Research
UNMOVIC	United Nations Monitoring, Verification and Inspection Commission (established Jan. 1990 to inspect Iraq for WMD)
UNSCOM	United Nations Special Commission, set up in 1991 to help the IAEA deal with Iraq's weapons programs
USEC	United States Enrichment Corporation
USSR	Union of Soviet Socialist Republics (Soviet Union)
VERTIC	Verification Research, Training & Information Centre (formerly the Verification Technology Information Centre)
WMD	weapons of mass destruction
WTO	Warsaw Treaty Organization

MORE ABOUT THE AUTHOR

Alexander DeVolpi has been an arms-control physicist, active in nuclear-arms policy and treaty-verification technology studies, for over 35 years. Now retired from Argonne National Laboratory, he writes from first-hand experience on most topics of this book..

On the subject of nuclear-weapons nonproliferation, DeVolpi is author or coauthor of many articles, two books (author of *Proliferation, Plutonium, and Policy* and co-author of *Born Secret: The H-bomb, The Progressive Case, and National Security* — both published by Pergamon), and a number of technical review articles (including "Fissile Materials and Nuclear Weapons Proliferation," *Annual Review of Nuclear and Particle Science*). He has been the lead author for the two volumes of *Nuclear Shadowboxing*.

Dr. DeVolpi has initiated numerous projects on the methodology and technology of treaty verification, including a technique for relatively unintrusive counting of nuclear-warhead multiplicity. Other proposals have pertained to nuclear-warhead detection and inspection on Earth and in space, fissile-material conversion, nuclear-facility monitoring, aerosol applications, weapons dismantlement, tagging and sealing, chemical-weapons verification, laser-brightness monitoring, cargo and luggage inspection, contraband-drug detection, and cooperative treaty-verification measures. He has given papers on these subjects at national and international conferences in Paris (1986), Japan (1989), Germany (1989), Austria (1990), Italy (1991), Canada (1991), Russia/Ukraine (1991), and Geneva (1993).

After earning an undergraduate degree in journalism from Washington and Lee University, DeVolpi served on active duty with the U.S. Navy in the mid-1950s. Later he received his Ph.D. in physics (and MS in nuclear engineering physics) from Virginia Polytechnic Institute, Blacksburg, Virginia.

DeVolpi has considerable experience in neutron physics and nuclear diagnostics, having served as principle investigator in a variety of research projects. He was manager of nuclear diagnostics in the Reactor Analysis and Safety Division at Argonne, and then became technical manager of the arms-control and nonproliferation program.

He holds a half-dozen patents, one of which is for the neutron/gamma hodoscope, a major instrument system used in the United States and France to image the motion of fissile material which was being tested under simulated accident conditions in special transient reactors.

In Argonne's Arms Control and Nonproliferation Program, he contributed to various technical arms-control projects, becoming Technical Manager for Physics and Engineering, and Principal Investigator for tamper-resistant tags and seals and assessment of foreign verification technology.

DeVolpi is recognized in *Who's Who in Frontiers of Science and Technology*, *American Men and Women of Science*, *Who's Who in Science and Engineering*, and other biographies. He was elected a Fellow of the American Physical Society for his contributions to arms-control verification and public enlightenment on the consequences of modern technology. He has been a member of the American Association for the Advancement of Science, the American Nuclear Society, and the Institute for Nuclear Materials.

Gaining the rank of Lieutenant Commander (retired) in the U.S. Naval Reserve as a result of 17 years on active duty and in the reserves, DeVolpi has had numerous assignments to the Naval Research Laboratory and the Radiological Defense Laboratory, where he participated in analysis of fissile-material safeguards and arms-control verification technology — including studies in 1969 regarding detection of ballistic-missile RV multiple-warheads, and monitoring for nuclear weapons on the seabed.

In a technical capacity, he has visited many national and international laboratories and has represented Argonne on various working and advisory groups — in the subjects of arms control, verification technology, radiation detection, tagging, on-site inspection procedures, and ground-based laser verification. He has been funded by the Departments of Energy and Defense to provide and participate in key briefings on related issues in Washington at government and contractor offices.

DeVolpi has also participated in major conferences on verification at Livermore, Sandia, Los Alamos, and Argonne; the DOD/DNA Arms Control Roundtables; American Physical Society, Association for the Advancement of Science, SAIC/CSNS verification workshops; DNA International Conference on Arms Control Verification Technology; and many international symposia. Because of long-term participation in treaty verification, he has had interactions with a wide spectrum of resource persons in and out of government, especially with arms-control organizations.

As a citizen-scientist, Alex has been active in public-interest arms-control issues since 1958. He was a participant and technical consultant in the FAS/NRDC joint project with the Soviets on nuclear-warhead dismantlement. He served as an elected member of the national council of the Federation of American Scientists in 1988-92. He was co-founder of Concerned Argonne Scientists, and a member of activist organizations and executive committees in the Chicago area. Some of his public-interest writings drew adverse reactions from DOE classification officials, resulting in a couple of publicized suspensions of his security clearance and the confiscation of his work computer. All of those actions were reversed after higher-level review.

Alex's personal life centers around his home at Carillon in Plainfield, Illinois. From a previous marriage, he had four children, now grown up, two of whom have their own families. Alex keeps in shape by playing handball, ping-pong, and swimming; and he enjoys — time permitting — poker, fishing, and woodworking as hobbies. Because of the time he has spent on writing about nuclear topics, two book projects about his family history and genealogy are backlogged.

[Independent Streak] (Involvement in Public Issues)

Not only because of the complexity of this book, but also because of its wide-ranging topics, it seems essential for the discriminating reader to have supplementary information about my qualifications and involvement, especially for this first volume.

I have benefitted from an unusual, but select and varied education in Virginia: military school at Staunton; BA in journalism from Washington and Lee University in Lexington; MS in nuclear engineering physics and PhD in physics from Virginia Polytechnic Institute, Blacksburg.

At Argonne, I started off with a more specialized nuclear-education upgrade at the International Institute for Nuclear Science and Engineering.

My formal academic progression was punctuated by commissioned military service in the U.S. Navy out of Norfolk and by subsequent participation in the Naval Research Reserve, with stints at the Naval Research Laboratory in Washington, DC, and the Naval Radiological Laboratory in San Francisco, CA. Despite this warriorlike background, I have frequently criticized military solutions to human problems. Still, I consider myself an activist, not a passivist

As a citizen-scientist, my background and experience has been focused on humanistic concerns, especially societal survivability and environmental health. This goal I partially attribute and credit to colleagues at Argonne National Laboratory, mostly members of its ground-breaking FAS Chapter, particularly David Inglis. Although the University of Chicago operated Argonne on behalf of the Department of Energy, it provided little, if any, shelter against government infringements.

Activism beyond the confines of an employer (Argonne, the University of Chicago, and the Department of Energy) requires risks and sacrifices beyond the norm. Anyway, following is a recollection of my public-interest activities or reactions, in partly chronological order, with particular relevance to Chapter IV.

[Paper Trail] (Professional Contributions Supported by Published Reports)

- Absolute and relative measurements of neutron sources and intensity
- Studies of fundamental nuclear fission parameters
- Radiation metrology
- Zero-power-reactor technical assistance
- Absolute measurements of fission fragments
- Gamma-ray measurements
- Neutron-beam measurements
- Neutron and gamma detector development
- Nuclear metrology
- Isotope half-lives, especially Mn-56 half-life
- Unexploded-ordnance (UXO) detection
- Evaluation of detecting nuclear weapons aboard ships and submarines
- Improvised explosive-device (IED) detection
- Dirty (radiological) bomb analysis and detection
- Laser-brightness verification
- Nuclear reactors and their safety
- Radiation parameters and Chernobyl safety
- Cineradiography
- Tags and Seals
- Nuclear-detection electronics
- Software for technical text processing
- Software for Fortran algorithms
- Cerenkov counting of radiation
- Manganese-bath absolute measurements of neutrons
- Statistical variance and correlation
- Fast-neutron and gamma radiography and tomography
- Electronic delays
- Dead-time corrections
- Radiation background measurements
- Instrumentation for nuclear-reactor safety experiments
- Time-resolved fast-neutron (hodoscope) radiography and diagnostics
- Detection of mass-casualty weapons
- Management, design, engineering, fabrication, installation, and operation of experiment apparatus
- Nuclear-reactor experiments
- Complex transient-data collection and analysis
- Proton-recoil proportional counters
- Fission-fragment detectors
- Water-level and fission-product monitoring for light-water reactors
- Fission-parameter analysis
- Data and analysis of nuclear-fuel-failure dynamics and loss-of-coolant simulations
- Image processing and resolution restoration
- Analysis of risks associated with possible diversion of nuclear materials
- Detector array design and implementation
- Verification analysis of START, SALT, CFE treaties
- MIRV nuclear-warhead detection
- Chemical and explosive identification; identification of dangerous substances
- Arms control, national security and counterterrorism
- Accidental-coincidence corrections
- Technology for arms control treaty verification
- Intrinsic-surface roughness tags
- Accommodation of on-site inspection at DOE facilities
- Reactor ex-vessel water-level detection experiments
- Accelerator-generated radiation
- Nuclear-fuel cladding experiments
- Time-resolved visualization of nuclear-fuel meltdown experiments
- Switchable radioactive neutron source
- Energy-policy decision making
- Chernobyl radiation effects
- Nuclear arms control and non-proliferation
- Sensitivity of plutonium to proliferation
- Nuclear-weapons characteristics and proliferation

Author's Credentials and Biography 467

[Cold War Professional Access]. Few who have written about the Cold War had an opportunity that I've had to visit relevant facilities and countries, or meet individuals, or knowledgeably assess their value.)

U.S. Laboratories
Argonne National Laboratory, Chicago, IL
Hanford site, WA
Idaho Nuclear Engineering Laboratory, Idaho Falls, ID
Kansas City Plant, KS
Los Alamos National Laboratory, Los Alamos, NM
Oak Ridge National Laboratory, Oak Ridge, TN
Pantex Ordnance Plant, Amarillo, TX
Sandia National Laboratories, Albuquerque, NM

Overseas Laboratories
Atomic Energy Canada Limited and NRL, Chalk River, Canada
Belgium, Sweden, Switzerland
BIPM, Sevres, France
Cadarache and Superphenix, France
Chelyabinsk-70 and Arzamas-16, USSR/RF
Atomic Energy Establishment, Dorset, England
Euratom, Ispra, Italy
Harwell and Aldermaston, England
IAEA Vienna and Siebersdorf, Austria
JAERI Tokyo and Osaka, Japan
Forschungszentrum Karlsruhe, Germany
Kurchatov Institute, ISTC, and Institute of Pulse Technics, USSR/RF
National Physical Laboratory, Teddington, England
ENEA, Ispra, Casccia, Rome, Pisa, and Bari, Italy
UKRATOMENERDOPROM, Kiev, Ukraine
VNIPIET, St. Petersburg, RF

Nuclear Experiment Facilities
ATSR, Argonne, IL
CP-5, Argonne
Juggernaut, Argonne
LANL low-power reactor, Los Alamos, NM
TREAT, Idaho Falls, ID
Various accelerators, Argonne
ZPRs and and ZPPR Argonne

NGOs
Bochum Verification Project, Germany
Brookings, Carnegie
FAS, ACA, and CDI
Heritage, American Enterprise Institute
INSTEAD, Netherlands
NRDC, SANE
SIPRI, Sweden
VERTIC, UK

U.S. Government Offices
ACDA, State Department, GAO
CIA
DARPA
DNA (DSWA)
DOE/AEC
Executive Office, House, and Senate
Pentagon
NIST

Foreign Government Offices
IMEMO, RF
MINATOM, USSR/RF
MOD, UK
Ukraine

U.S. Government Contractors
BDM
Mitre
RAND
SAIC
SRC
Titan
Lockheed
GeneralAtomics

Military Installations
Gibralter, Spain
Guatanamo Bay, Havana, and Santiago, Cuba
Naples, Palermo, and Taranto, Italy
Naval Radiological Defense Laboratory, San Francisco, CA
Navy Research Laboratory, Washington, DC
Rabat and Casablanca, French Morocco
Tripoli, Libya
U.S. naval shipyards and stations (many)
U.S. air-force bases (many)
U.S. missile-fabrication facilities
Valletta, Malta
Vieques, Roosevelt Roads, Ponce, and San Juan, Puerto Rico

[Public-Interest Activities –1]
- Participated with an Argonne group of mostly scientists meeting during our own free time, discussing current physics and technical developments, as well as nuclear public-policy issues. Many participants were also members of the FAS Chicago Chapter.
- One of the most prominent issues that the Argonne scientists group discussed and got involved in was the proposed siting of anti-ballistic missiles near cities, like Chicago. The group was heavily involved in defeating the proposed U.S. government siting plan. (My own involvement was two-fold: assessment of nuclear-safety issues for nuclear weapons stored close to population centers; and estimates of local fallout and radiation effects resulting from nuclear-warhead-armed missiles being exploded just above the cities in order to intercept incoming ballistic missiles.)
- Frequently called upon to speak in public forums about nuclear excess and concerns; participated in the group Chicagoans Against the ABM.
- Invented the fast-neutron hodoscope, after an earlier instrument developed by the Naval Research Laboratory failed; the hodoscope became a technical foundation for several arms-control verification applications that we proposed at a time when the President Reagan was claiming that arms-control measures were unverifiable.
- Having publically opposed the Air Force's plans for multiple warheads on ground-based missiles, I received a letter from the Navy requesting me to resign my naval commission (as Lieutenant Commander, U.S. Naval Reserve). Ironically, I preferred deployment of these missiles on Navy submarines. Calling a reporter that I knew on the *Washington Post* led to inquires that resulted in the Navy recanting, even asking me to send the original letter back.
- Often spoke at Hiroshima and Nagasaki annual memorial gatherings at the Henry Moore bronze mushroom-cloud evocative sculpture at the University of Chicago.
- Joined the Chicago Alliance to End Repression as a representative of our scientists' group because of constrictive actions by local and national security officials, including investigations and infiltration by the Chicago-police "Red Squad."
- Member of FAS, ACA, CDI, and other arms-control-related organizations.
- Wrote numerous technical articles and a book *Proliferation, Plutonium, and Policy* about still-prevailing misrepresentations of the weaponizability of reactor-byproduct plutonium.
- Was a recipient of an article alleged to be about hydrogen-bomb secrets, became a ACLU technical expert in support of its publication, supplied testimony for what became known as the *Progressive Case*, and was lead author of a book *Born Secret* about the case.
- Participated as a nuclear consultant in NRDC/FAS and foreign NGO arms-control-verification activities and experiments, despite concerns of DOE (but supported by Defense Nuclear Agency)

[Public-Interest Activities – 2]

- Was denied permission at the White House level to participate in the NGO "Black Sea" experiment aboard a Soviet nuclear-armed cruiser. However, government and intelligence agencies participated in my subsequent brokering of an unusual information dump from participants.
- Extensively investigated by federal agencies, as documented in CIA, FBI, Justice Department, DOE files released under FOIA requests. Numerous protocol conflicts with Argonne and DOE over participation in constitutionally protected speech and writing.
- Called on the carpet numerous times by Argonne supervisors or administrators for perceived infractions of Laboratory or DOE policy.
- DOE targeted me for alleged security infractions that proved to be unfounded. The Secretary of Energy personally called at home to overrule and apologize for precipitous DOE actions and confiscation of my office computer and disks.
- Co-founded the Concerned Argonne Scientists, an organization divorced from the FAS when MIT scientists rallied against the VietNam war.
- Publically wrote or spoke in opposition to many government positions, such as exaggerated radiation effects, oversimplified plutonium weaponizability, and treaty non-verifiability.
- Met with Soviet scientists during the Cold War (as reported in this book).
- Opposed high-power laser development for anti-ballistic-missile purposes.
- Marched and spoke in public demonstration regarding nuclear war, nuclear power, and the environment.
- In connection with, or as an adjunct to professional activities, visited many individuals, offices, laboratories, and individuals in connection with common arms-control and treaty-verification interests (see separate list of my professional access during the Cold War).
- Initiated Argonne's arms-control treaty-verification program and became its technical manager.
- Worked on arms-control and verification issues with Senators and their staff, particularly for Democratic Senator Paul Simon of Illinois
- Was a member of an advisory group appointed by Republican Congressman Harris Fawell of Illinois.
- Have written numerous professional and personal opinions published in public media regarding diverse technical and policy issues (e.g., Iraq, Iran, North Korea, gun control, nuclear-reactor policy, nuclear-weapons policy, plutonium weaponizability, global warming, and professional ethics).
- Still regularly meet and collaborate with Argonne colleagues after retirement.
- Engaged in informing the public and strengthening the historical record about the Cold War and nuclear weapons, reactors, and proliferation.

[Author's Summary of Professional Qualifications]. My official role at Argonne National Laboratory in arms-control and verification technology led me to relevant contracts with the Defense Nuclear Agency well before the beginning of formalized on-site inspection, including OSIA, as well as interactions with all the DOE weapons labs, with DOD, and at overseas laboratories. My volunteer activities allowed contribution of technical expertise to various NGO groups with which I collaborated, such as the FAS, NRDC, ACA, CDI, and others.

My professional activities at Argonne (and other laboratories) involved nearly 40 years of lab, field, and analytical activities in instrumentation, nuclear physics, nuclear engineering, reactor safety, radioisotopes, experiments, verification technology, and arms control. I have technical papers, review articles, and patents to back this up.

Besides being a technical consultant to the joint FAS/NRDC (Federation of American Scientists/Natural Resources Defense Council) verification project, I worked with European arms-control projects involving Soviet and Eastern European counterparts before the Cold War came to an end.
(www.NuclearShadowboxing.Info).

[Credential Overkill?]. Sorry if the presentation of these professional credentials seems like overkill or even boastfulness. In order to minimize misdirected challenges to the book's content, this supplement on Credential and Biography is clearly the place to substantiate credentials and to at least imply limitations.

Moreover, *Nuclear Insights* is derived not from singular personal experience and recollections, but incorporates those of my well-qualified *Nuclear Shadowboxing* coauthors.

In the world we find ourselves, full disclosure is often lacking: We were not institutionally encumbered in preparing either book; we are beholden to no one but our respective consciences.

TABLE OF CONTENTS FOR VOLUME 1

NUCLEAR INSIGHTS: THE COLD WAR LEGACY iii

VOLUME 1: NUCLEAR WEAPONRY (An Insider History) iii
   Dedication iv
   PRIMARY CONTENTS vi
   TABLE OF CONTENTS FOR PRELIMINARY PAGES vii
   CAVEAT EMPTOR ix
   CREDENTIALS xiii
   PREFACE xvii
   SYNOPSIS xxi
     Major Topics xxi
       Volume 1: Nuclear Weaponry (An Insider History) xxi
       Volume 2: Nuclear Threats and Prospects (A Knowledgeable Assessment) .. xxii
       Volume 3: Nuclear Reductions (A Technically Informed Perspective) xxii
     Major Themes xxiii
       Some Doubts xxiii
       Controversial Topics xxiii
       More Questions xxiv
       Russia's Legacy xxiv
       Shadowboxing xxv
       Recurring Themes xxv
   ORGANIZATION OF THE BOOK xxvii
   ACKNOWLEDGMENTS xxix
   DISCLAIMERS xxxi

   INTRODUCTION 1
     The *Trinity* Nexus 3
       Trinity minus 50: the Nuclear Genie Emerges 4
       ¡*Trinity!* Exclamation Mark for the 20 Century 5
       Deciding to Use the Bomb 6
         [Harry S. Truman's Announcement of the Dropping of an Atomic Bomb
          on Hiroshima] 7
       Demonstration? 9
       Motivations 10
       Potsdam Declaration 10
       Monday-Morning Quarterbacking 11
       Ending the War 11
         [Recollection by Wolfgang Panofsky] 12
       Nagasaki 12
       Postmortem 12
       Consequences 13
       Trinity plus 50: Getting Where We Are Now 14
       Figure 1: Full-Scale Replica of World's Largest Thermonuclear Bomb 15
     Problems for the New Millennium 16
       Reassessing Government Secrecy 17
       Nonproliferation and Counterterrorism Issues 18

Chapter I: THE COLD WAR (Two Scorpions in a Bottle) 21

A. NUCLEAR EVENTS THROUGH WORLD WAR II 23
[Enriching Uranium] 24
[The Frisch-Peierls Memorandum] 24
[Nuclear Chain Reaction] 25
The First Nuclear Weapons 25
[Nuclear-Weapon Configurations] 26
[German Atomic-Bomb Research] 26
[Soviet Pre-War Nuclear Research] 27
Figure 2: Mushroom Cloud After Atomic Bombing of Hiroshima 28
Hiroshima and Nagasaki 29

B. POST-WAR YEARS: *MANO A MANO* [1945–1960s] 30
[The Iron Curtain] 30
[Nuclear Secrecy] 31
[Dropshot] 34
Berlin 34
End of Nuclear Monopoly 36
[RDS-1] 36
[The Nuclear Aftermath] 36
NATO 37
Korean War 38
[Strategic vs. Tactical]. 38
Atoms for Peace 39
[The Smyth Report] 39
Testing Nuclear Weapons 40
[Nuclear-Weapon Testing and Secrecy]. 41
Developing Thermonuclear Weapons 41
[Boosted Weapons] 41
[Fusion Weapons] 42
Long-Range Bombers 44
Theater Weapons and the Neutron Bomb 45
Intercontinental Ballistic Missiles 46
Bomber and Missile Defenses 47
Surviving Nuclear War 48
Cuban Missile Crisis 49
[Events Leading Up to the Cuban Missile Crisis] 50
Vietnam War 52
[Vietnam War Narrative] 53
[Gulf of Tonkin Incident] 54
Changing Nuclear Balance 54

C. PHASES OF *DÉTENTE* [1970s-1989] 57
The 1970s 58
[ABM] 58
[MIRV] 58
[SALT I] 59
[CSCE/MBFR] 59
[Backfire] 60
[B-1] 60
[Cruise Missiles] 60
[SALT II] 61
Afghanistan 62
The 1980s 62
PD-59 63
[Neutron Weapons] 63

[INF] 65

D. COLLAPSE OF THE SOVIET UNION [1989-1991] 66
    [START I] 67
    [Dzerzhinsky Toppled] 67
    [Nunn-Lugar Act] 68
    Cold War Reprise 68
    Winners and Losers 69
    Russia as Successor 70

E. COLD WAR LEADERSHIP 71
    Superpower Leadership 73
    Superpower Leaders 73
    Non-Aligned Nations 74
    Treaties and Agreements 74
    [START II] 76

Chapter II: COLD WAR NATIONAL SECURITY
(Hypothermia and Hyperthermia) 79

A. NATIONAL DEFENSE 82
    Deterrence 83
        [X=Y≠Z] 86
        Figure 3: Mutual Assured Destruction 87
        Multiple Warhead Missiles 90
            Survivability 91
    Nuclear War Strategies 92
        U.S. Strategies 92
        Soviet Strategies 94
        [Hawks and Doves] 95
        British Global Strategy 97
        Germany's Nuclear-Capable Delivery Systems 98
        France and China 98
    Nuclear -Explosion Phenomena 99
        Fireball 100
        Blast 100
        Radiation 101
        Fallout 102
            [Radiation Units] 102
        EMP 103
        Examples of Nuclear Devastation 104
        [Nuclear-Explosion Effects] 105

B. THE WEAPONS 106
    Categories and Types of Weapon Systems 106
        PALs 107
    Strategic Systems 107
        Triad 108
        Bombers 109
        U.S. Intercontinental Ballistic Missiles (ICBMs) 110
        Soviet Strategic Modernization 111
        Submarine-Launched Missiles 111
        Shorter-Range Missiles 112
        Soviet Nuclear Weapon Systems 112
    Theater Weapons 113
        [Tactical Nuclear Weapons] 114
        U.S. Overseas Deployments 114

Soviet Deployments	116
INF Missiles	117
Enhanced-Radiation Warhead	117
British and French Deployments	118
Nuclear Warheads	119
Warhead Safety	119
Comparative Military Strength	120
British, French, and Chinese Nuclear Systems	121
C. DETERRENCE-SUPPORT TECHNOLOGY	122
Production of Nuclear Weapons	122
[Oak Ridge]	123
Nuclear weapons Testing	124
[A Sea Drill]	125
Nuclear Propulsion	126
U.S. Naval Vessels	126
Soviet Nuclear-Powered Vessels	127
Early Warning	127
Canceled Programs	128
Intelligence Collection and Satellites	129
U-2 Flights	129
[U-2 Memoir]	130
Satellite Reconnaissance	131
Command, Control, and Communication	133
C^3I	133
Command	134
[The Football]	135
Control	136
Communications	137
D. ARMS DEVELOPMENTS AND LIMITATIONS	139
Advances in Delivery Systems	139
Stealth	140
Technology Creep	140
Civil Defense	141
The 1960s	143
The 1970s	143
Emergency Management	143
Fallout Shelters	144
Strategic Defenses	145
Tallinn	147
Galosh	147
Sentinel	148
Safeguard	148
[Nixon's ABM Decision]	149
SDI	150
Directed-Energy Weapons	152
Antisatellite Technology	153
Overkill	154
[Nuclear Warheads Galore]	155
Improved Weapons Safety	156
Arms Control	156
SALT I	157
Nuclear Testing Limitations	158
SALT II	159
INF	160
START I	161

Chapter III: NUCLEAR LESSONS (Do We Learn?) ... 167

A. MYTHS AND REALITIES ... 170
Who Was Ahead? ... 171
- Bomber Gap ... 171
- Missile Gap ... 171
  - [Bang for the Buck] ... 172
- Window of Vulnerability ... 172
- Beam Gap ... 173

Initiatives for Peace and Disarmament ... 174
- Acheson-Lilienthal Plan ... 174

Arms Control Negotiations and Treaties ... 176
- Negotiations During the 1980s ... 177

Alleged Treaty Violations ... 179

Half-Truths and Misconceptions ... 180
- Half-Truths About the Arms Race ... 181
  - ??? Political rivalries do not matter — (the problem is runaway technology) ??? ... 181
  - ??? The arms race was accelerating ??? ... 181
  - ??? The arms race was accelerating ??? ... 181
- Misconceptions About the Arms Race ... 181
  - ??? Overkill exists in nuclear arsenals ??? ... 181
  - ??? The Soviet Union followed the U.S. lead ??? ... 182
- Fallacies About Arms Control ... 182
  - ??? Political instruments (arms control agreements) cannot constrain technology ??? ... 182
  - ??? Arm- control agreements inevitably benefit the Soviets ??? ... 182
  - ??? Technological innovations are always destabilizing and undesirable ??? ... 182
  - ??? The Soviets must be trusted ??? ... 183

Verification as a Red Herring ... 184
- OSI ... 185
- SLCMs ... 185
- ASAT ... 186
- MIRV ... 186
- Warheads ... 187

Perspective About Alleged Violations ... 188
- Treaty Experience ... 188
- Symmetry ... 188
- Vagueness ... 188
- Military Significance ... 188
- Balance of Terror ... 189
- Constraints on Treaties ... 189
- Verification Improvements ... 190
- Intangibles ... 190

Alternative Courses of Action ... 191
- NATO Nuclear Issues ... 191
- Offensive Strategic Forces ... 192
- Command & Control ... 193
- Nuclear Defense ... 193

The Role of Espionage ... 194
- Atomic Spies ... 195
- Thermonuclear Espionage ... 197

B. THE UNTHINKABLE ... 199
Strategic Assessments ... 199

```
        [Nostril to Nostril] ................................................ 201
   War Planning ........................................................... 202
        Second Thoughts ................................................... 205
   Security Paranoia ....................................................... 205
        McCloy-Zorin Agreement ........................................... 206
        Scowcroft Commission .............................................. 206
   Worst-Case Scenarios ................................................... 208
   Nuclear Bravado ........................................................ 209
        Nuclear Threats .................................................... 210
            [Nuclear Weapons Use Considered by Eisenhower] ............... 210
            [Lessons From the Cuban Missile Crisis] ....................... 211
            [DEFCON: Defense Condition] .................................. 212
        Nuclear Illusions ................................................... 214
            [Doomsday Machine] ........................................... 215
   National Security at Any Cost ......................................... 216
        Radiation Effects of Nuclear War .................................. 216
        Denver's Close Call ............................................... 217
        K-19: The Widowmaker ............................................. 218
        Deterrence Theory ................................................. 218
   Doomsday .............................................................. 219
   NATO's Nuclear Deterrent ............................................. 222
   Broken Arrows ......................................................... 225
   Accidental or Unauthorized Launch Risks ............................. 227
        Nuclear Safety and Authorization .................................. 228
   Radiological Warfare ................................................... 229

   C. THE NUCLEAR PRIESTHOOD ....................................... 232
   Think Tanks ............................................................ 233
        RAND ............................................................. 234
   Military-Industrial Complexes ......................................... 237
        Beltway Bandits ................................................... 238
        Armaments Lobby .................................................. 239
        Soviet Defense Establishment ...................................... 240
        Technological Innovation .......................................... 240
   The Scientific-Technological Elite ..................................... 241
        [Livermore] ....................................................... 243
        The Consummate Role of Secrecy .................................. 243
   JCAE ................................................................... 247
   Reagan's Intractability ................................................. 248
        The "Evil Empire." ................................................ 249
        The Revelation .................................................... 251
   Hardliners ............................................................. 252
        Committee on the Present Danger ................................. 253
        Perle .............................................................. 256
        Pipes .............................................................. 257
        Nitze .............................................................. 257
        Teller ............................................................. 259

   D. COST OF THE COLD WAR ......................................... 263
   Dollars for Defense .................................................... 263
        Foreign Aid ....................................................... 263
   Distorted Statistics .................................................... 265
   Economic Impact ....................................................... 266
        Stimulating the Economy .......................................... 267
        Infrastructure ..................................................... 267
        Productivity ...................................................... 267
        Jobs .............................................................. 268
```

Inflation and Taxes	268
Research and Development	269
Post-Cold War Impact	269
Nuclear Worthiness	270
Rubles for Defense	271
Vietnam War	273
Indirect Costs	273
Psychological Costs	274
Radical Compromises	275
[The Enemy of My Enemy]	275
Nuclear Weapons and Information Proliferation	275
[The Enemy of My Enemy (Getting to America)]	276
Nuclear Institutionalization	276
[The Enemy of My Enemy (In America)]	277
Who (What) Won the Cold War?	277
The Cold War as History	277
[CIA papers show agency knew where hunted Nazi was hiding]	278
The Cold War Was Won Before Reagan	278
Fallacies of Monopoly and Secrecy	280
Did Nuclear Deterrence Prevent War?	281
We All Lost the Cold War	282
Peace Works	283
[Personal Recollection About the Cold War *(Vadim A. Simonenko)*]	284
CIA Findings	285
Benefits of the Cold War	285
Who, Where, and When	287
What	288
Why	292
How	294
Chapter IV: PUBLIC INVOLVEMENT (Irritant or *Force Majeure*?)	299
A. PUBLIC-INTEREST ORGANIZATIONS	306
NGOs	307
FAS	308
SANE	310
Pugwash	310
Council for a Livable World	311
NRDC	312
[Our First Meeting]	313
UCS	313
SESPA	314
PSR	314
CAS	315
ACA	316
CDI	316
NOMOR	317
Mobe	317
Nuclear Freeze Campaign	318
Programmatic Organizations	319
Carnegie	320
Brookings	320
Hudson	321
Center for Security Policy	321
Henry L. Stimson Center	321
Nuclear Control Institute	322
Religious Associations	322

Extremists	323
Soviet Moderates	325
Soviet Academy of Sciences	325
Committee of Soviet Scientists	326
Sakharov	327
American Funding Foundations	329
W. Alton Jones Foundation	329
Ploughshares Fund	329
John D. and Catherine T. MacArthur Foundation	329
Educators	330
Individuals	330
MIT	332
Stanford University	333
Princeton University	334
University of Illinois	334
Monterey Institute	335
IGCC	335
University of Maryland	335
Bochum University	335
Notre Dame University	336
The Advanced International Studies Institute	336
Professional Associations	337
APS	337
JASON	338
AAAS	338
Academies of Science	339
International and Transnational Groups	340
IPPNW	340
UN	340
UNIDIR	341
SIPRI	342
International Institute for Strategic Studies	342
Canadian Institute for International Peace and Security	342
VERTIC	342
I N S T E A D Center for Verification Technology	343
Peace Research Institute Frankfurt	343
B A S I C	343
INESAP	343
Labor Unions	343
European Peace Movements	344
The Women of Greenham Common	345
The Greens of Germany	346
Networks	346
The Acronym Institute	346
INES	346
Abolition 2000	347
MoveOn	347
DontBlowIt	348
Institute for Global Communications	348
Safer World	348
Nuclear Threat Initiative	349
Fourth Estate	349
Scientific American	349
The *Progressive*	350
Dailies and Weeklies	350
The Silver Screen	351
[*On the Beach*]	351

[*The Day After*]	352
[*Dr. Strangelove*]	352
[*Seven Days in May*]	353
[*Fail-Safe*]	353

B. ARMS CONTROL CONTROVERSY ... 356
Fallout Shelters ... 356
Nuclear Testing ... 359
    [Impact of Public Opposition] ... 361
The ABM and Sentinel Debates ... 362
    Suburban Chicago Opposition ... 363
    Nationwide ... 364
    The Unraveling. ... 365
National Spending Priorities ... 366
Military-Industrial Complex ... 367
ACDA ... 368
SALT/MIRV ... 369
Proliferation ... 371
Euromissiles ... 371
Neutron Bombs ... 371
Freezing the Arms Race ... 373
SDI ... 376
ASAT ... 379
INF: Zero Solution/Option ... 380
Verification ... 382
MX ... 383
Missile Accuracy (and Reliability) ... 385
Nuclear Winter ... 387
B-1 and B-2 Bombers ... 389
Founding a Peace Institute ... 390
Conventional-Force Reductions ... 391
National Missile Defense ... 392

C. POLITICAL FORCES ... 394
U.S. Movers and Shakers ... 396
    Parties, Individuals, and Writers ... 396
        Joseph McCarthy ... 397
        Barry Goldwater ... 397
        Zbigniew Brzezinski ... 397
        Lyndon LaRouche ... 397
        Robert Welch and the John Birch Society ... 397
        William Buckley ... 397
        The Alsop Brothers ... 398
        Evans and Novak ... 398
        Berrigan Brothers ... 398
    Congress and Courts ... 398
    Paid Lobbyists and Advisors ... 399
    Private Institutions ... 400
        Cato Institute ... 400
        Heritage Foundation ... 401
        Hoover Institution ... 401
Western European Leaders ... 402
    Winston Churchill ... 402
    Charles DeGaulle ... 402
    Konrad Adenauer ... 402
    Willy Brandt ... 402
Soviet Policy Makers ... 402

Party Leaders ... 402
    Josef Stalin ... 402
    Nikita Khrushchev ... 403
    Alexei Kosygin ... 403
    Leonid Brezhnev ... 403
    Yuri Andropov ... 403
    Konstantin Chernenko ... 404
    Mikhail Gorbachev ... 404
Soviet Scholars and Organizations ... 404
    Petr Kapitsa ... 405
    Sergei Kapitsa ... 405
    Pugwash ... 405
    IMEMO ... 405
    The *Soviet Academy of Sciences* ... 405
    Andrei Kokoshin ... 405
Cultural Differences ... 406
    [Dead on Arrival] ... 406
Cultural Movements ... 409
    The Politics of Racism ... 410
    Gender Activism ... 411
    Student Radicalism ... 411
Religion vs. Atheism ... 412
Public Assemblies ... 413
    Conclaves ... 414
    Projects ... 414
    Conferences ... 415
    Demonstrations ... 416
    Summits ... 417
    Treaties ... 418

D. INTIMIDATION OF OPPOSITION ... 419
Prosecutorial Predominance ... 420
In the West ... 422
    Un-American Activities ... 423
    McCarthyism: Blacklisting ... 424
        [Korean Purges] ... 426
    Oppenheimer ... 426
    The Atomic Spies ... 428
    Pentagon Papers ... 429
    Red Squads ... 430
    The *Progressive* Case ... 431
In the East ... 432

POSTERIOR PAGES ... 439

COLD WAR REDUX ... 440
The Clash of Scorpions ... 440
Hypothermia and Hyperthermia ... 441
Learning Experiences ... 442
The Unthinkable ... 445
Cost of the Cold War ... 446
Loyal Opposition ... 447
Contentious Arms Control Issues ... 449

WHAT'S IN VOLUMES 2 AND 3? ... 451

GLOSSARY ... 453

ACRONYMS AND ABBREVIATIONS 457

MORE ABOUT THE AUTHOR 463
    [Independent Streak] 465
    [Paper Trail] 466
    [Cold War Professional Access] 467
    [Public-Interest Activities –1] 468
    [Public-Interest Activities –2] 469
    [Author's Summary of Professional Qualifications] 470
    [Credential Overkill?] 470

TABLE OF CONTENTS FOR VOLUME 1 471

INDEX FOR VOLUME 1 483

INDEX FOR VOLUME 1

(To find people: First look for last name [Bush], then for first name [George Bush], next for initials [G.W. Bush] , and finally for title [President Bush])

9/11 . . . . 213, 293, 424, 430, 435
AAAS 300, 338, 415, 457
Able Archer 221
ABM . . . xxiii, xxv, 47, 48, 58, 64, 65, 73, 74,
81, 90, 96, 113, 140, 146-
150, 152, 153, 156, 157,
159, 164, 179, 182, 190,
193, 194, 207, 235, 236,
238, 243, 251, 254, 256-
258, 272, 273, 291, 292,
301, 310, 311, 314, 315,
321, 331-333, 337-339, 349-
351, 360, 362-366, 369,
370, 377-379, 385, 392,
393, 396, 398, 399, 401,
403, 405, 409, 414, 416,
433, 443, 457, 468
    ABM bases 365
    ABM debate 235, 338, 366
    ABM deployment 147, 363, 393
    ABM sites 256, 363, 365, 366
    ABM Treaty . . xxiii, 58, 64, 65, 74, 148,
150, 153, 157, 159, 164,
182, 190, 193, 194, 207,
243, 254, 258, 339, 351,
369, 378, 392, 401, 409
        Safeguard 48, 58, 81, 146, 148-150,
235, 238, 256, 337, 363,
365, 366, 396
ABM Opposition
    ABM Opposition 362, 364
Abolition . xxiv, xxvi, 176, 177, 259, 301, 315,
346, 347, 374
    Abolition 2000 . . . . 301, 346, 347
    Nuclear Abolition . . . xxvi, 176, 259, 347
Abrams 253
Abyss 155, 270, 286, 361, 442
ACA 300, 316, 457, 467, 468, 470
Academies of Science 300, 339, 402
Accidents
    accidental nuclear war . . . 221, 236, 320
Accuracy . 15, 47, 60, 61, 75, 90-92, 107, 108,
112, 113, 122, 136, 138-
140, 143, 154, 158, 162,
181, 189, 193, 200, 203,
207, 223, 259, 266, 271,
272, 301, 369, 385-387
ACDA . 55, 188, 242, 253, 256, 258, 301, 331,
368, 369, 389, 390, 457,
467
ACDIS 334
Acheson 39, 168, 174, 307, 364

ACHRE 457
Ackerman 462
Acronym Institute 301, 346
activism . . 251, 302-304, 307, 308, 330, 337,
345, 346, 354, 376, 397,
398, 410, 411, 436, 465
ADA 400
Adelman 188, 253, 255
Adenauer 302, 402
Adequate Verification 184, 206
ADM 115, 457
Advanced International Studies 300, 336
AEC 33, 40, 42, 107, 196, 226, 246, 359,
423, 426, 467
AEDS 457
Afghanistan 22, 57, 61, 62, 69, 118, 120,
213, 216, 279, 282, 286,
287, 293, 356, 395, 412,
443, 444, 447
AFL 343, 344
Africa 14, 124, 411, 444
AFTAC 457
Aftergood 309
Agnew 360, 385
air defense . 109, 116, 130, 146, 207, 212, 233,
271, 272
Air Defense Emergency 212
Air Force . 34, 50, 54, 85, 86, 89, 91, 107, 109,
110, 112, 114, 119, 126,
127, 155, 171, 172, 185,
199, 200, 206, 225, 226,
233, 234, 236, 266, 287,
361, 386, 390, 429, 457
Air Force Generals 206
Aircraft Nuclear Propulsion . . . 126, 128, 457
Al Qaeda 213, 444
Alabama 277
Alaska 116, 124, 126, 127, 136, 180, 185,
229, 360, 457
Albuquerque 124, 225, 467
ALCMs 60, 112, 159
Aleksandrov 336
Aleutians 127
Algeria 125
Ali 375
Alibek 293
Alleged Violations 168, 188
Allen 254, 257, 331
        Richard Allen 254, 257
Allin 255, 256, 278, 280
all-out nuclear war . . xvii, 149, 186, 203, 217,
254, 328

Alperovitz	333
Alsop	302, 398
Alsop Brothers	302, 398
Alternative Courses	168, 191
Altmann	335
Altshuler	433
Alvarez	12, 426
Amaldi	415
Amarillo	124, 467
America Foundation	348
American Academy	332, 339
American allies	58, 233
American Civil Liberties Union	421, 430, 432
American Enterprise Institute	233, 254, 319, 467
American Friends Service Committee	308, 323, 389, 431
American leaders	323
American Minuteman	386
American Physical Society	153, 314, 337, 426, 457, 464
American Security Council	233, 319
Amnesty International	431
Amsterdam	343, 344, 417, 459
Andropov	73, 249, 302, 326, 380, 403
Angola	120, 264, 411
ANP	457
Antarctic	180, 183, 185, 188
Anthrax	180, 244, 293
anti-aircraft	226
antiballistic-missile	146
Antisatellite Technology	81, 153
Anti-Satellite Weapons	291
APS	300, 337, 338, 377, 426, 457
APS Council	337, 426
APS Forum	337
Arabian Sea	213
Arbatov	238, 240, 283, 326, 345, 405
Archer	221
Arctic	44, 112, 125, 218
Argentina	278
Argonne FAS	315
Argonne Scientists	315, 363, 367, 372, 430, 457, 464, 468, 469
Arizona	237
Arkin	312
Armageddon	229, 260, 288, 297, 442
Armaments Lobby	169, 239, 261
Arms Control	
Arms Control and Disarmament Agency	55, 177, 242, 253, 368, 370, 457
Arms Control Association	316, 457
Arms Control Measures	61, 160
Arms Control Today	316
Arms Reduction	16, 161, 175, 178, 320, 329, 394, 448
Arms-Control Issues	464
Arms-Control Verification	464, 468
Nuclear Arms Control	59, 176, 249, 308, 309, 311, 338, 346, 356, 388, 400, 417, 466
Arms Developments	xxii, 81, 139, 393
Arms Reductions	xxv, 16, 63, 345, 397, 462
Army	7-10, 12, 53, 62, 85, 95, 107, 112, 114-116, 118, 119, 121, 146, 147, 153, 155, 199, 223, 276, 277, 287, 293, 305, 323, 332, 362, 363, 425, 430-432, 440, 459
Article VI	18
Artsimovich	196, 326, 339, 405
ASAT	xxv, 138, 153, 154, 159, 168, 184, 186, 193, 301, 310, 312, 316, 326, 379, 380, 394, 414, 417, 448, 457
ASTOR	129
ASW	114, 121, 207, 457
Aswan	264
Atheism	302, 412
Atlantic	35, 37, 121, 124, 218, 277, 460
Atlas	46, 108, 110, 200
Atmospheric Testing	102, 304, 449
atomic	
A-bombs	106
Atomic Audit	94, 269, 270, 289, 290, 292, 294, 295, 320, 321, 358
Atomic Bomb	2, 5-7, 10, 11, 13, 14, 23-27, 29-33, 37-39, 41, 43, 82, 125, 145, 195, 196, 219, 241, 290, 292, 306, 308, 311, 333, 402, 410, 422, 423, 426, 453
atomic bombs	5, 9, 10, 13, 14, 17, 29-31, 34, 38, 39, 43, 56, 85, 93, 103, 107, 115, 121, 128, 175, 195, 209, 211, 261, 332, 340, 434
Atomic Energy Act	86, 93, 244, 246-248, 432
Atomic Energy Commission	33, 93, 150, 185, 247, 425, 426
Atomic Energy Detection System	185, 457
Atomic Energy Establishment	428, 467
Atomic Power	7, 8, 13, 122, 247
Atomic Scientists	32, 306, 307, 331-333, 349, 423
Atomic Scientists Association	307
Atomic-Bomb	6, 22, 25, 26, 31, 39, 86, 99, 124, 183, 330, 337, 352, 429
Atomic-Bomb Research	124
Atomic-Energy	4
Blast	7, 41, 45, 49, 63, 80, 91, 99-106, 108, 111, 117, 118, 142-144, 149, 156, 216, 230, 372, 388, 453
gravity bombs	37, 114, 223, 372
Atoms for Peace	22, 39, 175
Australia	10, 125
Austria	66, 219, 463, 467
Aviation Industry	240
Azerbaijan	68
BASIC	300, 343
Back from the Brink	347, 348

Index for Volume 1

Backfire 60, 109, 117, 176, 192, 272
bacteriological 34, 433
Badger 109
Baikonur 274
Bailey 243, 401
Baines 53
Baker 280
Baklanov 368
Balance of Terror . 84, 89, 168, 180, 189, 332
Balkhash 46
Ballistic Missiles ... xxiv, 2, 22, 30, 42, 44-47,
 51, 55, 57-59, 65, 75, 76,
 80, 85, 98, 107, 108, 110-
 112, 118, 125, 127, 128,
 139, 143, 145, 147, 151,
 152, 157, 171-173, 191,
 192, 203, 207, 227, 241,
 289, 291, 369, 371, 385,
 393, 404, 417, 445, 459,
 468
 antiballistic 116, 146, 150, 153, 204,
 209, 220, 256, 259, 362,
 365, 377, 457
Ballistic Missile Defense . 153, 207, 235,
 339, 354, 358, 362, 365,
 368, 378, 379, 384, 393,
 457
 ballistic-missile defense ... xxiii, 47, 77,
 146, 147, 151, 152, 163,
 174, 250, 251, 254, 256,
 257, 276, 296, 312, 331
 ballistic-missile submarines 59, 92,
 121, 157, 207
 Penaids 147
BAMBI 146, 457
Barash 290, 291
Barker 242, 321
Barkovski 195
Baruch 39, 174, 331
Batzel 360
Beam Gap 168, 171, 173, 174
Beijing 210, 415
Belarus 68, 116, 161, 404
Belgium 37, 115, 160, 345, 467
Beloretsk 135
Beltway Bandits 169, 238, 239, 261, 399,
 400, 453
Beneficial Impact 370
Beria 27, 36, 196
Berkeley 12, 26
Berkshire 345
Berlin .. 22, 30, 32, 34-36, 44, 50, 56, 57, 66,
 67, 71, 72, 77, 109, 210,
 279, 282, 286-288, 395,
 403, 443, 447
 Wall 35
Berrigan 302, 323, 398
Berryville 134
Bethe .. 26, 202, 330, 331, 333, 338, 350, 360,
 364, 426, 428
Big Oleg 368
Bikini 31, 43, 359, 360
Bin Laden 412
Biological 328

Biological Warfare 189, 333
biological weapons .. 179, 180, 293, 295,
 342, 349, 416, 457
Biological Weapons Convention 293,
 457
bioprogram 294
biowarfare 293
bioweapons 244, 293
Chemical and Biological ... 19, 107, 179,
 245, 286, 295, 333, 342
Birch 302, 323, 397
Birdie 156
Birmingham 24
Bison 44, 54, 171
Black Liberation Army 305
Black Mayors 318
Black Panthers 305
Black Sea . 125, 186, 312, 327, 333, 335, 414,
 469
Blackjack 109, 207
blacklisting 302, 424
Blair 320
Blechman 321
BMD .... xxiii, xxiv, 146-150, 194, 207, 378,
 379, 384, 401, 457, 461
BMEWS 136, 457
BNCT 457
Bochum Project 335, 391, 392
Bochum University 300, 335, 343
Bogolubov 405
Bohr 4, 174, 307, 427
Bomarc 225
Bomb Secrets 468
Bomber and Missile Defenses 22, 47
Bomber Gap 73, 128, 168, 171, 174, 197,
 236, 363
Bombers 22, 36, 43-45, 47-49, 51, 54, 57, 67,
 75, 80, 83, 85, 89, 92, 94,
 107-110, 113, 115, 116,
 118, 120, 121, 127, 128,
 135, 138-140, 146-148, 157,
 159-161, 164, 171, 173,
 176, 181, 189, 192, 193,
 200, 227, 229, 233, 263,
 266, 272, 292, 294, 301,
 353, 372, 389, 390, 404,
 435
Bonn 343, 344, 417
Bonner 327
border 66
Born Secret 432, 463, 468
Bosnia 391
Boston 314, 362
Bothe 26
Bradbury 360
Brandt 302, 402
Brennan 220, 309, 321, 332, 337, 339
Brent 178, 206, 352, 381
Brezhnev .. 61, 73, 74, 95, 159, 279, 284, 302,
 318, 345, 403, 404, 433
Brezinski 397
brinkmanship .. xxiv, 49, 141, 222, 261, 378,
 435, 442, 445
Britain ..... xvii, 7, 25, 37, 44, 62, 67, 88, 98,

486   *Nuclear Insights*

	114, 115, 118, 121, 125, 160, 175, 189, 215, 220, 233, 269, 275, 277, 293, 329, 336, 342, 344, 345, 359, 402, 428
British .. 5-7, 17, 24-26, 34-36, 74, 80, 95, 97, 98, 118, 121, 125, 129, 195-197, 204, 223, 244, 254, 256, 307, 311, 340, 343, 344, 359, 396, 428	
British American Security 343	
British Global Strategy 80, 97	
British Labour Party 344	
British Ministry 254	
Brodie 84, 234	
Broglie 4	
Broken Arrow 225	
Broken Arrows 169, 225, 226, 453	
Brookings .. 94, 270, 300, 320, 321, 429, 467	
Brown 206, 380	
George Brown 380	
Brussels 381	
Brzezinski 222, 252, 253, 302, 383, 397	
Buccaneer 121	
Buchanan 253, 372	
Buckley 253, 302, 351, 352, 397	
Buffalo 322	
Bukharin 334	
Bulganin 73, 403	
Bulletin of the Atomic Scientists 32, 306, 333, 349	
Bundy 10	
bunker....................... 103, 228	
Bunn 332	
Buran 274	
Bush	
Vannevar Bush 357	
Bush (G.W.)	
Bush Administration ... xxiii, 205, 327, 378, 392, 444, 459	
G.W. Bush . xxiii, 14, 110, 209, 327, 347, 348, 392, 401, 444	
George W. Bush 73, 206	
President Bush 68, 69, 76	
President G.W. Bush 110, 209, 347, 348, 401	
Bush (H.W.) 65, 71, 73	
H.W. Bush 378, 404	
President H.W. Bush............. 404	
Butler 192, 205, 239, 259, 281, 292, 294	
C3 .... 228, 229, 457	
C3I 81, 122, 133, 135, 457	
CAFF 330	
Cahn 271	
Cairo 213	
CALC 322	
Caldicott 314	
California Institute of Technology 339	
Calley 410	
Calogero 415	
Cambodia 53, 54, 309, 356	
Cambridge 288, 293, 295, 332	
Cambridge University 293	
Canada ... xiv, 25, 37, 59, 116, 127, 128, 210,	212, 252, 336, 342, 405, 463, 467
Canada Institute 405	
Canadian 6, 196, 300, 310, 342	
Canaveral 39	
Canceled Programs 81, 128	
Capitol Hill 94, 239, 312, 400	
Carlton 294	
Carnegie 300, 320, 467	
Carribean 52, 69, 77, 287, 447	
Carson 246, 360	
Carte Blanche 95	
Carter . 60-64, 69, 73, 93, 110, 143, 159, 160, 203, 220, 224, 246, 253, 256, 258, 262, 265, 280, 332, 358, 371, 383, 397, 431	
Ashton Carter 332	
Carthage 253	
CAS 300, 315, 316, 431, 457	
Casey 249, 253	
Castle Bravo 43, 44	
Castro 49, 50, 449	
Catholic Bishops 322	
Catholic Church 412	
Cato Institute 233, 302, 320, 400, 401	
Catonsville 398	
CDI .. 265, 300, 316, 317, 320, 329, 457, 467, 468, 470	
Center for Security Policy...... 254, 300, 321	
Central America 251, 279, 319, 436, 443	
Central Committee ... 86, 134, 368, 395, 402	
Central Park 375	
CEP 47, 91, 92, 108, 386, 458	
CFE 67, 187, 391, 414, 416, 458, 466	
CFE Treaty 67, 416	
CFE Verification 414	
Chadwick 4, 24	
Challenger 377	
Charges and Countercharges .. 179, 188, 200	
Charlottesville 329	
Chayes 365	
Chazov 340	
Cheating 256, 351	
Cheget 135	
Chelyabinsk .. xiv, 15, 36, 124, 130, 147, 312, 313, 467	
Chemical Warfare 183	
Chemical Weapons Convention 185, 458	
CW 187, 188	
CW Treaty 187	
CWC 458	
Cheney 204, 205, 255	
Chernenko 73, 302, 403, 404	
Chernobyl 144, 244, 377, 408, 466	
Cheyenne 134	
Chiapas 431	
Chicago . 5, 9, 25, 26, 47, 174, 209, 224, 225, 235, 243, 254, 301, 306, 314-318, 322, 329, 330, 337, 349, 350, 358, 362-364, 367, 370, 375, 389, 398, 430, 460, 464, 465,	

Index for Volume 1

467, 468
Chicago area ... 318, 330, 362, 363, 367, 389, 464
Chicago Area Faculty 318, 330
Chicago Committee 363, 367
Chicago Peace Council 430
Chicago Police 430
Chicago Seven 314, 337
Chicago Tribune 209, 225, 254, 318, 350, 367, 370, 375, 398
Chichi Jima 115
China .. xvii, xxiv, 17, 19, 38, 54, 58, 72, 80, 98, 99, 117, 121, 125, 129, 130, 148, 161, 210-213, 270, 294, 362, 393, 403, 412, 415, 424, 425, 443, 449, 450
Chinese Communists 86, 397, 423
Chinese Nuclear Systems 80, 121
Chomsky 333
Christian 12, 322, 397, 413
Christy 428
Chronology 210
Churchill 7, 30, 97, 302, 402
CIS 68, 117, 458
CISAC ... 326, 329, 333, 334, 339, 401, 458
CISSM 458
Civil Defense ... 48, 49, 57, 81, 96, 139, 141-145, 162, 164, 194, 202, 234, 260, 276, 291, 310, 321, 356-359, 394, 420, 435, 441
Civil Rights 304, 310, 330, 367, 398, 410, 411, 425
civilians . 48, 49, 95, 104, 118, 139, 141, 142, 162, 230, 236, 237, 254, 273, 293, 305, 316, 372, 440, 444
  civilian control 86, 349
  civilian use 346
Clamshell Alliance 324
Clark 243
Classified Information ... 244, 246, 256, 338, 390, 422
Clausewitz 286
Cleland 318
Clifford 243
Climatic 224, 387, 388
Clinton . 68, 73, 110, 209, 266, 331, 338, 347, 378, 392, 401, 460
Closed City 433
CND 344, 345, 458
CNS 335
Coates 398
Cochran 312, 313, 327, 377, 378
CODENE 345
Coffin 323
Cohen 63, 334
Colby 75
Cold War
  Cold War agreements 187
  Cold War as History 170, 277
  Cold War books 321
  Cold War dangers 414

Cold War developments 320
Cold War divide 323
Cold War environment 228, 389
Cold War excesses 303, 306
Cold War fears 352
Cold War history .... xvi, 303, 440
Cold War issues 337, 413
Cold War military ... xxii, 229, 256, 265, 267
Cold War policy 208
Cold War realities 412
Cold War Redux . vi, ix, xviii, xxvii, 439, 440
Cold War Reprise 23, 68
Cold War thinking 330
Cold War Was Won 170, 278
Cold War Weaponry .... ix, xiii
Cold Warriors . xxvi, 231, 256, 361, 411, 442, 444, 445, 450
Cost of the Cold War 169, 263, 439, 446
End of Cold War 437
Collaboration ..... x, xv, 287, 312, 380, 414
Collateral Damage .... xxiv, 63
Collateral Effects 104
Colorado Springs 134
Colson 410, 429
Columnists 216, 253, 358, 370, 372, 398
COMECON 264, 458
Comintern 425
Command .. 12, 16, 19, 32, 44, 60, 63, 81, 91, 93, 103, 108, 109, 113, 122, 129, 133-137, 153, 158, 159, 163, 168, 191-193, 199, 202, 204, 205, 212, 216, 220-222, 227, 228, 230, 239, 276, 279, 320, 358, 370, 381, 386, 390, 457, 460, 461
Committee of Soviet Scientists . 300, 313, 326, 354, 458
Committee on Nuclear Policy 329, 348
Commoner 324, 330
Communication .. xiv, xxxi, 81, 92, 129, 133, 135, 137, 138, 144, 193, 204, 211, 220, 231, 240, 279, 308, 344, 349, 379, 405, 428, 432, 446, 454
Communism ... 13, 14, 26, 31, 32, 50, 53, 54, 66-69, 71, 74, 76, 82, 93, 162, 163, 181, 206, 216, 231, 236, 255, 257, 261, 277, 279-282, 285, 304, 344, 395-397, 403, 406, 411-413, 417, 422, 428, 433, 443-446, 449
Communist Party . xiii, 33, 64, 66, 67, 86, 95, 249, 296, 352, 354, 395, 402, 404, 405, 419, 423, 424, 427, 428, 432, 458
Communists .. 32, 52, 86, 395, 397, 402, 411, 412, 423-425, 431, 434
Compliance 157, 159, 177, 179, 180, 183-185, 187-190, 382, 443

Comprehensive Test Ban Treaty 158, 159, 241, 338, 373, 458
Computer Hardware 70
computer technology 368
Conant 42
Concerned Argonne Scientists . 315, 367, 372, 430, 457, 464, 469
Concerned Scientists 308, 313, 324, 331, 363, 371, 462
Conclaves 302, 414
Condon 337
Conferences 16, 242, 302, 311, 318, 320, 340, 343, 383, 402, 414, 415, 418, 463, 464
Confidence-Building Measures 416
Conflict Resolution 334, 342, 353, 390
confrontationalism 418, 443
Congress
    Congressional .. 33, 50, 57, 72, 179, 233, 239, 247, 261, 266, 268, 304, 309, 311, 339, 360-362, 367, 378, 384, 385, 396, 398-401, 417, 445, 457
    Congressional hawks 362
    Congressional Research Service 268
    Congressional staff 401
    Congressional supporters 367
Constitution ... 346, 397, 422, 431, 433, 434
Consultants .... xi, xx, 200, 233
Containment ... xv, xxi, 31, 56, 61, 76, 77, 85, 86, 96, 200, 221, 255, 285, 288, 331, 396, 413, 441, 442, 449
Contamination ... 31, 48, 103, 105, 144, 217, 226, 245, 332
Contemporary Threats .... ix, xiii
Contractors . 72, 237, 238, 245, 316, 321, 368, 383, 421, 429, 432, 453, 467
Conventional Forces .. xxii, 55, 65, 67, 72, 88, 105, 172, 175, 191, 192, 200, 222, 224, 248, 270, 282, 391, 403, 405, 416, 451, 458
Conventional-Force Reductions 187, 301, 391
Cooper 321, 338
Cooperative Test-Ban 414
Cooperative Threat Reduction 458
Cornell 314, 330, 331, 333, 339
Cornell University 330, 331
CORONA 131, 291
Corporal 98
Cortright .. 283-285, 288, 289, 297, 310, 323, 336, 351, 358, 407
Cosmos 131, 228, 379
cost-effective 292
Council for a Livable World ... 300, 311, 319, 321, 363, 389, 392, 400
Countercharges 179, 188, 200
Counterforce .. 47, 63, 88, 89, 91, 92, 97, 104, 108, 138, 140, 143, 144, 162, 236, 255, 374, 385, 453
countermeasures 108, 145, 150, 153, 251, 291, 326, 365, 379, 394, 395, 448
counterproductive .... xxiv, 14, 19, 54, 75, 78, 87, 208, 245, 284, 361
Counterproliferation xxii
Counterterrorism 2, 18, 466
countervailing 65, 303
Countervalue 154, 453
Courts 302, 398, 432
CP-1 25
CPD 253-255, 458
CPSU 352, 458
Cranston 308, 347
Critical Mass 26, 217, 453
Criticality 156, 196, 217, 226, 453, 454
Cruise Missile 61, 64, 112, 118, 186, 193, 312, 327, 387, 457, 459, 461
CSCE 59, 458
CSS .. 312, 326, 327, 331, 380, 383, 405, 414, 458
CTBT 120, 158, 159, 242, 347, 361, 392, 393, 401, 416, 458, 459
Cuban Missile Crisis ... xix, 2, 22, 30, 49-52, 57, 61, 72, 96, 121, 138, 143, 175, 212, 219, 229, 311, 353, 359, 403, 442
Cultural Differences 302, 406, 420
Cultural Movements 302, 409, 410
Culver 216
Curie 4, 453
Current Policy .... xv, xxii
Curtis 10, 44, 49, 50, 205
Cyprus 213
Czech 32, 35, 336
Czech Republic 336
Czechoslovak 71, 432
Dana 278
Dangerous Illusions 216
Darmstadt 343
DARPA 177, 458, 467
Davis 375
Davy Crockett 114
Day
    Sam Day 350
Day Without Research 332, 416
De-Emphasis 403
    De-Alerting 320
Dean
    Arthur H. Dean 206
    Jonathan Dean 59, 391
Deaver 280
decapitation 89, 133, 136, 204
decisionmakers 48, 129, 283, 434
Declassification .... xviii, 17
Decoupling 418
Deep Cuts .... xxii, xxiii, 18, 175, 451
DEFCON 212
Defense
    Defense Analyses 238, 338
    Defense Budget 181, 273, 274, 400
    Defense Condition 212

Defense Department ... 91, 93, 154, 238, 259, 266, 321, 338, 362, 366, 367
Defense Emergency 212
Defense Industry 201, 240, 268, 397
Defense Information . 225, 226, 265, 316, 348, 457
Defense Monitor 317
Defense Nuclear Agency .. 243, 383, 458, 468, 470
Defense Priorities 316
Defense Programs ... 130, 265, 291, 314, 357, 360, 422
Defense Research 369
Defense Science Board 377
Defense Secretary . 55, 88, 172, 218, 237, 398
Defense Spending ... 65, 70, 72, 75, 173, 249, 252, 267-270, 272, 366, 367, 404
Defensive Systems 89
deficit 265, 266, 268, 274
DeGaulle 302, 402
Delivery Systems . x, xix, xxiv, 15, 17, 41, 44, 55, 61, 65, 77, 80, 81, 83, 98, 106, 107, 109, 112, 114-116, 118, 119, 138, 139, 158, 159, 162, 164, 178, 181, 189, 191, 199, 200, 211, 223, 227, 231, 232, 240, 245, 263, 287, 367, 370, 371, 383, 392, 402, 445
Dellinger 375
Dellums 375
Delta 112, 127, 377
Demilitarizing
  Demilitarization .... xxvi, 3, 16-19, 283, 376
  Demilitarized 16, 447
Democrat 69, 236, 262, 399, 411
Democratic .. 53, 54, 60, 68, 69, 71, 202, 237, 264, 283, 304, 314, 328, 337, 347, 400, 407, 413, 432, 469
Demonstrations ... 74, 94, 222, 251, 302, 317, 319, 344, 345, 355, 356, 359, 416-418, 436
denaturing 175
  Denaturing Plutonium 175
Denehey 353
Denmark 37
Denver ... 124, 217, 322, 385, 430, 431, 435
Denver's Close Call 169, 217
deployment of nuclear weapons . 30, 191, 319, 373, 374
Deployments ... 64, 66, 80, 94, 114, 116, 118, 132, 147, 182, 192, 293, 344, 417, 445
détente .... xxi, 22, 56-59, 61-64, 66, 77, 248, 249, 253, 271, 280, 403, 441
deterrence
  assured deterrence 106, 182

deterrence and compellence 282
deterrence strategy 202, 205, 292
Deterrence Theory 169, 218, 219
effective deterrence 84, 207
extended deterrence 98, 223
mutual deterrence .. xxiv, 42, 59, 85, 90, 108, 154, 198, 286
Nuclear Deterrence .... xxiv, 36, 44, 88, 141, 170, 211, 219, 221, 270, 281, 282, 292, 294, 329
Detroit 104, 305, 322, 362, 364, 430
Deutch 206
Deutsch 381
Devastation ... 2, 5, 14, 16, 29, 41, 56, 77, 80, 88, 99, 104, 121, 123, 145, 164, 170, 181, 203, 215, 222, 261, 329, 357, 440, 441
Development History 63
development of nuclear weapons . xix, 23, 77, 311, 376, 429
DeVolpi ... iii, v, xiii-xv, xx, 15, 313, 463, 464
DEW 127
DeWitt 243
Die Grönen 346
Dien Bien Phu 53, 70, 211
Directed-Energy Weapons .. 81, 152, 153, 338
Dirty Bombs xxvi
disarmament
  American Disarmament Study . 329, 339
  Disarmament Agency . 55, 177, 242, 253, 368, 370, 457
  Disarmament Agreements 326, 403
  Disarmament Diplomacy 346
  Disarmament Movement 323, 417
  disarmament negotiations 341, 344
  Disarmament Research ... 341, 347, 462
  Disarmament Studies 318, 341, 391
  Disarmament Week 318
Dissent ... 269, 280, 303, 305, 325, 354, 355, 357, 369, 371, 381, 382, 396, 414, 420, 422, 432-436, 444
Dissenters .... xxiv, 258, 356, 394, 420, 421, 438
Distorted Statistics 169, 265
Dobrynin 69, 284
DOD .. 86, 122, 137, 153, 226, 229, 239, 243, 369, 458, 464, 470
Dollar Cost 265
Dollars for Defense 169, 263
Dominici 348
DontBlowIt 301, 348
Doomsday .. xxii, 11, 15, 36, 89, 97, 169, 191, 214, 215, 219-221, 223, 231, 234, 296, 436, 453
Doomsday Machine ... 11, 97, 191, 215, 220, 221, 234, 453
Doty 339
Douglas 321, 353
Doves 95, 356, 396
Dr. Strangelove 220, 234
Drawbacks 140, 365

Drell 334, 364
Dresden 162
Dropshot 33, 34, 93, 422
DSWA 458, 467
Dubna 194
Dulles 34, 38, 87, 211, 235, 357
Duma 159, 393
Dunne 5
Dushanbe 186
Dzerzhinsky 67
Eagleburger 204
EAM 204
Early PALs 248
Early Warning . . . 81, 127, 128, 131, 228, 365,
379
East German 66, 210, 429, 432
Eastern Bloc 203, 264
Eaton 310
EcoNet 348
Economic
    Economic Impact 169, 266
Economy . . . 59, 62, 70, 82, 85, 169, 237, 264-
269, 271, 272, 278, 332,
379, 391, 395, 401, 403-
407, 422, 443, 449, 459
Educators 300, 330, 331, 345, 400
Egypt 74, 213, 264
Ehrlich 324, 330, 388
Einstein 4, 5, 32, 260, 307, 310, 371, 423
Einstein Manifesto 307, 310
Einstein Peace Prize 371
Eisenhower 10, 11, 34, 38-40, 73, 93, 94,
129, 130, 175, 210, 211,
231, 235, 237, 240, 247,
270, 337, 349, 359, 397,
403
Electronic News 364
Electronics Industry 240
ELF 137, 138, 458
ELINT 129
Ellsberg 234, 235, 429
Emelyanov 339
EMERGCON 212
Emergency Action Message 204
Emergency Committee 306, 331, 332
Emergency Management . . 81, 143, 144, 458
EMP 80, 103, 104, 124, 138, 150, 458
Encoding 179
encryption 160
Ending the War 2, 11, 12, 255
England . 4, 42, 115, 136, 224, 344, 345, 359,
403, 428, 457, 458, 467
English Channel 277
Enhanced-Radiation Warhead 80, 117
Enola Gay 6, 29
Enthoven 202
environment
    Environmental Remediation 263
    Environmental Studies 334
    Natural Resources Defense Council . 312,
324, 460, 470
Episcopal Church 322
Equal Rights Amendment 346
Erice 415

ERIS 146, 458
Erskine 362, 363
ERW 45, 63, 117, 118, 372, 458
Espionage . . 6, 17, 25, 27, 33, 34, 36, 44, 129,
132, 168, 185, 194-197,
201, 206, 244, 289, 293,
422, 423, 428, 429, 431,
434
Eurasian Studies 335
Europe
    East European 422
    Eastern Europe . . . 30, 33, 35, 60, 66, 71,
76, 82, 204, 252, 264, 277,
279, 282, 283, 286, 289,
345, 368, 395, 396, 402,
413, 425, 434, 436, 443,
447, 449
    Eastern European . 30, 70, 292, 417, 470
    Euromissiles . . . 301, 371, 380, 394, 448
    European allies 37, 255, 256, 351
    European Communism . . . 279, 280, 344
    European Peace Movements . . . 301, 344,
417
    European Union 344
    Southern Europe 345
    West European 255, 256
    Western Europe . . . 32, 37, 46, 55, 61, 72,
75, 85, 95, 97, 98, 112, 117,
161, 191, 203, 222-224,
232, 255, 264, 271, 274,
282, 289, 293, 343, 344,
372, 380, 392, 402, 416
    Western European . . 35, 37, 97, 264, 302,
402, 417
    Western European Leaders 302, 402
Evans 253, 302, 358, 398
Evans and Novak 302, 398
Evernden 242
Evil Empire . 64, 74, 169, 183, 205, 249, 278,
293
Evolution . . . . xxv, 3, 42, 127, 132, 140, 183,
208, 243, 255, 308, 414
Executive Branch . . . 116, 134, 150, 185, 235,
242, 247, 262, 351, 366,
382, 396, 399, 427, 446
Explorer 39, 132, 196, 277
Explosives
    Explosive Yields . . 16, 99, 110, 114, 154,
190, 372
    thermonuclear explosive 122
    thermonuclear explosives 393
Extremely Low Frequency 138, 458
Extremists . . . . x, 300, 319, 323-325, 410
Fail Safe 214, 220
Fail-Safe 291, 353
Fallacies . . . . xxii, 168, 170, 182, 280
fallout
    Fallout . . . xxii, 15, 24, 43, 48, 56, 80, 81,
91, 102-104, 106, 113, 114,
125, 128, 141-145, 162,
164, 215, 216, 236, 244,
301, 304, 308, 310, 328,
332, 350, 351, 354, 356,
357, 359, 360, 372, 385,

Index for Volume 1   491

	388, 468
Fallout Shelters ...	48, 81, 141-145, 162, 236, 301, 354, 356, 357
Local Fallout	102, 103, 114, 468
Farewell	237, 241, 247, 289, 368
Farewell Address	237, 241, 247
FAS ..	186, 300, 306-310, 312, 313, 315, 316, 320, 327, 329, 331, 333-335, 338, 358, 359, 363, 364, 377, 378, 380, 383, 385, 391, 392, 405, 414, 423-425, 431, 458, 464, 465, 467-470
FAS Chapter	313, 364, 465
FAS Chicago Chapter	363, 468
FAS Cooperative Research	320, 329
FAS Council	331
fascism	412
FBI ..	197, 322, 324, 423, 425, 427, 431, 469
Federal Civil Defense	142
Federal Emergency Management ...	143, 458
Federation of American Scientists ...	68, 306, 308, 313, 339, 348, 358, 423, 458, 464, 470
Feiveson	334, 374
Feld	306, 311, 333
FEMA	143, 144, 458
Fermi	4, 5, 9, 25-27, 42, 230, 311, 426
Fetter	335
Financial Cost	54, 273, 296, 421
Finlandization	255, 261
fireball	80, 99, 100, 102, 103, 118, 144
firestorm	100, 162
First Amendment	235, 430, 434
First Lightning	244
fission	
Fission .	xvii, xxi, xxv, 4-6, 14, 17, 23-27, 30, 32, 36, 41-45, 63, 99, 100, 102, 103, 107, 114, 115, 117-119, 122, 196, 197, 225, 230, 311, 372, 428, 443, 453-456, 466
Fission Chain Reaction	4, 5, 25, 119, 225, 453
fissionable ..	44, 186, 230, 382, 454, 455
Florida	49, 211
FOBS	140, 148, 182, 458
FOIA	423, 431, 458, 469
football	92, 108, 134, 135, 158, 204, 342, 404
Force Majeure	299, 303
Ford	59, 73, 74, 203, 252, 271, 329, 346
Ford Foundation	329, 346
Foreign	
Foreign Affairs	33, 60, 95, 116, 142, 162, 169, 235, 236, 261, 263, 264, 320, 335, 368, 388, 396-399, 406, 407, 410, 420, 424, 433, 436
Senate Foreign Relations Committee	
....	116, 235, 399
Former Soviet Union	xiv, 56, 68-70, 131, 244, 245, 267, 320, 335, 406, 450, 459, 460
FSU ..	161, 312, 313, 320, 329, 407, 450, 459
Forsberg	318, 346, 373, 375, 376
Fortress America	238
Foster ..	34, 38, 156, 235, 243, 342, 357, 362, 365, 368, 369, 437
Fourth Estate	301, 349
Fractional Orbital Bombardment System .	148
France ...	xvii, 37, 53, 67, 80, 88, 98, 99, 115, 118, 121, 125, 160, 175, 189, 220, 224, 233, 294, 325, 336, 339, 340, 344, 345, 402, 411, 464, 467
Franck	9, 32, 174, 306
Frankenheimer	353
Frankfurt	300, 343
Franklin...................	5, 73, 231, 339
Freedom House	233, 253, 319
Freeze ...	64, 88, 94, 157, 158, 204, 300, 310, 314, 317-319, 323, 330, 339, 344, 346, 373-376, 378, 380, 387, 399, 404, 414, 417
FREEZEPAC	318
Freije University	343, 459
French	4, 17, 34, 52, 53, 58, 80, 118, 121, 125, 211, 223, 235, 244, 250, 273, 289, 359, 360, 395, 402, 462, 467
Frisch	22, 24, 195
Froines	314, 337
FRP	459
Fuchs	25, 42, 195-197, 428
Fukuyama	288
Fulbright	238, 304, 338, 398, 399
Fulton	30
Funding Foundations	300, 329
Fusion	14, 15, 36, 41-44, 77, 86, 99, 115, 122, 260, 313, 372, 428, 453-456, 459
Fusion Weapons ..	14, 15, 36, 41, 42, 77, 115, 428
GAC	426, 427, 459
Gadget................................	5
Gaffney	253, 254, 262, 311, 321, 446
Galosh ...	48, 58, 81, 147, 148, 150, 272, 362
gamesmanship	218, 442
GAO	292, 459, 467
Gardner	351, 353
Garthoff	295
Garwin	331, 350, 360
GDP	268, 271
Gelb	429
Gender Activism	302, 411
General Accounting Office	177, 292
General Advisory Commission	247
General Assembly .	40, 74, 174, 175, 317, 341
General Machine Building	240, 272
General Secretary	6, 61, 64, 86, 134, 177, 249, 395, 402, 404, 437
Geneva	250, 314, 341, 416, 417, 463
Genie	2, 4, 14, 155, 451
genocide	233
GEO	132, 459

Georgetown University 414
Georgia 68, 411
geosynchronous 132, 186, 379, 459
German Democratic Republic 432
Germany .. 4-6, 23, 25, 34, 35, 37, 44, 67, 75,
95, 97, 98, 113, 115, 117,
160, 191, 195, 199, 219,
220, 223, 229, 239, 257,
264, 274-276, 278, 290,
301, 311, 323, 325, 331,
335, 336, 343, 344, 346,
347, 402, 411, 413, 417,
428, 432, 458, 463, 467
Ghana 74, 264
glasnost 57, 64, 244, 280, 361, 404, 454
Glassboro 403
GLCM 160, 384, 459
Global
    Global Communications 301, 348
    Global Conflict 335, 413, 459
    Global Consequences 415
    Global Positioning System 131, 137, 459
    Global Resource Action Center 348
    Global Responsibility 345-347
    Global Security 309, 320, 329
    Global Warming 198, 469
Glomar Explorer 132, 196
Glonass 131
Gold 110, 271
Goldberger 339
Goldhaber 314
Goldsboro 226
Goldschmidt 340
Goldwater . 237, 301, 323, 362, 368, 386, 397
Golf 132
Gorbachev .. 57, 62, 64-69, 73, 76, 153, 161,
177, 201, 221, 240, 265,
272, 274, 278-280, 284,
296, 302, 310, 312, 326-
328, 355, 360, 361, 368,
377, 381, 391, 403-405,
418, 433, 437, 455
Gore 338, 363, 385, 399
Gorgon Medusa 369
Gottfried 314
Government
    Government Expenditures 263
    Government Leaders ... 70, 71, 211, 285, 304, 326, 337, 360, 419, 446
    Government Operations 135, 424
    Government Scientists 17, 186
    government secrecy . xviii, xxv, 2, 17, 18, 31, 244-246, 304, 309, 431
    Soviet government 43, 99, 133, 244, 280, 285, 291, 325, 354, 391, 404, 405, 432
    U.S. government 34, 48, 50, 94, 104, 135, 175, 186, 188, 225, 236, 242, 244, 247, 285, 317, 335, 344, 353, 360, 368, 403, 411, 413, 422, 427, 433, 467, 468

GPS 131, 137, 141, 384, 459
Grand Forks 194
graphite 5, 25, 26, 290, 312
Gray 102
graybeards 252, 271
Grayson 432
Great Britain ... xvii, 7, 25, 44, 121, 125, 189,
215, 233, 277, 329, 336, 345
Great Powers ... xix, 43, 156, 174, 219, 281, 334
Greece 31, 37, 115, 213, 264
Greenbrier 134
Greenglass 27, 429
Greenham Common 301, 345
Greenland 116, 127, 136, 212, 226, 457
Greens 301, 346
Greider 238
Gross National Product 237, 332, 366
Ground Bursts 104, 144, 216
Ground Zero 48, 100-103, 142, 144, 162
ground-launched 117, 177
Groves 12, 13, 25, 31, 32, 39, 219, 428
GRU 129, 195, 196, 428, 459
Guam 115, 210
Guantanamo 50, 116
Gulf of Tonkin 52, 54
Gulf War 140, 213, 454
Hafemeister 331
Hahn 4
Haig 206, 284
Haldeman 430
Hall 27, 195, 196, 363, 428, 444
    Ted Hall 195
    Theodore Hall 27, 196
Halperin 191, 213, 233, 429
Hamburg 224
hardened .. 46, 47, 91, 94, 100, 101, 103, 104,
108, 110, 134, 140, 149,
178, 189, 203, 216, 235,
384, 385, 399
hardening 32, 85, 91, 207, 427
hardliner .. 180, 184, 188, 256, 257, 321, 370, 390, 397, 401
hardliners 64, 95, 96, 157, 158, 169, 172,
175, 179, 188, 197, 206,
218, 231, 249, 251-253,
255-259, 262, 271, 280,
281, 283, 288, 295, 296,
303, 311, 323, 324, 327,
351, 354, 355, 361, 374,
376, 378, 381, 383, 385,
386, 393, 395, 417-419,
433, 434, 441, 443-445,
447, 448
Harriers 121
Harriman 250
Harvard 173, 180-183, 188, 191-194, 257, 332, 339
    Harvard Study 428, 467
Hatch 309, 386
Hawaii 116
hawk 250
hazard

health hazard 120
  Nuclear Hazards 451
Heidelberg 26
Heisenberg 4, 25, 26
Helms 206
Helsinki 59, 280, 403
Helsinki Accords 403
HEO 131, 459
Hercules 47, 129, 146
Heritage Foundation . 233, 253, 302, 320, 376, 401
Herken 27, 31, 33, 280, 281
Hermitage 203
Hertz 253, 254
Hesburgh 336
hibakusha 274
High Frontier 173, 377
Highly Elliptical Orbit 131, 459
Higinbotham 306, 309
Hiroshima ... xxi, 5-10, 12, 13, 22, 23, 27-29, 39, 77, 82, 99, 100, 103, 105, 119, 126, 145, 155, 214, 215, 221, 255, 274, 297, 306, 317, 331, 340, 349, 404, 436, 440, 449, 468
Hitler 4, 5, 26, 219, 251, 260, 311
Holdren 332
Holloway 13, 181, 191, 237
Hollywood 351, 423, 424
Hollywood Ten 423, 424
Holocaust . xvii, 53, 68, 89, 90, 192, 221, 255
Holton 307
Homeland Security 424
Honecker 66
Hoover Institution ... 233, 253, 254, 302, 401
Hot Lines 138
House Science Council 253, 259
How Many .... xxiv, 2, 16, 133, 229
How Much 170, 245, 263, 361, 416
howitzer 114
HUAC 337, 423-425, 427, 459
Hubble 377
Hudson ... 220, 233, 300, 309, 320, 321, 332, 337, 357
Huerta 375
Hughes 132
Human
    Human Radiation Experiments 457
    Human Rights Committee 433
    Humanitarianism 160
    Humans 15, 18, 31, 56, 63, 100, 102, 129, 163, 214, 227, 245, 360
HUMINT 129, 459
Hungary 66, 252, 260
hybrid 18, 237
Hyde Park 317
hydrogen (weapons) ... 24, 33, 38, 41-43, 58, 85, 119, 197, 225, 226, 233, 235, 261, 289, 328, 331, 350, 364, 426, 432, 468
H-bomb .. 33, 41, 42, 125, 182, 235, 259, 281, 328, 350, 426, 431, 463
H-bombs 115, 215, 363
Hydrogen Bomb 33, 42, 43, 58, 197, 225, 226, 235, 289, 328, 364, 426, 432
hydrogen bombs ... 41, 43, 85, 197, 226, 233, 261, 331
Hyperthermia 1, 81, 165, 439, 441
Hypothermia 1, 81, 439, 441
I N S T E A D 300, 343, 416, 459, 467
IAEA 40, 56, 437, 459, 462, 467
IANUS 343
IBM 331
ICBM . xxv, 39, 46, 47, 51, 55, 58, 59, 61, 64, 91, 97, 104, 107, 108, 112, 130, 131, 133, 134, 136, 147, 150, 157, 159, 171, 172, 182, 183, 192, 194, 200, 207, 208, 212, 235, 267, 271-273, 275, 289, 291, 362, 365, 383, 384, 386, 443, 459, 460
ICBM accuracy 271, 272
ICBMs ... 44, 46, 47, 56, 57, 60, 61, 67, 72, 80, 91, 107-110, 113, 120, 131, 136, 138, 146-148, 157, 159, 161, 171-173, 179, 182, 192, 200, 201, 203, 207, 211, 228, 229, 255, 258, 272, 291, 292, 365, 377, 384, 389
Land-Based ICBMs ... 46, 91, 108, 258, 389
Iceland 37, 76, 116, 127, 404
ideological .. x, xxv, 69, 71, 81, 83, 160, 165, 198, 213, 263, 265, 278, 279, 297, 314, 355, 396, 399, 402, 407, 415, 428, 441, 443
IGCC 300, 335, 459
Iklé 202, 236, 248, 253, 262, 321, 446
Illinois 300, 317, 334, 385, 465, 469
Illusions ... 75, 169, 183, 214-216, 231, 234, 255, 275, 278, 446
imagery 131
IMEMO 302, 391, 405, 459, 467
Improvised .... xxvi, 193, 466
India . 14, 17, 19, 74, 126, 264, 281, 401, 450
Indian Ocean 125
Indianapolis 321
Indirect Costs 169, 273
Indochina 52, 53, 59
Indonesia 74, 264
INES 301, 346
INESAP 300, 343, 346, 459
INF ... 64, 65, 76, 80, 81, 116, 117, 160, 161, 163, 177, 178, 184, 192, 250, 251, 283, 284, 297, 301, 344, 376, 377, 380-382, 404, 406, 417, 418, 436, 459
INF Missiles 80, 117, 251, 284, 380
INF negotiations 250, 377, 382
INF Treaty .. 54, 169, 265, 268, 269, 273,

285, 324, 446
Information Bulletin 343
Information Centre 342, 462
Information Proliferation 170, 275
Infrastructure . xv, xxiv, 69, 94, 117, 122, 133, 139, 169, 177, 204, 267, 269, 273, 357
Inglis 358, 363, 465
Insecurity Trends 451
Insensitive HE 120
insiders .... xviii, xx, 29, 200, 231, 361, 382, 404, 418, 445
Institutional . xviii, xxii, 30, 81, 83, 165, 198, 214, 233, 262, 304, 315, 330, 357, 420, 421, 434, 441, 448, 451
   Nuclear Institutionalization ... 170, 276
Insurgencies 279
Intangibles 168, 190
Intelligence
   Air Force Intelligence 171
   CIA 50, 52, 62, 75, 86, 90, 122, 129, 131, 132, 170, 171, 173, 177, 196, 207, 209, 221, 222, 236, 242, 253, 257, 271, 275, 278, 285, 309, 316, 384, 386, 412, 422, 429, 431, 446, 458, 467, 469
   Defense Intelligence Agency .. 122, 129, 458
   DIA 177, 327, 458
   intelligence agencies . 147, 171, 177, 201, 252, 291, 313, 412, 469
   Intelligence Collection 81, 122, 129, 131, 138
   intelligence data . xviii, 43, 75, 149, 171, 174, 195, 197, 201, 290
   intelligence information ... 27, 133, 172, 291, 292
   intelligence officials 221, 278
   intelligence sources 171, 173, 196
   intelligence techniques 184
Internal Security Act 423, 424
International
   International Action Week 317
   international affairs 37, 59, 213, 264, 316, 338, 401
   International Civilian Leaders 259
   international controls 175, 331
   International Cooperation . 165, 283, 330, 335, 339, 348, 404
   International Development 32, 264
   International Inspections 175
   International Institute 300, 342, 465
   International League 308, 348, 363
   international markets 269, 270
   International Monetary Fund 264
   International Monitoring System ... 459
   international organization 39, 340
   International Peace .. 300, 320, 342, 366, 391, 461
   International Peace Research Institute
      .... 342, 461

International Physicians .. 340, 348, 355, 361, 459
international politics 157, 222, 440
international relations 285, 405, 459
International School 415, 459
International Security .. 17, 92, 326, 333-335, 339, 341, 342, 349, 355, 373, 458
International Studies . 233, 300, 319, 329, 333, 335, 336, 339
international system 206
Internet .. ix, x, 285, 308, 317, 330, 346, 348, 414, 460
Interuniversity Network 343, 459
Intimidation ... xxii, 125, 210, 213, 302, 305, 396, 419-422, 429, 435
Intimidation of Opposition 302, 419, 435
intrusive .. 177, 179, 187, 190, 198, 322, 393
   Intrusiveness 65, 187, 382
Inventories .. xxiv, 18, 19, 121, 286, 294, 445
Investigators 322, 425
IPPNW 300, 314, 340, 459
Iran ... xxiii, 19, 210, 251, 252, 382, 401, 469
Iraq 14, 85, 129, 140, 347, 348, 437, 444, 454, 455, 462, 469
   Iraq Invasion 455
IRBM 45, 459
Iron Curtain . iv, xiii, 30, 56, 77, 98, 251, 285, 287, 350, 395, 402, 412, 413, 436, 441, 447
irreversible cell damage 101
Isaacs 312, 321, 392
Islamists 279
ISODARCO 415, 459
Isotope 24, 123, 454-456, 466
Israel 14, 85, 212, 252, 256, 264, 333
ISTC 467
Italy . 37, 46, 50, 115, 160, 220, 339, 344, 345, 413, 415, 463, 467
Ivanov 406
Iwo Jima 115
Izvestia 405
Jackson . 61, 70, 253, 256, 284, 310, 323, 375
Japan .. xvii, xxi, xxii, xxv, 5-7, 9-14, 23, 25, 29-32, 44, 55, 82, 93, 105, 107, 114-116, 144, 158, 209, 245, 255, 274, 286, 304, 306, 325, 340, 410, 440, 443, 447, 463, 467
JASON 300, 334, 335, 338, 352
JCAE 169, 247, 248, 459
JCS 86, 202, 203, 353, 459
JDEC 459
Jewish 4, 26, 70, 284, 433
Jihad 286
Jim Crow 412
Johnson ... 52-54, 73, 89, 104, 148, 149, 231, 235, 238, 257, 290, 346, 362, 403, 411, 416
   President Johnson 52, 149, 238, 362, 403, 416
   Rebecca Johnson 346
Johnston Island 116
Joint

Index for Volume 1   495

Joint Chiefs . 86, 155, 200, 204, 210, 212, 237, 250, 362, 459
Joint Consultations 205, 312, 414
Joliot-Curie 4
Jones 257, 300, 329, 346, 348
Journalists .... xiii, xix, 6, 278, 287, 316, 350, 355, 377, 397
Jupiter 45, 49, 115, 277
Justice 232, 263, 305, 310, 348, 408, 423, 431, 438, 469
Justice Department 423, 469
Kahn . 202, 214, 215, 220, 234, 238, 309, 321, 357
Kaldor 289, 417
Kapitsa 302, 326, 339, 405
Kazakhstan .. 36, 68, 116, 125, 161, 312, 326, 327, 361, 377, 393
Kazbek 137
Kehler 318
Kelley 375
Kendall 216, 217, 314
Kennan 31, 85, 86, 221, 285, 371
Kennedy .. xix, 39, 41, 50-52, 55, 69, 73, 115, 171, 172, 175, 206, 226, 231, 239, 248, 289, 290, 317, 365, 368, 399
Kerry 304, 305, 410
Kevlar 110
Keyworth 376
KGB ... 36, 67, 122, 129, 196, 403, 428, 429, 459
Khariton 196
Khe San 211
Khrushchev . 41, 49-51, 73, 96, 206, 210, 284, 302, 307, 359, 360, 403
Khurchatov 196
Kidder 243
Killian 339
Kipper 213
Kirkpatrick 253-255, 262, 446
Kirov 127
Kirtland 225
Kissinger .. 58, 59, 74, 95, 140, 149, 202, 203, 206, 352, 381, 429, 430
Kistiakowsky 339, 370
Kohls 412
Kokoshin 302, 377, 391, 405
Koppel 351
Korea
    Korean War .. xiv, 22, 30, 33, 38, 49, 87, 114, 142, 210, 304, 442, 447
    North Korea   xxiii, 19, 38, 264, 281, 395, 426, 449, 450, 469
    South Korea ... 116, 264, 274, 426, 445
Kortunov 294
Kosygin 302, 403
Krasnoyarsk 65, 124, 179, 409
    Krasnoyarsk Radar 65, 179
Krass 331
Kremlin .. 32, 66, 88, 223, 236, 279, 284, 293, 311, 323, 377, 381, 405, 445
Kremlinologists 238, 395

Krepon 321
Kristol 255
Kroc Institute 336
Krokus 137
Kubrick 220, 352
Kurchatov .. 36, 124, 194, 196, 325, 328, 428, 467
Kuwait 67, 454
Kwajalein 116
Kyrgyzstan 68
Kyshtym 227, 312
Labor Unions 300, 343
Laboratory 98
    Argonne National Laboratory ... xiv, 27, 187, 306, 315, 362, 363, 457, 461, 463, 465, 467, 470
    Arzamas 36, 124, 467
    Brookhaven 421
    Hanford 459, 467
    Kansas City Plant 124, 467
    Kurchatov Institute 124, 325, 467
    Lawrence Livermore Laboratory 152
    Livermore . 123, 124, 152, 156, 200, 215, 242, 243, 254, 260, 270, 287, 289, 313, 334, 335, 360, 369, 377, 378, 382, 384, 421, 464
    Los Alamos National Laboratory . 26, 27, 42, 43, 124, 156, 195, 243, 246, 260, 287, 306, 311, 334, 335, 337, 360, 385, 421, 428, 429, 459, 464, 467
    Mound 124
    Oak Ridge 8, 25, 123, 124, 306, 467
    Sandia 124, 155, 243, 464, 467
Laird 148, 206, 243, 365, 369, 398, 430
Laity Concerned 322, 389
Lall 339
Lamb 335
Lancaster 353
Lance 114, 373
Lancer 110
landmines 37, 107, 191, 223
Laos 52, 54, 410
Lapp 226, 359, 369
LaRocque 226, 316
LaRouche 302, 397
Las Vegas 360
laser .. 141, 151, 152, 173, 186, 312, 313, 327, 377, 378, 380, 414, 422, 460, 463, 464, 466, 469
    Laser ASAT 186, 380, 414
Latin America 180, 188, 264
launch
    launch authority 136, 228
    Launch Risks 169, 227
    launcher 204, 462
    launchers 59, 67, 104, 157, 159-161, 176, 180, 193, 207, 208
Lawrence 9, 124, 152, 254, 313, 335, 352, 417
Leadership   . xv, xxii, 5, 23, 25, 49, 59, 61, 62,

71-73, 86, 89, 95, 135, 141,
161, 176, 203, 204, 213,
221, 243, 249, 272, 288,
308, 309, 314, 322, 323,
335, 339, 340, 346, 357,
358, 363, 383, 389, 392,
404, 415, 419, 426, 440
Lebow 282
Legacies .... ix, x, xxi-xxiii, xxvii, 6, 77, 252,
261, 287, 411, 440, 444,
451
    Legacies and Challenges .. ix, xxvii, 440,
451
Legacy ... i, iv, vii, x, xix, xxii, xxiv, 33,
69, 127, 275, 402
Legacy of Distrust 26
LeMay 10, 44, 45, 49, 50, 205, 206
Leningrad 147, 364
LEO ... 4, 131, 174, 202, 260, 306, 311, 312,
423, 460
Lessons . vi, ix, x, xiii-xv, xix, xxi, xxii, xxv, 3,
6, 52, 82, 167, 180, 194,
211, 287, 296, 297, 365,
437, 444, 450
    Nuclear Lessons .... vi, 167, 287
levitated 124
Lewis 196, 333, 341, 342, 374, 375, 383
    Patricia Lewis 341, 342, 383
Lexington 465
liberal .. xiii, 59, 95, 252, 278, 280, 283, 319,
336, 355, 366, 398, 400
    Liberals 252, 396, 410, 411
libertarians 235
Libya 401, 467
Lifton 214, 215, 281
Lilienthal 39, 168, 174, 307
limited nuclear war . 92, 95, 97, 143, 214, 215,
234, 322, 381
Limited Test Ban Treaty ... 74, 126, 158, 260,
304, 359, 360
Lithuania 67
Live Oak 210
LNT 460
Lobbying . 109, 183, 200, 239, 246, 253, 309,
321, 389
Lobbyists and Advisors 302, 399
Lockheed 368, 467
Logan Act 423
London .. 36, 37, 307, 342-344, 381, 415, 417
Long
    Frank Long 339
long-range
    Long-Range Bombers . 22, 43-45, 57, 85,
108, 113, 157, 159, 200
    Long-Range Missiles xxiii, 2, 17, 48, 61,
121, 126, 158
Lop Nur 125
Lopez 336
Los Angeles 112, 274, 324, 364, 425, 430
Lost the Cold War 69, 170, 282
Louisiana 95
Louisville 322
Lovins 324
Lown 340, 355, 377

Loyalty 353, 423, 426, 427
Luftwaffe 115
Luongo 334
Luxembourg 37
MacArthur Foundation ... 300, 329, 330, 335
Machine Building 122, 240, 272
Mack 362, 363
MAD .. xvii, xxi, 31, 56, 64, 84, 88, 89, 162,
164, 220, 236, 288, 386,
460
Madison 350
Maginot Line .... xxiv, 151, 450
Malenkov 38, 73, 403
Malmstrom 228
Managed Access 185
Manchuria 10, 82, 447
Manhattan District 39, 123, 195
    Manhattan Engineering District 123,
460
Manhattan Project . 5, 6, 9, 12, 13, 16, 17, 25-
27, 32, 39, 123, 124, 219,
221, 260, 289, 306, 311,
330, 333, 360, 420, 426-
429, 460
Marine Corps 114, 119
Mark 2, 5, 84, 171, 246, 360
    J. Carson Mark 246, 360
Markey 375, 399
Marshall . 10, 31, 85, 116, 124, 236, 264, 275,
310, 343, 424
    Andrew Marshall 236
Marshall Islands 116
Marshall Plan 31, 85, 264, 343
Martin 288, 314
Marx 412
    Marxism 278, 406, 412
Maryland 300, 335, 398, 458
Mass Casualty .... xxv, xxxi
Mass Destruction ... xi, 3, 38, 107, 174, 320,
329, 334, 335, 343, 347,
400, 451, 462
Massachusetts 313, 318, 332, 416, 460
Massachusetts Institute of Technology ... 313,
332, 416, 460
Matador 116
Maud 24, 195
Max Born 4
May
    Michael May 215, 270
Mayak 124, 196
MBFR 59, 391, 460
McCarran Act 423, 424
McCarthy .. 72, 301, 304, 397, 413, 424, 425,
427
McCarthyism 33, 72, 302, 304, 398, 411,
420, 424-426, 434, 448
McCloy 168, 175, 176, 206, 368
McGovern 332
McGuire AFB 225
McNamara .. 47, 48, 51, 55, 88, 89, 148, 156,
212, 218, 219, 236, 237,
243, 252, 281, 288, 352
McTaggart 360
MED 460

Medical . 14, 32, 102, 142, 143, 145, 216, 245,
267, 297, 309, 314, 315,
325, 340
Medicine 281
Mediterranean 226
Medium Machine Building 122, 240
Medvedev 433
Meese 254, 401
megaton .. 15, 36, 41, 42, 100-102, 104, 113,
125, 148, 149, 154, 156,
172, 200, 210, 215, 328,
350, 387, 399
megatonnage ... 113, 119, 121, 157, 173,
200, 258, 350, 351
Meitner 4
Melman 375
MEO 131, 460
Merck Fund 348
Messelson 309, 333
Met Lab 9, 26, 306
metaphysics 96
metastable 90, 198
Mexico 3, 5, 8, 26, 124, 241, 424, 431
Meyer 352
Mi Lai 305, 410, 412, 444
Miamisburg 124
Middle East 67, 161, 211, 288, 403
MILC 183, 460
Militarization 57, 304, 354, 381
Military
    Military Almanac 317
    Military Burden 265, 267
    Military Development .... xxii, 316, 328,
451
    Military Liaison Committee 247
    Military Officers .. xx, 75, 136, 137, 153,
200, 229, 233, 237, 263,
265, 316
    Military Policy 367
    Military Power ... 86, 96, 126, 172, 183,
207, 253, 287, 355, 389,
442
    Military Purposes . 13, 39, 308, 315, 366,
379, 437
    Military Significance 168, 188, 189
    Military Spending ... 253, 263, 265-270,
273, 290, 293, 295, 316,
371, 400, 408
    Military Strength . xxii, 58, 72, 75, 80-83,
94, 119, 120, 157, 171, 201,
366, 401, 443
    Military Technology 84, 241, 291
    MILLC 287
military-industrial
    Military-Industrial ... xxv, 18, 72, 76, 86,
169, 183, 237-241, 247,
252, 254, 261, 266, 271,
274, 278, 287, 289, 292,
296, 301, 304, 317, 319,
325, 332, 367, 368, 389,
395, 396, 398-400, 403,
440, 445, 460
    Military-Industrial Complex ... 237-240,
247, 252, 254, 266, 271,
274, 278, 289, 301, 367,
368, 398, 403
    Military-Industrial-Laboratory ... 18, 72,
183, 237, 261, 287, 296,
304, 317, 396, 445, 460
Miller 333
Millionshchikov 405
Milshtein 345
Milstein 95
MINATOM 122, 460, 467
Mining 123
Ministry of Defense 122, 201, 272, 460
Minkov xiii
Minuteman ... 46, 51, 90, 108, 111, 118, 148,
157, 172, 173, 200, 203,
208, 235, 273, 289, 291,
361, 369, 383, 385, 386,
399, 429
    Minuteman ICBMs 172, 200, 203
    Minuteman III 118, 273, 383
MIRACL 460
Mirage 121
MIRV .... xxv, 58-60, 90, 91, 106, 140, 147,
152, 154, 168, 186, 187,
190, 236, 272, 291, 292,
301, 316, 332, 364, 367,
369, 370, 384, 443, 460,
466
    MIRV technology 90, 91, 140
    MIRVed ICBMs 201, 207, 272
MIRVs .. 48, 58, 59, 61, 90, 91, 106, 139,
140, 147, 158, 159, 292,
310, 383, 384, 405
Misconceptions . 168, 180, 181, 214, 280, 281
Miss America 412
missile
    missile accuracy 193, 259, 301, 385,
386
    Missile Defense . xxiii, 3, 47, 77, 85, 146-
148, 151-153, 163, 174,
207, 220, 235, 243, 250,
251, 254, 256, 257, 276,
296, 301, 312, 315, 321,
331, 339, 347, 350, 354,
358, 362, 365, 366, 368,
378, 379, 384, 392-394,
401, 435, 441, 448, 457,
460
    Missile Gap 48, 55, 69, 73, 130, 142,
168, 171, 172, 174, 197,
236, 288, 291, 363, 398
    Missile Site Radar 149
    missile technology 107, 460
    Missiles Near Cities 468
Mississippi 185, 384, 414
Missouri 30, 116, 124, 390
Mistrust 310, 402, 441
MIT .. 202, 300, 314, 315, 332, 333, 338, 339,
364, 367, 385, 392, 460,
469
Mlad 27, 428
Mobe 300, 317, 318, 324
Modern World 249, 288
modernization 80, 97, 111, 176, 178, 192,

Moldova 68, 193, 201, 206-208, 250, 258, 273, 350, 384, 401
Moldova 68
Molnyia 131
Mongolia 264, 443, 450
Monopoly and Secrecy 170, 280
Montana .... xviii, 228
Monterey Institute 300, 329, 335
Moore 468
moratorium ... 40, 64, 65, 125, 260, 284, 315, 317, 323, 326, 356, 359-362, 367, 369, 373, 374, 376, 380, 393, 404, 417, 460
Morland 431
Mormon Church 323
Morocco 115, 467
Morrison 333
Morse 304
Moscow . xviii, 31, 32, 48, 52, 58, 62, 66, 96, 124, 135, 147, 150, 161, 192, 194, 204, 207, 209, 211, 213, 221, 228, 239, 250, 261, 272, 277, 284, 313, 320, 328, 334, 339, 362, 364, 391, 404, 418, 459
Motives 190
Mount Pony 134
Mount Weather 134
MoveOn 301, 347, 348
Movers and Shakers .... xx, 301, 396
Moynihan .. 90, 159, 173, 202, 262, 290, 310
MRV .... xxv, 58, 59, 90, 190, 443, 460
Mujahideen 62, 412
Mulroney 342
multilateral .. 55, 71, 188, 190, 230, 339, 391, 418, 458
Multilateralism 198, 414
multiple-warhead
 Multiple Warhead Missiles 80, 90
 multiple warheads 48, 132, 150, 158, 193, 468
 Multiple-Warhead Missiles 106
 multiplicity 104, 139, 158, 162, 186, 187, 190, 463
Muroroa 360
Muslim 62
mutual
 Mutual Assured Destruction .... xxi, xxv, 31, 48, 56, 64, 77, 80, 84, 87, 88, 129, 151, 188, 220, 230, 236, 255, 354, 386, 392, 433, 460
 Mutual Balanced Force Reduction .. 391, 460
 Mutual Economic Assistance .. 264, 458
 Mutual Freeze Resolution 317
 Mutual Security 415, 451
MX . 61, 63, 64, 75, 90-92, 97, 106, 108, 110, 118, 159, 173, 178, 193, 200, 208, 226, 249-251, 273, 301, 310, 316, 319, 323, 351, 383-386, 389, 394, 399, 436, 443, 448, 460
Myths .... xxi, xxii, 168, 170, 232, 288, 289, 296, 443
Nader 324
Nagasaki . xxi, 2, 5-7, 9, 12, 13, 17, 22, 29, 77, 100, 103, 105, 145, 221, 255, 274, 297, 340, 427, 468
Napoleon 251
NAS 326, 329, 334, 339, 377
 National Academy .. 177, 326, 331, 334
NASA 90, 275, 277, 460
National
 National Airborne Operations Center
  .... 134, 460
 National Council ... 318, 322, 460, 464
 National Defense . xxii, 80, 82, 174, 205, 263, 264, 366, 400, 401
 National Endowment 253
 National Ignition Facility 313
 National Intelligence Estimates 131, 271, 291
 National Military Command Authority
  134
 National Missile Defense ... 3, 146, 151, 301, 315, 392, 401, 460
 National Priorities 364, 367, 389
 National Security Studies . 269, 301, 366
 National Strategic Target Database ... 94
national security
 National Security Act 86
 national security advisor .. 184, 253, 377
 national security bureaucracy 233
 National Security Council .. 32, 86, 134, 135, 250, 322, 335, 422, 429, 430, 461
 National Security Policy 210
 national security staff 213
 National Security Studies 257
National Technical Means . 40, 133, 158, 178, 184, 190, 373, 461
Native American 431
NATO .. 22, 30, 35, 37, 40, 45, 48, 55, 62, 63, 65, 67, 72, 74, 75, 77, 95-98, 105, 112, 114-116, 118, 120, 147, 160, 168, 176, 184, 189, 191, 192, 203, 210, 213, 221-224, 228, 233, 250, 254, 257, 285, 289, 293, 343, 344, 346, 371-373, 380, 381, 388, 392, 402, 445, 460
 NATO Alliance 381, 402
 NATO Allies 37, 63, 293
 NATO Modernization 192
 NATO Nuclear Issues 168, 191, 343
Nature . 4, 9, 12, 14, 25, 55, 56, 60, 76, 90, 99, 122, 132, 140, 159, 180, 200, 206, 211, 213, 220, 255, 258, 268, 295, 305, 388, 390, 392, 395, 413, 419, 426, 443, 455
Nautilus 126

Naval .. xiv, 45, 50, 54, 80, 111-114, 116, 119, 122, 123, 125-127, 132, 138, 164, 182, 185, 186, 196, 199, 210, 218, 271, 464, 465, 467, 468
Naval Reactors 123, 127
Naval Vessels ... 80, 114, 119, 122, 126, 138, 164, 185
Navy ... xiv, 7, 25, 52, 86, 107, 110-112, 114, 119, 125, 127, 128, 132, 156, 171, 172, 200, 229, 287, 327, 442, 463, 465, 467, 468
Nuclear-Powered Submarines .. 108, 111, 126, 127, 292, 427
Nuclear-Powered Vessels 80, 127
NAVSTAR 137, 459
Naylor .. 69, 70, 183, 184, 252, 253, 256, 290
Nazi .. 4, 6, 23, 155, 170, 260, 275, 277, 278, 290, 402
NCI 322
NCRP 460
Neckarelz 343
Negotiations .. xiv, xxii, 18, 40, 52, 58-60, 67, 87, 89, 95, 110, 111, 139, 156, 157, 159, 160, 163, 168, 176-179, 184, 185, 187, 190, 192, 193, 211, 239, 249-251, 254, 256, 258, 284, 294, 295, 297, 307, 316, 327, 329, 334, 341, 342, 344, 346, 347, 359, 361, 368, 370, 372, 375, 377, 382, 384, 391, 393, 394, 404, 408, 416, 418, 441, 448, 451, 461
neoconservative 236, 255, 256, 259, 280, 288, 340, 354
neocon 255, 350, 380
Netherlands . 37, 115, 160, 336, 343, 344, 467
networks .. 129, 137, 158, 184, 193, 228, 301, 307, 346, 347, 355, 356, 404
neutron .. 4, 22, 26, 45, 60, 63, 117, 118, 127, 147, 223, 224, 249, 257, 260, 261, 301, 310, 316, 317, 344, 350, 371-373, 380, 381, 454, 455, 457, 458, 463, 464, 466, 468
neutron bombs .. 60, 224, 260, 301, 316, 371-373, 380, 381
neutrons ... 4, 24, 42, 45, 100, 101, 117, 118, 123, 312, 372, 453-455, 466
Neutron Weapons 63, 118, 257, 371, 373
Nevada . 40, 49, 111, 124-126, 150, 185, 200, 304, 312, 319, 360, 361, 386, 416, 417, 461
New Hampshire 331
New Jersey 225, 310, 403
New Left 324
New Mexico 3, 5, 8, 26, 124, 241
New Millennium .. x, xv, xxv, 2, 3, 16, 18, 19, 269, 296, 303, 304, 349, 401, 411, 436, 445, 451
New York . 36, 42, 47, 49, 141, 149, 153, 155, 224, 235, 250, 306, 310-312, 320, 322, 324, 327, 350, 358, 362, 371, 372, 375, 398, 403, 410, 417, 430, 444
New York City .. 49, 322, 358, 371, 375, 403, 417
New York Times . 42, 149, 153, 235, 250, 311, 327, 350, 372, 398, 410, 430, 444
New Yorker 219
Newsweek 28, 149, 226, 350, 368
NGO .. 53, 187, 312, 327, 334, 335, 342, 343, 349, 361, 380, 405, 415, 418, 437, 458, 460, 468-470
Nicaragua 264, 395
Nike . 47, 116, 146, 148, 149, 311, 362, 431
Nike-Ajax 146
Nike-Hercules 146
Nike-X 146
Nike-Zeus 146
Nitze ... 33, 60, 169, 173, 184, 257-259, 262, 362, 364, 446
Nixon 54, 57-59, 61, 73, 74, 88, 91, 148-150, 211, 231, 235, 243, 252, 254, 256, 271, 273, 294, 337, 356, 362, 363, 381, 383, 386, 400, 423, 429, 430, 444
NMD 392, 393, 460
Nobel Peace Committee 327
Nobel Peace Prize ... 296, 311, 402, 404, 431, 433
Nobel Prize 314, 340, 427, 433
Nolan 332
NOMOR ... 300, 316-318, 367, 372, 385, 460
Non-Proliferation
nonproliferation . x, xiii, xxii-xxiv, 2, 18, 19, 159, 246, 309, 312, 313, 320, 322, 333-335, 341, 343, 346, 371, 400, 401, 407, 437, 463, 464
Nonproliferation Review 335
Nonproliferation Treaty 341
Non-Proliferation Treaty .. 18, 40, 58, 74, 347, 418, 461
Nuclear Threat .. 65, 211, 213, 214, 301, 326, 348, 349, 352, 458, 462
Nuclear Threat Reduction Campaign
.... 348
noncompliance 184, 188
nonconformist 239, 433
nonmilitary .. 40, 69, 118, 126, 132, 154, 269, 307, 315, 363, 366-368
Nonviolent Action 345
NORAD 134, 460
Norris 312, 360
Norstad 206
North American Rockwell 109
North Dakota 194, 365

North Pacific 196
North Shore 318, 363
North Vietnam . . . 52, 54, 235, 264, 356, 395
Northwestern University 330, 389
Norway 2, 26, 37, 195, 229, 290
Norwegian
    Norwegians 201
Notre Dame University 300, 336
Nova Scotia xiv
Novak
    Robert Novak 253, 358, 398
NPR 460
NPT . . . 40, 56, 58, 70, 74, 99, 183, 347, 418,
437, 461
    NPT Conference 347
NRDC 300, 312, 313, 316, 326, 327, 331,
335, 378, 383, 414, 460,
464, 467, 468, 470
NSA 54, 177, 461
NSC . . . 32, 33, 86, 87, 93, 211, 257, 422, 461
    NSC 68 33
NSDD 204, 249
NTM 158, 160, 183-185, 249, 461
Nuclear
    Nuclear Age . . . xvii, xix, 217, 222, 308,
329, 330, 353, 369
    Nuclear Radiation 104, 105
    Nuclear Research 195, 363, 425
    Nuclear Systems . . xxv, 80, 121, 223, 443
    Nuclear Technology . . xiii, xvii, xxii, 39,
212, 275, 307, 451
    nuclear threats . vii, ix, xxi, xxii, 72, 169,
210, 211, 213, 282, 286,
347, 400, 436
    Nuclear Worthiness 169, 270
    nuclear-explosive 158
    Nuclear Arms . ix, x, xv, xxv, 3, 15, 16, 18, 19,
30-32, 36, 47, 56, 58, 59,
61, 62, 69, 71, 72, 74, 75,
77, 88, 162, 170, 174, 176,
177, 180, 181, 183, 191,
205, 209, 213, 217, 223,
224, 234, 240, 249, 254,
274, 289, 292, 294, 295,
303, 304, 308, 309, 311,
312, 317, 318, 321-323,
330, 331, 338, 340, 346,
349, 351, 355, 356, 371,
373-375, 384, 388, 400,
403, 406, 414, 417, 422,
430, 434, 436, 446, 448,
449, 466
    Nuclear Armaments . . . . xi, xxiii, 56, 68,
197, 211, 310, 312, 323,
338, 341, 353, 376, 392,
415, 435, 436
    Nuclear Arms Race . . . . ix, xxv, 3, 18, 19,
30-32, 36, 47, 61, 69, 71,
72, 74, 77, 88, 170, 174,
181, 205, 223, 234, 240,
274, 292, 295, 303, 304,
312, 317, 318, 330, 331,
340, 346, 349, 351, 355,
373-375, 417, 422, 434,
448
Nuclear Arsenals . . . xix, xxi-xxiv, 16, 18,
19, 33, 41, 56, 57, 76, 82,
87, 132, 146, 164, 168, 181,
202, 216, 231, 261, 288,
329, 371, 373, 387, 388,
404, 435, 441, 445, 447,
450, 451
Nuclear Balance . . 22, 54, 189, 250, 259,
451
Nuclear Bombs . 5, 36, 44, 45, 49, 76, 95,
115, 119, 121, 226, 227,
357, 373, 387, 404
Nuclear Capable 44, 77, 98
Nuclear Forces . . . 56, 63, 65, 68, 71, 89,
97, 116, 129, 135, 160, 172,
183, 191-193, 203-205, 210,
228, 250, 267, 270, 293,
320, 376, 380, 391, 418,
459
Nuclear Genie 2, 4, 14, 155
Nuclear Hazards . . . 6, 17, 132, 168, 175,
191, 251, 316, 332, 334,
335, 340, 343, 347, 351,
396, 435, 449
Nuclear Policy . . . 18, 29, 191, 231, 281,
303, 308, 310, 329, 348,
363, 420, 433, 436, 461
Nuclear Posture Review 460
Nuclear Powers . . . 82, 84, 126, 149, 164,
287, 290, 317, 345
Nuclear Priesthood . . xvii, 169, 232, 261,
445
Nuclear Reductions . vii, xviii, xxii, xxvi,
451
Nuclear War Strategies 80, 92
Nuclear-Arms Race xxv
Nuclear-Capable Missiles 49
Nuclear-Tipped . . . 17, 52, 121, 146, 147,
150, 197, 213
Nuclear Control Institute 300, 322
Nuclear Defense . . 55, 75, 145, 168, 191, 193,
268, 289
Nuclear Dilemma 315
Nuclear Events 22, 23, 410
Nuclear Explosion
    Nuclear Winter . 301, 316, 387, 388, 462
Nuclear Freeze . . 94, 204, 300, 310, 317, 318,
323, 346, 373-376, 378
    Nuclear Freeze Campaign . 300, 318, 373,
374
    Nuclear Weapons Freeze Campaign . 318,
319, 376
Nuclear Fuel
    Nuclear Fuel Cycle 246
Nuclear Materials . xxii, xxv, 18, 19, 174, 175,
244, 245, 349, 451, 457,
464, 466
Nuclear Power . . 14, 126, 245, 275, 315, 317,
324, 346, 373, 408, 451,
469
Nuclear Propulsion 25, 80, 122, 126-
128, 182, 457
Nuclear-Powered Aircraft 127, 128

Nuclear Reactors . x, 25-27, 32, 122, 324, 325, 466
 nuclear reactor ... xiv, 5, 24-26, 85, 290, 311, 334, 338, 455
 Power Reactors 123, 127, 144, 460
Nuclear Safeguards
 Safeguarding .... xxiii, xxxi, 17, 18
 safeguards . 16, 18, 40, 68, 227, 245-247, 434, 448, 464
Nuclear Safety.......... 156, 169, 228, 314
 Nuclear Warhead Safety 459, 461
Nuclear Security .... xxii, 213, 288, 334, 451
 Nuclear Secrets xviii
Nuclear Shadowboxing ... ix, xi-xiii, xv-xviii, xxii, xxv, xxvii, xxix, xxxi, 65, 128, 287, 288, 313, 403, 442, 452, 463, 470
 shadowboxing ... vii, ix, xi-xiii, xv-xviii, xxii, xxv-xxvii, xxix, xxxi, 65, 128, 287, 288, 292, 294, 296, 313, 403, 442, 444, 452, 463, 470
Nuclear Strategy 213, 234, 236
 Nuclear Bravado 169, 209, 219
 Nuclear Devastation ... 2, 5, 16, 80, 104, 222
 Nuclear Disarmament ... 3, 18, 174, 175, 259, 294, 308, 310, 322, 343-345, 359, 378, 381, 403, 416, 458
 Nuclear Doctrine 63
 nuclear umbrella .. 55, 98, 114, 223, 277, 282
 overkill . 80, 81, 114, 154, 155, 162, 163, 168, 181, 182, 317, 353, 439, 460, 470
Nuclear Testing . xxiv, 30, 40, 41, 75, 81, 125, 126, 158, 182, 261, 301, 307, 310, 315, 328, 332, 354, 357, 359-361, 368, 394, 397, 404, 416, 417, 435, 448
 Bravo 43, 44, 359
 George ... xiii, 10, 31, 41, 42, 65, 71, 73, 85, 86, 111, 173, 206, 219, 221, 253, 285, 332, 333, 336, 338, 339, 350-353, 363, 370, 371, 376, 378, 380, 424
 Mike 41-43, 197, 210, 398
 Mushroom Cloud 22, 28, 102
 Nevada Test Site 49, 125, 461
 Novaya Zemlya 44, 112, 125
 NTS.......................... 461
 Nuclear Explosion .. 12, 36, 48, 101, 103, 104, 128, 138, 142, 143, 149, 189, 217, 227, 229, 230, 242, 247, 328, 453, 458, 461
 nuclear test 34, 41, 52, 56, 211, 304, 312, 319, 361, 378, 397, 403
 Nuclear Test Ban Treaty 41, 403
 Nuclear Test Site 304, 319

Nuclear Testing Limitations 81, 158
Nuclear Tests 40, 64, 126, 175, 184, 185, 190, 241, 312, 326, 328, 359, 361, 373
Nuclear Weapons Testing .. 80, 124, 176, 318, 360
Pacific Proving Ground 124
PTBT 461
Semipalatinsk 36, 43, 125, 361, 416
Testing Moratorium 361, 393
Testing Nuclear Weapons ... 22, 40, 124, 210, 244, 260, 417
Nuclear Threat
 Nuclear Threat Initiative 301, 349
Nuclear Turning Point 320
Nuclear War .. xvii, xix, 22, 32, 38, 44, 48-51, 56, 57, 60, 63, 66, 69, 78, 80, 83, 88, 89, 92-95, 97, 102, 105, 108, 134, 141-143, 145, 149, 157, 159, 169, 176, 186, 194, 203, 205, 206, 212, 214-217, 220, 221, 223-225, 227, 228, 231, 232, 234, 236, 250, 251, 254, 257, 270, 272, 274, 276, 281, 282, 286, 293, 304, 308, 314, 315, 317, 318, 320-322, 326, 328-331, 333, 340, 346, 348, 351, 352, 355-359, 361, 367, 373-375, 381, 387, 388, 399, 404, 405, 412, 436, 446, 449, 459, 469
 Nuclear War Prevention 317
 Nuclear Warfare 9, 76, 88, 113, 155, 216, 229, 231, 234, 258, 336, 372, 387, 388, 446
 Unintentional Nuclear War 228
Nuclear Warheads .. xix, xxiii, xxiv, 3, 16, 34, 37, 44, 46, 47, 51, 58, 60, 67, 75, 80, 90, 93, 98, 106, 107, 113, 115, 116, 118-121, 125, 128, 132, 133, 146, 148, 150, 154, 155, 165, 186, 187, 189-191, 193, 199, 200, 218, 221, 227, 263, 382, 456
 Nuclear Assemblies 17
 nuclear-tipped 226
Nuclear Weapons
 Fat Man 5, 9, 17, 119, 427
 Honest John 98
 Little Boy 5, 9, 119
 Little John 98
 Neutron Bomb 22, 45, 63, 117, 118, 257, 261, 310, 317, 344, 350, 372, 373, 458
 Nuclear Weapons and Delivery 118, 367, 371, 402
 Nuclear Weapons and Materials xxv
 Nuclear Weapons Data 312
 Nuclear Weapons Production 315
 Nuclear Weapons Systems 75

nuclei 4, 42, 454
nucleus 4, 101, 236, 453-455
Nunn 68, 313, 348, 349
Ocean 58, 125, 132, 137, 156, 218, 294
Offensive Strategic Forces 168, 191, 192
Official Secrets Act 244
Ohio 61, 112, 124
Okinawa 10, 49, 115, 210
Olin 253
Omaha 205, 279
On-Site Inspection ... 65, 155, 162, 184-187,
327, 382, 461, 464, 466,
470
Openness 57, 329, 361, 404, 437, 454
Operation Crossroads 144
Operation Sage Brush 95
Oppenheimer ... 5, 9, 13, 26, 42, 43, 88, 174,
202, 230, 233, 235, 259,
260, 302, 426-428, 434
Orion 128
Orlov 404, 432
Orwellian 178
OSI 168, 185, 188, 461
Osirak 85
OSS 86, 461
Ostpolitik 402
OTA 377
outer space . 41, 126, 139, 146, 152, 175, 180,
182, 188, 380
Outer Space Treaties 152
overflights.................. 129-131, 291
Pacific . . 10, 29, 115, 124, 125, 196, 210, 222,
244, 345
Packard 254
Paid Lobbyists 302, 399
Paine 312, 313, 378
Pakistan .. 14, 17, 19, 62, 126, 264, 281, 401
PAL .. 107, 133, 136, 227, 228, 236, 248, 461
Palestinian
    Palestinians 226, 227
Panama Canal 397
Panofsky 260, 334
Pantex 119, 124, 467
paradigm .... xii, 234, 282, 370, 445
paradox 274
Paris 36, 53, 373, 381, 463
Parliament 75, 204, 328, 342, 404
Partial Test Ban Treaty 185, 461
particulates 196, 244
partisanship 209
Party Leaders 302, 402, 403, 419
pathogens 293
Pauling 332, 359
Pax Christi 323
Peace
    National Peace Academy 390
    National Peace Conversion Campaign
    389
    Peace Action 310, 348
    Peace and Disarmament .. 168, 174, 206,
    310, 328
    Peace Conversion Conference 389
    Peace Council 373, 381, 430
    Peace Institute 301, 390

Peace Movement ... 157, 252, 283, 285,
297, 304, 306, 323, 346,
359, 366, 374, 376, 380,
381, 394, 400, 407, 434,
436, 458
Peace Research Institute .. 300, 342, 343,
461
Peace Works 170, 283, 351
peaceniks 381, 434
United States Peace Institute 390
peaceful
Peaceful Uses of Nuclear Energy ... 307
PNE 60, 74, 158, 461
PNE Treaty 74, 158
Peacekeeper 64, 110, 385, 460
Peacemaker.................... 109, 340
Pearl Harbor 5, 7, 10, 29
Peierls 22, 24, 195
Pennsylvania..................... 134
Pentagon 52, 72, 146, 147, 180, 183, 194,
222, 226, 234, 235, 237,
238, 242, 250, 265, 266,
271, 281, 288, 289, 296,
303, 346, 365, 367, 370,
372, 377, 384, 387, 389,
390, 392, 410, 422, 429,
444, 467
Pentagon Papers 234, 235, 303, 410, 429
Penza 124, 293
perestroika ... 57, 64, 69, 280, 283, 328, 379,
404, 455
Perimeter Acquisition Radar 149
Perimetr 137
Perl 314
Perle .. 61, 169, 253, 254, 256, 257, 262, 321,
446
Permissive Action Link .. 107, 227, 236, 248,
461
Perry 206, 348
Perseus 428, 429
Peru 309
Petropavlovsk 218
Petrov 228
Pew 253
Pfaff 232
Philippines 116
Physics Today 374
Pickett 310
Pike 309, 377
Pilot 130, 186
Pintado 218
Pipes . 169, 173, 249, 252, 253, 257, 262, 271,
446
Pittsburgh 362
Planck 4
Plesetsk 179
Ploughshares Fund 300, 329, 348
PLUTO 128
Plutonium ... 4-6, 9, 12, 13, 16-18, 25-27, 32,
36, 120, 122-124, 175, 195,
196, 217, 225, 226, 230,
246, 247, 290, 312, 327,
455, 456, 460, 463, 466,
468, 469

Plutonium-Production 123, 195, 196, 327
Pu-239 .... 454, 456
Podgorny 403
Point Four Program 264
Poland 66, 67, 113, 252, 279, 282, 344
Polaris .. 51, 55, 111, 112, 121, 127, 148, 157, 200, 345
Policy Debate .... xxii, 451
policy-oriented 401
Politburo . 82, 86, 94, 223, 254, 326, 368, 395, 402, 443
Political Action . 308, 311, 314, 318, 364, 376, 461
Political Factors 418
Political Forces 301, 305, 394-396, 409
Politics of Racism 302, 410
Pollack 462
Pollution 77, 132, 267, 325, 330
Pope 279, 344
pork 247, 288, 293
Portugal 37
Poseidon 61, 111, 112, 116, 291, 427
Possible Violations 40, 179, 432
Postol 333, 350
Posture Review 460
Post-War Years 22, 30, 32, 39
Potemkin 278
Potsdam Conferences 402
Potsdam Declaration 2, 10, 11
Potter 335
Poverty 198, 277, 322, 398
Powaski 72, 76
Powell 283
Powers
    Gary Powers 130
Pragmatists 280
Prague Spring 432
Present Danger 19, 60, 65, 169, 173, 236, 252-254, 256, 259, 262, 320, 446, 458
President Roosevelt 7, 10, 32
Press .... xxvi, 6, 64, 76, 209, 223, 234, 235, 244, 277, 316, 318, 322, 336, 355, 365, 373, 379, 385, 397, 405, 429, 432
preventive 34, 86, 144, 206, 238, 336
Priesthood ..... xvii, 169, 232, 233, 261, 445
Primack 331, 365
Primakov 326
Princeton 300, 331, 334, 337, 364
Private Institutions 302, 400, 415
Production
    Production Capacity 122, 267
    production plant 26, 290
    Production Plants 7, 240
    Productivity 169, 267, 268, 388
Professional
    Professional Associations 300, 337
Program 300, 308, 319
Progressive .... xxv, 212, 245, 259, 297, 301, 303, 322, 333, 348, 350, 431, 432, 463, 468
Project Abolition 347

Project Orion 128
Project PLUTO 128
Proliferation
    Nuclear-Armed Nations .... xxii, 14, 56, 291, 451
propellant 51, 112, 120, 141, 229, 429
Propulsion 25, 27, 80, 122, 126-128, 138, 182, 457
protestors 194, 344, 356, 416
Proxmire 367
PSR 300, 314, 315, 358, 461
psychological ... 2, 5, 9, 11, 69, 78, 141, 169, 190, 205, 214, 216, 230, 252, 263, 270, 274, 340, 440, 441
Psychological ... 2, 5, 9, 11, 69, 78, 141, 169, 190, 205, 214, 216, 230, 252, 263, 270, 274, 340, 440, 441
Psychological Costs 169, 274
public
    Public Affairs 303, 335, 338
    Public Assemblies 302, 389, 413
    Public Debate ... 94, 145, 160, 180, 332, 356
    public dissent .. 269, 305, 325, 357, 369, 396, 434, 436, 444
    public information 184, 229, 359
    public interest ...... xxix, 309, 326, 343
    public involvement ... vi, 296, 299, 303, 356, 419, 435, 449
    public movement 52, 355, 373, 376
    public opposition ... xv, 41, 75, 77, 150, 159, 250, 257, 303, 304, 357, 361, 393, 436, 437, 448
    public policy ... 241, 309, 320, 332, 337, 338, 397, 400
    public protest .. 304, 319, 359, 362, 373, 385, 394, 414, 441
    public resistance . xxi, 49, 297, 344, 353-355, 360, 370, 448
    Public Sources xxxi
    Public-Interest Organizations ... xv, xxii, 68, 300, 305-307, 329, 330, 354, 448
Puerto Rico 116, 467
Pugwash .. 300, 302, 307, 310, 311, 329, 332, 333, 339, 354, 362, 388, 405, 415, 425, 433
    International Pugwash Group 339
    Pugwash Conference 311
    Soviet Pugwash 339, 405
Purex 123
Putin 73
Quaker 323
Quemoy 54
R&D .. 69, 140, 146, 148, 151, 157, 174, 194, 241, 269, 274, 289, 383, 416, 461
Rabi 42, 426
Rabinowitch 32, 306, 311, 349
Radiation
    Background Radiation 101

504 Nuclear Insights

Ionizing 101, 144, 461
Irradiation 217
Low Levels 175
Low-Level Radiation 275, 328
Radiation Doses 103, 218
Radiation Effects . . . 101, 105, 169, 216, 275, 453, 466, 468, 469
Radiation Exposures 31, 244
Radiation Legacies xxiii
Radiation-Exposure 15, 56
Radiophobia 14
radiophobic 245
Radio USA 253
radioactive
   Radioactive Contamination . . . 105, 217, 226
   Radioactive Materials . . 16, 56, 229, 230, 245
   radioactivity . . 4, 77, 102, 103, 123, 217, 218, 225, 229, 230, 359, 388
radiochemistry 27, 197
Radiography 466
radioisotope 25
radiological 6, 42, 169, 229, 230, 464-467
   Radiological Warfare . 42, 169, 229, 230
rainout 216
RAND   63, 169, 202, 215, 220, 233-236, 238, 248, 261, 288, 329, 334, 364, 429, 467
RANSAC 334
Rathjens 333, 338, 370
Raven Rock Mountain 134
RBE 102, 461
RDS 36, 43, 45
Reagan . 57, 61-67, 69, 73-76, 89, 91, 94, 110, 111, 118, 120, 126, 127, 150-153, 160, 170, 172, 177-180, 183, 184, 191-194, 198, 203, 204, 206, 213, 221, 222, 248-255, 259, 262, 266, 268, 271, 273, 274, 278-280, 282-285, 288-290, 293, 295-297, 311, 318, 321, 322, 326, 340, 344, 345, 351, 352, 358-360, 371, 373-379, 381, 383-387, 390, 394, 398, 399, 401, 403, 404, 407, 410, 417, 418, 426, 427, 436, 441, 442, 444, 446, 448, 460, 461, 468
   Reagan administration . . 64, 76, 89, 150, 153, 160, 178-180, 183, 184, 193, 203, 213, 222, 250-254, 259, 262, 266, 271, 273, 279, 284, 285, 288, 289, 293, 297, 321, 326, 351, 360, 371, 376-378, 385-387, 390, 394, 407, 417, 436, 441, 448
   Reagan Doctrine 204
Reaganites 262, 378
Reagan's Intractability 169, 248

Realities   x, xxi, xxii, 168, 170, 291, 394, 396, 412, 443, 448
Reciprocity 294
Recommendations   x, xxiii, 8, 33, 86, 98, 180, 206, 208, 248
Recurring Themes . . . . vii, xxv
Red Army 115, 199, 323
Red Herring 168, 184, 258, 370
Red Squad 430, 468
Redstone 98, 277
Reed . . 89, 155, 199, 203-205, 236, 244, 249-252, 266, 270, 271, 279, 280, 285, 289, 352, 361, 376, 427
Regulus 111
Reliability . . . xxiv, 46, 99, 113, 120, 124, 136, 140, 154, 158, 173, 255, 301, 359, 361, 365, 385, 386, 451
Religious Associations 300, 322
Reprocessing 123, 196, 246, 455
Republic . 50, 54, 98, 117, 121, 191, 212, 336, 404, 407, 412, 424, 432, 450, 458
Republican . . . 12, 57, 69, 148, 155, 254, 350, 362, 368, 469
Research
   Research and Development . . . 122, 164, 169, 172, 173, 177, 190, 207, 241, 242, 269, 343, 365, 390, 461
   Research and Test Reactors 461
Residue 29, 118, 217
Reuters 257
Reykjavik 65, 76, 184, 284, 404, 417
Rhodes 13, 27, 261, 282, 427
Rickover 127
Ridge 8, 25, 123, 124, 306, 467
Right-Wing 120, 177, 233, 252, 253, 289, 304, 319, 323, 324, 336, 340, 350, 351, 359, 370, 372, 376-378, 396-398, 400, 401, 418, 427
Ringo 363
riots 75, 274
risk of nuclear war 69, 176, 373
Roadmap 206, 452
Robinson 243
Rockefeller 141
Rockwell 109
Rocky Flats 124, 217
roentgens 105
Romania 67
Rome 344, 415, 417, 467
Roosevelt 5, 7, 10, 24, 32, 73, 260, 467
Rosenberg 290, 309, 429
Rosenbergs 27, 429
Rosenthal 253
Rossin 246, 247
Rostow 60, 253
Rotblat 4, 24, 27, 195, 219, 311, 359
Rowen 202, 249
Rowny 178
Royko 398

Rubles for Defense 169, 271	Satellite Reconnaissance ... 81, 128, 129, 131, 177, 379
Ruby 327, 363	Saturn 275
Rudman 349	Scaife 253, 254
Ruhr University 416	Schaerf 415
Ruina 332	Schell 219, 221, 224, 232, 233, 239, 262, 289
Rumsfeld 206, 254	Schiffer 363
Russell 307, 310	Schlesinger 143, 172, 206, 236, 360
Russia	Schmidt......................... 239
RF 339, 461, 467	Scholars and Organizations 302, 404
Russian Federation .. xxiv, 61, 160, 460, 461	Schroedinger 4
Russian Obstacles 117, 404, 407	Schultz 253
Rutherford........................... 4	Schwartz 314
SAC .. 44, 45, 49, 93, 94, 109, 203, 205, 212, 235, 461	Science Academies 326
Safer World 301, 314, 348, 349	Science Advisory Committee 331, 334
Safire............................ 253	Science and Technology ... xxv, 99, 195, 240, 309, 313-315, 331, 347, 464
Sagan 328, 352, 387, 388, 405, 462	scientific
Sagdeev ... 196, 201, 223, 240, 312, 326, 377, 405	scientific advisors 364
Sage 95	Scientific Advisory Group 203
Sahara 24	Scientific American . 301, 331, 349, 350, 363, 370
SAIC 233, 238, 464, 467	scientific community 151, 365, 415, 427, 436
Saigon 53	Scientific Freedom 338
Sakharov ... 43, 153, 196, 284, 300, 327-329, 360, 432, 433	scientific organizations 358
SALT ... 58-61, 63-65, 74, 81, 111, 140, 157-161, 163, 176, 178-180, 183, 189, 190, 192, 208, 253, 254, 256, 258, 260, 273, 292, 301, 334, 344, 351, 362, 368-370, 378, 383, 384, 397, 403, 405, 417, 461, 462, 466	scientific research 240, 241, 340
	Scientists Against Proliferation ... 343, 459
	scorpions 21, 43, 78, 88, 439, 440
	Scott 352, 353
	Scoville 90, 316, 384-386
	Scowcroft 168, 178, 206, 207, 352, 381
SALT I. 58, 59, 61, 74, 81, 111, 157-160, 163, 176, 190, 256, 260, 351, 369, 370, 384, 403, 405, 461	SDI . 64, 76, 81, 146, 150-153, 178, 179, 184, 251, 254, 256, 258, 259, 266, 272, 278-280, 291, 301, 310, 314, 316, 326, 331, 338, 365, 368, 376-381, 384, 393, 394, 397, 404, 417, 418, 422, 436, 448, 461
SALT I negotiations 111, 190, 370	
SALT I Treaty 351, 370, 403	
SALT II . 59-61, 63-65, 74, 81, 159, 160, 176, 178-180, 183, 189, 192, 208, 253, 254, 258, 260, 344, 368-370, 378, 383, 384, 397, 417, 461	
	SDS 305
	Seabed .... xix, 58, 180, 182, 188, 464
	Seabed Arms Control Treaty 182
SALT II negotiations 59, 258	Seattle 362, 363, 423, 430
Samuelson 269, 270	Second Stage 455
San Francisco 243, 465, 467	secrecy
Sandinista	Secrecy . xviii, xxiii, xxv, xxvi, xxxi, 2, 6, 8, 9, 16-18, 29, 31, 56, 77, 78, 94, 115, 116, 169-171, 174, 181-183, 198, 213, 216, 231, 243-248, 275, 280, 294-296, 304, 309, 330, 351, 356, 390, 408, 420, 422, 431, 434, 461
Sandinistas 251	
SANE 220, 300, 308, 310, 319, 322, 323, 359, 363, 364, 375, 417, 461, 467	
Sane Nuclear Policy 308, 310, 363, 461	
Santa Monica 233, 234	
Sapwood 46	
Sarov 124, 328	Secret ... xviii, xxvi, xxxi, 4, 6, 9, 15-17, 24, 25, 27, 29, 31, 32, 34, 37, 41, 42, 44, 51, 56, 62, 83, 94, 114-116, 124, 130, 140, 172, 179, 195, 196, 199, 202, 209-213, 218, 229, 231, 234, 235, 240, 244-247, 249, 275, 281,
Sary Shagan ... 146, 152, 186, 312, 327, 377	
satellites ... 46, 55, 64, 81, 90, 104, 111, 126, 129, 131, 132, 136, 137, 139-141, 146, 153, 154, 158, 172, 182, 185, 186, 190, 193, 277, 292, 308, 379, 380, 385, 450, 459	

```
                   293, 328, 332, 356, 377,     Silver Screen . . . . . . . . . . . . . . . . . . . . 301, 351
                   378, 389, 390, 402, 422,     Simon . . . . . . . . . . . . . . . . . . . . . . . . . 330, 469
                   424, 428, 429, 431, 432,          Paul Simon . . . . . . . . . . . . . . . . 330, 469
                   435, 463, 468                Simonenko . . . . . . . . . . xiii-xv, 170, 284, 313
Secrets .. xviii, 155, 185, 196, 202, 244,      Simpson . . . . . . . . . . . . . . . . . . . . . . . . . . 306
                   246, 333, 383, 429, 468      Sinclair . . . . . . . . . . . . . . . . . . . . . . . . . . . 355
Secretary of Defense .. 13, 47, 85, 88, 89, 118,   Single Integrated Operational Plan . . . 94, 202,
                   143, 146, 148, 156, 203,                                                   461
                   206, 219, 243, 253, 256,     SIOP . . . . . . . . . . . 38, 94, 154, 202-205, 461
                   257, 271, 288, 364, 365,     SIOP Targets . . . . . . . . . . . . . . . . . . . . . . . 94
                   369, 424                     SIPRI . . . . . . . . . . . . . . . 300, 342, 461, 467
Secretary of Energy . . . . . . . . . . . . . 360, 469   Sixth Fleet . . . . . . . . . . . . . . . . . . . . . . . . 212
Security                                        Skobeltzen . . . . . . . . . . . . . . . . . . . . . . . . 405
     Security Council . . . . 32, 70, 86, 93, 134,   Skolnikoff . . . . . . . . . . . . . . . . . . . . . 332, 333
                   135, 175, 233, 250, 319,     Slava . . . . . . . . . . . . . . . . . . . . . . . . . 186, 327
                   322, 335, 422, 429, 430,     SLBM . . . . . . . . . . . . 111, 112, 193, 458, 461
                   461                               SLBMs . . . 47, 55, 85, 97, 107-112, 116,
     Security Issues . 314, 316, 321, 331, 337,                   120, 126, 127, 136, 138,
                   342, 345, 363, 396                             140, 159, 161, 173, 193,
     Security Paranoia . . . 168, 205, 219, 422                   197, 207, 229, 292
     Security Policy .. 49, 210, 254, 300, 321,   SLCM . . . . . xxv, 161, 186, 327, 382, 443, 461
                   402, 406, 424                Sloika . . . . . . . . . . . . . . . . . . . . . . . . . 43, 328
     Security Studies . . . . 257, 335, 392, 458   Smaar . . . . . . . . . . . . . . . . . . . . . . . . . . . 335
     security umbrella . . . . . . . . . . . . . . . 87   Smith . . . . . . . . . . . . . . . . . . . . . . . . 396, 398
Seignious . . . . . . . . . . . . . . . . . . . . . . . . 370   Smith Amendment . . . . . . . . . . . . . . . . . . 396
Seismic Monitoring . . . . . 312, 360, 393, 414   Smithsonian . . . . . . . . . . . . . . . . . . . . . . . . 6
seismology . . . . . . . . . . . . . . . . . . . . . . . 242   Smoke . . . . . . . . . . . . . . . . . . . . . . . . 387, 388
Selden . . . . . . . . . . . . . . . . . . . . . . . . . . . 246   Smyth . . . . . . . . . . . . . . . . . . . . . 22, 39, 195
Senate . . . 58, 61, 74, 116, 159, 160, 235, 239,   Smyth Report . . . . . . . . . . . . . . . . . . 39, 195
                   248, 304, 305, 318, 333,     Snezhinsk . . . . . . . . . . . . . . . . . . xiv, 124, 130
                   369, 390, 392, 393, 397-     SNTRA . . . . . . . . . . . . . . . . . . . . . . . . . . 462
                   399, 409, 423-425, 458,      Social Responsibility . . . . 314, 319, 348, 358,
                   467                                                                 363, 461
     Senate Armed Services Committee .. 235,    Socialist . xix, xxv, 31, 50, 52, 66, 67, 71, 165,
                   398                                            286, 295, 297, 325, 406,
     Senate Committee . . . . . . . . . . . 369, 424                      426, 441, 462
     Senate Internal Security Subcommittee     Socialist Labor . . . . . . . . . . . . . . . . . . . . . 325
                   . . . . . . . . . . . . . . . 333, 423   Societal Risks . . . . . . . . . . . . . . . . . . . . . 427
Sentinel ABM . . . . . . 148, 243, 257, 362, 365   Soddy . . . . . . . . . . . . . . . . . . . . . . . . . . . . . 4
Sentry . . . . . . . . . . . . . . . . . . . . . . . . . . . 146   Sokov . . . . . . . . . . . . . . . . . . . . . . . . . . . 201
Serbia . . . . . . . . . . . . . . . . . . . . . . . . . . . 219   solid fuel . . . . . . . . . . . . . . . . . . . . . . . 46, 189
Sergeant . . . . . . . . . . . . . . . . . . . . . . . . . . 98   Solidarity . . . . . . . . . . . . . 66, 279, 344, 346
SESPA . . . . . . . . . . . . . . . . . . . 300, 314, 461   Solzhenitsyn . . . . . . . . . . . . . . . . . . . 252, 433
Shakers . . . . . . . . . . . . . . . . . . . . . xx, 301, 396   South Africa . . . . . . . . . . . . . . . . 14, 124, 411
shibboleths . . . . . . . . . . . . . . . . . . . . . 163, 191   South Asia . . . . . . . . . . . . . . . . . . . . . 161, 334
Shipbuilding Industry . . . . . . . . . . . . . . . . 240   South Carolina . . . . . . . . . . . . . . . . . . . . . 322
Shipping . . . . . . . . . . . . . . . . . . . . . . . . . 267   South Dakota . . . . . . . . . . . . . . . . . . . . . . 150
shortage . . . . . . . . . . . . . 198, 241, 266, 355   South Pacific . . . . . . . . . . . . . . . . . . . . . . 125
shortcomings . . . xxi, 120, 198, 410, 431, 441   South Vietnamese . . . . . . 53, 54, 356, 410, 412
shortcuts . . . . . . . . . . . . . . . . . . . . . . 271, 276   Southeast Asia . . . . . . . . . . . . . . . . . . . 53, 54
Shorter-Range Missiles . . . . . . . . . . . . 80, 112   Southern Christian Leadership Council . . 322
shortfall . . . . . . . . . . . . . . . . . . . 173, 266, 271   Southern Hemisphere . . . . . . . . . . . . . . . . 388
Shultz . . . . . . . . . . . . . . . . . . . . . . . . 65, 376   Soviet
Shyster . . . . . . . . . . . . . . . . . . . . . . . . . . . 46        Soviet ABM . . . . . . . 147, 150, 207, 364
Siberia . . . . . . . . . . . . . . . . . . . . . . . . . . . 36        Soviet Academy . . . . 238, 300, 302, 307,
Sicily . . . . . . . . . . . . . . . . . . . . 345, 381, 415                          312, 325, 327, 339, 388,
Sickle . . . . . . . . . . . . . . . . . . . . . . . . . . 279                          405
Sierra . . . . . . . . . . . . . . . . . . . . . . . . . . . 185        Soviet allies . . . . . . . . . . . . . 49, 233, 368
SIGINT . . . . . . . . . . . . . . . . . . 129, 132, 461        Soviet Army . . . 10, 62, 85, 95, 276, 432
Signal . . . . . . . . . . . 135, 137, 140, 141, 228        Soviet ASAT . . . . . . . . . . . 186, 326, 379
Signatories . . . . . . . . . . . . . . . . . . . . . . . . 18        Soviet atomic bomb . . . . . . . . . . . . . 402
significant influence . 195, 234, 286, 355, 402,        Soviet attack . . . . 94, 129, 148, 149, 192,
                   447                                                 204, 216, 222, 362, 372,
Silicon Valley . . . . . . . . . . . . . . . . . . . . 347                                    385
```

Soviet A-bomb 27, 33, 336
Soviet bloc 48, 264, 280, 352
Soviet bombers 171, 189
Soviet cheating 351
Soviet Communism 50, 68, 74, 277, 281
Soviet Defense Council 134
Soviet Defense Programs 265
Soviet Deployments 80, 116
Soviet economy ... 59, 70, 265, 271, 278, 379
Soviet espionage 34, 289
Soviet first-strike 369, 398
Soviet hydrogen bomb 328, 432
Soviet H-bomb 328
Soviet ICBM 55, 130, 136, 171, 207, 272, 384
Soviet Jews 284
Soviet leaders 82, 83, 205, 207, 221-223, 254, 282, 283, 285, 340, 367, 378, 404, 405
Soviet military ... xiv, 32, 55, 72, 89, 94, 96, 97, 105, 111, 130, 135, 171, 201, 227, 253, 265, 266, 271, 272, 285, 289, 316, 325, 363, 368, 391, 403
Soviet Military Power 96, 253
Soviet MIRV 60, 369
Soviet missiles ... 2, 136, 203, 383, 386
Soviet Moderates 300, 325
Soviet moratorium 125, 360, 380
Soviet Navy ..... xiv, 111, 112, 127, 327, 442
Soviet nuclear arsenal 77, 350
Soviet nuclear test 34, 56, 361
Soviet nuclear weapon 80, 112, 327
Soviet officials . 195, 218, 251, 282, 309, 323, 354, 448
Soviet policies 340, 433
Soviet Policy Makers 302, 402
Soviet Power 278
Soviet Program 195, 293
Soviet protest 130
Soviet reformers 251
Soviet reformists 391
Soviet response .. 32, 106, 291, 293, 387
Soviet satellite 204, 273, 379
Soviet scientists ... 5, 27, 195-197, 295, 300, 310-313, 325, 326, 339, 349, 354, 359-361, 383, 388, 405, 415, 437, 458, 469
Soviet Secrecy 182, 351
Soviet sites 327
Soviet Socialist Republics .. xix, xxv, 66, 67, 71, 165, 295, 297, 441, 462
Soviet Strategic Forces 94, 132, 134, 135
Soviet Strategic Warheads 94
Soviet Strategies 80, 94, 96
Soviets 206
Soviet-American .. 31, 33, 329, 331, 339

Soviet-bloc 336, 380
Soviet defense ... 75, 134, 169, 201, 221, 240, 265, 271, 272
Soviet Defense Establishment 169, 240
Soviet hardliners 296, 378
Soviet Missiles
    SS-9 236
space
    Space Research Institute 240, 326
    spacecraft 128, 132, 274
Space Exploration 277
Spain 116, 226, 467
Spartan 47, 146, 149, 150, 363, 365
SPC 238
Spencer 401
Spending .. xix, xxii, 65, 70, 72, 75, 173, 242, 245, 246, 249, 252, 253, 263-274, 285, 290, 293, 295, 296, 301, 316, 361, 366, 367, 371, 389, 392, 400, 403, 404, 408, 414, 451
Spent Nuclear Fuel 230
Spies . xviii, 6, 17, 25, 33, 168, 195, 278, 281, 303, 420, 426, 428-430
Spirit .... xxxi, 110, 390, 421
Spock 310, 375
Sputnik ... xxiii, 39, 46, 48, 55, 77, 128, 146, 153, 171, 183, 277, 337, 362, 403
Spying 184, 421-423, 428-430, 432, 434
SSBN 111, 462
SSBNs 112, 137, 200
Stagg Field 25
Stalin ... 6, 11, 13, 27, 31, 38, 69, 73, 82, 191, 195, 196, 244, 264, 289, 302, 402, 408, 423, 425, 433, 447
Stalinism 412, 448
Standards .. 16, 184, 188, 247, 258, 268, 337, 382, 383
Standing Consultative Commission . 254, 461
Stanford .... xiii-xv, 300, 333, 334, 363, 401, 458
    George Stanford 363
Star Wars ... 64, 146, 250, 326, 377, 378, 394, 461
START .. 63, 67, 68, 161, 163, 178, 184, 185, 208, 321, 382, 462, 466
    START I .. 16, 63, 67, 81, 155, 157, 161, 179, 404
    START II 76, 157, 294
    START negotiations 178, 184
    START Treaty 185, 382
State Department .. 31, 34, 39, 160, 249, 250, 361, 369, 424, 467
State-Sponsored 3
Statistics 169, 231, 265, 445
Stealth .. 60, 81, 110, 140, 162, 193, 249, 377, 384, 389, 390
stealth technology 110
Stein 282
Steinbruner 335
Stevenson 330, 385

Stewart 398
Stimson Center 300, 321
Stimulating the Economy 169, 267
Stockholm 342, 461
Stockpile
    Stewardship 313, 401, 403
    Stockpile Stewardship 313, 401
Stone
    Jeremy Stone 377
Strassman 4
Strategic
    Strategic Air Command 32, 44, 109,
        199, 205, 212, 279, 461
    Strategic Armaments 62, 176
    Strategic Arms Limitation Treaty . 58, 59,
        61, 159, 461
    Strategic Arms Reduction Treaty ... 161
    Strategic Assessments 168, 199, 233
    strategic bombers .. 36, 47, 51, 108, 113,
        120, 121, 139, 157, 159,
        176, 189, 233, 389, 404,
        435
    Strategic Defense Initiative .. 64, 76, 146,
        150, 250, 376, 461
    Strategic Defenses ... 81, 145, 207, 257
    Strategic Forces ... 85, 94, 97, 108, 132-
        135, 159, 160, 168, 178,
        181, 191-193, 202, 206,
        208, 220, 223, 228, 235,
        236, 273, 320, 339, 441
    Strategic Implications 257
    Strategic Modernization ... 80, 111, 176,
        201
    Strategic Nuclear Forces 205
    Strategic Rocket Force 201
    Strategic Rocket Forces 96, 288
    Strategic Studies 236, 300, 342
    Strategic Systems . 59, 80, 107, 158, 207,
        366
    Strategic Warheads ... 94, 109, 110, 112,
        118, 121, 161
    Strategic Weapons 88, 95, 103, 114-116,
        119, 132, 157, 159-161,
        207, 224, 291, 387, 455
strategists ... 83, 87, 200, 220, 221, 272, 293,
    294, 444
Stratofortress 109
Stratojet 109
stratosphere 102, 388
Strauss 196
Student Activism 330
Student Peace Union 304
Student Radicalism 302, 411
submarine . 2, 52, 59, 80, 85, 94, 99, 107-109,
    111, 112, 118, 120, 121,
    126-129, 132, 133, 136,
    148, 156, 157, 159, 180,
    196, 201, 207, 211, 213,
    218, 220, 229, 241, 289,
    345, 351, 373, 386, 457,
    461, 462
    submarines .. xix, 51, 55, 59, 61, 75, 89,
        91, 92, 94, 108, 109, 111-
        113, 118, 120, 121, 126,
        127, 132, 135-138, 146,
        148, 151, 157, 159, 189,
        203, 207, 211, 212, 292,
        294, 372, 427, 466, 468
    Submarine-Launched Missiles .. 80, 111,
        136
suborbital 112
subsurface 138, 244
Suburban Chicago Opposition 301, 363
Suez 70, 211, 216
Sullivan 335, 425
Sun 7, 42, 123
superbomb 42, 125, 197, 284
Superfortress 109
superpower ... xv, xvii, xix, xxi, xxiii, 19, 23,
    36, 56, 62, 65, 67-69, 71-75,
    82, 88, 89, 95, 116, 126,
    154, 155, 163, 174, 179,
    263, 271, 280, 282, 285-
    287, 289, 291, 292, 296,
    304, 326, 340, 344, 374,
    410, 412, 414, 417
superpowers .. xix, 2, 19, 58, 61, 74, 75,
    78, 109, 125, 133, 136, 138,
    145, 159, 163, 165, 176,
    178, 183, 187, 188, 193,
    197-199, 207, 211, 228,
    230, 232, 264, 272, 274,
    281, 287, 290, 291, 295,
    296, 393, 418, 442, 445,
    446, 448, 451
Supreme Court 235, 262
Supreme Soviet 135, 175, 409
surface
    surface burst 102
    surface ships 127, 212
    surface vessels 126
Survivability ... xiv, 80, 91, 99, 108, 191, 208,
    358, 384, 465
Surviving Nuclear War 22, 48, 359
Survivors Envy 234
Sverdlovsk 124, 130, 180, 227
Sweden .... xvii, 467
Switzerland .... xvii, 341, 467
SWORD 154, 450
Symington 116, 171
Symmetry 168, 188
Syndrome 191
Syria 401
Systematic Assessment x
Szilard . 4, 5, 9, 13, 174, 202, 260, 306, 307,
    311, 312, 423
Tactical
    battlefield nuclear weapons 372
    tactical nuclear weapons .. 37-39, 45, 51,
        55, 63, 65, 75, 114-116,
        121, 164, 189, 200, 202,
        203, 211, 223, 224, 257,
        292, 372, 404
    Tactical Weapons 42, 113, 117, 185
Taiwan 116, 212
Tajikistan 68
Taliban 213, 447
Tallinn 81, 147, 291

Index for Volume 1 509

Talmadge 411
Tamm 43, 326, 328, 405
Taylor 243, 246, 309
Team B 173, 257, 271, 272
technocrat 155, 252, 273, 361
Technological Innovation 169, 240
technology
    Technology Assessment .. 177, 343, 377, 459
    Technology Creep 81, 140
Teller .. 4, 26, 42, 43, 152, 156, 169, 173, 197, 214, 215, 233, 239, 241, 243, 252-254, 259-262, 328, 331, 364, 376-378, 401, 415, 426, 427, 446
Tennessee 8, 25, 123, 241, 322, 363, 399
Terkel 355
Terror .. 84, 89, 135, 168, 180, 189, 293, 332
    Terrorism .... x, xi, xxiii, xxv, 3, 17, 183, 198, 286, 410, 424, 430, 437, 451
    Terrorists .... xxiv, 3, 273, 323
terrorism
    international terrorism 183
Texas 124, 239
Thatcher 262, 345
The Day After 75, 306, 351, 352
Theater
    Theater Weapons 22, 45, 80, 113
Thermonuclear ... xiii, xiv, xxv, 2, 15, 22, 30, 33, 36, 41-46, 56, 77, 99-101, 107, 110, 111, 119, 120, 122-124, 144, 154, 156, 157, 162, 168, 175, 191, 194, 197, 210, 215, 234, 241, 244, 260, 284, 310, 321, 357, 359, 383, 393, 426, 427, 443, 455, 456
    Thermonuclear Bomb 2, 15, 42, 197, 359
    thermonuclear bombs 41, 157
    thermonuclear detonation 100
    Thermonuclear Espionage 168, 197
    thermonuclear war 234, 321, 357
    Thermonuclear Warheads ... 42, 46, 110, 120, 383
    Thermonuclear Weapon 42, 43, 144, 154, 194, 455, 456
    Thermonuclear Weapons .... xiv, xxv, 22, 30, 41-45, 99, 101, 156, 162, 175, 191, 194, 197, 215, 241, 260, 310, 359, 443
Third World .. 14, 76, 86, 264, 281, 376, 411, 437
Thomas . 69, 89, 155, 183, 199, 203-205, 236, 244, 249, 252, 266, 270, 279, 285, 289, 312, 317, 352, 361, 376, 427, 462
Thor 45, 115
thorium 123
Threat Reduction 348, 458, 462
Three Mile Island 373, 377

Threshold Test Ban 59, 60, 120, 158, 180, 185, 360, 393, 462
Throw-Weight 16, 67, 73, 113, 160, 161, 173, 176, 200, 207, 258, 259, 456
Thule 226, 227
Thumper 146
thyroid 144, 459
Titan 108, 110, 172, 238, 377, 467
Tlatelolco 188
TNT 7, 113, 119, 156, 163
Tokyo 6, 9, 10, 29, 411, 467
Tomahawk 60, 63, 112
Tomsk 124
Topchiev 307, 339
Topol 251
Tornado 109, 121
torpedo . 45, 52, 112, 117, 129, 132, 213, 462
Torture 426, 432
Toxic 244
Trade Act 284
Transnational ... xi, xxiii, 251, 286, 300, 314, 340, 345, 354, 355, 360, 361, 391, 393, 402, 405, 418, 436, 451
    Transnational Groups 300, 340
    Transnational Movement 361, 393
Treaty
    Treaties and Agreements 23, 74, 157, 175, 176, 350, 392
    Treaties and Negotiations ..... xxii, 451
    Treaty Experience 168, 180, 188
    treaty negotiations ... 110, 239, 295, 359
    Treaty Verification .. 157, 184, 187, 241, 243, 256, 316, 339, 409, 414, 435, 441, 457, 463, 464, 466
    Treaty Violations 168, 179, 183, 188, 190, 197
Trenton 225
Triad 38, 80, 91, 108, 109, 126, 138, 173, 200, 220, 260, 271, 272, 384, 389
Trident ... 2, 61, 92, 108, 112, 118-120, 159, 189, 193, 226, 229, 377, 458
Trinity 2-6, 9, 13, 14, 18, 27, 31, 124, 334
Tritium 41-43, 122
Truman ... 6, 9, 11, 12, 29, 31, 33, 38, 42, 43, 52, 72, 73, 85, 93, 142, 196, 210, 231, 235, 264, 400, 410, 423
Truman Doctrine 31, 85, 264
Trutnev 43
Tsipis 333
TTBT 59, 157, 158, 179, 334, 462
Tupolov 44, 326, 405
turbofan 141
turboprop 171
Turkey .. 31, 37, 46, 49-51, 77, 115, 213, 264, 287, 447
Turkmenistan 68
Turner 54, 349
Turning Point 42, 320, 428

U.S.
   U.S. allies 178
   U.S. Army .. 10, 107, 112, 114, 116, 118,
                 146, 147, 155, 199, 276,
                 425, 430
   U.S. bombers 107, 115, 192, 272
   U.S. Military Spending ... 263, 268, 269
   U.S. Radars 179, 457
   U.S. Strategic Forces 220, 223, 235
   U.S. Strategies 80, 92
UCS . 300, 313-315, 329, 338, 364, 377, 380,
                 392, 462
UK ... 24, 121, 160, 250, 262, 345-349, 434,
                 462, 467
Ukraine 68, 116, 161, 313, 336, 404, 463,
                 467
Ulam 43, 260
ultraconservative 252, 397
UN ... 9, 40, 70, 74, 129, 174, 175, 210, 244,
                 255, 300, 302, 337, 340,
                 341, 353, 359, 397, 403,
                 418, 423, 437, 454
   UN General Assembly 40, 74, 175
UNIDIR 300, 341, 462
Unitarian 318
Unitarian Universalists 318
United Campuses 308
United Kingdom 40, 45, 55, 98, 128, 225,
                 241, 339, 462
United Methodist Church 322
United Nations .. 39, 158, 206, 230, 264, 340,
                 341, 370, 375, 397, 437,
                 462
United Nations Institute 341, 462
United Presbyterian Church 318, 322
United World Federalists 308
Universalists 318
University of Illinois 300, 334
University of Maryland 300, 335, 458
UNSCOM 462
Unthinkable . xxii, 88, 96, 168, 199, 202, 295,
                 296, 439, 445
Un-American Activities 302, 337, 423
Update xxii
Urals .... xiv, 135
uranium
   Depleted Uranium 16
   HEU 123, 459
   Highly Enriched Uranium . 25, 123, 126,
                 349, 459
   LEU 460
   U-235 .... 123, 454-456
   U-238 454-456
   Uranium Committee 5, 27
   uranium-235 .... 24, 26, 123
   uranium-238 123
USSR .. xv, 31, 35, 38, 46, 49, 50, 59, 61, 62,
                 70, 73, 122, 134, 171, 175,
                 199, 205, 244, 248, 249,
                 260, 271, 279, 280, 282,
                 284, 285, 290, 295, 309,
                 312, 313, 325, 328, 349,
                 352, 353, 361, 366, 368,
                 384, 389, 391, 395, 396,
                 401-405, 407, 409, 415,
                 418, 437, 442, 455, 462,
                 467
Ustinov 221
Utah 111, 200, 323, 386
Uzbekistan 68
U-2 .... 34, 81, 129, 130, 171, 229, 291, 379
Vagueness 168, 188
Van der Graaf 343
Vance 93
Vannik 70
Vanunu 333
VEGA 274
VELA 185
   VELA Hotel 185
   VELA Sierra 185
   VELA Uniform 185
Velikhov .. 312, 325-327, 339, 361, 377, 388,
                 391, 405
Verifiable .... xxvi, 19, 68, 75, 112, 177, 178,
                 184, 258, 265, 318, 345,
                 351, 361, 373, 380, 383
Verification
   Verification Improvements 168, 177,
                 190
   Verification Issues 184, 469
   Verification Technology .. 184, 198, 300,
                 342, 343, 374, 383, 441,
                 462-464, 470
   Verification Technology Information
                 Centre 342, 462
VERTIC ... 300, 341-343, 383, 416, 462, 467
Very Low Frequency 138
Vessey 250
Vested Interests .... x, 19, 209, 233, 239, 272,
                 276, 296, 304
Vienna 416, 467
Viet Cong 53, 63, 273, 356, 447
Viet Nam 273, 274, 367, 429, 434
Vietcong 53
Vietminh 53
Vietnam . 22, 50, 52-54, 57, 58, 62, 63, 68-70,
                 87, 89, 120, 169, 211, 216,
                 234, 235, 237, 250, 264,
                 273, 285-288, 304, 305,
                 310, 314, 315, 322, 323,
                 330, 332, 337, 343, 345,
                 348, 356, 366, 368, 395,
                 397-399, 410, 412, 416,
                 429, 430, 442-444, 446,
                 447, 469
Vinogradov 326, 405
Virginia 134, 329, 463, 465
Vladivostok 176
VLF 138
Von Hippel 309, 312, 327, 331, 334, 365,
                 374, 377, 378, 391
von Neuman 234, 260
Vulnerability .. 47, 65, 72, 75, 85, 91, 97, 109,
                 134, 135, 145, 164, 168,
                 171-174, 192-194, 197, 228,
                 235, 251, 255, 261, 365,
                 385, 443
VVAW 304, 305

Index for Volume 1

Wald 332, 363
Walesa 279
Walker 288
Wall Street 237
war
   War Crimes 332
   War Game 88, 95, 132
   War Planning . 93-95, 168, 202, 219, 358
   War Resisters League 308
   War Without Winners 317
   warfare .. 7, 9, 23, 32, 33, 41, 42, 45, 49,
               76, 83, 84, 86, 88, 89, 108,
               112-114, 117, 122, 148,
               155, 169, 171, 179, 182,
               183, 188, 189, 193, 207,
               214, 216, 229-231, 234,
               244, 258, 260, 272, 317,
               322, 333, 336, 372, 380,
               387, 388, 446, 457
   warfighting 163, 224, 374
   warhawks .. 95, 361, 441, 442, 444, 445
   warlike 66, 262, 263, 324, 325, 379,
                  394
Warhead
   W76 120
   W88 .... 112, 119, 120, 229
   Warhead Dismantlement .. 123, 382, 464
   Warhead Production 94, 124, 146
   Warhead Safety 80, 119, 459, 461
Warnke 258, 368
Warren Air Force Base 91
Warsaw 30, 37, 62, 65, 74, 121, 184, 192,
               223, 224, 388, 392, 404,
               416, 462
Warsaw Pact .. 30, 65, 74, 184, 192, 223, 224,
               388, 392, 404, 416
Washington .. xxix, 6, 8, 39, 47, 51-53, 77, 92,
               111, 134, 150, 177, 194,
               202, 209, 211, 233, 235,
               239, 242, 243, 249, 251,
               278, 293, 305, 309, 312,
               316, 318, 320, 327, 329,
               336, 337, 343, 350, 362,
               363, 378, 398, 410, 414,
               416-418, 423, 441, 453,
               463-465, 467, 468
Washington Post 235, 327, 350, 398, 468
Washington Summit 177, 418
Watergate 59, 68
Wattenberg 253
Wayne State University 362
We All Lost 170, 282
Weapon
   weaponeers 123, 158, 243, 246
   Weaponizability 468, 469
   Weaponization ..... xvii, 24, 26, 36, 56
   weaponry ..... iii, vii, ix, xi, xiii, xix-xxi,
           xxxi, 18, 23, 68, 76, 82, 92,
               141, 145, 149, 154, 155,
               171, 181, 231, 257, 261,
               277, 281, 321, 324, 353,
               355, 365, 373, 398, 433,
               436, 445-447
Weapons

Weapons Delivery 370
Weapons of Mass Destruction .... xi, 38,
              107, 320, 329, 334, 335,
              343, 347, 400, 451, 462
Weapons Safety 81, 125, 156
Weapons Systems . 75, 88, 159, 226, 238,
              240, 241, 294, 316, 338,
              370, 374, 394, 401
weapons technology ... 3, 152, 162, 183,
              371
Weapons-Grade 9, 16, 25, 27, 123,
              175, 217, 247, 408, 455,
              456
Weapons-Usable 294
Weinberger 85, 89, 118, 253, 254, 271
Weiss 375
Weisskopf 332
Welch 302, 397, 425
Werner 4, 25, 26
West Germany 95, 97, 98, 113, 115, 223,
              239, 257, 278, 323, 346,
              402, 411, 458
West Suburban Concerned Scientists 363
West Virginia 134
Western Allies 23, 25, 30, 35, 77
Wheeler 364
White .. 37, 94, 134, 147, 210, 246, 252, 253,
              259, 279, 280, 284, 314,
              318, 322, 323, 331, 346,
              351, 360, 377, 378, 381,
              382, 397, 404, 411, 469
White House . 37, 94, 134, 147, 210, 246, 252,
              253, 259, 280, 284, 314,
              318, 322, 331, 351, 360,
              377, 378, 381, 382, 397,
              404, 469
White House Science Council 253, 259
Whiteman Air Force 390
Wicker 283
Widowmaker 169, 218
Wiesel 352
Wiesner 364, 365
Wigner 260, 337, 357, 364
Will
   George Will 253, 351
William Broad 398
Williams 352
Window of Vulnerability ... 65, 168, 171-174,
              197, 251, 255, 385
Wisconsin 138, 350
Wittner 417
WMD 329, 444, 462
Wohlstetter 202, 235, 364
Woodruff 378
Woolsey 200, 206, 207, 381
World Bank 264
World Federalist Association 206
World Forum 259
World War I 27, 154, 219, 232, 281, 401,
              444
World War II ... xiv, xxii, xxiv, xxv, 5, 14, 22,
            23, 27, 29-32, 36, 44, 46,
            53, 56, 67, 72, 74, 77, 78,
            82, 86, 95, 96, 98, 109-111,

World Wide Web 308
Worst-Case Scenarios 168, 208
WTO .. 37, 45, 67, 95, 96, 117, 120, 203, 222, 380, 381, 462
Wyoming 91
Xinjang 125
Yamantau 135
Yates 363
                                   141, 145, 146, 162, 174,
                                   190, 194, 195, 209, 226,
                                   229-231, 233, 237, 263,
                                   266, 275, 277, 278, 281,
                                   282, 286, 289, 297, 306,
                                   321, 330, 332, 333, 340,
                                   343, 398, 402, 419, 423,
                                   428, 434, 441, 447

yellowcake 123
Yeltsin 2, 19, 67, 68, 70, 73, 76, 229, 279, 368, 404
Yevtushenko 292
Yield-to-Weight 16, 36
Yom Kipper 213
Young 367, 397, 401
Zero Option 380, 381, 418
Zero Solution 301, 380, 381
Zeus 116, 146, 362
Zimmerman 332
Zlatoust 124
Zorin 168, 175, 176, 206
Zuckerman 215, 223, 241, 254, 257, 340

www.ingramcontent.com/pod-product-compliance
Lightning Source LLC
Chambersburg PA
CBHW071957150426
43194CB00008B/902